Few-
Body
Systems

Suppl. 1

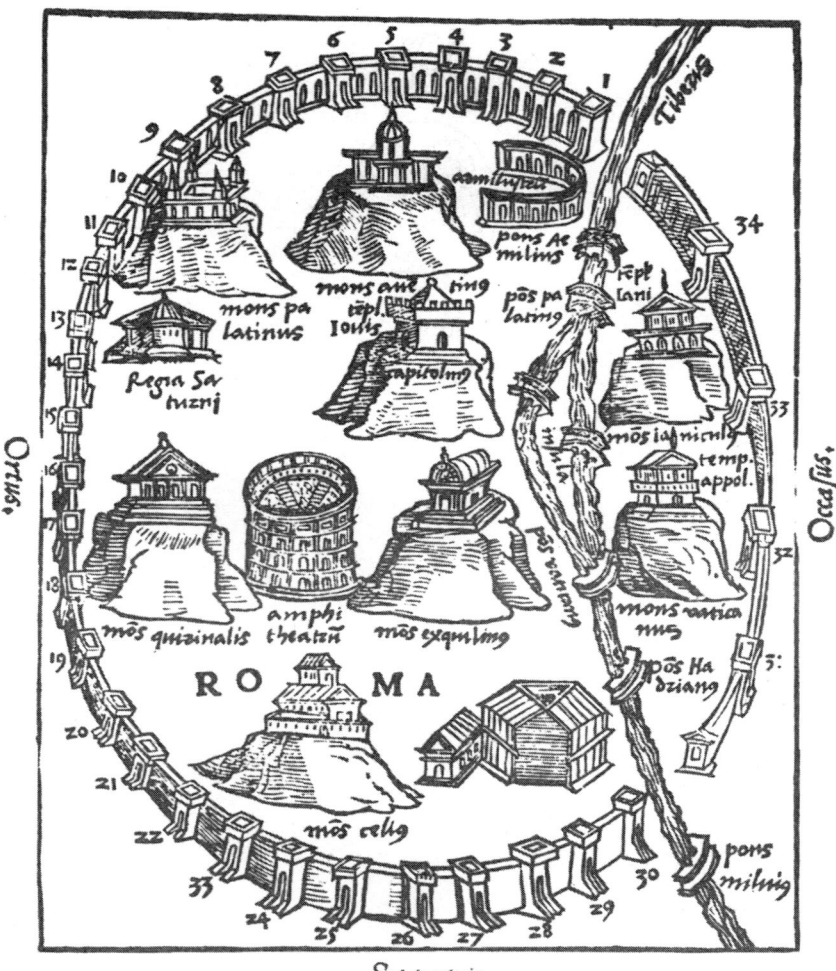

Meridies.

Roma

Septentrio.

1 Porta Laurentina	10 Porta Latina	18 Porta Minutia	28 Porta Rhomanula
2 Porta Ostiensis	11 Porta Labicana	19 Porta Exquilina	29 Porta Rhutamena, siue Veientana
3 Porta Trigemina	12 Porta Gabina	20 Porta Salutaris	
4 Porta Trigonia	13 Porta Querquetula= na, siue Cælimontana	21 Porta Piacularis	30 Porta Flumentana, siue Flaminea
5 Porta Naualis		22 Porta Lauernalis	31 Porta Valeria
6 Porta Sancualis	14 Porta Triumphalis, siue sacra	23 Porta Viminalis	32 Porta Septimiana
7 Porta Neuia		24 Porta Quirinalis	33 Porta Aurelia
8 Porta Radusculana	15 Porta Prenestina	25 Porta Catullaria	34 Porta Portuensis.
9 Porta Camena, siue Appia, siue Capena	16 Porta Tyburtina	26 Porta Colluna	Quatuor
	17 Porta Mutia	27 Porta Collatina	

1538 - Roma antica edita da SEBASTIANO MÜNSTER

Few-Body Systems

Editor-in-Chief: H. Mitter, Graz   Associate Editor: W. Plessas, Graz

Supplementum 1

Theoretical and Experimental Investigations of Hadronic Few-Body Systems

Proceedings of the European Workshop
on Few-Body Physics, Rome, October 7–11, 1986

Edited by
C. Ciofi degli Atti, O. Benhar, E. Pace, and G. Salmè

Springer-Verlag Wien New York

Prof. Dr. Claudio Ciofi degli Atti
Dr. Omar Benhar
Dr. Emanuele Pace
Dr. Giovanni Salmè
Istituto Nazionale di Fisica Nucleare
Sezione Sanità
Roma, Italy

ISSN 0177-8811
ISBN-13:978-3-7091-8899-6   e-ISBN-13:978-3-7091-8897-2
DOI: 10.1007/978-3-7091-8897-2

Local Organizing Committee

C. Ciofi degli Atti
(Workshop Chairman)

O. Benhar

E. Pace

G. Salmè

International Advisory Committee

V. Belyaev
Dubna

B. L. Berman
Washington

C. de Vries
NIKHEF

M. M. Giannini
Genova

B. F. Gibson
Los Alamos

F. Gross
CEBAF

J. Morgenstern
Saclay

G. Pisent
Padova

W. Plessas
Graz

D. Prosperi
Roma

S. Rosati
Pisa

W. Sandhas
Bonn

T. Sasakawa
Sendai

P. U. Sauer
Hannover

C. Schaerf
Roma

E. Schmid
Tübingen

Yu. A. Simonov
Moscow

H. Tanaka
Sapporo

W. T. H. van Oers
TRIUMF

Sponsored
by
Istituto Nazionale di Fisica Nucleare (INFN)

Organized
by
INFN—Sezione Sanità
with support of Physics Laboratory, Istituto Superiore di Sanità

Site
Istituto Superiore di Sanità
Roma—Viale Regina Elena, 299

Editorial

By the beginning of 1986 the former journal "Acta Physica Austriaca" was transformed to its new format "Few-Body Systems". One of the principal motives was to provide an active forum for research on few-body problems in the various domains of physics and neighbouring fields. Few-body systems, understood as consisting of a small number of constituent structures, play an important role in such areas like particle, nuclear, atomic, molecular, and condensed-matter physics as well as statistical and mathematical physics, astrophysics, chemical and biological physics. The new journal should above all serve the purpose of bringing together at one place relevant work hitherto dispersed in the literature, thereby fostering research done on related problems in different areas of natural science.

Since its first appearance "Few-Body Systems" has obtained welcoming approval and it has been obvious from the very beginning that it is meeting the actual needs of few-body systems research. The publication of the original papers in the first few issues already received considerable attention; also the journal's regular News Section containing information, both of scientific and practical nature, about latest developments in the field of few-body physics has been much appreciated. Thus the journal "Few-Body Systems" is on the best way towards providing a lively platform for the communication of few-body systems research.

It is in the same spirit that we are now initiating the publication of relevant conferences in connection with the new journal. We hope that with the help of "Few-Body Systems Supplementa" we will be able to contribute to a fruitful exchange of the most recent results as typically presented at conferences on the subject of few-body physics and at corresponding topical meetings. This will constitute the appropriate addendum to the publication of original research work in the form of usual articles, letter-type notes, rapid communications etc. in the regular issues of the journal "Few-Body Systems".

The Rome Workshop, to the proceedings of which the first Supplementum is devoted, provided a good starting point for this new series. The meeting was excelling for its well-balanced program, its high scientific standard, and an informative and stimulating atmosphere, what is also reflected in the contributions collected in the present volume. These are, of course, just the features we will have to observe as essential requirements also for future publications in the series of "Few-Body Systems Supplementa".

Graz, Jülich
December 1986

H. Mitter
W. Plessas

Preface

This volume collects the papers given at the European Workshop "Theoretical and Experimental Investigations of Hadronic Few-Body Systems" which, adhering to an invitation of the European Few-Body-Physics Research Committee, was organized in Rome on October 7–11, 1986. All papers presented at the workshop appear in the volume, plus two papers which could not be presented orally because their authors were at the last moment unable to attend. The list of contents closely follows the programme of the workshop.

The workshop, attended by 128 American, European, and Japanese physicists from 60 different institutions and universities, was sponsored by the Italian National Institute for Nuclear Physics (INFN) and was organized by the INFN Section located at the Istituto Superiore di Sanità (ISS), which kindly provided the venue for the meeting and many related facilities. The goal of the workshop was to summarize the present situation and the future perspectives concerning the theoretical descriptions of strongly interacting few-body systems and their experimental investigation by electromagnetic and hadronic probes, mainly at intermediate energies. To this end, representatives from most international groups working within different theoretical methods and with different experimental facilities, were invited and asked to illustrate their latest results and future research programs; the intention was to provide, by this way, an impartial and broad information which could be useful to whom is actively working in few-body physics, as well as to young students entering this field of research. The workshop, entirely based upon plenary sessions and invited papers, started with a self-contained introductory session, aimed at a presentation of the main features and open problems in theoretical methods and computational techniques, and went over to the discussion of more specific subjects. Special emphasis was devoted to such topics as realistic descriptions of nucleon-deuteron scattering, polarization experiments, three-body forces, quark and relativistic effects, nucleon and nuclear form factors at high momentum transfer, momentum distributions, spectral functions, final-state interactions, exclusive and inclusive electrodisintegration, y-scaling. Particular attention was also paid to a wide presentation of the most relevant projects of new particle accelerators, whose construction is expected to have a strong impact on research activity in few-body physics. A short session was also devoted to the highlights of the Sendai XI International IUPAP Conference on "Few-Body Systems in Particle and Nuclear Physics" (August, 1986) and the Washington Symposium on "The Three-Body Force in the Three-Nucleon System" (April, 1986); the aim of this session was to inform that part of the audience (young students and many European physicists) which was unable to attend these two important events.

The workshop was opened by Professor Nicola Cabibbo, President of INFN, and the closing remarks were given by Professor Franz Gross; to both of them I would like to express my sincere appreciation.

The participation of the leading international experts together with many young students, the large amount of information that circulated during the workshop, the presentation of many yet unpublished experimental and theoretical results and, finally, the highly qualified and live discussions, make me confident that the workshop has fully reached its goal.

Much of the success of the workshop was due to the considerable help of our sponsor and the extensive efforts of many people. I would like to express my hearthfelt thanks to Professor Francesco Pocchiari, Director of ISS, for his kind hospitality and support; to the members of the International Advisory Committee, for their advice on the programme; to all session chairmen and discussion leaders, for their expertise and skillness; to Mrs. Bruna Ceccarelli, who acted as administrative secretary of the workshop; to the staff of the Physics Laboratory of ISS, particularly to Mr. A. Grisanti and Mr. G. Monteleone, for their continuous assistance in smoothing out many logistical difficulties; to the Service for Cultural Activities and to the Editorial Section of ISS, for their constant support; to Mrs. Paola De Castro-Pietrangeli, Ms. Alessandra Raponi, Mrs. Paola Tacchi-Venturi and Ms. Giovanna Vacri, for running the secretarial work during the workshop.

Finally, it is my pleasant duty to record the extraordinary energy, efforts and skillness of my co-organizers and collaborators Omar Benhar, Emanuele Pace and Giovanni Salmè without whom the workshop would not have been possible; to them I am deeply indebted.

Roma, December 1986   Claudio Ciofi degli Atti

Contents

Session I
"Theoretical Frameworks in Few-Body Physics"

Chairman: *E. Schmid*

Chairman: *J. L. Friar*

Chairman: *W. Plessas*

Panel Session
"Reports on Progress in Realistic Calculations for Three- and Four-Nucleon Systems"

Chairman: *S. Rosati*   Discussion leader: *P. U. Sauer*

Session II
"Nucleon-Deuteron Scattering"

Chairman: *W. Sandhas*

Session III
"Few-Body Clusters and Reactions"

Chairman: *G. Pisent*

Session IV
"Polarization Effects"

Chairman: *S. Boffi*

Session V
"Final State Interaction, Meson-Exchange Currents, Scaling"

Chairman: *M. M. Giannini*

Panel Session
"Reports on Progress in Electron Scattering Experiments"

1. Latest Experimental Results

Chairman: *B. L. Berman*   Discussion leader: *B. Frois*

2. Research Programs

Chairman: *J. Mougey*

Session IX
"Future Experimental Developments"

Chairman: *C. de Vries*

Chairman: *G. Backenstoss*

THE NUCLEON-NUCLEON INTERACTION

J.J. de Swart, T.A. Rijken, J.R.M. Bergervoet,

P.C.M. van Campen, W.M.G. Derks, and W.A.M. v.d. Sanden

Institute for Theoretical Physics,

University of Nijmegen, The Netherlands

A review is given of some of the theoretical ideas about the NN-interaction. Topics covered are: a discussion of the Bonn (1986), Paris (1980), and the Nijmegen (1978) potentials, a new phase shift analysis of the pp-data giving the preliminary value for the $pp\pi^0$ coupling constant $g_0^2/4\pi = 13.15 \pm 0.2$, and a discussion of the charge symmetry and charge independence of the NN-scattering lengths.

1. Introduction

There are many different models for the nucleon-nucleon (NN) interaction. When one tries to judge these various approaches, then there are two important criteria:

(i) What is the philosophy behind the model?

(ii) How well does the model fit the experimental data?

Let us discuss each of these criteria in somewhat more detail.

(i) Philosophy behind the model

The first questions that require an answer are: "Why did the authors construct this specific model?", "What are they trying to test?", or "What kind of applications do they have in mind?" Are they just trying to give a good description of the low energy data or do they want a model that is also valid in the neighborhood of the $N\Delta$-threshold? Do the authors think about extending their model to all other baryon-baryon channels (like ΛN, ΣN, ΞN, $\Lambda\Lambda$, etc.) or are they more concerned with applications to nuclei or infinite nuclear matter?

A second series of questions concerns more their specific approach. Are the authors using a well-established theory or are they just following a fashionable trend? This last question can be phrased often in a quite different way. It sounds suddenly quite differently when one asks: Are the authors following an old-fashioned or a modern approach? This question is especially important when one tries to compare models based mainly on "meson theory" and models based on quarks and QCD. The answer can and will depend on the specific topic. For example, it is clear that the one-pion-exchange potential (OPEP) is well-established and cannot be considered as old-fashioned.

This first criterium is clearly not easy to apply and the conclusions will depend strongly on the person trying to judge. A difficulty is, that most work sounds plausible, is often highly technical, and is very hard to judge for outsiders (and often even for insiders). Still everybody should make up his mind and try to judge if the philosophy behind a certain model is sound.

(ii) How well does the model fit the experimental data?

To be fair one must first determine if the model is meant to give a qualitative description of the data, or do the authors try to be more ambitious and have they tried to give a quantitative description of the data. This criterium is much easier to handle. When the fits to the data are presented in figures only, then almost certainly only a qualitative fit is presented. When one tries to present quantitative fits to scattering data, then one must rely on statistical criteria, like χ^2, and one should present the χ^2 per datapoint or per degree of freedom. However, that is not totally sufficient. One should also know and compare with the minimum χ^2, reached in a phase shift analysis for the same set of data.

2. Different Approaches

Let us start with the more modern approach (or fashionable trend) based on quarks and QCD. It is probably fair to say that no reliable, direct information about the NN-force is obtained this way. Some people claim that it explains the repulsive core in the NN-interaction, but even that is not clear. One thing, however, is clear. Nucleons and mesons are extended objects made of quarks and antiquarks. In all treatments one should keep this picture in mind [1].

Let us next consider the meson exchange models. The OPEP is well-established. For example in NN-phase shift analyses one can determine the coupling constant and even the mass of the pion. However, there are still uncertainties. What is the precise value of the coupling constant g^2 and which form factor $F(k^2)$ should be used? This form factor is clearly an expression of the spatial extension of the nucleons and the pion. One could also wonder if OPEP is a good representation for all values of r or if OPEP is only valid for r > 2R with R the nucleon radius.

Let us next consider the two-pion-exchange potential (TPEP) and connected with it the one-boson-exchange (OBE) potentials of the $J^{PC} = 0^{++}$ mesons ε (~ 700), S (975), ε'(1300), ...; of the $J^{PC} = 1^{--}$ mesons ρ (770), ρ', ...; and of the $J^{PC} = 2^{++}$ mesons f (1270), f'(1525), etc. The situation here is unclear and muddy.

Consider first the uncorrelated TPEP given by the contribution of the planar- and crossed box graphs minus the twice iterated OPEP (Fig. 1). How well can one calculate this contribution? One first must decide how to

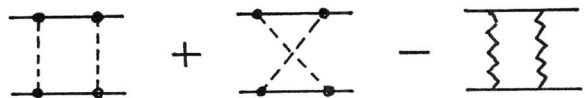

Fig. 1: Uncorrelated 2π-exchange potential

treat the πNN vertices. Should one use PS-coupling (γ_5), or PV-coupling ($\gamma_\mu \gamma_5$), or even more complicated couplings? What choice must one make for the form factor? Some examples are $F(k^2) = [\Lambda^2/(\Lambda^2 + k^2)]^n$ and $F(k^2) = \exp(-k^2/\Lambda^2)$, but many more forms are possible. Because in the calculation of TPEP one has to integrate over one internal momentum, the answer will depend strongly on the choice of coupling and form factor. The uncorrelated TPEP is strongly model dependent!

But we are not there yet. What to do with N or Δ-resonances in the intermediate states, or with interactions between the two exchanged pions (Fig. 2)? Are these diagrams related or are they independent?

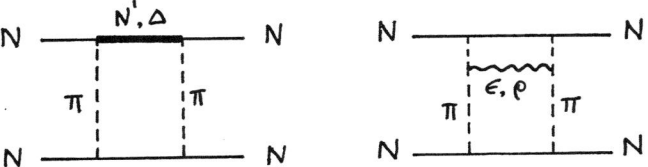

Fig. 2: Other contributions to TPEP

Another way to approach this problem is to look directly at the correlated TPE-potential. In the t-channel one could decompose the system in angular momentum states (J = 0, 1, 2, ...). Next one could hope, that when one tries to calculate the J = 0 contribution to TPEP, that the OBE-potentials corresponding to ε(700), S(975), etc. exchange give a fair representation of this J = 0 contribution. Analogously one could hope that the case J = 1 is already well represented by ρ-exchange, the case J = 2 by f and f' exchange, etc.

Some authors like to represent the complete TPEP as the sum of the uncorrelated TPEP and the correlated TPEP as represented by OBE-potentials. A problem of this latter approach is that certain exchanges are counted twice. Let us call this double counting of the first kind. It is practically certain that this way a lot of double counting is done. How much this double counting is and how to avoid it in a practical way, is not known.

However, there is another and much more subtle way of double counting. Let us call that double counting of the second kind. Around 1968-1969 the concept of duality became very important in high-energy physics [2]. At that time it became clear that the description of the πN-scattering amplitude via an infinite sum of s-channel resonances is in a sense equivalent with the description of the same amplitude via the t-channel exchange of an infinite sum of Regge trajectories. In pictorial form this is shown in Fig. 3.

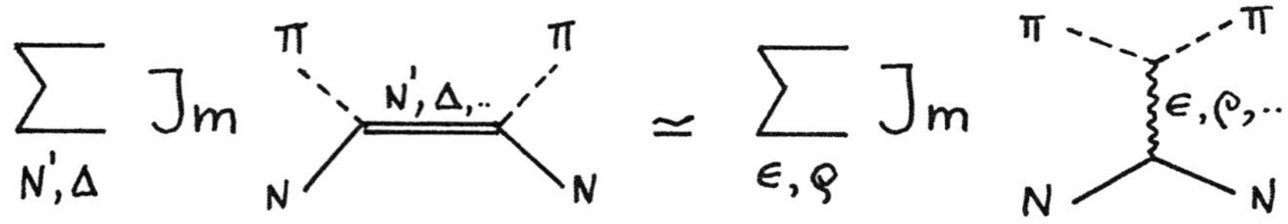

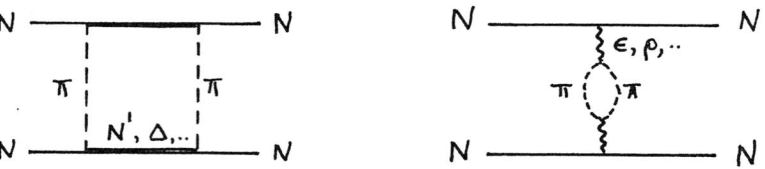

Fig. 3: Duality for πN-scattering amplitude

Applied to NN-scattering this duality shows that the TPE-diagrams of Fig. 4a with N and Δ-resonances as intermediate states are at least partially contained in the OBE-diagrams of Fig. 4b representing the exchange of broad mesons like ε, S, ρ, f, f', etc.

Fig. 4a: TPE-diagrams   b. Broad OBE-diagrams

Also here the situation is not very clear cut. When one tries to describe the scattering in the neighborhood of the NΔ-threshold, then it is obvious that one should include these channels. When one tries then to describe the potentials with OBE-exchanges like ε, S, ρ, f, f', etc, then clearly a double counting of this second kind is done.

Finally let us come back to the quark picture. Nucleons and mesons are extended objects with radii R_N and R_M. In classical language meson exchange is thus only allowed for distances larger than $2(R_N + R_M)$, which is of the order of 2-3 fm. This makes it difficult to understand the importance of heavy meson exchange (like ρ, ω, etc.). How to reconcile meson-exchange with the quark picture is unclear.

3. Some NN-Potentials

In this section we will review quickly some of the meson theoretical potentials [3], available at this moment.

The Bonn (1986) potential [4]

The Bonn group tried to improve on their older NN-potentials [5] by constructing a totally new model, in which they carried the meson-exchange picture to the utmost extreme. They include the exchanges of one, two, three, and even four pions. Correlated exchange is included via the ω and ρ mesons and via a fictitious σ'-meson. Also they discuss the $\pi\omega$, $\pi\rho$, and $\pi\sigma'$ exchange as well as nucleon and Δ-resonances as intermediate states. They do omit the exchange of the well-established mesons as η, η', ϕ, and S, and also they do not include Pomeron exchange. It appears that they do all kinds of double counting.

Important to note is, that it is only an np-potential. Therefore we will not confront it with the pp-data. The comparison with the experimental data is only done in some figures, where they show the fit to certain np (and even pp?) observables. No χ^2 is ever mentioned.

The Paris (1980) potential [6]

The Paris group tried to avoid the difficulties with the computation of the complete TPE-meson theoretical potential by using the dispersion approach. Next to this dispersion theoretical 2π-potential they include OBE-contributions due to π, ω, and A_1. The inner region of this potential is treated totally phenomenologically [7]. After this potential was constructed they presented a parametrized version in which they adjusted some of these parameters in order to improve their fit with the data. They present a pp as well as an np-potential.

The Nijmegen (1978) soft-core potential [8]

This potential is based on Regge theory [9] and is a generalized OBE-potential. Included are the exchanges of all non-strange mesons of the $J^{PC} = 0^{-+}$, 1^{--}, and 0^{++} nonets, like π, η, η'; ρ, ω, ϕ; $\varepsilon(700)$, $S(975)$, and $\delta(983)$. Of the 2^{++}-nonet (A_2, f, f') in Regge language only the "J = 0 component" is included. An important contribution comes from Pomeron exchange. The uncorrelated TPEP is totally neglected in order to avoid double counting. Let us point to two very nice theoretical features of this model.

(i) At the vertices the form factor $F(k^2) = \exp(-k^2/\Lambda^2)$ is used. In high energy scattering and low momentum transfers one sees this form factor [10]. It is clear that the exponential form is a good representation of this form factor in that momentum transfer region.

(ii) The potential contains a contribution due to Pomeron exchange. The elastic scattering amplitude at high energies and low momentum transfer is totally dominated by Pomeron-exchange [11]. This contribution is then analytically extrapolated to lower energies, where it gives rise to a Gaussian, repulsive, central potential and some contributions to the spin-orbit and quadratic spin-orbit potentials. The parameters of this potential, representing the strength and the range, were fitted to the low energy NN-data. These parameters can also be deduced from the Regge fits to the high energy data. The agreement is really excellent. How should one look at this Pomeron-exchange potential? In high-energy scattering the Pomeron can be explained as a two-gluon (or multigluon) exchange effect [12]. Therefore the Pomeron contribution to the Nijmegen potential can be viewed as a phenomenological way to incorporate a two-gluon exchange potential.

4. Phase Shift Analyses

Instead of trying to improve the Nijmegen (1978) soft core potential [8] we decided to study first carefully the NN-scattering data and see if this way some things could be learned. At present we have finished the study of the 0 - 3 MeV [13] and of the 0 - 30 MeV pp-data [14]. The analysis of the 0 - 350 MeV pp-data is in progress and a study of the np-data just has been started.

Our improvements over older analyses [15] are especially important in the low energy region, where we had many more data to analyze. We used a thorough treatment of the electromagnetic interactions and a very careful treatment of the various complications due to OPEP.

The electromagnetic interaction

In low energy pp-scattering it has been shown that not only the Coulomb interaction (order α), but also the vacuum polarization (VP) potential [16] (order α^2) is important. In our analysis of the 0 - 30 MeV data the VP is a 10 s.d. effect. When one wants to improve on this standard description of the electromagnetic interaction, one needs to take account of the fact that the charges are moving (instead of fixed as in the static Coulomb potential) and one needs to incorporate also two-photon exchange (which is also of order α^2). Some time ago we constructed such an improved Coulomb potential [17]. In our language it is really a one-photon-exchange potential, where the two-photon-exchange contributions are generated by off-shell effects. This improved treatment of the electromagnetic interactions shows itself in two ways.

(i) The Coulomb parameter $\eta = \alpha/v$, where v is the relative velocity in the cm system, must be replaced by $\eta' = \alpha/v_L$, where v_L is the laboratory velocity. This replacement is well-known [18] and is in our 0 - 30 MeV phase shift analysis a $4\frac{1}{2}$ s.d. effect.

(ii) The Coulomb phase shift gets an extra contribution. This effect is not visible in an improved χ^2, but it influences the precise values of the phase shifts.

Treatment of the various partial waves

The phases in the higher partial waves must always be predicted by theory. Mostly one calculates in Born approximation (BA) the phases due to OPEP, neglecting the influence of the repulsive Coulomb interaction. We calculated in Coulomb distorted wave (CDW) BA the phases due to the OPEP, the VP-potential and the improved Coulomb potential. This guarantees a quite accurate treatment of the higher partial waves, which is very important when one wants to determine the $pp\pi^0$-coupling constant in the analysis.

The lower partial waves are always searched for in phase shift analyses. In order to describe properly their energy dependence one needs a good parametrization. It is standard to use for this some form of effective range expansion. These effective range functions have lefthand cuts starting at $T_L = - 9.7$ MeV due to OPEP. The influence of this lefthand cut has always been neglected or treated in an approximate way. After several years of struggling with these problems we found a very obvious as well as simple solution, which allows us in _every_ partial wave to take account _exactly_ of the long range tails of the improved Coulomb potential, the VP-potential, as well as the OPEP.

We use a P-matrix description [19] where we specify the P-matrix at b = 1.4 fm. Outside this radius we use the OPEP, the improved Coulomb potential, and the VP-potential. The energy dependence of this P-matrix is parametrized. This allows now for a unified treatment of all partial waves.

The waves with an intermediate value of ℓ were treated in a special way. Using conformal mapping techniques and the knowledge of the analytic structure of the amplitudes [20] we can extrapolate from low ℓ to higher values of ℓ the deviation of the phases from their OPEP-value. In these optimal polynomial expansions [21] one uses as input the lower partial

waves and one obtains as output the higher partial waves. In practice:
from $^1D_2$ and $^1G_4$ one predicts $^1I_6$ and $^1L_8$,
from $\qquad$ $^3F_2$ one predicts $^3H_4$ and $^3K_6$,
from $^3P_1$ and $^3F_3$ one predicts $^3H_5$ and $^3K_7$, and
from $^3P_2$ and $^3F_4$ one predicts $^3H_6$ and $^3K_8$.

In the 0 - 350 MeV analysis the inclusion of this optimal polynomial expansion is a 10 s.d. effect.

The pion-nucleon coupling constant

The most important result of our 0 - 350 MeV phase shift analysis (still in progress) is a very precise determination of the π^0pp-coupling constant. We find (preliminary result)

$$g_0^2/4\pi = 13.15 \pm 0.2 \qquad .$$

This value must be compared with the coupling constant g_c of the charged pions to the nucleons as determined in the precise analyses of the accurate πN-scattering data [22]. There one finds

$$g_c^2/4\pi = 14.3 \pm 0.2 \qquad .$$

5. Charge Symmetry and Charge Independence

The charge symmetry breaking in the pp- and nn-scattering lengths was studied [23] in a potential model, where the Nijmegen (1978) soft-core model [8] was used and where we included the full electromagnetic interaction treating the nucleons as extended sources [24].

We started with the construction of a series of $^1S_0$ pp-potentials (each with a different value of g_0), which all fit exactly the low energy parameters as determined in the 0 - 3 MeV phase shift analysis [13]. To guarantee a proper high energy behavior we required the phase shift at 250 MeV to be equal to the value in the original Nijmegen potential [8], which agrees with the multienergy phase shift analysis from 0 - 350 MeV.

Next we assumed the various coupling constants to be charge symmetric. This allowed us to construct a series of $^1S_0$ nn-potentials (including e.m. interactions) and to calculate a_{nn} and r_{nn} as a function of g_0^2. The results are given in Fig. 5.

Using the experimental value [25] $a_{nn} = 18.45 \pm 0.45$ fm, we obtain

$$g_0^2/4\pi = 13.3 \pm 0.9 \qquad \text{and} \qquad r_{nn} = 2.823 \pm 0.003 \text{ fm} \qquad .$$

We think that this result is rather model-independent.

Next we assumed all potentials to be charge independent, except the OPEP. This allowed us to construct a series of $^1S_0$ np-potentials (including

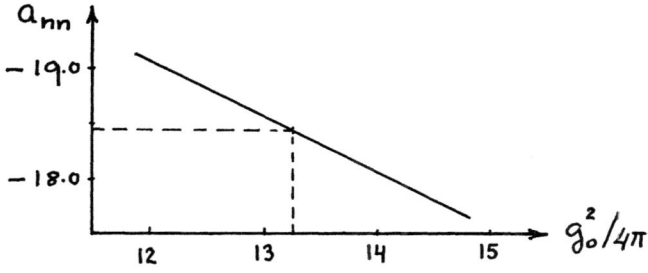

Fig. 5: a_{nn} as function of $g_0^2/4\pi$

e.m. interactions). For the OPEP we write

$$V_\pi = 2 \frac{g_c^2}{4\pi} V_c - \frac{g_0^2}{4\pi} V_0 \quad .$$

In V_c (or V_0) we used the $\pi^\pm$-mass (or π^0-mass). We can now calculate a_{np} as a function of g_0, keeping g_c fixed at a reasonable value. It turned out that the value of a_{np} varies only slightly between -21 fm to -22 fm, but never reached the experimental value $a_{np} = -23.75$ fm. The reason that a_{np} is so insensitive to the value of g_0 is understandable. The OPEP in momentum space is

$$V_\pi(k^2) = g^2 \frac{k^2}{k^2 + m^2} F(k^2) \quad . \tag{1}$$

In coordinate space the volume integral of the OPEP is

$$\int d^3r \, V_\pi(r) = V_\pi(k^2 = 0) = 0$$

independent of the form factor. Because changes in a_{np} are related to changes in this volume integral [26] we can understand the insensitivity of a_{np} on g_0.

However, we could look at this problem a different way. We write (1) as

$$V_\pi(k^2) = g^2 \left\{ 1 - \frac{m^2}{k^2 + m^2} \right\} F(k^2) \quad . \tag{2}$$

The first term $g^2 F(k^2)$ represents a repulsive, short range, Gaussian potential and the second term an attractive, long range, Yukawalike potential. We then assume that the short range interactions cannot be described properly by meson exchange, but come from much more complicated quark-quark interactions. This means that the short-range interactions are then described totally phenomenologically. If we assume these short range interactions to be charge independent, we must vary only the long range, Yukawalike part of OPEP. In that case the calculated value of a_{np} is very sensitive to the precise value of g_0. The experimental value of a_{np} can be obtained when

$$2 \frac{g_c^2}{4\pi} - \frac{g_0^2}{4\pi} \simeq 15.2 \quad .$$

A solution is $g_0^2/4\pi = 13.2$ and $g_c^2/4\pi = 14.2$. We find this way a very simple and very natural explanation for the charge dependence of the NN-scattering lengths.

6. Comparison of Different pp-Potentials

In our 0 - 350 MeV phase shift analysis we have constructed a handy parametrization of the χ^2-surface for the pp-scattering data. This allows us to compare rather easily the χ^2 of different pp-potentials. Because the Bonn potentials [4,5] and the Argonne potential [27] are only np-potentials, we refrained from calculating their total χ^2 with respect to the pp-data. We did this only for the Nijmegen (1978) and the Paris (1980) potentials [6,8]. The results are displayed in Fig. 6. The distance along the horizontal axis we took proportional to the number of data in the interval.

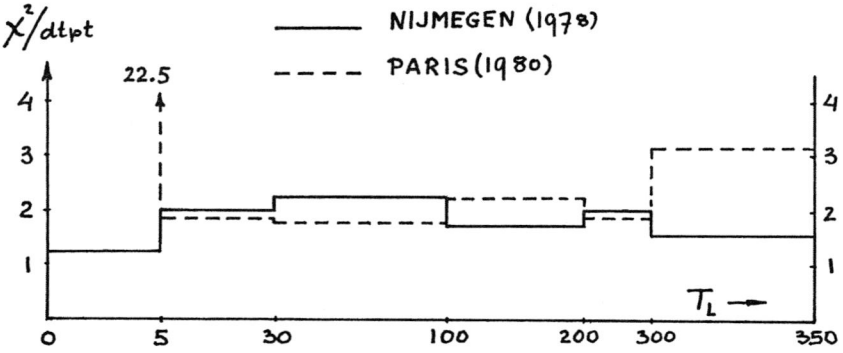

Fig. 6: χ^2-contributions of the Nijmegen (1978) and the Paris (1980) potentials

We see that at the beginning (0 - 5 MeV) and the end (300 - 350 MeV) of the energy range the Paris potential has a rather large χ^2. In Table 1 we give for the Nijmegen and for the Paris potential the total χ^2 and the χ^2/datapoint for three energy intervals.

Table 1: χ^2 for the Nijmegen (1978) and Paris (1980) potentials with respect to the pp-data.

T_L (MeV)	N_{data}	Total χ^2		χ^2/dtpt	
		Nijm	Paris	Nijm	Paris
0 - 350	1345	2433	7026	1.81	5.22
5 - 350	1146	2188	2540	1.91	2.22
5 - 300	860	1739	1646	2.02	1.91

Acknowledgements

Part of this work was included in the research program of the Stichting voor Fundamenteel Onderzoek der Materie (F.O.M.) with financial support from the Nederlandse Organisatie voor Zuiver-Wetenschappelijk Onderzoek (Z.W.O.).

References

1. De Swart J.J., and Nagels M.M., Fortschr.Phys. $\underline{28}$, 215 (1978).

2. Dolen R., Horn D., and Schmid C., Phys.Rev. $\underline{166}$, 1768 (1967); Schmid C., "Duality and Exchange Degeneracy", in: "Subnuclear Phenomena", part A (ed. Zichichi A.), Academic Press, 1970; Frampton P., "Dual Resonance Models", Benjamin Books, 1973.

3. De Swart J.J., van der Sanden W.A., and Derks W., Nucl.Phys. A $\underline{416}$, 299c (1984).

4. Machleidt R., Holinde K., and Elster Ch., Bonn preprint (1986).

5. Holinde K., and Machleidt R., Nucl.Phys. A $\underline{247}$, 495 (1975); A $\underline{256}$, 479 (1976); A $\underline{280}$, 429 (1977).

6. Lacombe M., Loiseau B., Richard J.M., Vinh Mau R., Côté J., Pirès P., and de Tourreil R., Phys.Rev. C $\underline{21}$, 861 (1980).

7. Lacombe M., Loiseau B., Richard J.M., Vinh Mau R., Pirès P., and de Tourreil R., Phys.Rev. D $\underline{12}$, 1495 (1975).

8. Nagels M.M., Rijken T.A., and de Swart J.J., Phys.Rev. D $\underline{17}$, 768 (1978).

9. Rijken T.A., Ph.D. Thesis, University of Nijmegen (1975), unpublished; Ann.Phys. (NY) $\underline{164}$, 1 and 23 (1985).

10. Barger V.D., and Cline D.B., "Phenomenological Theories of High Energy Scattering", W.A. Benjamin, Inc. (NY), 1969.

11. Barger V., and Olsson M., Phys.Rev. $\underline{146}$, 1080 (1966); ibid $\underline{148}$, 1428 (1966).

12. Low F.E., Phys.Rev. D $\underline{12}$, 163 (1975); Nussinov S., Phys.Rev.Lett. $\underline{34}$, 1286 (1975).

13. Van der Sanden W.A., Emmen A.H., and de Swart J.J., THEF-NYM-83.11 (1983).

14. Bergervoét J.R.M., van Campen P.C.M., and de Swart J.J., in preparation.

15. Arndt R.A., et al., Phys.Rev. D $\underline{28}$, 97 (1983); Naisse J.P., Nucl.Phys. A $\underline{278}$, 506 (1977); Noyes H.P., Lipinski H.M., Phys.Rev. C $\underline{4}$, 995 (1971).

16. Durand L., Phys.Rev. $\underline{108}$, 1597 (1957).

17. Austen G.J.M., and de Swart J.J., Phys.Rev.Lett. $\underline{50}$, 2039 (1983).

18. Breit G., Phys.Rev. $\underline{99}$, 1581 (1955).

19. Jaffe R.L., and Low F.E., Phys.Rev. D $\underline{19}$, 2105 (1979).

20. Cutkosky R.E., and Deo B.B., Phys.Rev. $\underline{174}$, 1859 (1968); Ciulli S., Nuov.Cim. $\underline{61A}$, 787 (1969); ibid $\underline{62A}$, 301 (1969).

12

21. Rijken T.A., and Signell P., "A New Optimal Polynomial Theory for NN Scattering" (1985), to be published; Rijken T.A., Lect.Notes in Physics 236, 196 (1985).
22. Dumbrajs O., et al., Nucl.Phys. B 216, 277 (1983).
23. Derks W., van der Sanden W.A.M., and de Swart J.J., in preparation.
24. Derks W., van der Sanden W.A.M., and de Swart J.J., in preparation.
25. Slaus I., Akaishi Y., Tanaka H., Phys.Rev.Lett. 48, 993 (1982).
26. De Swart J.J., and Dullemond C., Ann.Phys. (NY) 19, 458 (1962).
27. Wiringa R.B., Smith R.A., Ainsworth T.L., Phys.Rev. C 29, 1207 (1984).

NUCLEON-NUCLEON INTERACTION ABOVE PION THRESHOLD

Peter U. Sauer
Theoretical Physics, University Hannover
3000 Hannover, Germany

Abstract

A force model with Δ-isobar and pion degrees of freedom is presented. It is noncovariant. It accounts for elastic two-nucleon scattering and for its unitarily coupled inelastic channels up to 500 MeV in the c.m. system with satisfactory accuracy. The force model forms a microscopic basis for a unifying description of nuclear phenomena at low and intermediate energies.

1. Introduction

The active degrees of freedom, which one sees in nucleon-nucleon scattering at intermediate energies and which any trustworthy force model has to incorporate, are those of the Δ-isobar and the pion (π) besides the nucleonic (N) one. In addition, however, one had hoped to also find dibaryon resonances in this energy domain, i.e., two-baryon states of exotic nature irreducible into those single interacting baryons considered anyhow. Experimentally, no pronounced resonance has been detected, which could be identified with them. On the theoretical side, force models can describe the existing data without the dibaryon degree of freedom rather well. The present paper adds further evidence to this fact.

A force model is considered with the traditional nucleon, Δ-isobar and pion degrees of freedom. It uses a hamiltonian approach within the framework of noncovariant quantum mechanics. It is able to account

(1) for two-nucleon scattering below and above pion threshold up to
    500 MeV in the c.m. system,
(2) for the coupling of the two-nucleon system to the pion-deuteron
    channel and
(3) for elastic pion-deuteron scattering

with satisfactory accuracy. The break-up channel has not been considered
yet. The fact that the present force model achieves a unifying description
of two-nucleon scattering and its inelasticities by the same hamiltonian
extends the validity of the potential parametrization for the two-nucleon
interaction to the energy domain above pion threshold without destroying
its success at low energies.

2. Force Model with Δ-Isobar and Pion Degrees of Freedom

The inelasticity in two-nucleon scattering up to at least 500 MeV in
the c.m. system, i.e., far beyond the two-pion threshold, is predominantly
single-pion production and occurs essentially in the isospin-triplet partial waves. These experimental facts are borne out in Fig. 1 and suggest
pion production in intermediate-energy two-nucleon scattering to proceed
through single-Δ excitation. Processes like double-Δ excitation with subsequent production of two pions appear suppressed in the considered energy
region.

A force model [2,3], with Δ-isobar and pion degrees of freedom added,
is diagrammatically defined in Fig. 2. It acts in the isospin-triplet two-
nucleon partial waves. Its two-nucleon part is not any more the traditional
two-nucleon potential. There, processes due to Δ-excitation and pion propagation are contained in the instantaneous potential, whereas they are
treated explicitly in the present model. In the model of Fig. 2 the coupling to pionic states is mediated by the Δ-isobar, which is excited in a
two-nucleon interaction. The one-baryon πNΔ vertex then yields the subsequent decay of the Δ-isobar into a pion and a nucleon. Thus, pion production and absorption is described as two-stage event built up from the two
processes of Figs. 2(b) and 2(e). A direct step from nucleonic states to
states with pions is not provided by the force model. The isospin-singlet
partial waves are taken over unchanged from a purely nucleonic potential.

The structure of the force model pays special tribute to the fact
that pion-nucleon scattering up to 300 MeV pion lab energy is dominated by
the P_{33} partial wave and that the P_{33} partial wave itself is dominated by
the Δ-resonance in this energy domain. The nonresonant pion-nucleon interactions are accounted for by potentials. This difference in describing re-

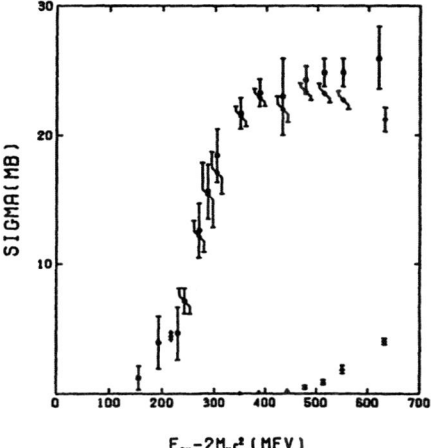

Fig. 1

Proton-proton total inelastic cross section (full dots) and contributions from single-pion production (open squares) and from two-pion production (open triangles). The data are from Ref. [1].

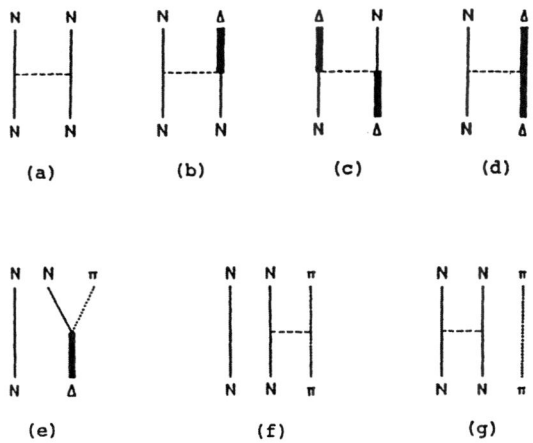

Fig. 2

Building blocks of the force model with Δ-isobar and pion degrees of free-dom. The Hermitian adjoint pieces corresponding to the processes (b) and (e) are not shown. The Δ-isobar is a bare particle. Process (e) yields the physical P_{33} pion-nucleon resonance in the one-nucleon Hilbert space by iteration. Process (f) stands for the nonresonant pion-nucleon interaction, whose partial waves up to orbital angular momentum $\ell = 2$ are included. For process (g), i.e., for the two-nucleon potential in the presence of the pion, only the $^3S_1 - {}^3D_1$ partial wave is retained. The extended force model acts in the isospin-triplet partial waves only. In the isospin-singlet par-tial waves the force model is purely nucleonic; the Paris potential [4] is used as parametrization.

sonant and nonresonant pion-nucleon partial waves provides the natural framework for a physically well-motivated approximation scheme, which keeps resonant P_{33} pion-nucleon scattering exactly, but treats the nonresonant contributions perturbatively.

The force model is identical in spirit to the one of T.-S.H. Lee [5]. Conceptually, it exploits the successful phenomenology of purely nucleonic potentials and adds pion production and absorption as a correction. This is the strategy suggested years back by A.M. Green [6]. In contrast, the complementary approach of Ref. [7] considers the three-body πNN dynamics as basic physics, which is coupled to the two-nucleon system through the nucleon pole of P_{11} pion-nucleon scattering. The latter procedure attempts to derive the two-nucleon interaction without any reference to the existing corresponding phenomenology. It is therefore clearly more ambitious, but also less successful in its account of experimental data.

3. Results

The force model is tested in two-nucleon scattering above pion threshold. It is employed to describe elastic scattering together with some of its unitarily coupled channels. The present calculation uses pion- and rho-exchange for the Δ-excitation of Fig. 2(b) and puts the diagonal nucleon-Δ potential of Fig. 2(d) to zero, reserving its parameters for a later quantitative calibration. At this time no attempt is made to fit the force model to data.

The experimental data for pion production are the basis for the special features of the force model. Though the important break-up channel has not yet been investigated in detail, its overall coupling to the two-nucleon channel is calculated and found to yield the correct strength according to Fig. 3. In the following characteristic examples for the obtained results are given.

(1) Elastic two-nucleon scattering:

Phase shifts and inelasticities for all isospin-triplet partial waves with total angular moemtnum $J \leq 4$ are computed. In all partial waves the reproduction of experimental data is as good as for the Paris potential [4] at lower energies and in most better at intermediate energies. Fig. 4 shows the results for the $^1D_2$ and $^3F_3$ partial waves. In the ^3P_J partial waves, not documented here, the theoretical phase shifts show systematic deficiencies at higher energies.

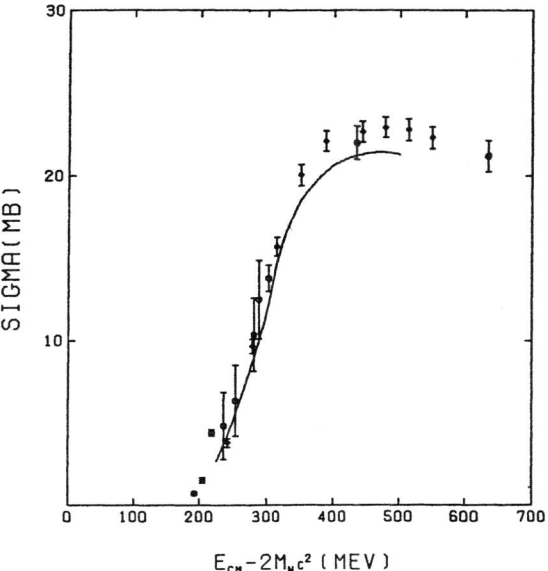

Fig. 3

Total inelastic cross section pp→πNN. In contrast to Fig. 1 the inelasticity does not contain the πd channel. The prediction of the force model of Fig. 2 is compared to data [1].

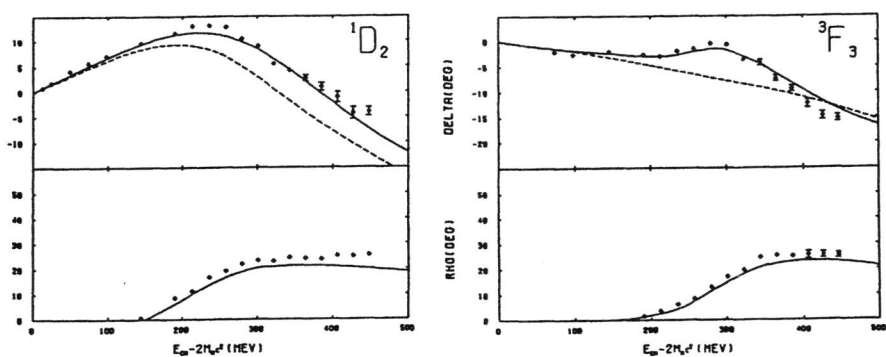

Fig. 4

Nucleon-nucleon phase shifts and inelasticities for two selected isospin-triplet partial waves at intermediate energies. The predictions of the force model of Fig. 2 (solid curve) and of the purely nucleonic Paris potential (dashed curve) are compared to data [8].

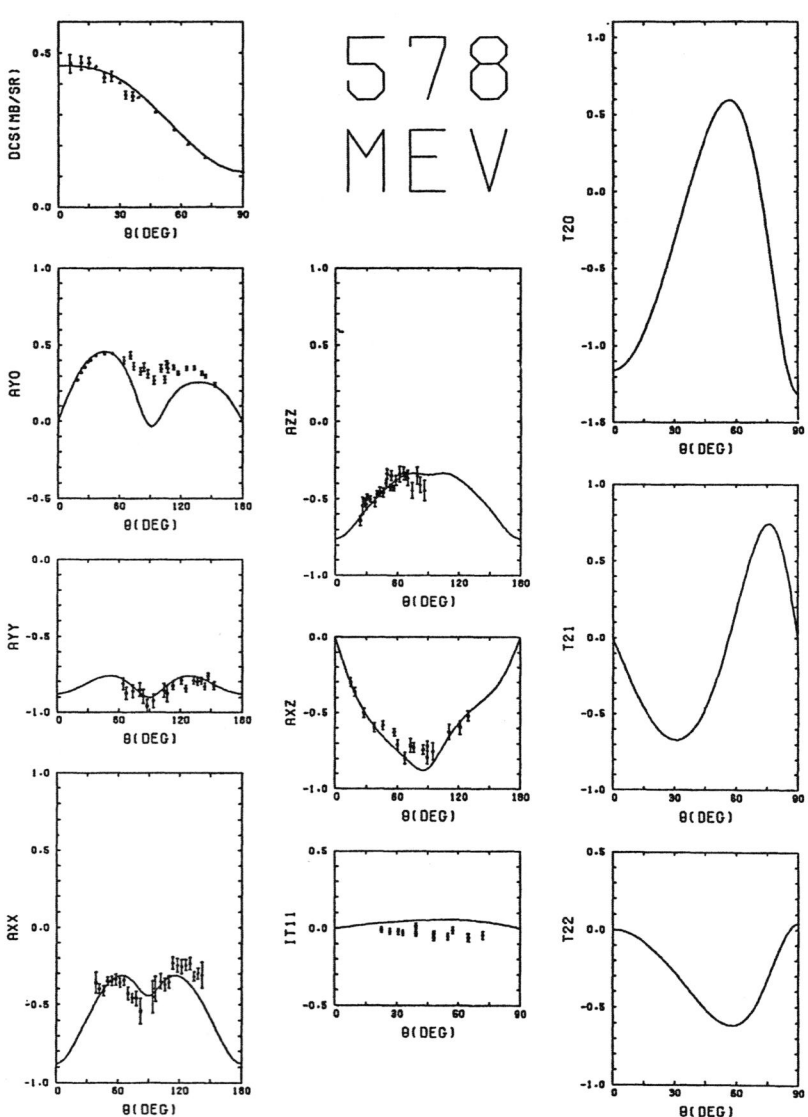

Fig. 5

Observables for the pp→π⁺d reaction at 578 MeV proton lab energy as a
function of the pion c.m. angle θ. The prediction of the force model of
Fig. 2 is compared to existing data [9].

(2) Inelastic pp↔ πd process:

Due to the large number of observables results are presented in Fig. 5
only for one selected two-nucleon lab energy, i.e., 578 MeV, for which a
large amount of experimental data is available. The spin-averaged differential cross section and the spin observables are qualitatively reproduced.
The analyzing power A_{yo} has attracted special interest. Though our result
shows too strong an oscillatory structure, the order of magnitude is in
satisfactory agreement with the data. The correct description of the coupled elastic two-nucleon channel is found vital for this agreement as Fig.
6 proves. We think this is the reason why other calculations [11] fail
badly for A_{yo}.

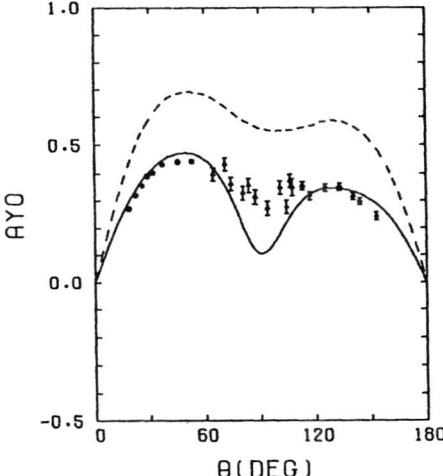

Fig. 6

Analyzing power A_{yo} for the pp→π⁺d reaction at 578 MeV proton lab energy as
a function of the pion c.m. angle. THe prediction of the force model of
Fig. 2 (solid curve) is compared to a model (dashed curve) in which artificially the account of two-nucleon scattering is spoilt: The purely nucleonic part of the interaction is decreased by an overall factor of 0.5.

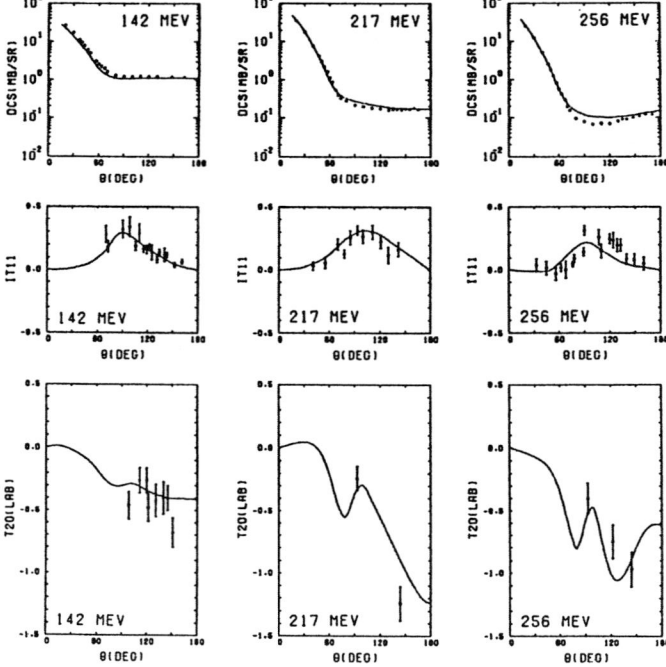

Fig. 7

Observables for elastic pion-deuteron scattering at three pion lab kinetic energies as a function of the pion c.m. scattering angle. The prediction of the force model of Fig. 2 is compared to data [12,13].

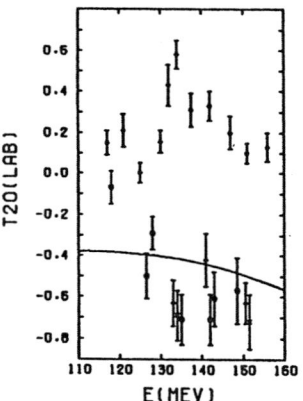

Fig. 8

Tensor polarization t_{20}^{lab} of elastic pion-deuteron scattering at 150° pion c.m. scattering angle as a function of pion lab energy. The data refer to Ref. [13] (squares) and to Ref. [14] (other symbols).

(3) Elastic pion-deuteron scattering

Fig. 7 shows the results for the spin-averaged differential cross section and for the spin observables it_{11} and t_{20}^{lab} at four selected pion lab kinetic energies. The spin averaged differential cross section is overestimated with increasing energy in the angular region between 80° and 150° pion c.m. angle, a deficiency shared by all comparable force models. With respect to the tensor polarization t_{20}^{lab} of Figs. 7 and 8, our prediction is compatible with the measurements which find negative values. The positive and strongly varying data of Ref. [14] can clearly not be reproduced.

4. Conclusion

The force model of Fig. 2 is applied to the two-nucleon system above pion threshold. A qualitatively satisfactory and for most observables even a quantitative agreement between theoretical description and experimental data is achieved. The existing discrepancies are not so severe that they required a calibration of the force model at this stage or that they even called for more exotic degrees of freedom.

The force model is not tested in the two-nucleon system for its own sake. It is constructed to also be applied later on in many-nucleon systems and to provide the microscopic basis for nuclear structure and nuclear reactions at intermediate energies [2,15]. In particular, the presented force modell yields corrections for the classic theory of nuclear phenomena in terms of nucleons only. Examples for the corrections [15] are three-nucleon forces, electromagnetic and weak exchange currents, pion absorption and pion production and the excitation and propagation of the Δ-resonance in the nuclear medium. In nuclear structure, the Δ-isobar and the pion appear as constituents of bound many-nucleon systems, whereas they contribute to the reaction mechanism in nuclear reactions with hadronic and leptonic probes. The considered force model gives a consistent microscopic description for the different roles which the Δ-isobar and the pion play and for the different corrections which they yield in nuclear phenomena. The present discussion of the force model in two-nucleon scattering above pion threshold lays the basis for these applications.

Acknowledgement

The talk is largely based on work done in collaboration with H. Pöpping and Zhang Xi-Zhen. Financial support by funds of the BMFT, project MEP 0234 HAA, is gratefully acknowledged.

References

1. Flaminio, V., et al.: CERN-HERA 84-01 data compilation (1984)
2. Sauer, P.U.: Prog. Part. Nucl. Physics 16, 35 (1986)
3. Pöpping, H., Sauer, P.U., Zhang Xi-Zhen, to be published
4. Lacombe, M.,et al.: Phys. Rev. C21, 861 (1980)
5. Lee, T.-S.H.: Phys. Rev. C29, 195 (1984)
6. Green, A.M.: Rep. Prog. Phys. 39, 1109 (1976)
7. Mizutani, T., et al.: Phys. Lett. 107B, 177 (1981)
   Blankleider, B., Afnan, I.R.: Phys. Rev. C24, 1572 (1981)
   Ueda, T.: Phys. Lett. 119B, 281 (1982)
   Rinat, A.S., Starkand, Y.: Nucl. Phys. A397, 381 (1983)
8. Arndt, R.A., et al.: Phys. Rev. D28, 97 (1983)
9. Bugg, D.V.: J. Phys. G, Nucl. Phys. 10, 717 (1984)
10. Laptev, A.B., Strakovsky, I.I.: A Collection of Experimental Data for the pp$\leftrightarrow\pi^+$d Process, Leningrad Nuclear Physics Institute, Leningrad 1985
11. Seth, K.K., et al.: Phys. Lett. 126B, 164 (1983)
12. Gabathuler, K., et al.: Nucl. Phys. A350, 253 (1980)
    Smith, G.R., et al.: Phys. Rev. C29, 2206 (1984)
13. Holt, R.J., et al.: Phys. Rev. Lett. 47, 472 (1981)
    Ungricht, E., et al.: Phys. Rev. C31, 934 (1985)
    Shin, Y.M., et al.: TRIUMF preprint 1985
    Smith, G.R., et al.: TRIUMF preprint 1986
14. König, V., et al.: Proceedings of the Few Body X Conference, ed. B. Zeitnitz, Amsterdam: North-Holland 1984, Vol. II, p. 169
15. Sauer, P.U., Lecture Notes in Physics 260, eds. B.L. Berman and B.F. Gibson, Berlin: Springer 1986, p. 107

SKYRME SOLITONS, EFFECTIVE LAGRANGIANS AND STATIC PROPERTIES OF BARYONS

R. Vinh Mau

Division de Physique Théorique[::], Institut de Physique Nucléaire, 91406 Orsay
and L.P.T.P.E.[::], Université Pierre et Marie Curie, 75252 Paris Cedex 05.

Predictions of the static properties of *baryons* from a chiral invariant
effective Lagrangian constructed from *pions* and the low mass *mesons*, are
discussed. The agreement with experiment is quite satisfactory. This gives strong support to the idea that baryon physics can be deduced from
meson physics via *solitons*.

1. Introduction

The conjecture of Skyrme [1] that a nucleon can be approximately regarded as a soliton in the pion and sigma fields is fairly well supported by
the phenomenological success in the prediction of the static properties of
baryons [2]. These encouraging results suggest that baryon physics can be
deduced from meson physics via solitons [3]. In such a general program the
Skyrme model is probably too crude for a simultaneous fit of the meson physics and the baryon physics since it contains only the pion field. On the
other hand a basic theory for meson physics is still missing. One way to
get around this shortcoming to construct effective Lagrangians implementing
the known phenomenological features of meson physics at energies below about
1 GeV. In particular, any minimal phenomenological description of meson physics must take into account the observed low mass mesons such as the ρ meson, the ω meson, the ε meson (which is responsible for the enhancement at
around 1 GeV of the $\pi\pi$S wave).

[::] Laboratoire Associé au C.N.R.S.

In this talk, I shall discuss the work along this line by Lacombe et al. [4]. References on other related works can be found in this reference. In their work [4], Lacombe et al. proceed as follows

i) an effective Lagrangian from the non linear σ model along with a chiral singlet scalar meson (the ϵ meson), a chiral singlet vector meson (the ω meson) and isovector vector mesons (the ρ meson and its chiral partner the A_1 meson) is constructed in a chirally invariant way.

ii) the parameters of this effective Lagrangian are determined in the meson sector.

iii) Skyrme type soliton solutions of this effective Lagrangian are investigated and from these are derived not only the properties of skyrmions (which is not too hard) but also the static properties of the physical baryons (which is more difficult). Predictions on the baryon-baryon interaction can be also made.

2. The model

The effective Lagrangian density considered in ref. [4] is of the form

$$\mathcal{L}(x) = \mathcal{L}_\epsilon(x) + \mathcal{L}_\omega(x) + \mathcal{L}_{\pi\rho A_1}(x) \tag{1}$$

where

$$\mathcal{L}_\epsilon(x) = \frac{1}{2}(\partial_\mu \epsilon)^2 - \frac{1}{2} m_\epsilon^2 \epsilon^2 - \delta_\epsilon \, \mathrm{Tr}(\partial_\mu U \partial^\mu U^+) \tag{2}$$

$$\mathcal{L}_\omega = -\frac{1}{4}(\partial_\mu \omega_\nu - \partial_\nu \omega_\mu)^2 + \frac{1}{2} m_\omega^2 \omega_\mu^2 + \beta \omega_\mu B^\mu \tag{3}$$

with

$$B_\mu = \frac{1}{24\pi^2} \epsilon_{\mu\alpha\beta\gamma} \, \mathrm{Tr}(U^+ \partial^\alpha U \, U^+ \partial^\beta U \, U^+ \partial^\gamma U) \tag{4}$$

are the ϵ and ω meson terms.

The third term,

$$\mathcal{L}_{\pi\rho A_1}(x) = -\frac{1}{8} \mathrm{Tr}(X_{\mu\nu}^2 + Y_{\mu\nu}^2) + \frac{1}{4} m_\rho^2 (X_\mu^2 + Y_\mu^2) + f\mathrm{Tr}[D_\mu U(D^\mu U)^+] \tag{5}$$

results from the introduction of the isospin-1 vector meson (ρ and A_1) as gauge fields in the non linear σ model through the transformations

$$D_\mu U = \partial_\mu U + ig(X_\mu U - U Y_\mu) \tag{6}$$

$$
\begin{aligned}
X_{\mu\nu} &= \partial_\mu X_\nu - \partial_\nu X_\mu + ig[X_\mu, X_\nu] \\
Y_{\mu\nu} &= \partial_\mu Y_\nu - \partial_\nu Y_\mu + ig[Y_\mu, Y_\nu]
\end{aligned}
\tag{7}
$$

with

$$
\begin{aligned}
X_\mu &= \vec{X}_\mu \cdot \vec{\tau} \\
Y_\mu &= \vec{Y}_\mu \cdot \vec{\tau} \\
U &= e^{2i\vec{\tau}\cdot\vec{\phi}/F_\pi} , \quad \vec{\phi} \text{ is the pion field.}
\end{aligned}
\tag{8}
$$

In eqs. (5)-(7), in place of the ρ and A_1 fields, the left-handed $\vec{X}_\mu$ and right-handed $\vec{Y}_\mu$ fields are used. They are more convenient for the construction of chirally invariant Lagrangians.

In the meson sector, the above Lagrangian describes the low energy dynamics of pions and the other mesons. When the U field acquires the soliton configuration, and is identified as a baryon, it also accounts for the coupling of these mesons with baryons. These couplings are constructed in a chiral invariant way. Expressing the physical ρ and A_1 meson fields in terms of $\vec{X}_\mu$, $\vec{Y}_\mu$ and $\partial_\mu\vec{\phi}$ and normalizing the pion field as usual, yield the relations $m_A^2 = m_\rho^2 + 8fg^2$ and $16f = F_\pi^2 m_A^2/m_\rho^2$ which determine the parameters f and g in terms of the known quantities m_ρ, m_A and F_π.

If, as usual, the hedgehog ansatz is assumed for the static field $U(\vec{r}) = e^{i\vec{\tau}\cdot\hat{r}\theta(r)}$, and for the space components of the X and Y fields

$$
\begin{aligned}
X_i &= \alpha(r)(\hat{r}\wedge(\vec{\tau}\wedge\hat{r}))\cdot\hat{i} + \beta(r)(\vec{\tau}\cdot\hat{r})(\hat{r}\cdot\hat{i}) + \gamma(r)(\vec{\tau}\wedge\hat{r})\cdot\hat{i} \\
Y_i &= -\alpha(r)(\hat{r}\wedge(\vec{\tau}\wedge\hat{r}))\cdot\hat{i} - \beta(r)(\vec{\tau}\cdot\hat{r})(\vec{\tau}+\hat{i}) + \gamma(r)(\vec{\tau}\wedge\hat{r})\cdot\hat{i}
\end{aligned}
\tag{9}
$$

The time components of the X and Y fields are zero in the static limit as are the space components of the ω field.

The static energy of the system is then

$$
\begin{aligned}
E = 4\pi \int dr\, r^2 \Big\{ &-\frac{1}{2}(\omega_o'^2 + m_\omega^2\omega_o^2) + \frac{\beta_\omega}{2\pi^2}\omega_o\,\frac{\theta'\sin^2\theta}{r^2} \\
&+ \frac{1}{2}(\epsilon'^2 + m_\epsilon^2\epsilon^2) - 2\delta_\epsilon\epsilon(\theta'^2 + \frac{2\sin^2\theta}{r^2}) +
\end{aligned}
$$

$$+ m_\rho^2(2\alpha^2 + \beta^2 + 2\gamma^2) + 4(\frac{\gamma}{r} - g(\alpha^2 + \gamma^2))^2 \tag{10}$$

$$+ 2(\gamma' + \frac{\gamma}{r} - 2g\alpha\beta)^2 + 2(\alpha' + \frac{1}{r}(\alpha-\beta) + 2g\beta\gamma)^2$$

$$+ 2f(\theta' + 2g\beta)^2 + 4f(\frac{\sin\theta}{r} + 2g(\alpha\cos\theta - \gamma\sin\theta))^2\}$$

Stationary values of E with respect to θ, ϵ, ω_o, α, β and γ give the classical Euler-Lagrange equations for these variables. Soliton solutions exist with the boundary conditions $\theta(0) = \pi$ and $\theta(\infty) = 0$. It is then straightforward to calculate the classical soliton mass M by inserting these solutions into equ.(10).

However, the interesting physical quantities are rather the baryon masses, radii, magnetic moments, etc... To calculate them, one has to quantize the classical solutions. This can be done by introducing rotational dynamics through a unitary transformation [2] A(t) applied to the U, X_i and Y_i fields.

This rotation induces non trivial field configurations for the space components of the ω field [5], [6]

$$\omega_i(r) = \frac{\beta_\omega}{2\pi^2 m_\omega^3} \frac{I(r)}{r^3} \epsilon_{ij\ell} \kappa^j r_\ell \tag{11}$$

and for the time components X_o, Y_o of the gauge fields

$$X_o = a(r)(\hat{r}\wedge(\vec{\tau}\wedge\hat{r}))\cdot\vec{\kappa} + b(r)(\vec{\tau}\cdot\hat{r})(\hat{r}\cdot\vec{\kappa}) + c(r)(\vec{\tau}\wedge\hat{r})\cdot\vec{\kappa}$$
$$Y_o = a(r)(\hat{r}\wedge(\vec{\tau}\wedge\hat{r}))\cdot\vec{\kappa} + b(r)(\vec{\tau}\cdot\hat{r})(\hat{r}\cdot\vec{\kappa}) - c(r)(\vec{\tau}\wedge\hat{r})\cdot\vec{\kappa} \tag{12}$$

with

$$\vec{\kappa} = -\frac{i}{2} Tr(A^+ \frac{dA}{dt} \vec{\tau}) \tag{13}$$

The variables I, a, b and c are again determined by a variation principle applied to the effective Lagrangian which now reads $L = -M + \lambda\ Tr(dA/dt\ dA^+/dt)$.

The moment of inertia λ for the soliton in the present model is far more difficult to calculate than its mass M. This has been done for the first time in ref. [4] and

$$\lambda = \frac{2\pi}{3} \int_0^\infty dr \; r^2 \left\{ -16 \, \delta_\epsilon \, \epsilon(r) \sin^2\theta + \frac{\beta_\omega^2}{2\pi^4 m_\omega^3} \theta' \frac{\sin^2\theta}{r^3} I(r) + \right.$$

$$+16f \, [\sin\theta + g(a \sin\theta + c \cos\theta)]^2$$

$$+ 2[\frac{a-b}{r} + 2\gamma(1+gb)]^2 + 2[\frac{c}{r} - 2\alpha(1+gb)]^2$$

$$+ 2[\frac{a-b}{r} + 2 \, (gc\alpha - \gamma(1+ga))]^2 \qquad (14)$$

$$+ 2[\frac{c}{r} - 2(gc\gamma + \alpha(1+ga))]^2$$

$$+ 2[a' + 2gc\beta]^2 + 2[c' - 2\beta(1+ga)]^2$$

$$\left. + b'^2 + m_\rho^2 \, [2(a^2 + c^2) + b^2] \right\}$$

The nucleon and delta masses are given in terms of the soliton mass M and its moment of inertia λ by $m_N = M + 3/8\lambda$ and $m = M + 15/8\lambda$. Other static properties of single baryons such as mean square radii, magnetic moments, and the axial charge are also computed in ref. [4]. They are given by

$$<r^2>_{I=0} = -(2/\pi) \int_0^\infty r^2 \theta' \, \sin^2\theta \, dr \qquad (15)$$

for the isoscalar mean-square radius,

$$<r^2>_{M,I=0} = [\int_0^\infty r^4 \theta' \, \sin^2\theta \; dr] \, / \, [\int_0^\infty r^2 \theta' \, \sin^2\theta \; dr] \qquad (16)$$

for the isoscalar magnetic mean-square radius, and

$$\vec{\mu} = 2m_N \vec{\sigma}[\; -(1/6\pi\lambda) \int_0^\infty r^2 \theta' \, \sin^2\theta \; dr + (4\pi/9)\tau_3 \int_0^\infty F(r)r^3 dr] \qquad (17)$$

for the nucleon magnetic-moment operator in units of nuclear magnetons, with

$$F(r) = 4f \sin\theta \; [(\sin\theta)/r + 2g(\alpha \cos\theta - \gamma \sin\theta)]$$

$$- 2[\beta\alpha' + (\alpha\beta - \beta^2 - 2\gamma^2)/r + 2g\gamma(\alpha^2 + \beta^2 + \gamma^2)]$$

$$+ (\beta_\omega/2\pi^2)\omega_0 \theta'(\sin^2\theta)/r - 4 \, \delta_\epsilon \epsilon(\sin^2\theta)/r \qquad (18)$$

The axial coupling constant is

$$g_A = 4\pi \int_0^\infty G(r) \, r^2 \, dr \tag{19}$$

with

$$G(r) = - (2f/3)[(\sin 2\theta)/r + \theta' + 4g \cos\theta(\alpha \cos\theta - \gamma \sin\theta) + 2g\beta]$$

$$+ \frac{2}{3} \{(\alpha + \beta)(\gamma' + 2\gamma/r) - \alpha'\gamma - 2g[(\alpha + \beta)(\alpha^2 + \beta^2 + \gamma^2) - \beta^3]\}$$

$$- (\beta_\omega/12\pi^2)[\theta'(\sin 2\theta)/r + (\sin^2\theta)/r^2] \, \omega_0 +$$

$$+ (2\delta_\epsilon/3)[\theta' + (\sin 2\theta)/r]\epsilon \tag{20}$$

3. Results

Among the parameters of the model, the mesons masses m_ω, m_ϵ, m_ρ, m_A and the pion decay constant F_π are known, β_ω can be identified with the ωNN vector coupling constant which is known to lie between 10 and 15. One can also get an upper bound $\beta_\omega < 25$ from the ω meson width [5] Finally, one can attempt to determine δ_ϵ from the $\pi\pi S$ wave phase shifts. It is preferable, for the time being, to leave δ_ϵ as a free parameter. Actually as it was already found in previous work [6] δ_ϵ is bounded by the stability condition for the solutions of the Euler Lagrange equations. However, taking the meson masses, the pion decay constant, and β_ω at their physical values i.e. m_ω = 782.6 MeV, m_ϵ = 800 MeV, m_ρ = 769 MeV, m_A = 1275 MeV, F_π = 186 MeV and β_ω = 12, it is found that with the limited possibility of varying δ_ϵ, the nucleon and delta masses were too high. The best values are m_N = 1308 MeV and m_Δ = 1651 MeV. It should be, however, noted that the mass difference between the delta and nucleon agrees better with experiment. This is more important than the values for absolute masses. In view of a comparison with the original Skyrme model one can try to adjust F_π and δ_ϵ to fit the baryon static masses This can be achieved with the meson masses kept at their physical values, β_ω = 13.65 and δ_ϵ = 27.3 MeV but with F_π = 142.4 MeV somewhat smaller than its known value. This value of F_π is however closer to experiment than that required to fit the nucleon and delta masses in the Skyrme model [2].

The results along with those of the original Skyrme model and experimental data are listed in Table I .

Table I

Static properties of baryons

Observable	Results of this calculation	Results with the Skyrme model[2]	Experiment
m_N (MeV)	input	input	938.9
m_Δ (MeV)	input	input	1232.0
F_π (MeV)	142.4	129	186
$\langle r^2 \rangle^{1/2}_{I=0}$ (fm)	0.66	0.59	0.72
$\langle r^2 \rangle^{1/2}_{M,I=0}$ (fm)	0.83	0.92	0.81
μ_p ($e/2m_N$)	2.33	1.87	2.79
μ_n ($e/2m_N$)	-1.63	-1.31	-1.91
$|\mu_p/\mu_n|$	1.42	1.43	1.46
g_A	0.62	0.61	1.23

Apart from the axial coupling constant g_A, the agreement with experimental data is quite satisfactory. It is generally better than that given by the original Skyrme model. It is worthmentioning that one can fit the experimental values of g_A and of F_π by increasing β_ω but at the expense of having too high baryon masses.

It was shown in ref. [6] that, in the approximation where the kinetic terms of the ε and ω mesons are neglected in the field equations by comparison with their mass terms, the ε and ω fields can be eliminated in the lagrangians $\mathcal{L}_\varepsilon$ and $\mathcal{L}_\omega$ to yield terms in powers of derivatives of the pion field,

$$\mathcal{L}_\varepsilon(x) \rightarrow \frac{\delta^2_\varepsilon}{m^2_\varepsilon} (\text{Tr } \partial_\mu U \partial^\mu U^+)^2 \tag{21}$$

$$\mathcal{L}_\omega(x) \rightarrow \frac{\beta^2_\omega}{m^2_\omega} B_\mu B^\mu \tag{22}$$

Likewise, if the same approximation is applied to the ρ and A_1 mesons, $\mathcal{L}_{\pi\rho A_1}$ becomes formally identical to the Skyrme Lagrangian

$$\mathcal{L}_{\pi\rho A_1}(x) \rightarrow \frac{F^2_\pi}{16} \text{Tr } \partial_\mu U \partial^\mu U^+ + \frac{1}{32e^2} \text{Tr}[\partial_\mu U U^+, \partial_\nu U U^+]^2 \tag{23}$$

with now a coupling constant given by

$$e^2 = 2g^2 (1 - m_\rho^4/m_A^4)^{-2} \qquad (24)$$

This shows that effective Lagrangians resulting from local expansion in derivatives of the pion field, like the original Skyrme model, can be derived as approximations of those involving other observed mesons.

The nucleon and delta masses were also calculated with the same set of parameters as above but in this approximation and the good fit is spoilt (m_N = 1010 MeV, m_Δ = 1168 MeV). This confirms previous finding [6] that the local expansion of the effective Lagrangian, keeping only a few powers of derivatives of the pion field alone, is of limited accuracy.

The contribution of $\mathcal{L}_\varepsilon$ and $\mathcal{L}_\omega$ to the nucleon-nucleon interaction have been derived in ref. [6]. The complete calculation of the contribution from $\mathcal{L}_{\pi\rho A_1}$ is currently underway [7].This contribution can however be readily estimated approximately as above. The results show that the model is able to provide the medium range attraction in the central NN potential which is lacking in the original Skyrme model.

It must be stressed that the good description of baryons provided by the Lagrangian (1) has been obtained with very little freedom in its form and in the choice of the parameter values and the results obtained give strong support to the idea that one can predict the low energy physics of baryons to a good approximation from an effective Lagrangian constructed with meson fields alone.

References

(1)   T.H.R. Skyrme, Proc. Roy. Soc. A260, 127 (1961), Nucl. Phys. 31, 556 (1962).

(2)   G.S. Adkins, C.R. Nappi and E. Witten, Nucl. Phys. B228, 552 (1983). M. Rho, A.S. Goldhaber and G.E. Brown, Phys. Rev. Lett. 51, 747 (1983) ; A.D. Jackson and M. Rho, Phys. Rev. Lett. 51, 751 (1983),

(3)   E. Witten, in Solitons in Nuclear and Elementary Particle Physics, edited by A. Chodos, E. Hadjimichael, and C. Tze (World Scientific, Singapore, 1984), p. 306.

(4)   M. Lacombe, B. Loiseau, R. Vinh Mau and W.N. Cottingham, Phys. Rev. Lett. 57, 170 (1986).

(5)   G.S. Adkins and C.R. Nappi, Phys. Lett. 137B, 251 (1984).

(6)   M. Lacombe, B. Loiseau, R. Vinh Mau and W.N. Cottingham, Phys. Lett. B169, 121 (1986).

(7)   M. Lacombe, B. Loiseau, R. Vinh Mau and W.N. Cottingham, to be published.

RESONATING GROUP METHOD AND PAULI REPULSION OF CLUSTERS

H. Walliser

Institut für Theoretische Physik, Universität Tübingen,
D-7400 Tübingen, Federal Republic of Germany

In few body physics, a major task of the resonating group method is to provide effective interactions between composite particles or clusters. These interactions contain information about the internal structure of the clusters, the forces between their constituents and most importantly about the antisymmetrization of these constituents. Conventional resonating group calculations usually span, because of so-called forbidden states, a restricted relative motion space. However, these forbidden states appear only in a very singular case. The consideration of the complete relative motion space leads to effective interactions describing the short ranged Pauli repulsion more appropriately. It will be shown that this improvement may have decisive consequences in few body systems, taking the 3α system as an example.

1. Introduction

The interaction of clusters composed by fermions should reflect the Pauli repulsion for small relative distances. Two rare gas atoms may feel for example the repulsive part of a local Morse potential. Because the Born-Oppenheimer approximation is no longer valid for hadronic clusters, the situation there becomes more involved. The appropriate microscopic theory is the resonating group method (RGM) [1]. Because the RGM observes the antisymmetrization of the underlying fermions rigorously, it should describe the Pauli repulsion properly. However, in conventional RGM calculations, relative motion states appear which are forbidden by the Pauli principle. This leads to the truncation of that part of the relative motion space which is most relevant for the description of the Pauli repulsion. It should be noted that this truncation does not follow from the dynamics of the system but rather is imposed by a specific assumption about the internal cluster wave functions used frequently in the RGM. Namely, forbidden states appear only if all clusters are described by lowest configurations in harmonic oscillator wells with the same frequency. Any other description will allow for the full relative motion space and all amplitudes are determined dynamically [2].

It is this observation which leads to refined effective potentials acting in the complete relative motion space and describing the Pauli repulsion more appropriately. In contrast to the rare gas case the Pauli repulsion of hadronic clusters turns out to be manifestly non-local.

In the following I will mainly refer to the $\alpha\alpha$ system although the findings can easily be applied to many other systems as well. As an application I will report about effective $\alpha\alpha$ interactions used in 3α calculations.

2. Formulation

We assume a system of two clusters e.g. two α particles, described by internal wave functions ϕ_1 and ϕ_2 . Starting point is the well-known single channel RGM ansatz [1]

$$|\Psi\rangle = |Au\phi_1\phi_2\rangle \tag{1}$$

for the many body wave function Ψ . For clarity in presentation angular momentum indices are suppressed. The antisymmetrizer A exchanges particles between the two clusters. The ansatz (1) leads to the RGM equation

$$[H-EN]|u\rangle = o \tag{2}$$

for the relative motion function u . Matrixelements of the norm kernel N and the hamilton kernel H , e.g. in the basis $\{\varphi_n\}$ of harmonic oscillator relative motion states

$$\langle\varphi_n|N|\varphi_{n'}\rangle = \langle\Phi_n|\Phi_{n'}\rangle$$
$$\langle\varphi_n|H|\varphi_{n'}\rangle = \langle\Phi_n|H - E_{int}|\Phi_{n'}\rangle \tag{3}$$

are readily expressed in terms of the corresponding many body states

$$|\Phi_n\rangle = |A\varphi_n\phi_1\phi_2\rangle , \tag{4}$$

where H is the microscopic hamiltonian and E_{int} the cluster internal energy. Many authors [2–8] now agree that the mathematically equivalent normalized RGM equation

$$[\tilde{H}-E]|\tilde{u}\rangle = o , \quad \tilde{H} = N^{-1/2}HN^{-1/2} , \quad |\tilde{u}\rangle = N^{1/2}|u\rangle \tag{5}$$

is accessible to physical interpretations more directly. For the derivation of an effective potential V_{eff} used in a Schrödinger type of equation it is even a necessity to consider the normalized eq. (5), i.e.

$$\tilde{H} \equiv T + V_{eff} = T + V_D + \tilde{V} . \tag{6}$$

T is the kinetic energy operator of the relative motion. The effective potential V_{eff} consists of the local direct or folding potential V_D, obtained by folding the microscopic interaction of the constituents over the cluster densities, and a non-local part $\tilde{V}$ which contains all antisymmetrization effects of the two cluster system.

So far everything seems to be straightforward. However, the step from the original RGM eq. (2) to the normalized eq. (5) can only be carried out if the norm kernel N is non-singular. In conventional RGM calculations where the clusters ϕ_1 and ϕ_2 are represented by oscillator shell model ground states with the same frequency this is generally not the case. So-called redundant or forbidden states appear and the transition from eq. (2) to eq. (5) involves a zero divided by zero.

In order to understand this in more detail, let us consider a simple analytical model. We assume a microscopic oscillator hamiltonian $H = H_{osc}$ and internal oscillator wave functions ϕ_1 and ϕ_2 as discussed above, all with the same frequency. Then, the many body states (4) become orthogonal eigenstates of H_{osc} [1,9]

$$[H_{osc} - E_{int}]|\Phi_n> = \varepsilon_n |\Phi_n> \; , \; \varepsilon_n = (n + \frac{3}{2})\hbar\omega \; , \tag{7}$$

and both, the norm matrix and hamilton matrix (3) are diagonal. This is due to the special quadratic form of the oscillator hamiltonian H_{osc} which can be separated into internal and intercluster coordinates. For the oscillator shell model states Φ_n of eq. (7), a minimum number of oscillator quanta is required from the Pauli principle, e.g. $4\hbar\omega$ for the $(0s)^4(0p)^4$ state of ^8Be (Fig. 1). In the cluster picture these quanta must be carried by the relative motion function φ_n if the clusters are assumed

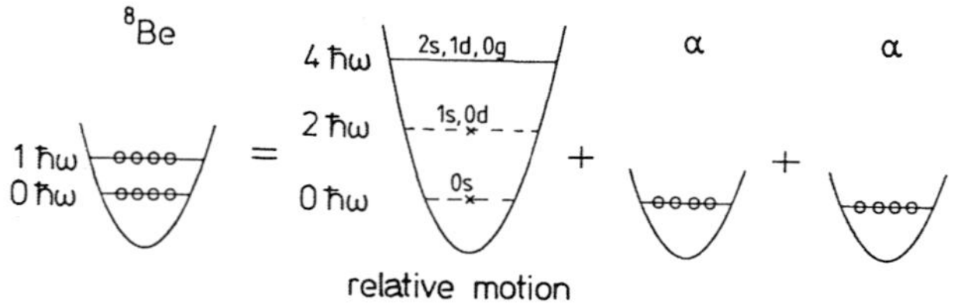

Figure 1

Illustration of the equivalence of shell model states and cluster states in the oscillator model for the $\alpha\alpha$ system. From counting oscillator quanta it is immediately concluded that the 0s, 1s and 0d relative motion states of two α-particles are forbidden.

to be unexcited. The norm of states Φ_ν with less quanta in the relative
motion function φ_ν vanishes by antisymmetrization e.g. in the $\alpha\alpha$ case
$\nu = 0s,1s,0d$ (Fig. 1). Because of these zero norm states, the norm kernel
(3) becomes singular and the transition to the normalized RGM eq. (5) is
no longer possible in a rigorous mathematical sense.

Orthogonality Condition Model

One way out is to exclude the space of Pauli forbidden states,
$<\varphi_\nu|\tilde{u}> = o$ for ν redundant. In the remaining relative motion space $\tilde{H}$ can
be evaluated without any problem. Because in the analytical model with
$H = H_{osc}$ the norm matrix and hamilton matrix are already diagonal, we
readily obtain

$$\tilde{H}_{osc} = \sum_n \varepsilon_n |\varphi_n><\varphi_n| = T + V_{osc} \; , \quad <\varphi_\nu|\tilde{u}> = o \; , \qquad (8)$$

where V_{osc} is the local relative motion oscillator potential. For practical calculations V_{osc} can be replaced by a direct potential V_D derived
from a realistic microscopic interaction, or fitted to scattering data.
This leads to the familiar orthogonality condition model (OCM) [3]

$$\tilde{H} = T + V_D \; , \quad <\varphi_\nu|\tilde{u}> = o \; , \; \nu \; red \; . \qquad (9)$$

In the OCM all antisymmetrization effects are contained in the orthogona-
lity conditions $<\varphi_\nu|\tilde{u}> = o$.

Finite Pauli Repulsion Model

Remembering that forbidden states appear only in the special case described above, the wilful removal of a part of the relative motion space
is certainly not the best solution. Any small deviation, e.g. slightly
different oscillator frequencies for different clusters, or Saxon-Woods
shell model functions for identical clusters will leave all many body
states Φ_n with a finite norm and the transition to the normalized eq. (5)
is performed without difficulty. Even in the limit when this deviation
goes to zero, the result is stable and quite different from the OCM.

We will see that the result cannot be influenced much by the particular
deviation chosen, which we may characterize by a parameter ε

$$|\Phi_n^\varepsilon> = |A\varphi_n\phi_1^\varepsilon\phi_2^\varepsilon> \; . \qquad (10)$$

To remain as close as possible to the analytical oscillator model, we consider the limit $\varepsilon \to o$. Numerical calculations have shown that this limit
is a fair approximation of the realistic case of finite ε .

While the allowed states Φ_n^ε remain unchanged in the limit $\varepsilon \to o$ leading
still to eq. (8), the remainder of the Pauli forbidden states Φ_ν^ε surviving

36

antisymmetrization must gain additional oscillator quanta. Thus, after
orthogonalization (diagonalization of the norm matrix) we get, in lowest
non-vanishing order of ϵ , oscillator states again, but with an additional
energy a_ν necessary to push these states above the Fermi level. Independently from the particular deviation chosen, this must be an excited level
for all p-shell compound systems (Fig. 2), because the non-degenerate

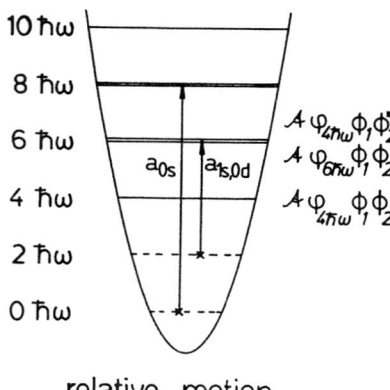

relative motion

Figure 2

Formation of Pauli resonances in the $\alpha\alpha$ system. The
forbidden states (compare Fig. 1) experience an energy
shift of $a_{0s} = 8\hbar\omega$ and $a_{1s} = a_{0d} = 4\hbar\omega$ and are pushed
above the Fermi level ($4\hbar\omega$ level). On the $2\hbar\omega$ excited
level we thus find the allowed states and additional
states of $\alpha\alpha^*$ structure. The latter and the corresponding $4\hbar\omega$ excited state appear as Pauli resonances
in a realistic calculation.

ground state is already contained in the allowed basis. In the $\alpha\alpha$ system we
obtain two states of $\alpha\alpha^*$ structure at the $2\hbar\omega$ excited level and one state
at the $4\hbar\omega$ excited level.

For $\tilde{H}$ we now get the result

$$\tilde{H}_{osc} = T + V_{osc} + \sum_{\nu,red} a_\nu |\varphi_\nu><\varphi_\nu| \ , \tag{11}$$

which is rigorous for harmonic forces in the limit $\epsilon \to o$ [2]. The shift
energies a_ν are determined for the respective system under consideration.
As in the OCM, V_{osc} is again replaced by a direct potential V_D of a realistic microscopic interaction

$$\tilde{H} \approx T + V_D + \sum_{\nu,red} a_\nu |\varphi_\nu><\varphi_\nu| \ , \tag{12}$$

Alternatively, V_D and the constants a_ν which determine positions of compound resonances may be fitted to scattering data.

To distinguish this model from the OCM, we call it finite Pauli repulsion model (FPRM), because all antisymmetrization effects are contained in a few separable repulsive terms of finite strengths [2,10]. Note, in this context that the OCM is obtained from eq. (12) for $a_\nu \to \infty$.

The difference, when working in the full respectively truncated relative motion space demonstrated in the analytical oscillator model (FPRM resp. OCM), will of course also show up in numerical RGM calculations

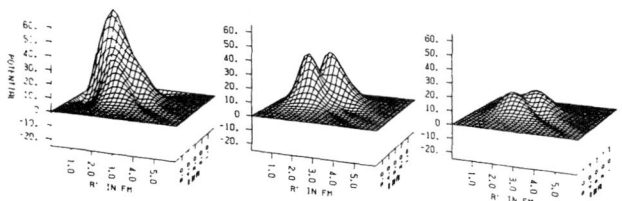

Figure 3a

Nonlocal potential $\tilde{V}$ (coordinate representation) of a realistic RGM calculation in the complete relative motion space for the α-t system [11]. The $\ell=0,1,2$ partial waves (from left to right) are dominated by Pauli repulsion terms similar to those of eq. (11). These are not appearent in calculations done in the truncated space.

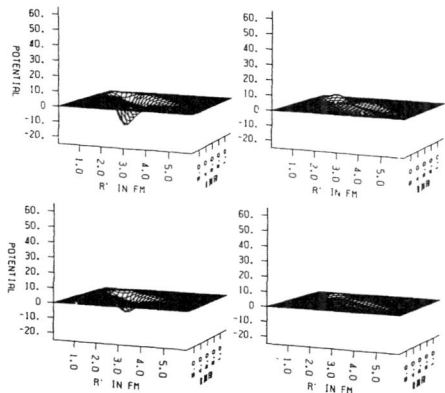

Figure 3b

Same as Fig. 3a for the higher partial waves. These potentials are relatively flat because there are only contributions from the allowed states and no Pauli repulsion terms. The odd-even effect ($\ell=3,5$ ℓ.h.s., $\ell=4,6$ r.h.s.) is nicely recognized.

using realistic microscopic forces. The non-local potential $\tilde{V}$ of eq. (6) will then contain additional contributions from the allowed states. While $\tilde{V}$ is relatively flat when working in the truncated space, it is dominated by Pauli repulsion terms similar to those of eq. (11) when the complete relative motion space is considered (Fig. 3).

3. Comparison of various models

Taking the $\alpha\alpha$ system as an example, we will compare three models which treat the Pauli repulsion quite differently. These are the local Ali-Bodmer potential [12], and the OCM and FPRM discussed above. All three models are adjusted to reproduce the 2α data approximately, i.e. they are approximately on-shell equivalent. Their off-shell behaviour at small relative distances is, however, quite different, due to the apparently different description of the short ranged Pauli repulsion. This is reflected most distinctively in the 2α ground state wave functions at small relative distances (Fig. 4) [10]. While the wave function of the local AB potential

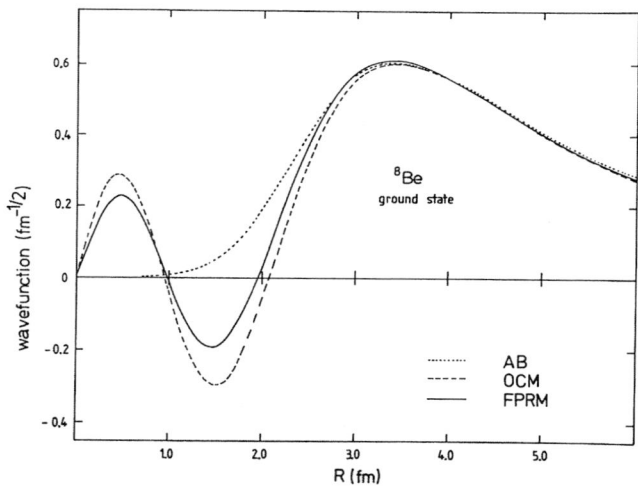

Figure 4

2α ground state wave functions without Coulomb, normalized to 1. The wave functions are those of the local Ali-Bodmer potential (AB), the OCM and FPRM derived from the RGM. The wave functions differ decisively for small relative distances reflecting the different treatment of the Pauli repulsion.

does not even have a node, that of the OCM exaggerates the nodal behaviour due to the orthogonality conditions imposed. For the FPRM these oscillations are dampened appreciably because this wave function has finite amplitudes in the "forbidden" space.

If we calculate quantities which are sensitive to the wave function resp. effective potential at small relative distances, the three models will lead to different results. In the $\alpha\alpha$ case, such a quantity is the binding energy of the 3α system calculated with pairwise $\alpha\alpha$ potentials [10]. This is because the 3α system is relatively compact compared to the 2α system.

The results of a variational 3α calculation are given in Table 1. While

Table 1

3α binding energies and root mean square charge radii obtained from a variational 3-body calculation with various $\alpha\alpha$ models. The Coulomb interaction is included.

	AB	OCM	FPRM	EXP
3α binding energy (MeV)	-1.4	-16.1	-10.0	-7.274
3α r.m.s. radius (fm)	2.8	2.1	2.3	2.460

the local AB potential gives too little binding, the OCM overbinds the system considerably. The latter is a common feature of all models working in the truncated space, even of a full three cluster RGM calculation [13]. Considering the simplicity of the model, the FPRM result is quite satisfying. The main result of this calculation is that the so-called forbidden space is important for the 3α binding problem. In my opinion this problem will not be resolved in the truncated space.

4. Conclusions

Effective interactions of clusters should be deduced from the normalized resonating group equation. In doing so, the so-called forbidden space should be taken into consideration, yielding a potential which operates in the complete relative motion space. In a simple analytical model the Pauli repulsion is then described by a few separable terms of finite strengths (FPRM), rather than by orthogonality conditions (OCM). A 3α calculation shows that the difference may become important for quantities calculated typically in few body physics.

References

1. Wildermuth, K., Tang, Y.C.: A Unified Theory of the Nucleus. Braun-schweig: Vieweg 1977.

2. Fliessbach, T., Walliser, H.: Nucl. Phys. A377, 84 (1982).

3. Saito, S., Okai, S., Tamagaki, R., Yasuno, M.: Prog. Theor. Phys. 50, 1561 (1973).

4. Fliessbach, T.: Z. Phys. A272, 39 (1975).

5. Buck, B., Friedrich, H., Wheatley, C.: Nucl. Phys. A275, 246 (1977).

6. Saito, S.: Prog. Theor. Phys. Suppl. 62, 11 (1977).

7. Fiebig, H.R., Stingl, M., Timm, W.: J. Phys. G4, 1291 (1978).

8. Schmid, E.W.: Z. Phys. A297, 105 (1980).

9. Horiuchi, H.: Prog. Theor. Phys. Suppl. 62/III (1977).

10. Walliser, H., Nakaichi-Maeda, S.: preprint 1986, to be published in Nucl. Phys. A.

11. Brown, R.E., Tang, Y.C.: Phys. Rev. 176, 1235 (1968).

12. Ali, S., Bodmer, A.R.: Nucl. Phys. 80, 99 (1966).

13. Tohsaki-Suzuki, A., Kamimura, M., Ikeda, K.: Prog. Theor. Phys. Suppl. 68, 359 (1980).

MESON THEORETICAL MODELS OF THREE-NUCLEON FORCES

Sidney A. Coon
Department of Physics
University of Arizona
Tucson, Arizona 85721 USA

Abstract. The analogy between photon-exchange and meson-exchange many-body forces is examined with the aid of dispersion theory and used to draw conclusions about the role of baryon resonances in meson-exchange three-nucleon potentials. The importance of low-energy theorems and soft pion theorems in the three-nucleon potential is stressed.

1 Introduction

There is a close analogy between the three-body forces of rare gas systems and those currently being investigated in nuclear physics. This analogy can be instructive in organizing our thoughts about the many-body forces between particles with substructure such as helium atoms or nucleons [1]. Furthermore, the analogy is not only qualitative but, as we shall see for two-body forces, can be made quantitative. In this paper, I shall review the calculation of two- and three-body potentials of long range. The potentials considered are of the $1/r^n$ type due to photon exchange between neutral spinless particles, and of the Yukawa type due to meson exchange between nucleons. The tool used will be the dispersion approach, which is based on the ideas of Lorentz invariance, analyticity, unitarity, and crossing symmetry. This approach, reaching a maturity before the acceptance of QCD as the underlying theory of strong interactions, is particularly helpful in establishing the degree of model-independence of the resulting potentials in each system. With the results in hand, one can then address often-asked questions such as: "What is the degree of consistency between a two-body and three-body force in a given formalism?" "What is the importance of excited states of the composite particle in the three-body force?"

We begin by examining the many-body forces between the neutral rare gas atoms of helium, neon, argon, etc. at such a long range that their electron clouds do not overlap. The instantaneous interaction between the electrons in their orbits causes a mutual polarization of the charge clouds to occur. The interaction energy can be calculated in

Rayleigh–Schrödinger perturbation theory with the instantaneous interaction potential regarded as a perturbation. A general term in the interaction energy possesses two characteristics -- a purely geometrical factor depending on the mutual positions of the interacting atoms (*structure*) and an interaction constant which depends on the atomic species involved (*strength*). In second order, the *strength* and *structure* are displayed by the well known London–Van der Waals potential $V(r) = -C/r^6$, where C is a positive constant related to the static polarizability of the atoms. In third order of perturbation theory, one encounters two-body contributions consisting of sums of simple pair interactions, as well as three-body contributions in which the coordinates of all three atoms are coupled together. The *structure* of the Axilrod–Teller–Muto force [2], the earliest such three-body force, is nicely illustrated by some three-dimensional plots by Varandas [3]. The *strength* of the triple-dipole ATM force is also due to the polarizability of the atoms which can be completely described in terms of dipole oscillator strengths for all possible transitions from the ground state [4]. This description of long-range forces is thoroughly reviewed in Ref. 5.

This nonrelativistic formalism is incomplete, however, as shown by Casimer and Polder [6], who quantized the electromagnetic field to show that the true long range force between two atoms at very large distances falls off as $1/r^7$ (as indicated by experiment). The $1/r^6$ behavior is recovered at distances large compared to atomic dimensions, but not large compared to largest wavelength associated with electric–dipole excitations of the atomic ground state. *The strength, however, is still determined by the static polarizability α of the atoms.*

These features of the potential do not depend on the atomic model of Ref. 6, but follow from the very general principles of dispersion theory. In the language of Feynman diagrams, the Van der Waals interaction arises from the exchange of two photons between two atoms (Fig. 1) and the exchange of three photons between three atoms (Fig. 2). Note that no long-range force of electromagnetic origin can exist to order e^2 between spinless, neutral systems; hence the topology of the figures. From these diagrams one could surmise that the *structure* of a potential would be determined by the topology of the diagram and the *strength* by the blobs, which appears to be related to a Compton scattering amplitude. The work of Feinberg and Sucher [7,8] and of Bernabeu and Tarrach [9] demonstrates the *strength* hypothesis. They showed that the two-system potentials are uniquely determined by the charge, magnetic moment, and electric and magnetic polarizabilities of both systems. These model-independent potentials incorporate

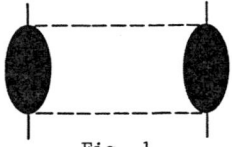

Fig. 1.

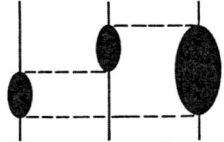

Fig. 2.

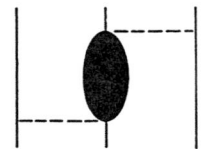

Fig. 3.

the link between Compton scattering polarizabilities (response to real photons) and classically defined polarizabilities (response to a static electromagnetic field). The two definitions of α were shown to be equivalent for neutral spinless systems [9]. It is this demonstration, based on dispersion theory, which allows the analogy between photon exchange and meson exchange to be made quantitative for the two-body potential [10] and indicative in the three-body potential.

Next we turn to the meson exchange theory of nuclear forces, expected to be valid at distances greater than nucleon dimensions. In practice, one draws Feynman diagrams like Figs. 1 and 3 and then defines potentials in terms of the depicted S-matrix elements. That is, one obtains a *structure* from the potential approximation and from covariant perturbation theory. One can use dispersion theory on Fig. 1 to show that if a spin-independent two-body potential exists, it will have the asymptotic form $V(r) \sim r^{-n} e^{-m_0 r}$, where m_0 is the lowest mass that can be exchanged between the particles [8]. The *strength* is then given by Compton scattering of photons ($m_0 = 0$) from atoms or by Compton-like scattering of mesons from nucleons (e.g., elastic πN and $\pi\pi \to N\bar{N}$ amplitudes [10]). One could determine the *strength* of the potential of Fig. 3 by approximating the blob by excited states of the nucleon [11]. The nonrelativistic methods of these models would correspond to the nonrelativistic approach which relates the strength of the Van der Waals force to the dipole excitations of the neutral atom [4,5]. Alternatively, one could obtain the *strength* by a dispersion relation in energy which amounts to integrating over the meson-nucleon scattering cross section [12]. This corresponds to a dispersion relation over the total Compton cross section of photon-atom scattering. These approaches to the construction of molecular and nuclear three-body forces are diagrammed in Fig. 4. Finally, one could use dispersion theory to derive the *strength* of potentials [13-15]. For Fig. 3, one could exploit the low-energy theorems (including soft-pion theorems), which constrain the Compton-like scattering of hadronic currents from nucleons [13]. The pion is related via PCAC (Partially Conserved Axial Vector Current) to the SU(2) axial vector current A_μ^i, and the vector mesons ρ and ω are related to the SU(2) vector current j_μ^i by the current-field identity (vector dominance) [16].

In the electromagnetic case, the massless nature of the exchanged particle makes the model independence exact as the two-body potential is expressed in terms of the low-energy Compton amplitude involving on-mass-shell photons. In contrast, neither the two-body [17] nor three-body nuclear potential can be expressed in terms of the physical meson-nucleon amplitude. Because the mesons have mass, the needed amplitude is always in an unphysical region and, for the three-body potential, off-the-pion-mass-shell. Even in this case the low-energy theorems of these Compton-like amplitudes impose a degree of model independence on the derived three-body potentials.

One of the important points of this paper is that the three alternatives for obtaining the strength of a three-body force are not always of equal validity. The dispersion

Molecular force strength	**Nuclear force strength**

i)   sum over dipole excitations

$$\alpha = 2e \sum_I \frac{|\langle I|\vec{r}\cdot\vec{\epsilon}|0\rangle|}{E_I - E_0}$$

i)   sum over nucleonic excitations

$$\Delta(1232)$$

$$N(1440) \qquad [11]$$

ii) off-shell expansion via
     dispersion relation

$$\alpha \sim \int \frac{\sigma_{tot}(k^2)dk}{k^2}$$

ii) off-shell expansion via
     dispersion relation

$$C_p \sim \int \frac{\sigma_{33}(k^2)dk}{(k^2+\mu^2)} \qquad [12]$$

$$C_s \sim \tfrac{1}{3}(a_1+2a_3) \sim 0$$

iii) Compton scattering

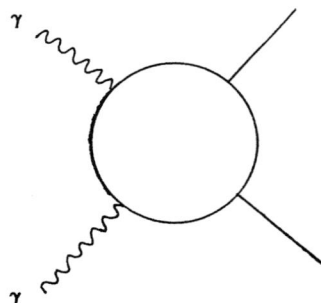

iii) Compton-like scattering
     of hadronic currents

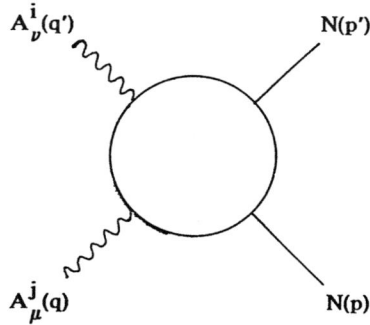

low-energy theorem
on the amplitude

$$\lim_{k^2 \to} T \longrightarrow \alpha k^2 \vec{\epsilon}\cdot\vec{\epsilon}\,'$$

low-energy theorems
on the amplitude

$\pi\pi$ – PCAC and soft-pion
    theorems   [13,14]
$\pi\rho$ – PCAC and Kroll-Ruderman
    theorem   [15]
$\rho\rho$ – Thirring-Gell-Mann-Low
    theorem   [15]

Fig. 4. Methods used to set the scale of a three-body force.

theory calculation of the electromagnetic interaction yields an impeccable foundation for the strength parameter of the two- and three-body forces of molecular physics. That is, the polarizability obtained by dipole excitations, total cross-section integrals, and measurement is the same as the polarizability of the Compton amplitude. *This confluence of methods is not true for meson exchange potentials.* The strength parameter (Fig. 4) of the blob in Fig. 3 is solely determined neither by excited states of the nucleon (Δ isobars), nor by the low-energy π-N cross-section (dominated by the (3,3) resonance). The amplitudes provided by these models do not satisfy the low energy theorems demanded by current conservation and the breaking of chiral symmetry. One is then obliged to accept only the potentials of the dispersion theoretic method as an adequate representation of the situation. The three-body forces built solely on models of nucleon isobars are wrong. The question remaining is "Are they wrong in an important numerical way?"

2 Dispersion Theoretic Derivations of 2π Exchange Forces

The two-pion exchange two-nucleon potential of Fig. 1 is defined by a Laplace-like transform (in the variable of $\sqrt{t}$) of the spectral function $\rho(t)$ obtained from a πN amplitude for time-like values of $t = (q-q')^2$ so that the potential is a superposition of Yukawas. The dispersion theoretic two-nucleon potentials utilize πN amplitudes $\pi^j(q) + N(p) \rightarrow \pi^i(q') + N(p')$ in which all four particles are on-mass-shell. These amplitudes are crossed into the $N\overline{N} \rightarrow \pi\pi$ amplitudes in the pseudo physical region above the $\pi\pi$ and below the $N\overline{N}$ threshold. The energy dependence of these potentials comes mainly from the box diagrams (the blobs of Fig. 1 are replaced by nucleon poles) and is nearly eliminated by the subtraction of iterated one-pion exchange [20]. (The one-photon exchange between neutral spinless systems is of the range of an atomic dimension and so is its iterate; hence no analogous subtraction is needed for the asymptotic potential [8]. The one pion exchange potential, however, exists, is of order g^2, and does have a long-range iterate.)

In contrast, the blob of Fig. 3 represents a meson-nucleon scattering amplitude with the term corresponding to the iterated one-meson exchange (or forward-propagating Born term, FPBT) subtracted. The subtraction is made to avoid double-counting from three-nucleon cluster contributions. The two mesons, unlike those of Fig. 1, are off-mass-shell and spacelike (four momenta $q^2, q'^2 \leq 0$). The kinematics of the scattering amplitude expected in a nuclear medium are $q_0^2 \sim 0$ and $\vec{q}^2 < 10\ \mu^2$ [18]. (Only very recently has retardation, i.e., $q_0^2 \neq 0$, been handled in a consistent manner [19]. The crossing variable $\nu \equiv (s-u)/4m = (q'+q)\cdot(p'+p)/4m$ is of the order μ^2/m, where μ is the pion mass and m the mass of the nucleon. The three-body potential is essentially the Fourier transform of the off-mass-shell amplitude and, hence (if the amplitude is expanded around $q^2, q'^2 = 0$), has the form of simple Yukawas and derivatives of Yukawas in coordinate space.

In spite of these differences, which are not so apparent in a nonrelativistic formalism, we will later make a limited comparison of the two- and three-body potentials which will help answer questions of consistency.

3 Off-Mass-Shell πN Amplitudes

The off-mass-shell amplitude needed for the three-body force is written as

$$T = T_B - T_{FPBT} + \Delta T + \overline{T} \quad , \tag{1}$$

where T_B is the nucleon pole (Born) term and ΔT is added to T_B so that $T_B + \Delta T$ satisfy low-energy theorems (LETs). The axial-vector LETs lead to the soft ($q^2 \to 0$) pion theorems associated with the current algebra (CA) Ward identity program for πN scattering and pion photoproduction [16]. In order to motivate (1), we briefly review the main features of this program for πN scattering, setting down equations only as they are needed later. To begin with, we consider the Compton-type process with SU(2) axial vector currents "scattering" off of target nucleons as in Fig. 4. The Ward identity is obtained by taking the double divergence of the amplitude and expressing it as equal time commutators (ETCs) plus an integral over derivatives of axial vector currents. The algebra of currents relates the ETCs to the isovector nucleon charge form factors (see (5)), and, in the important isospin-even forward amplitude, to the sigma term defined as

$$\langle N|[Q_5^i, i\partial \cdot A^j]|N\rangle = \delta^{ij} f(t) \, \sigma \, \overline{u}(p') u(p) \quad . \tag{2}$$

Then we use PCAC in the matrix element form $\langle o|\partial \cdot A^i|\pi^j\rangle = \delta_{ij} f_\pi \mu^2$ and let the axial vector current be dominated by the pion pole to turn the Ward identity into a relation with the sum of a pion-nucleon amplitude and a background axial-vector-nucleon amplitude on the left-hand side and the ETCs on the right-hand side. Finally, the nucleon poles are subtracted from both the amplitudes to give the current algebra representation of the background (nucleon pole removed) pion-nucleon amplitude as an ETC and the background axial-Compton amplitude with both pion and nucleon poles removed. The complete current algebra amplitude is found by adding the (pseudoscalar coupling) nucleon pole terms evaluated according to dispersion theory. The result is a representation of the four invariant πN amplitudes defined by the even and odd t-channel isospin decomposition

$$T^{ij} = T^{(+)} \delta^{ij} + T^{(-)} i \, \epsilon^{ijk} \tau^k \tag{3}$$

and by

$$T^{(\pm)} \equiv \overline{u}(p') \left\{ F^{(\pm)} - \frac{1}{4m} B^{(\pm)} [\gamma \cdot q', \gamma \cdot q] \right\} u(p) \quad , \tag{4}$$

where the forward combination $F = A + \nu B$ reduces to the non-spin flip amplitude of a nonrelativistic formalism.

Of the four amplitudes, only the forward-even $F^{(+)}(\nu,t)$ and spin-flip odd $B^{(-)}(\nu,t)$ are needed in the low order q^2/m^2 expansion currently used in three-body force. We first display them on–pion–mass–shell:

$$F^{(+)} = F_B^{(+)} + \overline{F}^{(+)} \; ; \quad \overline{F}^{(+)}(\nu,t) = f(t)\,\sigma/f_\pi^2 + C^{(+)}(\nu,t)$$

$$B^{(-)} = B_B^{(-)} + \overline{B}^{(-)} \; ; \quad \overline{B}^{(-)}(\nu,t) = (F_1^V(t) + F_2^V(t))/2f_\pi^2 - g^2/2m^2 + D^{(-)}(\nu,t) \quad ,$$

(5)

where $f_\pi = 93$ MeV is the on-shell value of the pion decay form factor, and $g^2/2m^2$ is the non-nucleon-pole background difference between the axial-Compton scattering and pion scattering for the πNN coupling $g\gamma^5$. The background axial-vector-nucleon amplitudes $C(\nu,t)$ and $D(\nu,t)$ correspond to the pion-nucleon amplitudes F and B.

To test this current algebra representation with πN data, one must determine $C(\nu,t)$ and $D(\nu,t)$. They are dominated by the Δ isobar. A dispersion theoretic analysis allows one to use the (unique) on-shell spin 3/2 projection operator at resonance [21]. The appropriate kinematic region to compare these amplitudes with data is, as emphasized by Höhler et al. [22], in the sub-threshold region with $\nu < \mu$ and $t < 4\mu^2$ (Fig. 5). Near the point $\nu = t = 0$, the amplitudes vary rapidly because of the nearby nucleon poles. Once the nucleon poles are subtracted, however, the background amplitudes $\overline{F}$ and $\overline{B}$ can be expanded in a power series around $\nu=t=0$ in the crossing symmetric variables t and ν^2. The data is then expressed as expansion coefficients, which are compared to those of theoretical models. In this region the full current algebra amplitude compares very well [14]. Approximating the background amplitudes by the Δ dominated terms $C(\nu,t)$ and $D(\nu,t)$ alone, however, gives a terrible representation of the subthreshold data [22,23].

For the reader who considers data extrapolated to the subthreshold region somehow suspect, I remark that the full current algebra amplitude also maps out the threshold scattering lengths and the low-energy non-resonant phase shifts in a satisfactory manner [24]. I further note that this subthreshold region is the closest on-shell to the off-shell values of $\nu \cong 0$, $0 \leq q^2$, $q'^2 \leq -10\,\mu^2$ expected to be dominate in the three-body force. Indeed, consider a Fermi gas of nucleons so that the pion only scatters forward ($t = 0$) from an on-shell nucleon. Then the nucleon poles at $t = 2\mu^2$, $\nu = 0$ near the center of Fig. 5 are precisely the appropriate starting points for the off-mass-shell extrapolation needed [18]. This extrapolation, at least for interacting nucleons but restricting still to $t = 0$, is from the nucleon pole $s = (p+q)^2 = m^2 \sim 50\,\mu^2$ down to $\sim 40\,\mu^2$. One should have qualms about the extrapolations which instead start from threshold $s = (m+\mu)^2 \sim 64\,\mu^2)$ [12]. An assumed dominance of the Δ pole up at $s = (m+2\mu)^2 \sim 78\,\mu^2$ down in the region of the three-body force would seem even more suspect [11], considering the terrible description Δ dominance gives of the subthreshold data, even if the field-theoretic ambiguities of the off-shell Δ propagator were addressed (which they often are not). In any event, these extrapolation prescriptions do not satisfy the soft-pion theorems to which we now turn.

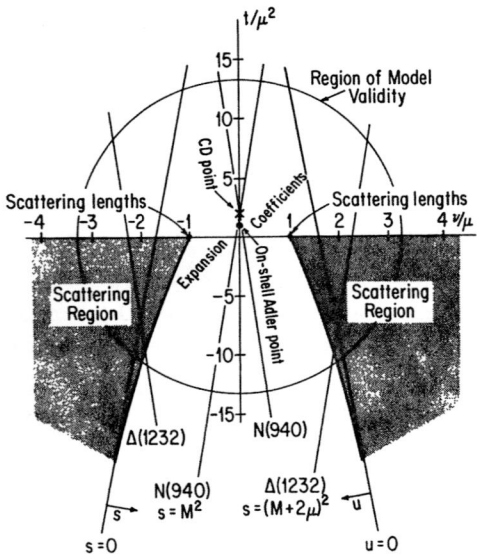

Fig. 5. Kinematics of on-mass-shell πN scattering.

The πN amplitudes needed in the 2π exchange three body force must be extrapolated to the subsoft region $q'^2,q^2 < 0$. The extrapolation of the (residual) nucleon pole term is accomplished by the πNN vertex function [25] and that of the background amplitude by PCAC. The matrix element version of PCAC mentioned before is equivalent to a soft pion version of PCAC: once the rapidly varying pion and $O(1/q)$ bremmstrahlung-type poles are removed from the axial-vector amplitude, the background is a smoothly varying function of pion momentum squared [16]. Tests of the extrapolation from soft pions ($q^2 = 0$) to physical pions ($q^2 = \mu^2$) are shown on Fig. 6, which depicts the projection onto the hyperspace $\nu = 0$ of the coordinates (ν,t,q^2,q'^2) needed to describe a fully off-shell amplitude. The extrapolations shown are PCAC corrections to the amplitude $\overline{F}^{(+)}$ from the Adler zero A (one pion soft, one pion on shell) and the current algebra point CA (both pions soft) to the on-mass-shell line keeping ν and t fixed. The extrapolation is of the order of 10% per pion, indicating that PCAC is a reliable hypothesis. An additional test of PCAC is the measurement of the on-shell analogue of the Adler-Weisberger soft-pion theorem on $\overline{F}^{(-)}$. Again the on-shell extrapolation shows a PCAC correction of 10%. The on-shell values were determined for these tests via interior dispersion relations [26].

Given this proven validity of the pion PCAC extrapolation from $q^2 = 0$ to $q^2 = \mu^2$, the forward even amplitudes at the on-shell points $\overline{F}^{(+)}(0,\mu^2)$ and $\overline{F}^{(+)}(0,2\mu^2) = \sigma/f_\pi^2$ can be used to obtain the off-shell amplitude $\overline{F}^{(+)}(0,t,q^2,q'^2)$ needed for (the dominant portion of) the three-body force. That is, this amplitude becomes

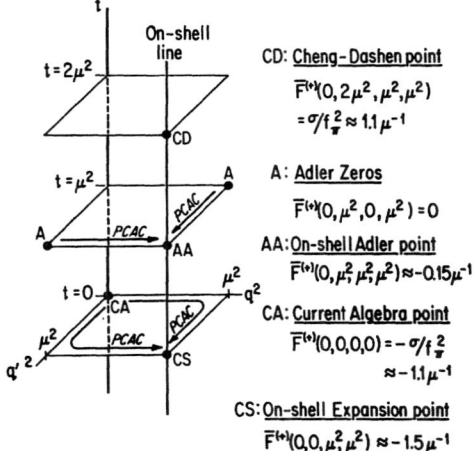

Fig. 6. The geometry of off-mass-shell πN amplitudes for $\nu = 0$.

$$\overline{F}^{(+)}(0,t,q^2,q'^2) = \left(\frac{t}{\mu^2} - 1\right)\frac{\sigma}{f_\pi^2} + (q'^2+q^2-t)\frac{\overline{F}^{(+)}(0,\mu^2,\mu^2,\mu^2)}{\mu^2} + O(q^4/m^4) \quad . \tag{6}$$

Thus to order q^2/m^2, the off-shell background amplitude is determined by measurable on-shell quantities and is then model-independent.

The physical content of (6) can be seen by inspection of Fig. 6. Note that $\overline{F}^{(+)}(0,t,q^2,q'^2)$ to order q^2/m^2 satisfies the Adler conditions (points A), the current algebra point (CA), recovers the on-shell Adler point (AA), and is symmetric in q^2 and q'^2. These four conditions suffice to determine the power series up through q^2/m^2. The scale of the off-shell amplitude is given by the sigma term determined, for example, by the on-shell Cheng-Dashen point (CD) of Fig. 6. Note that a model of the off-shell amplitude $\overline{F}^{(+)}$, which only incorporates a nonrelativistic delta isobar somehow extrapolated off-shell can never satisfy the soft pion theorems of Fig. 6 and (1). Even the background axial-vector nucleon amplitude $C(\nu,t)$ has the structure of a double divergence and therefore vanishes as q or $q' \to 0$, so cannot be non-zero at the current algebra point (Fig. 6).

The second amplitude $\overline{B}^{(-)}$ multiplies the invariant $[\gamma \cdot q, \gamma \cdot q']$, which is of second order in q^2/m^2 already, so only the constant term of the on-shell amplitude is needed. This constant amplitude can be obtained either by the empirical expansion coefficients or by a dispersion-theoretic model of the Δ contribution to $D^{(-)}$ as the numbers agree. In any event, to this order, no off-shell extrapolation is necessary.

This model-independent derivation has the disadvantage that one must go back to the models of $\overline{T}$ if one finds it necessary to include higher order terms in the potential.

This was in fact found necessary in the derivation of the ρ exchange three-nucleon potentials [15] following the same current algebra-PCAC program. They therefore have a greater portion of model-dependent Δ-isobar contributions than the π-exchange potentials reviewed here.

The full amplitude T of (1) requires evaluation of the term $T_B - T_{FPBT}$, which brings back in the nucleon pole, but only the part which is not the iterate of the one-pion-exchange two-body potential. This can be done by subtraction techniques and a subsequent nonrelativistic reduction while ignoring retardation ($q_0^2 \neq 0$) in the pion lines [14]. An alternative technique makes a Foldy-Wouthuysen reduction procedure to define pion-nucleon vertices and uses time-dependent perturbation theory to tie together the nonrelativistic nucleons with pions [19]. The latter technique does not depend on the small-momentum expansion of the invariant πN amplitude and includes retardation. It is gratifying that the two approaches agree on the local terms once an ambiguity parameter has been identified in the former approach. The role of overlapping, retarded pion exchanges in the three-body force was finally clarified by the latter technique. This result is a step toward full consistency of wave function calculations with consistent two- and three-body potentials and relativistic corrections such as exchange currents.

4 Discussion

Now it is time to reexpress (1) in the form familiar to readers of three-body-force calculations in order to continue the discussion. Setting $q = (0,\vec{q})$ and rearranging (6) in powers of $\vec{q}$ and $\vec{q}'$, we obtain an amplitude which, with the attendent pion propagators and vertex functions $F_{\pi NN}$, can be Fourier transformed into coordinate space to obtain a potential. The amplitude is then

$$T = F_{\pi NN}(-\vec{q}^2)\, F_{\pi NN}(-\vec{q}'^2)\, [a + b\vec{q}\cdot\vec{q}' + c(\vec{q}^2 + \vec{q}'^2) + \dots] \quad . \tag{7}$$

where

$$a = \sigma/f_\pi^2 \qquad\qquad\qquad = (1.13 \pm 0.12)\ \mu^{-1}$$

$$b = -2/\mu^2 \left[\sigma/f_\pi^2 - \overline{F}^{(+)}(\nu=0, t=\mu^2) \right] \qquad = -(2.58 \pm 0.33)\ \mu^{-3} \tag{8}$$

$$c = \sigma/(\mu^2 f_\pi^2) - g^2/4m^3 + F'_{\pi NN}(0)\sigma/f_\pi^2 \qquad = (1.05 \pm 0.10)\ \mu^{-3} \quad .$$

The final coefficient $d_3 + d_4 = -0.75\ \mu^{-3}$ is the leading term in the expansion of the isospin-odd, spin-flip πN amplitude $\overline{B}^{(-)}$. The forward odd and spin-flip even amplitudes do not contribute to the TBF to $O(q^2/m^2)$.

The relative importance to the 2π TBF of the terms in (1) is obscured by the presentation in (8), which uses on-shell data to parametrize the sum $\Delta T + \overline{T}$. The pair term $T_B - T_{FPBT}$ ($\propto g^2/4m^3 \approx 0.15\ \mu^{-3}$) plays a small role in the 2π TBF (15% of c and 20%

of $d_1 + d_2$) because of "pair suppression" enforced by the Adler zero. The Δ-isobar dominates $\overline{T}$ and therefore contributes to $\overline{F}^{(+)}(0,\mu^2) \cong 0.15 \ \mu^{-1}$. The Δ-isobar cannot, however, contribute to the πN sigma term σ which, as is evident in (8), dominates the 2π TBF. That is, the ΔT term, which remains in $\overline{F}(\nu,t)$ as both pions go soft, is exactly σ/f_π^2, a measure of chiral symmetry breaking [16]. The empirical value of σ/f_π^2 is determined by an extrapolation of πN data via dispersion relations to the unphysical but on-shell Cheng-Dashen point (Fig. 6). The most recent such analysis [27] utilized the Karlsruhe-Helsinki phase-shift analysis to find $\sigma/f_\pi^2 = 1.02 \pm 0.13 \ \mu^{-1}$, confirming the determination [26] quoted in (8). Very low-energy πN measurements [28], in a kinematical region never explored experimentally before, agree with the prediction of this phase-shift analysis and thus support this value of the sigma term. In summary, the pair term is small, the small isobar term is not isolated, and the chiral-symmetry-breaking term is dominant in the forward even πN amplitude. The coefficient $d_1 + d_2$ of the smaller spin-flip amplitude is 20% d_3 from the pair term, 30% from the current commutator term, and 50% from the isobar contribution $D^{(-)}$.

From (7) and (8), we conclude that off-mass-shell, the s-wave πN scattering (strengths a and c) are equally as important as the p-wave $\vec{q} \cdot \vec{q}'$ term (strength b). Only the latter term would appear in a model amplitude based on the lowest excited nucleon state [11], or extrapolated from threshold [12] where the s-wave scattering lengths are accidently near zero. However, all three strength parameters of the off-shell $\overline{F}$ are due primarily to the sigma term, which receives almost no contribution from the Δ isobar. It does contribute about half of the much smaller spin-flip amplitude. We must conclude that the Δ isobar plays a rather small role in the 2π three-nucleon force.

Proposals have been made to construct a "super" background amplitude by subtracting Δ poles from both sides of the Ward identity. Then one could add back in the Δ pole to the πN amplitude in a "dynamical" sense in the form of coupled channel equations that include Δ isobar states in the nuclear wave function. It is unclear to me how one would merge a three-body operator obtained properly as a nonrelativistic reduction of a dispersion theoretical calculation with a strictly nonrelativistic picture of resonances and nucleons interacting via potentials. One can draw diagrams like those of Figs. 1-3, but the nonrelativistic interpretation is quite different. In any event, the Δ isobar of $C(\nu,t)$ plays a very small role in the potential of Fig. 3, so the proposal, if enacted, would find a rather small change in the 2π three-body force.

An often asked question is "Why isn't the same amplitude used to calculate the two-nucleon force?" The answer should be clear from Section 2. However, in pursuit of the analogy [10] with the molecular force, the on-shell $\overline{F}(0,t)$ of (5) was extrapolated via the empirical expansion coefficients into the region $t > 4\mu^2$ near the two-pion threshold. It appears [10] that the $\pi\pi \to N\overline{N}$ helicity amplitude $f_+^0(t)$ obtained from $\overline{F}(0,t)$ can be used a little above threshold by the same analytic expression as the real amplitude, which is correct for $t < 4\mu^2$ (see Fig. 5). In this application, as in the three-body-force amplitude, the pions were $q_0^2 = 0$, $\nu = 0$, and the linear approximation in t was also used. The helicity amplitude for $t < 15\ \mu^2$ thus obtained is not so different from that used in the Paris potential. The spin and isospin-averaged NN potential obtained with this amplitude is then good for $r \geq 3$ fm. It does have a qualitative agreement with the purely empirical Hamada-Johnston potential at large r. Thus in this limited way, the three-nucleon potential off-shell amplitude of (6) and (7), brought back on-shell, is indeed consistent with the very long-range spin-independent two-nucleon potential.

References

1. Coon, S. A.: Invited talk at International Nuclear Physics Conference, Harrogate, England, August 25-30, 1986, to appear in the Proceedings
2. Axilrod, B. M., Teller, E.: J. Chem. Phys. 11, 299 (1943); Muto, Y.: Proc. Phys. Math. Soc. (Japan) 17, 629 (1943)
3. Varandas, A. J. C.: Mol. Phys. 49, 817 (1983)
4. Leonard, P. J., Barker, J. A.: Theoretical Chemistry, Advances and Perspectives 1, 117 (1975)
5. Klein, M. L. and Venables, J. A. (eds.): Rare Gas Solids. Academic Press - London 1976
6. Casimir, H. B. G., Polder, D.: Phys. Rev. 73, 360 (1948)
7. Feinberg, G., Sucher, J.: Phys. Rev. 139, B1619 (1965); A2, 2395 (1970); D20, 1717 (1979)
8. Sucher, J.: In: Cargese Lectures in Physics Vol. 7, Levy, M. (ed.). Gordon and Breach - London 1977
9. Bernabeu, J., Tarrach, R.: Phys. Lett. 55B, 183 (1975); Ann. of Phys. 102, 323 (1976)
10. Tarrach, R., Ericson, M.: Nucl. Phys. A294, 417 (1978)
11. Green, A. M.: Rep. Prog. Phys. 39, 1109 (1976); Muther, M.: Prog. Part. Nucl. Phys. 14, 123 (1985); Sauer, P. U.: this workshop
12. Miyazawa, H.: Phys. Rev. 104, 1741 (1956); Fujita, J., Miyazawa, H.: Prog. Theor. Phys. 17, 360 (1957); Loiseau, B. A., Nogami, Y.: Nucl. Phys. B2, 470 (1967); Sato, M., Akaishi, Y., Tanaka, H.: Prog. Theor. Phys. (Supp.) 56, 76 (1974)
13. Coon, S. A., Scadron, M. D., Barrett, B. R.: Nucl. Phys. A242, 467 (1975)
14. Coon, S. A., Scadron, M. D., McNamee, P. C., Barrett, B. R., Blatt, D. W. E., McKellar, B. H. J.: Nucl. Phys. A317, 242 (1979)
15. Ellis, R. G., Coon, S. A., McKellar, B. H. J.: Nucl. Phys. A348, 631 (1985)
16. Scadron, M. D.: Rep. Prog. Phys. 44, 213 (1981)
17. Epstein, G. N., McKellar, B. H. J.: Phys. Rev. D10, 1005, 2169 (1974); Cottingham, W. N., Lacombe, M., Loiseau, B., Richard, J. M., Vinh Mau, R.: Phys. Rev. D8, 800 (1973); Chemtob, M., Durso, J. W., Riska, D. O.: Nucl. Phys. B38, 141 (1972)
18. Brown, G. E., Green, A. M., Gerace, W. J.: Nucl. Phys. A115, 435 (1968).
19. Coon, S. A., Friar, J. L.: Phys. Rev. C34, 1060 (1986)
20. Partovi, M. H., Lomon, E. L.: Phys. Rev. D2, 1999 (1970)
21. Scadron, M. D., Thebaud, L. R.: Phys. Rev. D9, 1544 (1974)
22. Hohler, G., Jakob, H. P., Strauss, R.: Nucl. Phys. B39, 237 (1972)
23. See Table 2.4.7.1 in Höhler, G.: Landolt-Bernstein, Vol I/9B, Schoppor, H. (ed.): Pion-Nucleon Scattering. New York - Springer-Verlag 1983
24. Wilde, B. H., Coon, S. A., Scadron, M. D.: Phys. Rev. D18, 4489 (1978); Wilde, B. H., unpublished
25. See, for example, Coon, S. A., Scadron, M. D.: Phys. Rev. C23, 1150 (1981)
26. Hite, G. E., Jacob, R. J., Scadron, M. D.: Phys. Rev. D14, 1306 (1976)
27. Koch, R.: Z. Phys. C15, 161 (1982)
28. Goring, K. et al.: to appear in Proc. of the Second Conference on the Intersections of Particle and Nuclear Physics, Lake Louise, Canada, May 24-31, 1986

FADDEEV EQUATIONS FOR BOUND AND SCATTERING STATES

G. L. Payne
Department of Physics and Astronomy
The University of Iowa
Iowa City, Iowa 52242, U.S.A.

Abstract. The numerical solution of the Faddeev equations to obtain accurate binding energies and wave functions for the trinucleon system is reviewed. The methods used to include the effects of a three-body force are discussed.

I. Introduction

Much of the recent theoretical and experimental work in the three-nucleon problem has been directed at understanding certain low-energy properties of the trinucleon system. The detailed comparisons between the calculations and the experimental data serve as a test of the traditional approach to nuclear physics. This approach is formulated in terms of non-relativistic nuclear Hamiltonians which contain realistic two-nucleon interactions and possibly three-nucleon interactions. The fundamental question is whether this model which neglects subnuclear degrees of freedom, such as mesons or quarks, can provide an adequate model of the atomic nucleus. The trinucleon system plays an extremely important role in the investigation of the validity of the standard model; for this system the solution of the model Hamiltonian may be numerically solved "exactly." That is, the numerical solution can be obtained with any desired accuracy; therefore, these calculations serve as a stringent test of the theoretical models. Any disagreement between the calculated and experimental value must be due to a failure of the model. For more complicated systems the present numerical methods can not yield "exact" solutions, and one can not determine whether any discrepancies between theory and experiment are due to the model or to the numerical approximations used. In this paper the present status of the calculations for the bound state of the trinucleon system is reviewed. Since the numerical calculations are not trivial, it is important that several different groups solve the calculations for the same model Hamiltonian. This has been done for the bound state, and the results of these independent calculations yield bound-state energies which agree to within 10 keV. These numerical results will be presented by the various groups in separate papers.

Accurate calculation of the energy eigenvalues and wave functions for the bound-state system has proved to be exceptionally difficult. Only a few years ago different calculational methods applied to the same model Hamiltonian gave different results. These differences were the result of the various numerical approximations which were used. As mentioned above, more recent calculations which use these approximations give virtually identical results. This agreement has greatly increased the level of confidence in our calculational ability. However, this is not the case for the trinucleon scattering problem. While there have been many calculations for n-d and p-d scattering with realistic potentials, there has not been a detailed comparison between the different groups solving the same problem. Until this comparison has been made, it will be difficult to judge our ability to solve the scattering problem. Hopefully, this goal will be achieved within the next few years.

In the next section we review the Faddeev equations for the bound state, and in Section III we discuss the Faddeev equations for the scattering problem in momentum and coordinate space. Finally, in Section IV we present our conclusions.

II. Faddeev Equations: Bound States

Given the model Hamiltonian for the trinucleon system with two-body interactions V and a three-body force W:

$$H = H_0 + V(\vec{x}_1) + V(\vec{x}_2) + V(\vec{x}_3) + W(\vec{x}_1, \vec{x}_2, \vec{x}_3) \qquad (II.1)$$

$$= H_0 + V_1 + V_2 + V_3 + W \quad ,$$

where H_0 is the free-particle kinetic energy operator and $\vec{x}_i$ is the vector between particles j and k, we want to solve the Schrödinger equation

$$(H - E)|\Psi> = 0 \quad . \qquad (II.2)$$

In (II.1) the Hamiltonian has both a two-body interaction V and a three-body interaction W. For the bound-state system one can solve the

Schrödinger directly using a method such as the Rayleigh-Ritz Variational Principle [1,2]. Also, one could write the Schrödinger equation as a differential equation and use a numerical procedure such as a hyperspherical expansion [3] to solve the equation. An alternate procedure [4] is to introduce the three Faddeev amplitudes to replace the Schrödinger equation by the three coupled Faddeev equations. Since there exist several techniques for including the three-body force in Faddeev equations, we first consider the case with only two-body interactions. For this case the total wave function is decomposed into three parts

$$|\Psi> = \sum_{i=1}^{3} G_0 V_i |\Psi> = \sum_{i=1}^{3} |\Psi_i> \quad , \qquad (II.3)$$

where

$$|\Psi_i> = G_0 V_i |\Psi> \quad , \qquad (II.4)$$

is the Faddeev amplitude, and

$$G_0(E) = (E - H_0)^{-1} \qquad (II.5)$$

is the free-particle resolvent. Introducing the resolvent operator

$$G_i(E) = [E - (H_0 + V_i)]^{-1} \quad , \qquad (II.6)$$

$$= (E - H_i)^{-1} \quad ,$$

where $H_i = H_0 + V_i$, equation (II.3) can be written in the form:

$$|\Psi_i> = G_i V_i [|\Psi_j> + |\Psi_k>] \quad , \qquad (II.7)$$

for $i = 1,2,3$. The i,j,k in (II.7) are taken to be cyclic.

The three coupled equations in (II.7) comprise the Faddeev equations, and the solution of these equations is equivalent to the solution of the Schrödinger equation. To solve these equations one must choose a particular representation, and different groups have chosen different representations. Possible choices are the momentum space representation [5,6], the coordinate space representation [7], and a combination of the two representations [8]. For the momentum space calculations, it is customary to introduce the two-body t-matrices via

$$G_i V_i = G_0 t_i \quad , \tag{II.8}$$

and to write the bound-state Faddeev equations in the form:

$$|\Psi_i> = G_0 t_i [|\Psi_j> + |\Psi_k>] \quad . \tag{II.9}$$

In both representations the Faddeev amplitudes are projected onto a complete set of channel wave functions which represent the orbital angular momentum and spin-isospin variables. These channel basis states are normally defined for channel i as

$$|\alpha_i> = | [(\ell_\alpha, s_\alpha) j_\alpha, (L_\alpha, S_\alpha) J_\alpha)] JM; (t_\alpha, T_\alpha) TM_T > \quad , \tag{II.10}$$

where

ℓ_α = relative orbital angular momentum of particles j and k   ;

s_α = total spin angular momentum of particles j and k   ;

j_α = total angular momentum of particles j and k   ;

L_α = orbital angular momentum of particle i relative to
   particles j and k   ;

S_α = spin of particle i $\left(S_\alpha = \frac{1}{2} \right)$;

J_α = total angular momentum of particle i   ;

J = total angular momentum of the system $\left(J = \frac{1}{2} \right)$;

t_α = total isospin of particles j and k   ;

T_α = isospin of particle i $\left(T_\alpha = \frac{1}{2} \right)$;

T = total isospin of the system $\left(T = \frac{1}{2} \right)$.

In addition, the two-body potential is replaced by a potential projected onto the channel functions. The potential for the Faddeev calculations has the form

$$V_i = \sum_{\alpha=1}^{N_c} \sum_{\alpha'=1}^{N_c} |\alpha_i'> v_{\alpha'\alpha} <\alpha_i| \quad , \tag{II.11}$$

where N_c is the number of channels and $v_{\alpha'\alpha} = <\alpha_i|V(\vec{x}_i)|\alpha_i>$. The channel functions form a complete set; however, in practice the set must be truncated in an actual calculation. The difficulty of the numerical calculations increases as the number of channels is increased. In order to obtain an accurate solution to the Faddeev equations, one must use enough channel functions to accurately describe both the Faddeev amplitudes and the potential. In practice, the number of channels is increased until the solution has converged to the desired accuracy, where the convergence is determined by the value of the bound-state energy. For the Schrödinger equation the binding energy increases monotonically as the number of basis states is increased. This is not the case for the Faddeev equations. The Faddeev operator is not a Hermitian operator; consequently, the equation can not be derived from a variational principle. However, the Faddeev amplitudes can be used to construct the total wave function, and this wave function can be used to determine a variational upper bound for the binding energy [9].

Typically, the number of channels is chosen by choosing the maximum value of j_α and the parity of the interacting pair (j,k). The initial calculations were performed using the channels with $j_\alpha \leqslant 1$ and positive parity; this case has five channels which corresponds to using a two-body potential which has only ${}^1S_0$ and ${}^3S_1$-${}^3D_1$ interactions. Using a modern computer, calculations with $j_\alpha \leqslant 2$ and both positive and negative parities can be done; this case has 18 channels and yields binding energies within 100 keV of the converged result. In order to obtain a result for the binding energy with an error less than 10 keV, it is necessary to retain all channels with $j_\alpha \leqslant 2$, which corresponds to 34 channels.

To solve the Faddeev equations, one first projects the equations on the the channel basis functions. Since the three nucleons are treated as identical particles, all three Faddeev amplitudes will have the same functional form, and it is only necessary to solve one of the equations. The projection onto N_c channel functions yields N_c coupled equations for the channel wave functions. In the momentum space calculations, the resulting equation is an integral equation which contains the unknown eigenvalue for the energy in a nonlinear form. To find the eigenvalue, equation (II.9) is rewritten in the form:

$$|\Psi_i> = \lambda G_0 t_i [|\Psi_j> + |\Psi_k>] \quad , \tag{II.12}$$

where the parameter λ has been added to the right-hand side of (II.12). This equation can be solved as an eigenvalue problem for λ with a fixed value of the energy. If the value assumed for the energy is the correct value, then λ will be unity. In practice, the value of the energy is varied until a value is found for which $\lambda = 1$. Normally, an iterative procedure is used to determine the eigenvalue λ.

In coordinate space the Faddeev equation is written as a differential equation. This is done by multiplying (II.7) by G_i^{-1}. The resulting equation is

$$[E - (H_0 + V_i)]|\Psi_i> = V_i[|\Psi_j> + |\Psi_k>] \quad , \qquad (II.13)$$

where in coordinate space the kinetic energy operator is a differential operator. Introducing the N_c channel basis functions leads to N_c coupled differential equations for the channel wave functions. The energy E appears as a linear term in (II.13), and the equations can be rewritten as an eigenvlaue problem for the energy. However, for many channels the numerical calculations become prohibitive, and a procedure similar to the momentum space calculations is used. Equation (II.13) is rewritten as

$$[E - (H_0 + V_i)]|\Psi_i> = \lambda V_i[|\Psi_j> + |\Psi_k>] \quad , \qquad (II.14)$$

and the value of E is varied until the eigenvalue λ is unity.

If the model Hamiltonian includes a three-body force, there exists a variety of methods for including the three-body potential in the Faddeev equations. Three-body forces derived from meson exchange models can be written as a sum of three terms, that is

$$W = W_1 + W_2 + W_3 \quad , \qquad (II.15)$$

where physically W_i corresponds to the process in which particle i exchanges a one meson with each of particles j and k. One method of including a force of this type is to define

$$H_i' = H_0 + V_i + W_i \quad , \qquad (II.16)$$

and to follow the same procedure as for the case with only two-body forces. This leads to the modified form of the Faddeev equations

$$|\Psi_i> = G_i'(V_i + W_i)[|\Psi_j> + |\Psi_k>] \quad , \qquad (II.17)$$

where the resolvent G_i' includes W_i. Introducing the three-body T-matrices via

$$G_i'(V_i + W_i) = G_0 T_i \quad , \qquad (II.18)$$

the momentum space integral equation can be written as

$$|\Psi_i> = G_0 T_i [|\Psi_j> + |\Psi_k>] \quad . \qquad (II.19)$$

Different groups use variations of this procedure for the many channel calculations being done today. The Sendai [10] method consists of writing the Faddeev equations in the form

$$|\Psi_i> = G_i[V_i(|\Psi_j> + |\Psi_k>) + W_i(|\Psi_i> + |\Psi_j> + |\Psi_k>)] \equiv U|\Psi_i> \quad , \quad (II.20)$$

where the operator U includes the permutation operators which generate Ψ_j and Ψ_k from Ψ_i. Then through an iterative procedure a continued fraction expansion is generated for the Faddeev amplitude. In this method the energy is a parameter, and its value is varied until a matrix identity is satisfied to the desired accuracy. Given any initial state vector, this procedure can be used to find the wave function for the bound state.

The Bochum [11] group starts with the solution of the Faddeev equations with two-body forces and uses perturbation theory to calculate the additional binding energy due to the three-body force. The initial perturbation calculations [12] used only first-order perturbation theory; however, it has been shown [13] that the higher-order corrections are not negligible. In order to include these higher-order terms, the Bochum group has developed a perturbation expansion for arbitrary orders in the three-body force. A straightforward perturbation expansion of equation (II.19) leads to problems with the amount of comptuer storage required. To develop an efficient method, a new form of the Faddeev equations for three-body forces was introduced. Instead of writing the total wave function as the sum of three Faddeev amplitudes, a fourth Faddeev component was added for the three-body force. The total wave function is written as

$$|\Psi\rangle = G_0 \sum_{i=1}^{3} V_i |\Psi\rangle + G_0 W |\Psi\rangle = \sum_{i=1}^{3} |\Psi_i\rangle + |\Psi_4\rangle \quad . \qquad (II.21)$$

Using the Lippmann-Schwinger equation for the scattering by W alone

$$T_4 = W + W G_0 T_4 \quad , \qquad (II.22)$$

a set of coupled integral equations can be derived. The equations have the form

$$|\Psi_i\rangle = G_0 t_i [\, |\Psi_j\rangle + |\Psi_k\rangle + |\Psi_4\rangle \,] \quad , \qquad (II.23a)$$

and

$$|\Psi_4\rangle = G_0 T_4 [\, |\Psi_i\rangle + |\Psi_j\rangle + |\Psi_k\rangle \,] \quad . \qquad (II.23b)$$

These equations were used to derive a perturbation expansion around the unperturbed solution (W = 0) of the Faddeev equations. Starting with an 18-channel solution for the Reid-soft-core and Paris potentials, it was found that the second-order corrections were large and that it was necessary to keep terms up to fifth order. This explains why the early first-order perturbation calculations gave conflicting results.

The coordinate-space calculations [13] for the three-body force are done by writing (II.17) in the form

$$[E - (H_0 + V_i)] \Psi_i\rangle = \lambda [V_i (\, |\Psi_j\rangle \qquad (II.24)$$

$$+ |\Psi_k\rangle) + W_i (\, |\Psi_i\rangle + |\Psi_j\rangle + |\Psi_k\rangle) \,] \quad ,$$

using the same techniques as for the two-body case. Another choice for the coordinate-space Faddeev equations is

$$[E - (H_0 + V_i)] |\Psi_i\rangle = \lambda [V_i (\, |\Psi_j\rangle + |\Psi_k\rangle) \qquad (II.25)$$

$$+ \frac{W}{3} (\, |\Psi_i\rangle + |\Psi_j\rangle + |\Psi_k\rangle) \,] \quad .$$

Both equations give the same result for the binding energy if the number of channels is large enough. However, the results for the 5-, 9-, and

18-channel cases are different. This is a further indication that only
the converged calculation should be compared with the experimental values
or with results obtained by other numerical techniques.

Also, it is important to note that most of the Faddeev calculations
use a projected three-body force. That is, following the same procedure
as for the two-body force, the three-body force is written as

$$W_i = \sum_{\alpha=1}^{N_c} \sum_{\alpha'=1}^{N_c} |\alpha_i'> w_{\alpha'\alpha} <\alpha_i| \quad , \qquad (II.26)$$

where $w_{\alpha'\alpha} = <\alpha_i|w(\vec{x}_1,\vec{x}_2,\vec{x}_3)|\alpha_i>$ and the number of channels is truncated.
Therefore, to obtain an accurate solution to the equations, it is neces-
sary to keep enough channels to accurately represent both the Faddeev
amplitude and the potentials.

III. Faddeev Equations: Scattering States

For the scattering problem the Faddeev equations are considerably
more difficult to solve. The source of this difficulty is the fact that
the resolvent operator is no longer bounded. In momentum space this
results in singularities in the kernel of the integral equation, and the
numerical integration of the equation becomes more complicated. In
coordinate space the asymptotic boundary conditions are more difficult to
formulate in a form convenient for numerical calculations.

The Faddeev equations for the scattering problem are similar to the
bound-state case, except for the addition of the incident wave. The
Faddeev amplitudes are still defined as in (II.3), and the equations have
the form

$$|\Psi_i> = |\Phi_i> + G_i V_i [|\Psi_j> + |\Psi_k>] \quad , \qquad (III.1)$$

where $|\Phi_i>$ is the incident wave. Using (II.8) this equation is rewritten
as

$$|\Psi_i> = |\Phi_i> + G_0 t_i [|\Psi_j> + |\Psi_k>] \quad . \qquad (III.2)$$

Using the channel basis states, one obtains a set of coupled integral
equations which must be solved numerically. The integrals in these equations are two-dimensional integrals, and the numerical solution of the
equations for a realistic interaction is a large numerical calculation,

even for a modern computer. Nevertheless, computer programs to solve these equations have been developed [14], and these codes are being used to test the current model Hamiltonians. Also, various approximation methods have been used to solve the equations. A common procedure is to approximate the two-body t-matrix by a separable expansion, then the two-dimensional integrals are reduced to one-dimensinal integrals; these integral equations are usually easier to solve numerically.

In coordinate space the differential form of the Faddeev equations is the same as the form for the bound state. However, as mentioned above, the boundary conditions are different, and finding an exact form for the boundary conditions which is convenient for numerical calculations has proved to be very difficult. The initial calculations [7,15] used an approximate form derived using the stationary phase method. The errors introduced by the use of approximate boundary conditions were difficult to estimate, and recent studies [16] have indicated that the stationary approximation is only valid at distances which are prohibitive for numerical calculations. These questions about the boundary conditions must be resolved before any serious calculations can be done for the Faddeev equations in coordinate space.

IV. Conclusions

We have now reached the point where the Faddeev equations can be solved "exactly" for the bound-state three-body problem, either in momentum space or in coordinate space. These solutions have become an important tool which can be used to test our understanding of the nuclear system. The three-body scattering problem is more difficult, but much progress has been made during the past few years. However, the scattering problem has not reached the point where different groups find the same results for a given realistic interaction. Hopefully, within the next few years this goal will be achieved, and detailed comparisons with the experimental results can be made.

Acknowledgments

This work was supported, in part, by the U.S. Department of Energy. The author is grateful for the help of his collaborators J. L. Friar, B. F. Gibson, and W. N. Polyzou.

References

1. Carlson, J., Pandharipande, V. R., Wiringa, R. B.: Nucl. Phys. A 401, 59 (1983).

2. Ciofi delgi Atti, C., Smith, S.: Phys. Rev. C 32, 1090 (1984).

3. Ballot, J. L., de la Ripelle, M. Fabre: Ann. Phys. (NY) 127, 62 (1980).

4. Glöckle, W.: The Quantum Mechanical Few-Body Problem - New York: Springer-Verlag 1983, and references therein.

5. Harper, E. P., Kim, Y. E., Tubis, A.: Phys. Rev. Lett. 28, 1533 (1972); Glöckle, W., Hasberg, G., Neghabian, A. R.: Z. Phys. A 300, 217 (1982).

6. Hadjuk, C., Sauer, P. U.: Nucl. Phys. A 322, 329 (1979).

7. Merkuriev, S. P., Gignoux, C., Laverne, A.: Ann. Phys. (NY) 99, 30 (1976); Payne, G. L., Friar, J. L., Gibson, B. F., Afnan, I. R.: Phys. Rev. C 22, 823 (1980).

8. Ishikawa, S., Sasakawa, T. Sawada, T., Ueda, T.: Phys. Rev. Lett. 53, 1877 (1984).

9. Wiringa, R. B., Friar, J. L., Gibson, B. F., Payne, G. L., Chen, C. R.: Phys. Lett. 14 B, 273 (1984).

10. Sasakawa, T., Ishikawa, S.: Few-Body Systems 1, 3 (1986).

11. Bömelburg, A.: Phys. Rev. C 34, 14 (1986).

12. Bömelburg, A., Glöckle, W.: Phys. Rev. C 28, 2149 (1983).

13. Chen, C. R., Payne, G. L., Friar, J. L., Gibson, B. F.: Phys. Rev. C 33, 1740 (1986)

14. Takemiya, T.: Prog. Theor. Phys. 74, 301 (1985).

15. Payne, G. L.: Nucl. Phys. A 353, 61c (1981).

16. Levin, F. S.: International Workshop on Few-Body Approaches to Nuclear Reactions - Singapore: World Scientific Publishing Company, to be published.

INTEGRAL EQUATION APPROACH TO FEW-BODY COLLISION PROBLEMS

W. Sandhas

Physikalisches Institut, Universität Bonn, 5300 Bonn, West Germany

Abstract: The Faddeev approach, its applications to break-up and photodisintegration processes, and its extension to higher particle numbers is reviewed, with the emphasis on structural aspects. Recent four-nucleon break-up and photodisintegration calculations are discussed.

I. Introduction

The basic entities of multichannel collision theory are most naturally defined by means of time limits. For their practical calculation integral equations represent a particularly powerful tool. In going over to such equations one has to make sure that the information contained in the original definitions is fully preserved. From this point of view the three-body Faddeev equations play an exceptional role [1,2]. Their uniqueness has been proved in a mathematically rigorous (rather technical) manner [2] and, moreover, can be understood quite directly as an immediate consequence of the typical Faddeev coupling scheme [3].

In the present report we are going to consider some further characteristic features of this theory. Based on its modified formulation introduced by Alt, Grassberger, and Sandhas (AGS) [4], a striking similarity between the three-body and the two-body relations could be established [5,6]. Emphasizing these structural aspects we describe in the following the application of the Faddeev concept to break-up and photodisintegration processes and its extension to higher particle numbers.

In accord with this strategy, sec. II is devoted to those features of the two-body theory which are of relevance when going over to the three-body formalism. This transition is performed in sec. III by exhibiting the singu-

larity properties of the entities involved. Algebraical structures are discussed, and the operator identities for break-up and photodisintegration
processes are given in sec. IV. In view of the two- and three-body analogies
thus established, it is a priori clear that the reduction of the three-body
operator identities by separable subsystem expansions leads to effective
matrix equations of two-body Lippmann-Schwinger (LS) form (sec. V). The
correspondence between the two- and the three-body theory, moreover, allows
one to go step by step to higher particle numbers. As shown in sec. VI, the
extension to four particles can be based on AGS-like equations, but with the
respective three-body operators as subsystem input in their kernels.This
also holds true for break-up and photodisintegration processes. Recent results obtained by means of this four-body theory are reviewed in sec. VII.

II. Two-Body Scattering Theory

Let us consider a two-body situation with a Hamiltonian H composed
of a (free) kinetic energy part H_0 and a short-ranged potential V.
The usual steps leading from the basic concepts of time dependent collision
theory to Abelian limits then provide us with the well-known definitions of
the scattering states,

$$|\vec{p}\rangle^{(\pm)} = \lim_{\varepsilon \to 0} (\pm i\varepsilon) \, G(E \pm i\varepsilon)|\vec{p}\rangle , \qquad (2.1)$$

and of the S-matrix,

$$S(\vec{p}',\vec{p}) = \lim_{t \to \infty} \lim_{\varepsilon \to 0} e^{i(E'-E)t} \, i\varepsilon \, \langle\vec{p}'|G(E+i\varepsilon)|\vec{p}\rangle . \qquad (2.2)$$

In view of these results it is a priori clear that the momentum representa-
tion $\langle\vec{p}'|G(z)|\vec{p}\rangle$ of the resolvent of H must be rather singular. Other-
wise the ε-factor and the time oscillations would lead to zero when performing the above limits.

This can be made explicit by extracting the free resolvent G_0 from the
full resolvent according to

$$G(z) = G_0(z) + G_0(z) \, T(z) \, G_0(z) , \qquad (2.3)$$

a relation which in momentum space reads

$$\langle \vec{p}' | G(E+i\epsilon) | \vec{p} \rangle$$

$$= \frac{\delta(\vec{p}'-\vec{p})}{E+i\epsilon-p^2/2\mu} + \frac{1}{E+i\epsilon-p'^2/2\mu} \langle \vec{p}' | T(E+i\epsilon) | \vec{p} \rangle \frac{1}{E+i\epsilon-p^2/2\mu} . \quad (2.4)$$

From the alternative definition

$$T(z) = V + V G(z) V \quad (2.5)$$

of the transition operator we infer that in $\langle \vec{p}' | T(z) | \vec{p} \rangle$ the full resolvent appears between <u>normalizable</u> states $V|\vec{p}\rangle$. Hence, like any resolvent of a self-adjoint operator considered in Hilbert space, this expression is a meromorphic function, its poles and cuts being associated with the discrete and continuous spectrum of H, i.e., with the dynamics of the problem. The additional singularities, which typically occur if $G(z)$ is sandwiched between the <u>improper</u> vectors $|\vec{p}\rangle$, therefore are exhibited explicitly by (2.4). These singularities are of decisive relevance when going over from (2.1) and (2.2) to

$$|\vec{p}\rangle^{(\pm)} = [1 + G_0(E \pm io)T(E \pm io)]|\vec{p}\rangle = \Omega^{(\pm)}|\vec{p}\rangle \quad (2.6)$$

and

$$S(\vec{p}',\vec{p}) = \delta(\vec{p}'-\vec{p}) - 2\pi i\delta(E'-E) \langle \vec{p}' | T(E+io) | \vec{p} \rangle . \quad (2.7)$$

<u>Integral equations:</u> According to (2.1) and (2.2) the scattering states and the S-matrix are in principle determined by the solutions of the resolvent identity

$$G(z) = G_0(z) + G_0(z) V G(z) \quad (2.8)$$

or of its half-on-shell restriction, the Lippmann-Schwinger (LS) equation,

$$|\vec{p}\rangle^{(\pm)} = |\vec{p}\rangle + G_0(E\pm io)V|\vec{p}\rangle^{(\pm)} . \quad (2.9)$$

Due to the kinematical singularities of $\langle \vec{p}' | G | \vec{p} \rangle$, eq. (2.8) is inadequate when being considered as an integral equation in momentum space. This short-coming can be easily cured by splitting off these singularities, i.e., by inserting (2.3) on both sides of (2.8) and multiplying from the left and from the right with G_0^{-1} . We, then, arrive quite naturally at the transition operator LS-equation

$$T = V + V G_0 T \; , \tag{2.10}$$

and thus at the well behaved equation

$$T(\vec{p}',\vec{p};E) = V(\vec{p}',\vec{p}) + \int d^3k \; V(\vec{p}',\vec{k}) \; \frac{1}{E+io-k^2/2\mu} \; T(\vec{k},\vec{p};E) \; . \tag{2.11}$$

In view of (2.6) and (2.7) its solutions are most directly related to the scattering states and the S-matrix.

Photodisintegration: The Born amplitude for the photodisintegration of a two-nucleon bound state $|\psi_n\rangle$ is given by

$$B(\vec{p}) = \langle\vec{p}|H_{em}|\psi_n\rangle \; , \tag{2.12}$$

with H_{em} denoting the operator of the electromagnetic interaction. In order to take into account the final state interaction of the outgoing nucleons, $\langle\vec{p}|$ has to be replaced by the scattering state $^{(-)}\langle\vec{p}|$. Using eq. (2.6), the full disintegration amplitude, therefore, reads

$$M(\vec{p}) = {}^{(-)}\langle\vec{p}|H_{em}|\psi_n\rangle = \langle\vec{p}|[1 + T(E+io)G_0(E+io)]H_{em}|\psi_n\rangle \; . \tag{2.13}$$

This representation not only serves to define an off-shell extension of $M(\vec{p})$, but immediately leads to an integral equation for this amplitude. Writing the operator LS equation (2.10) in the form

$$(1 + TG_0) = 1 + VG_0(1 + TG_0) \tag{2.14}$$

and sandwiching this relation with $\langle\vec{p}|$ and $H_{em}|\psi_n\rangle$, we in fact arrive at

$$M(\vec{p}) = B(\vec{p}) + \int d^3k \; V(\vec{p},\vec{k}) \; \frac{1}{E+io-k^2/2\mu} \; M(\vec{k}) \; . \tag{2.15}$$

It is remarkable that the kernel of this integral equation coincides with the one of (2.11). Thus, in going over from nucleon-nucleon scattering to two-nucleon photodisintegration, we only have to replace the inhomogeneous term of (2.11) by the Born amplitude (2.12).

III. Three-Body Collisions

The transition to the three-body case can be performed in close structural analogy with the above formalism. We only have to have in mind that there are three two-fragment partitions consisting of an elementary particle $\alpha = 1, 2$ or 3 and of the respective subsystem $(\beta\gamma)$ of the other two particles. As usual we denote the subsystem potential $V_{\beta\gamma}$ complementarily by V_α and the corresponding two-body bound states and energies by $|\psi_{\alpha n}\rangle$ and $\hat{E}_{\alpha n}$. Denoting, moreover, the relative momentum between the clusters by $\vec{q}_\alpha$, the "free" channel states and the energies of this situation are given by $|\Phi_{\alpha n}\rangle = |\psi_{\alpha n}\rangle|\vec{q}_\alpha\rangle$ and $E_{\alpha n} = q_\alpha^2/2M_\alpha + \hat{E}_{\alpha n}$, respectively.

The total interaction in the full Hamiltonian H and, hence, in the resolvent $G(z)$ is now a sum of the three potentials V_α, V_β, V_γ and instead of one free resolvent G_0 there are three channel resolvents

$$G_\alpha(z) = (z - H_\alpha)^{-1} = (z - H_0 - V_\alpha)^{-1} \quad , \tag{3.1}$$

G_β and G_γ. The natural generalization of (2.3), consequently, is given by

$$G(z) = \delta_{\beta\alpha} G_\alpha(z) + G_\beta(z) U_{\beta\alpha}(z) G_\alpha(z) \quad . \tag{3.2}$$

This, in particular, means that the single channel T-operator is automatically replaced by the 3×3 transition operators $U_{\beta\alpha}$ introduced by Alt, Grassberger, and Sandhas (AGS) [4]. As in sec. II it is easily seen that $\langle\Phi_{\beta m}|U_{\beta\alpha}(z)|\Phi_{\alpha n}\rangle$ has only dynamical singularities corresponding to the spectrum of H, while additional kinematical singularities occur in $\langle\Phi_{\beta m}|G(E_{\alpha n}+i\varepsilon)|\Phi_{\alpha n}\rangle$, a fact exhibited similarly to (2.4) by sandwiching (3.2) between the channel states. It is, therefore, not astonishing that we end up with the following generalizations of (2.6) and (2.7) for the two-fragment scattering states and the re-arrangement S-matrix:

$$|\psi_{\alpha n}^{(\pm)}\rangle = [\delta_{\beta\alpha} + G_\beta(E_{\alpha n} \pm io) U_{\beta\alpha}(E_{\alpha n} \pm io)]|\Phi_{\alpha n}\rangle = \Omega_\alpha^{(\pm)}|\Phi_{\alpha n}\rangle \tag{3.3}$$

$$S_{\beta m,\alpha n} = \delta_{\beta\alpha}\delta_{mn} \delta(\vec{q}_\beta' - \vec{q}_\alpha) - 2\pi i\delta(E_{\beta m}' - E_{\alpha n})\langle\Phi_{\beta m}|U_{\beta\alpha}(E_{\alpha n}+io)|\Phi_{\alpha n}\rangle \tag{3.4}$$

Integral equations: Due to the occurrence of three channel resolvents, we now have instead of the single relation (2.8) three resolvent identities

$$G = G_\beta + G_\beta \sum_{\gamma\neq\beta} V_\gamma G \quad , \tag{3.5}$$

with $\beta = 1, 2$ or 3. In sec. II the kinematical singularities of the solutions of (2.8) were split off by means of (2.3). The same step now consists in inserting (3.2) on both sides of (3.5). Here, however, we are confronted with the additional freedom of how to choose the value of β. It appears to be quite natural to insert (3.2) on the left-hand side as it stands, but on the right-hand side with β being replaced by γ. Then the indices of the subsystem potential and of the channel resolvent coincide, which allows us to make use of the relation $V_\gamma G_\gamma = T_\gamma G_0$. Multiplying from the left and from the right with G_β^{-1} and G_α^{-1}, we then arrive at the three-body analogue of the transition operator LS equation (2.10) [4],

$$U_{\beta\alpha} = \bar{\delta}_{\beta\alpha} G_0^{-1} + \sum_\gamma \bar{\delta}_{\beta\gamma} T_\gamma G_0 U_{\gamma\alpha} , \tag{3.6}$$

where $\bar{\delta}_{\beta\alpha} = (1 - \delta_{\beta\alpha})$ denotes the anti-Kronecker symbol. It should be emphasized that these equations combine the two-body transition operators determined by

$$T_\gamma = V_\gamma + V_\gamma G_0 T_\gamma \tag{3.7}$$

and the three-body re-arrangement operators $U_{\beta\alpha}$. They, in other words, represent a most direct relationship between entities which play the same role on the two- and on the three-body level .

IV. Structures, Break-up, Photodisintegration

The transition from the three resolvent equations (3.5) to the coupled set of AGS equations (3.6) has been performed in complete analogy to the derivation of the two-body operator LS equation (2.10). This anticipates that (3.6) can be read as a matrix equation of the same structure. Introducing the transition operator matrix, the potential matrix and the matrix Green function

$$T = \{U_{\beta\alpha}\} \tag{4.1}$$

$$V = \{\bar{\delta}_{\beta\alpha} G_0^{-1}\} \tag{4.2}$$

$$G_0 = \{\delta_{\beta\alpha} G_0 T_\alpha G_0\} \tag{4.3}$$

respectively, eq. (3.6) indeed takes the form [5,6]

$$T = V + V G_0 T . \tag{4.4}$$

A full resolvent matrix G, moreover, may be defined via (2.3), with T and G_o being replaced by (4.1) and (4.3). This, of course, implies that also the resolvent identity (2.8) has its counterpart in the present matrix formulation of the three-body problem. Explicitly written the latter relation turns out to be the set of equations derived by Faddeev in [2] for the operators $M_{\beta\alpha} = \delta_{\beta\alpha} V_\alpha + V_\beta G V_\alpha$. Going over to the half on-shell restriction of the matrix resolvent identity, the original Faddeev equations [1] for the components $|\psi_{\alpha n}^\beta\rangle = G_o V^\beta |\psi_{\alpha n}^{(+)}\rangle$ are obtained. For details we refer to [5] and [6].

The above formulation of the Faddeev approach, therefore, represents a two-fragment theory of the structure of the genuine two-body problem. Within this language the AGS equations are the natural generalization of the T-operator Lippmann-Schwinger equation (2.10), while the Faddeev equations for the operators $M_{\beta\alpha}$ or the components $|\psi_{\alpha n}^\beta\rangle$ represent the three-body analogues of the resolvent equation (2.8) or of the LS equation (2.9). The relevance of this analogy becomes most striking when studying questions of unitarity [5,6] or when going over to effective two-body equations (see next section).

Break-up: The advantage of working in momentum space with the two-body T-operator equations (2.10) has been pointed out above, and the same holds true for their three-body analogue, the AGS equations. These relations have the additional advantage that for $\beta = 0$ they provide immediately the break-up operators

$$U_{o\alpha} = G_o^{-1} + \sum_{\gamma=1}^{3} T_\gamma G_o U_{\gamma\alpha} . \tag{4.5}$$

It should be noticed that the $U_{o\alpha}$ are determined algebraically by the rearrangement operators $U_{\gamma\alpha}$, i.e., by the solutions of the integral equation (3.6). For alternative formulations of this problem and early applications see [7] and [8].

Photodisintegration: Generalizing eq. (2.13), the photodisintegration of a three-body bound-state $|\Psi_{III}\rangle$ into a free two-fragment configuration $\langle\Phi_{\beta m}|$ is given by

$$M_{\beta m}(\vec{q}_\beta) = \langle\psi_{\beta m}^{(-)}|H_{em}|\Psi_{III}\rangle = \langle\Phi_{\beta m}|\Omega_\beta^{(-)\dagger} H_{em}|\Psi_{III}\rangle$$

$$= \langle\vec{q}_\beta|\langle\psi_{\beta m}|[\delta_{\beta\alpha} + U_{\beta\alpha}(E_{\beta m}+io)G_\alpha(E_{\beta m}+io)]H_{em}|\Psi_{III}\rangle . \tag{4.6}$$

Here, the representation (3.3) of the scattering states has been used.

As in the two-body case we immediately infer from (4.4)

$$(1+TG_0) = 1 + V G_0(1+TG_0) . \qquad (4.7)$$

Inserting the explicit expressions (4.1) - (4.3) we find, after some slight
modifications which make use of $G_\alpha = G_0 + G_0 T_\alpha G_0$,

$$[\delta_{\beta\alpha} + U_{\beta\alpha} G_\alpha] = 1 + \sum_\gamma \bar{\delta}_{\beta\gamma} T_\gamma G_0[\delta_{\gamma\alpha} + U_{\gamma\alpha} G_\alpha] . \qquad (4.8)$$

This generalization of (2.14) anticipates that the final integral equation
for the photodisintegration of a three-body bound state into two fragments
is given by replacing the inhomogeneous term of the scattering equation
by the photodisintegration Born amplitude, as in the transition from (2.11)
to (2.15). Originally the basic relation (4.8) was derived [9,10] without
making use of the structural equivalence between two- and three-body theory
emphasized in our approach. It should be noticed, however, that it is just
this equivalence which suggests the four-body generalization of the present
result, as being sketched in sec. VI.

V. Effective Two-Body Equations

In the two-body case the momentum representation of (2.5) is obtained
by inserting there the explicit expression

$$G_0(E+io) = \int d^3k |\vec{k}\rangle \frac{1}{E+io-k^2/2\mu} \langle\vec{k}| \qquad (5.1)$$

of the free resolvent, and by multiplying from both sides with the free states
$\langle\vec{p}'|$ and $|\vec{p}\rangle$, respectively. In the three-body case the generalization of
this step consists in expanding the resolvent matrix (4.3) analogously

$$G_0 = \{\delta_{\beta\gamma} G_0 T_\gamma G_0\} = \{\delta_{\beta\gamma}\int d^3q''_\gamma \sum_{r,s} |\vec{q}''_\gamma\rangle G_0 |\gamma r\rangle \Delta_{\gamma,rs} \langle\gamma s| G_0 \langle\vec{q}''_\gamma|\}.$$

$$(5.2)$$

Here we have made use of the fact that for short-ranged potentials the two-
body transition operator T_γ can be expanded into a series of separable terms.
In order to simplify the notation the energy dependence of $\Delta_{\gamma,rs}$, of the
form factors $|\gamma r\rangle$ or $\langle\gamma s|$ and of G_0 has been suppressed. But it should
be recalled that the usual energy shift $(z - q_\alpha^2/2M_\alpha)$, due to the existence

of the third particle of momentum $\vec{q}_\alpha$ and of reduced mass M_α, has to be performed in these entities.

Inserting (5.2) in the AGS equations (3.6) or (4.4) and multiplying from both sides, as in the two-body case, with the free states which now are the $\langle\vec{q}_\beta'|\langle\overline{\beta m}|G_0$ and $G_0|\alpha n\rangle|\vec{q}_\alpha\rangle$, we obtain the effective two-body matrix equation (compare in this context also [11])

$$\tilde{T} = \tilde{V} + \tilde{V}\,\tilde{G}_0\,\tilde{T} \quad . \tag{5.3}$$

The amplitudes, potentials and Green functions occurring in this relation are given in momentum space by

$$\tilde{T} = \{\langle\vec{q}_\beta'|\,\langle\overline{\beta m}|\,G_0\,U_{\beta\alpha}\,G_0\,|\alpha n\rangle|\vec{q}_\alpha\rangle\} \tag{5.4}$$

$$\tilde{V} = \{\delta_{\beta\alpha}\,\langle\vec{q}_\beta'|\langle\overline{\beta m}|G_0|\alpha n\rangle|\vec{q}_\alpha\rangle\} \tag{5.5}$$

$$\tilde{G}_0 = \{\delta_{\beta\gamma}\,\delta(\vec{q}_\beta' - \vec{q}_\gamma)\Delta_{\gamma,rs}\} \quad . \tag{5.6}$$

In view of the analogy between (5.1) and (5.2) it was a priori clear that the LS structure of the AGS equations exhibited in (4.4) would be preserved when going over to the effective equation (5.3).

Break-up: Applying the same steps to (4.5) we obtain on the energy shell

$$\tilde{T}_{0,\alpha n}(\vec{q}_\beta',\vec{p}_\beta';\vec{q}_\alpha) = \sum_{\gamma,rs} \langle\vec{p}_\gamma'|\gamma r\rangle\,\Delta_{\gamma,rs}\,\tilde{T}_{\gamma s,\alpha n}(\vec{q}_\gamma',\vec{q}_\alpha) \quad , \tag{5.7}$$

which shows that the break-up amplitudes are essentially given by the off-shell re-arrangement amplitudes, i.e., by the solutions of (5.3). In addition, only the two-body entities $\Delta_{\gamma,rs}$ and $\langle\vec{p}_\gamma'|\gamma r\rangle$ are involved, a result which reflects explicitly the general aspects discussed after (4.5).

Photodisintegration: Inserting the expansion (5.2) of T_γ in (4.8) and multiplying from the right with $H_{em}|\Psi_{III}\rangle$ and from the left with $\langle\vec{q}_\beta|\langle\overline{\beta m}|G_0$, the effective two-body equation

$$\tilde{M} = \tilde{B} + \tilde{V}\,\tilde{G}_0\,\tilde{M} \tag{5.8}$$

for the photodisintegration of a three-body bound state into two fragments is obtained. Here

$$\tilde{M} = \{<\vec{q}_\beta|<\overline{\beta m}|G_0[\delta_{\beta\alpha} + U_{\beta\alpha}G_\alpha]H_{em}|\Psi_{III}>\} \qquad (5.9)$$

and

$$\tilde{B} = \{<\vec{q}_\beta|<\overline{\beta m}|G_0 H_{em}|\Psi_{III}>\} \qquad (5.10)$$

are off-shell extensions of the full amplitude (4.6) and of the corresponding Born approximation, respectively. Hence, as anticipated by the structural equivalence of (2.14) and (4.8), the integral equations (5.3) and (5.8) for re-arrangement collisions and photodisintegration differ only in their inhomogeneous terms. In other words, also this feature of the genuine two-body problem [see eq. (2.11) and (2.15)] is preserved in the effective two-body equations of the three-body problem.

The advantages of the Faddeev approach to three-nucleon photodisintegration, and the present status of this problem have recently been reviewed by D. R. Lehman [12].

VI. Four Body Equations

The structural analogies between the two- and three-body formulations exhibited above represent the basis of the extension of the Faddeev-AGS approach to higher particle numbers as proposed by Grassberger and Sandhas (GS) [13].

Having combined two particles in one fragment $\alpha = (ij)$, the original three-body equations are reduced algebraically (or, after separable approximation, also literally) to effective two-body equations. In the four-body case this <u>first step</u> reduces the original problem only to an effective three-body problem. Equation (5.3) now plays the role of a subsystem relation like the two-body LS-equation (3.7) within the genuine three-body theory. Thus, an index τ [corresponding to γ in (3.7)] is needed to label the subsystem to which this relation belongs

$$\tilde{T}^\tau = \tilde{V}^\tau + \tilde{V}^\tau \tilde{G}_0 \tilde{T}^\tau . \qquad (6.1)$$

After this first step we are in the same position as in the genuine three-body case. We only have to keep in mind that instead of V_γ the interaction is given by the exchange potential $\tilde{V}^\tau$ defined in (5.5). It, moreover, should be realized that there are now six values of the subsystem indices $\alpha = (ij)$ and that α and β do not necessarily belong to the same three-body subsystem.

Apart from these additional complications, which are discussed in detail in [13], we may proceed as in the three-body case. Starting from the three-fragment analogue of the three-body AGS equations (3.6), equations (3.6)

$$\tilde{U}^{\sigma\rho} = \tilde{\delta}_{\sigma\rho} \tilde{G}_o^{-1} + \sum_\tau \tilde{\delta}_{\sigma\tau} \tilde{T}^\tau \tilde{G}_o \tilde{U}^{\tau\rho} . \qquad (6.2)$$

and expanding the subsystem T-operators into a series of separable terms [cf. eq. (5.2)]

$$\tilde{T}^\tau = \sum_{\mu,\nu} |^{\tau\mu}\rangle \tilde{\Delta}^{\tau,\mu\nu} \langle \overline{\tau\nu}| . \qquad (6.3)$$

We again end up, after this second reduction step, at an equation of the form (5.3)

$$\tilde{\tilde{T}} = \tilde{\tilde{V}} + \tilde{\tilde{V}} \tilde{\tilde{G}}_o \tilde{\tilde{T}} . \qquad (6.4)$$

The effective amplitudes, potentials, and Green functions are defined as in (5.4) - (5.6), but of course with the free resolvent G_o, the transition operators $U_{\beta\alpha}$ and the form factors $|\alpha n\rangle$ being replaced by the corresponding entities occurring in (6.2) and (6.3).

Break-up_and_Photodisintegration: Since we have treated the four-body problem like an effective three-body problem, with one of the particles being composite, all further conclusions can be taken over almost without any modification from sec. V. The analogue of the break-up relation (5.7) is of course valid [14]. We, moreover, find that the photodisintegration of four-body bound states into three fragments follows from (6.4) by replacing its inhomogeneous term by an off-shell extension $\tilde{B}$ of the Born-amplitude

$$B \sim \langle\vec{q}|\langle\psi_{III}|H_{em}|\psi_{IV}\rangle \qquad (6.5)$$

of this process. Analogously to (2.15) and (5.8) we, consequently, have instead of (6.4) [15-17]

$$\tilde{\tilde{M}} = \tilde{\tilde{B}} + \tilde{\tilde{V}} \tilde{\tilde{G}}_o \tilde{\tilde{M}} . \qquad (6.6)$$

Algebraic_procedure: In the above treatment we have simultaneously performed the algebraic re-orderings and the separable expansions which reduce the original four-body relations to effective two-body equations. Bearing in

mind the structural aspects exhibited above, we could have performed the
same steps first purely algebraically. The expansion (5.2), and the expansion (6.3) of the resulting operators, then lead again to the effective
equations (6.4) [18]. Proceeding in this way changes only the sequence of the
steps necessary to arrive at the final effective two-body equations. This
procedure, however, has at least the advantage that it enables comparison
with the alternative purely algebraical treatment of the problem by Yakubovsky [19].

What is trivially seen is that the coupling scheme of the GS treatment
and the Yakubovsky approach are identical. But the structural analogies between the two- and the three-body theory exhibited above have not been realized
in the latter access to the problem. Neither were the Yakubovsky equations
written for the transition operators nor, what is more essential, were their
kernels built up by the two-body and the three-body transition operators.
Hence, first attempts to reduce the Yakubovsky equations to an appropriate
effective theory were not successful. Only after taking over the above two-
step reduction scheme practical equations had also been obtained within this
framework. These, however, are identical to the ones achieved immediately in
the GS approach, a fact anticipated by the proof of equivalence given in
[18].

N-body_equations: It is a characteristic feature of the GS approach that the
original equations are step by step reduced to effective equations of a
lower number of fragments in a structurally equivalent manner. Having described this reduction procedure in detail for $N = 3$ and 4 its generalization is, at least in principle, trivial. For $N = 5$ the first step, e.g.,
reduces the original equations to effective four-body relations of the
form known from the genuine four-body case. The following two reduction steps
then can be performed as explained above.

VII. Recent Four-Nucleon Results

Adequate separable expansions (5.2) and (6.3) of the genuine and of
the effective subsystem amplitudes T_γ and $\widetilde{T}^\tau$ are required in the above
two-step reduction procedure. In this context one has to have in mind that
the effective potentials $\widetilde{V}^\tau$ defined by (5.5) are energy dependent. This
does not cause any real complications as long as (energy dependent) Sturmian

functions are used. The much simpler unitary pole expansion (UPE), successfully used in expanding energy independent two-body potentials V_γ, however,
has to be modified when being applied on V^τ. Two respective proposals, the
energy dependent pole expansion (EDPE) [20] and the generalized unitary pole
expansion (GUPE) [21] play an essential role in recent calculations. Since
earlier four-nucleon calculations have been summarized in various review
talks, we restrict ourselved to some hints on these recent calculations (for
more details see the report by H. Fiedeldey [22]).

Below the break-up threshold converged calculations have been achieved
by Fonseca with only two or three EDPE terms [23]. The convergence of the
EDPE and the GUPE above the threshold was studied by Sofianos et al. within
the K-matrix approach developed in [24], taking into account, however, additional principal value parts of the effective resolvents [25]. In spite of
dramatic improvements, in particular below threshold, there remain remarkable
discrepancies.

Break-up: In recent calculations of the break-up reactions ^{3}He(p,pd)^{1}H and
^{2}H(d,nd)p by Mdlalose et al. a high number of spin-isospin combinations and
of GUPE terms had to be taken into account in order to reach convergence.

Reasonable agreement between theory and experiment could be achieved
in the first case both with respect to shape and magnitude. Moreover, a clear
distinction between the contributions originating from quasi free scattering
(QFS) and from final state interaction (FSI) could be given, without adjust-
ing any parameters as done in phenomenological models [26].

In the second case, i.e., for deuteron-deuteron initial states, the
shape is well reproduced in kinematical regions governed by QFS mechanisms.
The magnitudes, however, had to be divided by factors 2 or 2.87, respectively [27]. Kinematical regions where both mechanisms compete have been considered most recently by the same authors [28]. There, a slight shift of the QFS
peak is being observed and the theoretical curves now have to be multiplied,
e.g., by a factor of 1.75. In spite of the rather satisfactory agreement of
the shape of the experimental and theoretical curves, it therefore appears to
be necessary to go beyond the lowest order K-matrix approximation.

Photodisintegration: In the energy region between 20 and 40 MeV it turned
out that the Born approximation is unable to reproduce the data even after
incorporating the mesonic degrees of freedom via Siegert's theorem. The full
solution of the integral equation (6.6) with the Siegert-modified Born term
was necessary to get agreement between theory and experiment. When the
corresponding calculations were presented first [16], this was by no means
obvious. At the same time, however, new experimental data had been published

which were in astonishing agreement with these theoretical predictions (cf. [17]).

For energies between 40 and 140 MeV the pure Born approximation without any mesonic corrections came already close to experiment in the calculations of [15]. It was, therefore, not unexpected that incorporating these corrections would lead to results above experiment [17]. Recently Sofianos et al. have repeated these calculations by applying EDPE and GUPE up to convergence [29]. In this way a considerable reduction of the pure Born approximation was achieved, so that incorporation of Siegert's theorem led to almost perfect agreement between theory and experiment. This indicates that the final state interactions should be small. Further investigations concerning this question are in progress.

References

1. L.D. Faddeev, Sov. Phys. JETP 12, 1014 (1961).
2. L.D. Faddeev, Mathematical Aspects of the Three-Body Problem in the Quantum Scattering Theory (Davey, New York, 1965).
3. W. Sandhas, in Few-Body Nuclear Physics, edited by G. Pisent, V. Vanzani, and L. Fonda (IAEA, Vienna, 1978), p. 3.
4. E.O. Alt, P. Grassberger, and W. Sandhas, Nucl. Phys. B2, 167 (1967).
5. W. Sandhas, in Elementary Particle Physics, edited by P. Urban [Acta Physica Austriaca, Suppl. IX, 57 (1972)].
6. W. Sandhas, in Few-Body Methods: Principles and Applications, edited by T.K. Lim et al. [World Scientific, Singapore, 1986], p. 3.
7. R. Aaron and R.D. Amado, Phys. Rev. 150, 857 (1966); R.T. Cahill and I.H. Sloan, in Three Body Problem, edited by J.S.C. McKee and P.M. Rolph (North-Holland, Amsterdam, 1970), p. 265.
8. W. Ebenhöh, Nucl. Phys. A191, 97 (1972).
9. I.M. Barbour and A.C. Phillips, Phys. Rev. C1, 165 (1970).
10. B.F. Gibson and D.R. Lehman, Phys. Rev. C11, 29 (1975).
11. C. Lovelace, Phys. Rev. 135, B1225 (1964)
12. D.R. Lehman, in Three-Body Force in the Three-Nucleon System, edited by B.L. Berman and B.F. Gibson [Lecture Notes in Physics, Vol. 260, Berlin, 1986], p. 287; invited talk presented at the Eleventh International Conference on Few Body Systems in Particle and Nuclear Physics, Tokyo and Sendai, 1986.
13. P. Grassberger and W. Sandhas, Nucl. Phys. B2, 181 (1967).
14. W. Sandhas, talk presented at the International Workshop on Few-Body Approaches to Nuclear Reactions in Tandem and Cyclotron Energy Regions,

Tokyo, 1986.

15. W. Böttger, A. Casel, and W. Sandhas, Phys. Lett. B92, 11 (1980).

16. A. Casel and W. Sandhas, in Proceedings of the Ninth International Conference on Few Body Problems, Vol. I, edited by M.G. Moravczik, Eugene, 1980, p. 33

17. A. Casel and W. Sandhas, Czech. J. Phys. B36, 300 (1986).

18. E.O. Alt, P. Grassberger, and W. Sandhas, JINR Report E4-6688, Dubna, 1972; and in Few Particle Problems in the Nuclear Interaction, edited by J. Slaus et al. (North Holland, Amsterdam, 1972), p. 299;
W. Sandhas, in Progress in Particle Physics, edited by P. Urban [Acta Physics Austriaca, Suppl. XIII, 679 (1974)]; Czech. J. Phys. B25, 251 (1975).

19. O.A. Yakubovsky, Sov. J. Nucl. Phys. 5, 937 (1967).

20. S.A. Sofianos, N.J. McGurk, and H. Fiedeldey, Nucl. Phys. A318, 295 (1979).

21. A. Casel, H. Haberzettl, and W. Sandhas, Phys. Rev. C25, 1738 (1982).

22. H. Fiedeldey, invited talk presented at the Eleventh International Conference on Few Body Systems in Particle and Nuclear Physics, Tokyo and Sendai, 1986.

23. A.C. Fonseca, Few Body Systems 1, 68 (1986).

24. E.O. Alt, P. Grassberger, and W. Sandhas, Phys. Rev. C1, 85 (1970).

25. S.A. Sofianos, H. Fiedeldey, and W. Sandhas, Phys. Rev. C32, 400 (1985).

26. T.E. Mdlalose, H. Fiedeldey, and W. Sandhas, Phys. Rev. C33, 784 (1986).

27. T.E. Mdlalose, H. Fiedeldey, and W. Sandhas, Nucl. Phys. A457, 273 (1986).

28. T.E. Mdlalose, H. Fiedeldey, and W. Sandhas, contribution to the Eleventh International Conference on Few Body Systems in Particle and Nuclear Physics, Tokyo and Sendai, 1986.

29. S.A. Sofianos, H. Fiedeldey, and W. Sandhas, invited contribution to the Eleventh International Conference on Few Body Systems in Particle and Nuclear Physics, Tokyo and Sendai, 1986.

COULOMB EFFECTS ON FEW-BODY SCATTERING STATES

E.O.Alt

Institut für Physik

Johannes Gutenberg-Universität

Staudinger Weg 7, Postfach 3980

D-6500 Mainz, West Germany

Abstract

Modifications of stationary momentum space scattering theory, necessitated by the presence of Coulomb forces, are described, both in the formalism which uses unscreened Coulomb potentials and in the screening and renormalization approach. Thereby, emphasis is laid on exposing the conceptual differences, as well as the different, presently achieved status of applicability. Some of the unresolved problems in both methods are enumerated.

I) Introduction

Despite of the fact that the Coulomb potential is the prototype of a local interaction, the first systematic investigations of how to include it in stationary few-body scattering theory have been performed in momentum space, [1-4]. One of the reasons was certainly the great success of momentum space integral equations in describing processes with neutral composite particles. Another one was the realization [5,6] that in this approach the amplitudes for two-cluster arrangement reactions can be easily and exactly made to satisfy a coupled set of equations independent from those pertaining to break-up and other more complicated processes, thereby making it possible to study each of them separately.

Already for two charged elementary particles the momentum space treatment of scattering reveals some very interesting facts. Let me mention just a few of them:

(i) the unfamiliar relation between S- and T-matrix [7-10];

(ii)   the lack of a smooth on-shell limit of T-matrix elements [11,12] and momentum space wave functions [13], thus rendering useless the Lippmann-Schwinger (LS) equations for their calculation;

(iii) the non-existence of partial-wave (p.w.) projections of the on-shell scattering amplitude, respectively the divergence of the p.w. series for the p.w. amplitudes [14];

(iv)   if the Coulomb potential is being screened, the non-existence of a smooth zero-screening limit in interesting physical quantities [14-17].

The cause for all these disturbing results resides in the infinite range of the Coulomb potential which affects the particle's motion even asymptotically far away from the scattering center, thereby preventing it from ever becoming free. Consequently, conventional scattering theory which is based on a comparison of the motion under the full Hamiltonian with the one under the free Hamiltonian, can not be applied without alterations to the present situation (an obvious exception is the bound state problem in which all particles are confined to a finite region of space. For this case no modifications are required. Thus, this topic is not being dealt with here.) In momentum space this infinite range gives rise to a singularity of the potential which occurs even at the on-shell point. It can, therefore, coincide with the corresponding pole of the free Green's function, making the LS equation untractable by standard techniques. (For an exhaustic review see ref. [18].)

From the physical point of view it is to be expected that all these problems show up also in charged-composite particle scattering. Of course, additional ones are expected to arise due to the complexity of the latter (see, e.g., ref. [19]).

In the following I briefly describe some important results obtained in the integral equations approach to three-charged particle scattering, both by using unscreened as well as screened Coulomb potentials. For more details, and references to possible generalizations to arbitrary particle numbers, I refer, e.g., to ref. [20].

II) Scattering by unscreened Coulomb potentials

1) Notation

Consider three particles with charges e_α, masses m_α and momenta $\vec{k}_\alpha$, $\alpha = 1,2,3$. From the latter we construct the Jacobi variables $(\alpha \neq \beta \neq \gamma \neq \alpha)$:

$$\vec{p}_\alpha = \mu_\alpha \ (\vec{k}_\beta/m_\beta - \vec{k}_\gamma/m_\gamma) \ , \qquad (II,1)$$

the relative momentum between particles β and γ, and

$$\vec{q}_\alpha = M_\alpha \ (\vec{k}_\alpha/m_\alpha - (\vec{k}_\beta+\vec{k}_\gamma)/(m_\beta+m_\gamma)), \qquad (II,2)$$

the momentum of particle α relative to the center of mass of the pair $(\beta+\gamma)$. $\vec{P} = \vec{k}_1+\vec{k}_2+\vec{k}_3$ is the total momentum. Herein, $\mu_\alpha^{-1} = m_\beta^{-1}+m_\gamma^{-1}$ and $M_\alpha^{-1} = m_\alpha^{-1} + (m_\beta+m_\gamma)^{-1}$ are the corresponding reduced masses.

The total interaction between the three particles is (here and in the following we employ the familiar cyclic notation)

$$V = \sum_{\alpha=1}^{3} V_\alpha = \sum_{\alpha=1}^{3} (V_\alpha^S + V_\alpha^C) \ , \qquad (II,3)$$

whereat we considered only pair potentials consisting of a short-ranged (V_α^S) and a Coulombian part (V_α^C). The latter is simply

$$V_\alpha^C(\vec{r}_\alpha) = e_\beta e_\gamma/|\vec{r}_\alpha| \ , \qquad (II,4)$$

with $\vec{r}_\alpha$ denoting the co-ordinate canonically conjugate to $\vec{p}_\alpha$. The total and the channel Hamiltonians are $H = H_o + V$ and $H_\alpha = H_o + V_\alpha$, $\alpha = 1,2,3$, respectively. The break-up channel ($\alpha = 0$) is included provided we define $V_o \equiv 0$. From now on we always assume that the trivial center-of-mass motion has been extracted from the free Hamiltonian H_o. The two-fragment channel states $|\Phi_{\alpha m}\rangle = |\Psi_{\alpha m}\rangle |\vec{q}_\alpha\rangle$ are product states of an internal wave function $|\Psi_{\alpha m}\rangle$ of the bound pair $(\beta+\gamma)$ with quantum numbers "m" and energy $\hat{E}_{\alpha m}$, and a plane wave describing its free motion relative to particle α, and are eigenfunctions of H_α to the energy $E_{\alpha m} = \vec{q}_\alpha^2/2M_\alpha + \hat{E}_{\alpha m}$. Similarly for the three-fragment states $|\Phi_o\rangle = |\vec{p}_\alpha\rangle |\vec{q}_\alpha\rangle$ which are eigenfunctions of H_o to the energy $E_o = \vec{q}_\alpha^2/2M_\alpha + \vec{p}_\alpha^2/2\mu_\alpha$ (in this case there is no preferable choice for the index α; we, therefore, omit it altogether whenever possible).

2) Definition of scattering amplitudes

A convenient definition of the three-body transition operators $U_{\beta\alpha}$ in terms of the full resolvent $G(z) = (z - H)^{-1}$ is [5]

$$G(z) = \delta_{\alpha\beta} \ G_\alpha(z) + G_\beta(z) \ U_{\beta\alpha}(z) \ G_\alpha(z) \ , \quad \alpha,\beta = 0,\ldots,3. \qquad (II,5)$$

Here, $G_\alpha(z) = (z - H_\alpha)^{-1}$ is the channel resolvent.

If there were only short-ranged forces acting, the physical reaction amplitudes $T_{\beta n,\alpha m}(\vec{q}'_\beta , \vec{q}_\alpha)$ describing the transition from an incoming configuration $(\alpha m, \vec{q}_\alpha)$ to an outgoing one $(\beta n, \vec{q}'_\beta)$, would be determined by

$$T_{\beta n, \alpha m}(\vec{q}_\beta', \vec{q}_\alpha) = \lim_{\varepsilon \to 0} \langle \Phi_{\beta n} | U_{\beta\alpha}(E_{\alpha m} + i\varepsilon) | \Phi_{\alpha m} \rangle \Big|_{E_{\beta n}' = E_{\alpha m}}, \tag{II,6}$$

with the operators $U_{\beta\alpha}$ obtained as solutions of the familiar three-body equations. Analogously for the $2 \to 3$ and $3 \to 3$ amplitudes $T_{0,\alpha m}(\vec{p}', \vec{q}'; \vec{q}_\alpha)$ and $T_{00}(\vec{p}', \vec{q}'; \vec{p}, \vec{q})$, respectively.

In the presence of Coulomb forces, however, the desired scattering amplitudes can not be calculated as described above. The reason is the occurrence of singularities, caused by the infinite range of the former, which prevent the on-shell limit (II,6) to be taken.

In the two-particle case, these singularities have been extracted explicitly so that the physical Coulomb scattering amplitude could be recovered from the momentum space representation of the Green's function [21], respectively of the transition operator [22]. A similar result was proven in [7,23] for three-particle scattering with Coulomb potentials, by making use of the existence of Møller operators [24]. Thus, instead of (II,6) we have the relations

$$T_{\beta n, \alpha m}(\vec{q}_\beta', \vec{q}_\alpha) = \lim_{\varepsilon \to 0} Z_\beta^{-1/2}(q_\beta'; \varepsilon) \langle \Phi_{\beta n} | U_{\beta\alpha}(E_{\alpha m} + i\varepsilon) | \Phi_{\alpha m} \rangle Z_\alpha^{-1/2}(q_\alpha; \varepsilon) \Big|_{E_{\beta n}' = E_{\alpha m}} \tag{II,7}$$

$$T_{0,\alpha m}(\vec{p}', \vec{q}'; \vec{q}_\alpha) = \lim_{\varepsilon \to 0} Z_0^{-1/2}(\vec{p}', \vec{q}'; \varepsilon) \langle \Phi_0 | U_{0\alpha}(E_{\alpha m} + i\varepsilon) | \Phi_{\alpha m} \rangle Z_\alpha^{-1/2}(q_\alpha; \varepsilon) \Big|_{E_0' = E_{\alpha m}} \tag{II,8}$$

$$T_{00}(\vec{p}', \vec{q}'; \vec{p}, \vec{q}) = \lim_{\varepsilon \to 0} Z_0^{-1/2}(\vec{p}', \vec{q}'; \varepsilon) \langle \Phi_0 | U_{00}(E_0 + i\varepsilon) | \Phi_0 \rangle Z_0^{-1/2}(\vec{p}, \vec{q}; \varepsilon) \Big|_{E_0' = E_0}. \tag{II,9}$$

The singular "renormalization" factors $Z_\alpha^{1/2}$ and $Z_0^{1/2}$ are given by

$$Z_\alpha^{1/2}(q_\alpha; \varepsilon) = \Gamma(1 - i\bar{\eta}_\alpha(q_\alpha)) \left(\frac{2q_\alpha^2/M_\alpha}{\varepsilon} \right)^{-i\bar{\eta}_\alpha(q_\alpha)} \tag{II,10}$$

$$Z_0^{1/2}(\vec{p}, \vec{q}; \varepsilon) = \Gamma(1 - \sum_{\alpha=1}^{3} \eta_\alpha(p_\alpha)) \prod_{\alpha=1}^{3} \left(\frac{2p_\alpha^2/\mu_\alpha}{\varepsilon} \right)^{-i\eta_\alpha(p_\alpha)}. \tag{II,11}$$

Herein, we have introduced the Sommerfeld parameters for the two elementary particles β and γ, $\eta_\alpha(p_\alpha) = e_\beta e_\gamma \mu_\alpha / p_\alpha$, and for the two fragments consisting of particle α and the bound pair $(\beta+\gamma)$, $\bar{\eta}_\alpha(q_\alpha) = e_\alpha(e_\beta+e_\gamma) M_\alpha / q_\alpha$. Furthermore, in (II,11) the notation is such that, after having chosen the momenta $(\vec{p}, \vec{q})$ as $(\vec{p}_\alpha, \vec{q}_\alpha)$ with some fixed α, the momenta $\vec{p}_\beta$ with $\beta \neq \alpha$ on the r.h.s. have to be expressed as linear combinations of $\vec{p}_\alpha$ and $\vec{q}_\alpha$. $\Gamma(.)$

is the Gamma-function. Note the close similarity [25] of these renormalization factors with the corresponding distortion factors occurring in the time-dependent approach of ref. [24].

The above-mentioned proof of additional singularities of the off-shell T-matrix elements when the energy shell is approached, has been rather indirect. A more direct one for two-cluster reactions has been found in [26-28].

3) Equations for physical quantities

An immediate consequence of relations (II,7) - (II,9) is that the conventional integral equations for $U_{\beta\alpha}$ are not appropriate for the calculation of physical scattering amplitudes, unless some method is developed for extracting the singular factors described above in a numerically stable way. However, it appears to be more promising to try to isolate them analytically. This goal can be achieved by means of a two-potential approach, at least for arrangement scattering. It entails the introduction of the so-called Coulomb-modified short-range transition operators which are free from these singularities and are, therefore, better suited for practical applications [26-28,20]. Yet there are still problems with the integral equations for these quantities for energies above the break-up threshold, ([19], compare also [29]). One possibility for overcoming them lies in a thorough investigation of the matrix elements between the channel states of the occurring driving terms [30]. For the three-charged particle break-up a promising attempt at isolating the most dangerous singularities has been undertaken in ref.[31]. Up to now there have been no scattering calculations based on this method.

III) Scattering by screened Coulomb potentials

An intrinsically different approach to charged particle scattering consists in screening the Coulomb potentials. From the physical point of view, this is motivated by the fact that in Nature unscreened Coulomb potentials do _not_ exist. It also has the formal advantage that now all occurring potentials are short-ranged and, therefore, the conventional scattering formalism is applicable.

In such a procedure two questions arise. Firstly, can those scattering quantities which are directly related to experimental observables be introduced in such a way that they become independent of the precise form of the screening as well as of the size of the screening radius R, both of which are not normally known in praxi, for such large but finite values of the latter as they are realized in actual experimental situations. Only if answered assertively does the application of the screened theory make sense. Secondly, do these suitably defined quantities, in the limit of infinite R, go over into the corresponding expressions as introduced in the approach which uses unscreened Coulomb potentials. The affirmative answer to this question provides the justification for the interpretation of experimental data with the latter theory.

1) Notation

The Coulomb potentials (II,4) are now replaced by

$$V_\alpha^R(\vec{r}_\alpha) = V_\alpha^C(\vec{r}_\alpha)\ g_\alpha(r_\alpha/R) \qquad \text{(III,1)}$$

where $g_\alpha(\cdot)$ is assumed to be a smooth function with the properties $g_\alpha(x) \to 0$ for $x \to \infty$ and $g_\alpha(x) \to 1$ for $x \to 0$ (an example is provided by exponential screening, $g_\alpha(x) = \exp(-x)$). The resulting dependence of various quantities on the screening radius R will be indicated by a superscript R.

2) Definition of scattering amplitudes

Since, for finite R, the V_α^R are short-ranged the conventional definitions (II,5) (with all quantities adorned by a superscript R) for the three-body transition operators $U_{\beta\alpha}^{(R)}$ yield, when inserted into formulas like (II,6), the physical screened amplitudes for all interesting reactions. The values of the latter, however, vary strongly with R. But this unwanted dependence can, for large screening radii, be factorized such that the remainder amplitude becomes independent of it [4,26,32,33]. The same conclusion has been reached in time-dependent scattering theory [34,35], namely that the zero-screening limit does exist for suitably renormalized screened amplitudes, independently of the specific functional form of $g_\alpha(\cdot)$ (under mild assumptions on the latter).
That is, we have

$$\lim_{R\to\infty} Z_\beta^{-1/2}(q_\beta';R)\ T_{\beta n,\alpha m}^{(R)}(\vec{q}_\beta',\vec{q}_\alpha)\ Z_\alpha^{-1/2}(q_\alpha;R)\Big|_{E_{\beta n}'=E_{\alpha m}} = T_{\beta n,\alpha m}(\vec{q}_\beta',\vec{q}_\alpha) \qquad \text{(III,2)}$$

$$\lim_{R\to\infty} Z_o^{-1/2}(\vec{p}',\vec{q}';R)\ T_{o,\alpha m}^{(R)}(\vec{p}',\vec{q}';\vec{q}_\alpha)\ Z_\alpha^{-1/2}(q_\alpha;R)\Big|_{E_o'=E_{\alpha m}} = T_{o,\alpha m}(\vec{p}',\vec{q}';\vec{q}_\alpha)$$

$$\text{(III,3)}$$

$$\lim_{R\to\infty} Z_0^{-1/2}(\vec{p}',\vec{q}';R) \; T_{00}^{(R)}(\vec{p}',\vec{q}';\vec{p},\vec{q}) \; Z_0^{-1/2}(\vec{p},\vec{q};R) \Big|_{E_0'=E_0} = T_{00}(\vec{p}',\vec{q}';\vec{p},\vec{q}),$$

(III,4)

the limits on the r.h.s. coinciding with the corresponding reaction amplitudes defined in the no-screening approach of Sect. II, cf. Eqs. (II,7) - (II,9). The renormalization factors can be calculated for any given $g_\alpha(\cdot)$. For the special case of exponential screening, $g_\alpha(x) = \exp(-x)$, one obtains in the limit of large R

$$Z_\alpha^{1/2}(q_\alpha;R) = (2q_\alpha R)^{-i\bar{\eta}_\alpha(q_\alpha)} \; e^{i\bar{\eta}_\alpha(q_\alpha)C}$$

(III,5)

$$Z_0^{1/2}(\vec{p},\vec{q};R) = \prod_{\alpha=1}^{3} (2p_\alpha R)^{-i\eta_\alpha(p_\alpha)} \; e^{i\eta_\alpha(p_\alpha)C} .$$

(III,6)

In Eq. (III,6) the same notation has been employed as in (II,11). $C = 0.5772 \ldots$ is the Euler number.

3) Equations for physical quantities

The relations (III,2) - (III,4) suggest an algorithm for the direct evaluation of unscreened reaction amplitudes. Namely, calculate for some R the desired screened amplitude by any method known in short-range scattering theory and multiply the result by the appropriate renormalization factors. Thereupon repeat the procedure for increasing values of the screening radius until independence on it is achieved. The resulting quantity is eventually to be interpreted as the corresponding unscreened transition amplitude.

However, for such an approach to be meaningful it must be ascertained that the limits in (III,2) - (III,4) hold in a pointwise sense. In contrast to the claims in ref. [34] this could be shown to be correct [4,26,32,33] for relation (III,2), and for (III,3) in the special case that one of the three particles be neutral. For three charged particles the interpretation of the limits in (III,3) and (III,4) is still an open problem.

The method described above has been applied extensively to the calculation of proton-deuteron phase parameters. For a detailed account of the corresponding results confer refs. [36].

Let me add two remarks:

(i) Results like Eqs. (III,2) - (III,4) prove that the unscreened theory follows from the screened one in the limit $R \to \infty$, as expected.

(ii) The question, for what values of the screening radius independence on it is actually achieved for the renormalized scattering amplitudes, has been investigated numerically. For proton-deuteron scattering, the corresponding R appears empirically to be determined by $(L+1)/q_\alpha R \sim$ const., with q_α being the Compton wave length of the projectile and L the total angular momentum.

IV) Summary and Conclusions

Starting from stationary momentum space scattering theory we have first discussed the problem of definition for charged-composite particle reaction amplitudes, in the formulations which use unscreened as well as screened Coulomb potentials. Then the question of how to actually compute them has been approached. In this connection several as yet unresolved problems have been encountered.

Concerning applications, for arrangement processes the screening and renormalization approach has been most successful up to now. However, the wide and interesting field of calculating Coulomb corrections to nuclear scattering observables has just been opened up. A lot of work needs still to be done concerning both the development of more efficient methods (e.g., by improving the analytical treatment of the long-ranged parts of the interaction) and their application to many interesting physical phenomena.

References

1. Noble, J.V.: Phys.Rev. 161,945(1967).
2. Faddeev, L.D.: in Mc Kee, J.S.C., Rolph, P.M. (eds.): Three Body Problem in Nuclear and Particle Physics, p.154. Amsterdam: North-Holland 1970. Veselova, A.M.: Theor.Math.Phys. 3,542(1970).
3. Prugovecki, E., Zorbas, J.: Nucl.Phys. A213,541(1973); Zorbas, J.: J.Math.Phys. 18,1112(1977).
4. Alt, E.O.: in Mitra, A.N., Slaus, I., Bhasin, V.S., Gupta, V.K. (eds.): Few Body Dynamics, p.76. Amsterdam: North-Holland 1976.
5. Alt, E.O., Grassberger, P., Sandhas, W.: Nucl.Phys. B2,167(1967).
6. Grassberger, P., Sandhas, W.: Nucl.Phys. B2,181(1967).
7. Veselova, A.M.: Theor.Math.Phys. 13, 368(1972).
8. Rosenberg, L.: Phys.Rev. D8,1833(1973).
9. Herbst, I.W.: Commun.Math.Phys. 35,181(1974).
10. Bajzer, Z.: Z.Physik A278,97(1976).
11. Ford, W.F.: Phys.Rev. 133,B1616(1964).
12. van Haeringen, H., van Wageningen, R.: J.Math.Phys. 16,1441(1975).

13. Guth, E., Mullin, C.J.: Phys.Rev. $\underline{83}$,667(1951).

14. Taylor, J.R.: Nuovo Cim. $\underline{B23}$,313(1974);
    Semon, M.D., Taylor, J.R.: Nuovo Cim. $\underline{A26}$,48(1975).

15. Gorshkov, V.G.: Sov.Phys. - JETP $\underline{13}$,1037(1961).

16. Ford, W.F.: J.Math.Phys. $\underline{7}$,626(1966).

17. Goodmanson, D.M., Taylor, J.R.: J.Math.Phys. $\underline{21}$,2202(1980).

18. van Haeringen, H.: Charged-Particle-Interactions. Leiden: Coulomb
    Press Leyden 1985.

19. Veselova, A.M.: in Few Particle Problem in Nuclear Physics, p.326.
    Dubna: JINR 1980.

20. Alt, E.O.: in Lim, T.K., Bao, C.G., Hou, D.P., Huber, S. (eds.):
    Few Body Methods: Principles and Applications. Singapore: World
    Scientific 1986.

21. Schwinger, J.: J.Math.Phys. $\underline{5}$,1606(1964).

22. Okubo, S., Feldman, D.: Phys.Rev. $\underline{117}$,292(1960).

23. Bajzer, Z.: in Pisent, G., Vanzani, V., Fonda, L. (eds.): Few-Body
    Nuclear Physics, p.365. Trieste: ICTP 1978.

24. Dollard, J.D.: J.Math.Phys. $\underline{5}$,729(1964); Rocky Mtn. J.Math.
    $\underline{1}$,5(1971).

25. Zorbas, J.: Lett.Nuovo Cim. $\underline{10}$,121(1974).

26. Alt, E.O., Sandhas, W.: Phys.Rev. $\underline{C21}$,1733(1980).

27. Thalheim, H.O.: Diploma thesis, University of Mainz, 1984: unpub-
    lished.

28. Veselova, A.M.: Theor.Math.Phys. $\underline{35}$,180(1978).

29. Kok, L.P., van Haeringen, H.: Phys.Rev. $\underline{C21}$,512(1980).

30. Kharchenko, V.F., Shadchin, S.A., Zepalova, M.L.: J.Phys.
    $\underline{B18}$,949(1985).

31. Avakov, G.V., Ashurov, A.R., Levin, V.G., Mukhamedzhanov, A.M.:
    J.Phys. $\underline{A17}$,1131(1984).

32. Alt, E.O., Sandhas, W., Ziegelmann, H.: Phys.Rev. $\underline{C17}$,1981(1978).

33. Alt, E.O.: in Pisent, G., Vanzani, V., Fonda, L. (eds.): Few Body
    Nuclear Physics, p.271. Trieste: ICTP 1978.

34. Zorbas, J.: J.Phys. $\underline{A7}$,1557(1974).

35. Alt, E.O., Sandhas, W.: in Zingl, H., Haftel, M., Zankel, H. (eds.):
    Few Body Systems and Nuclear Forces, Vol.I, p.373. Berlin: Sprin-
    ger 1978.

36. Alt, E.O., Sandhas, W., Ziegelmann, H.: Nucl.Phys. $\underline{A445}$,429(1985);
    Alt, E.O.: in Bencze, Gy., Doleschall, P., Revai, J. (eds.): Dy-
    namics of Few Body Systems. Budapest: KFKI 1986.

RECENT DEVELOPMENTS IN HYPERSPHERICAL HARMONIC METHOD

M. Fabre de la Ripelle

Division de Physique Théorique[*], Institut de Physique Nucléaire

F-91406 Orsay Cédex

Abstract - The infinite system of coupled differential equations of the hyperspherical harmonic expansion method is transformed in a single equivalent two variables integrodifferential equation. This equation is identical for 3 bosons in S state to the Faddeev equation written by Noyes for S state projected local potentials. The integrodifferential equation can be solved by an Adiabatic method. The Adiabatic eigenpotentials become asymptotically constant for large hyperradii. Each one is the total binding energy related to a definite asymptotic channel where clusters are at rest. Each eigenpotential is related to bound and/or scattering states. This method enables one to solve scattering.

The Potential Harmonic Expansion Approach

One possible method of solving the many-body problem is to write the wave function as the sum of partial waves

$$\Psi(\vec{x}) = \sum_{i < j \leq A} \psi_{ij}(\vec{x}) \tag{1}$$

for a system of A identical particles and to assume that $\psi_{ij}(\vec{x})$ is the product [1]

$$\psi_{ij}(\vec{x}) = H_{[L]}(\vec{x}) \, F(r_{ij}, r) \tag{2}$$

of a suitably symmetrized Harmonic Polynomial caracterized by $[L]$, the set of quantum numbers defining the "state", and a function of $r_{ij} = |\vec{x}_i - \vec{x}_j|$ and $r = \left[2/A \sum_{i < j \leq A} r_{ij}^2 \right]^{1/2}$, where $\vec{x}$ is the set of coordinates $(\vec{x}_1, \vec{x}_2, \cdots, \vec{x}_A)$ of the particles. We assume a local central two-body interaction $V(r_{ij})$. The partial waves must be a solution of

$$(T-E)H_{[L]}(\vec{x})F(r_{ij},r) = -V(r_{ij})H_{[L]}(\vec{x}) \sum_{k<\ell\leq A} F(r_{k\ell},r) \qquad (3)$$

in order that Ψ be a solution of the Schrodinger equation obtained by summing (3) over all pairs (i,j).

Our problem is now to find a function of r_{ij} and r only which satisfy (3). It can be solved by performing an expansion of $H_{[L]}(\vec{x})F(r_{ij},r)$ on the Potential Harmonic basis [1,2] complete for the expansion of any function of r_{ij}

$$H_{[L]}(\vec{x})F(r_{ij},r) = \sum_{K} P^{o}_{[L+2K]}(\Omega_{ij})u_K(r)/r^{\mathcal{L}+1} \qquad (4)$$

$$, \mathcal{L} = L + 3A/2 - 3,$$

and by projecting (3) on the same basis generating the system of coupled differential equations :

$$\{\frac{\hbar^2}{m}[-\frac{d^2}{dr^2} + \frac{\mathcal{L}_K(\mathcal{L}_K+1)}{r^2}] - E\} u_K(r) = -\sum_{K'} f^2_{K'} V^{K'}_K(r)u_{K'}(r) \qquad (5)$$

where $\mathcal{L}_K = \mathcal{L} + 2K$
which determine the radial waves $u_K(r)$ in (4). The potential matrix

$$V^{K'}_K(r) = \int P^{o}_{[L+2K]}(\Omega_{ij})V(r_{ij})P^{o}_{[L+2K']}(\Omega_{ij})d\Omega$$

is an integral taken over the surface of the unit hypersphere. The coefficient f^2_K is the overlap integral

$$f^2_K = \int P^{o}_{[L+2K]}(\Omega_{ij})[\sum_{k<\ell\leq A} P^{o}_{[L+2K]}(\Omega_{k\ell})]d\Omega \qquad (6)$$

This method can be applied to any systems with an arbitrary number of particles. For either large systems (A large) and/or for strongly repulsive core potentials a large number of partial waves must be included in the expansion and consequently the same number of coupled equations (5) have to be solved to obtain a good accuracy. This difficulty can be overcome by transforming the infinite system (5) into an equivalent single integrodifferential equation which determines $F(r_{ij},r)$.

The Integrodifferential Approach

Starting again from (3) we introduce the new variable

$$z = \cos 2\phi = 2r^2_{ij}/r^2 - 1 \quad , \text{ (i.e. } \cos\phi = r_{ij}/r \text{), and we write}$$

the partial wave

$$\psi_{ij}(\vec{x}) = H_{[L]}(\vec{x})\, P(z,r)/r^{L+1} \tag{7}$$

In order to obtain an equation for $P(z,r)$ we premultiply (3) by $r^{-L}\, H^{*}_{[L]}(\vec{x})$ and we integrate over all variables to generate the kinetic energy operator

$$\frac{\hbar^2}{m}\left[-\frac{\partial^2}{\partial r^2} + \frac{L(L+1)}{r^2} - \frac{4}{r^2}\frac{1}{W_{[L]}(z)}\frac{\partial}{\partial z}(1-z^2)W_{[L]}(z)\frac{\partial}{\partial z}\right] \tag{8}$$

where the integral in the weight function

$$W_{[L]}(z) = 2^{-3(A-1)/2}(1-z)^{3A/2-4}(1+z)^{\frac{1}{2}}\int_{(\vec{r}=1)}|H_{[L]}(\vec{x})|^2\, d\Omega_1$$

is taken for $r=1$ over all variables but z at the surface of the unit hypersphere $r=1$. The second part of (3) is calculated through a Potential Harmonic expansion of (7) for all pairs followed by a projection on the $\vec{r}_{ij}$ space. In expansion (4) $u_K(r)$ is given by

$$u_K(r) = \int_{-1}^{1} P(z',r)P_K^{[L]}(z')W_{[L]}(z')dz' \tag{9}$$

where the "Potential Polynomials" $P_K^{[L]}(z)$ are <u>associated</u> with the weight function $W_{[L]}(z)$ and where

$$P^{\circ}_{[L+2K]}(\Omega_{k\ell}) = r^{-L}\, H_{[L]}(\vec{x})P_K^{[L]}(z') \quad \text{for} \quad z' = 2r^2_{k\ell}/r^2 - 1 \tag{10}$$

are the Potential Harmonics for the pair (k,ℓ).

We introduce (9) in (4) and (4) in (3) for all pairs (k,ℓ), we premultiply by $r^{-L}\,H^{*}_{[L]}(\vec{x})$, we integrate over all variables but z and we exchange the sum over K with the integral over z' to obtain the projection of the second part of (3) leading to the integrodifferential equation [3]

$$\{\frac{\hbar^2}{m}\left[-\frac{\partial^2}{\partial r^2} + \frac{L(L+1)}{r^2} - \frac{4}{r^2}\frac{1}{W_{[L]}(z)}\frac{\partial}{\partial z}(1-z^2)W_{[L]}(z)\frac{\partial}{\partial z}\right] - E\} P(z,r) \tag{11}$$

$$= -V(r\sqrt{\frac{1+z}{2}})\,[P(z,r) + \int_{-1}^{1}f_{[L]}(z,z')P(z',r)dz']$$

where

$$f_{[L]}(z,z') = W_{[L]}(z')\sum_{K}(f_K^2-1)P_K^{[L]}(z)P_K^{[L]}(z') \tag{12}$$

is called the "Projection Function".

The serie in (12) which depends on the state $[L]$ cannot be in general summed up analytically, but when we have to deal with bosons in ground state where $L = 0$, $H_{[0]}$ is a constant and the weight function $W_{[0]}(z) = (1-z)^\alpha (1+z)^{1/2}$ is associated with the properly normalized Jacobi polynomials $P_K^{\alpha, 1/2}(z)$, where $\alpha = 3A/2 - 4$ and the sum can be calculated. The result is

$$\int_{-1}^{1} f_{[0]}(z,z') P(z',r) dz' = (A-2) \, \Gamma(\lambda + \frac{1}{2})/\sqrt{\pi} \{2/\Gamma(\lambda) \int_{-1}^{1} (1-v^2)^{\lambda-1} \quad (13)$$

$$\times P[(\sqrt{1+z} \, \cos \delta + v\sqrt{1-z} \, \sin \delta)^2 - 1, \, r][1+v\sqrt{\frac{1-z}{1+z}} \, tg \, \delta] dv$$

$$+ (A-3)/\Gamma(\lambda-1) \int_{-1}^{1} (1-v^2)^{\lambda-2} \, P[(1-z)v^2 - 1, \, r] \, v^2 \, dv \}$$

where $\lambda = (3A-7)/2$ and $\delta = \pi/3$. When $A = 3$ and $L = 0$ eq.(11) can be reduced to the well known Faddeev equation written by Noyes[4] for 3 bosons in S state when the variable θ is substituted for $z = \cos\theta$ and $\sqrt{r} \, U(r,\theta)/\sin\theta$ for $P(z,r)$ in (11)[5]. Then our integrodifferential equation (11) extend to any number of particles the Faddeed equation for two-body correlations in S state.

Adiabatic Approximation Method

In the Adiabatic Approximation we assume that the orbital and the radial motions of the point $\vec{x}$ in the $D = 3(A-1)$ dimensional space are nearly decoupled in such a way that the wave function can be written as the product

$$\psi(\vec{x}) = B_\lambda^{[L]}(\Omega,r) u_\lambda(r)/r^{L+1} \quad (14)$$

where $B_\lambda^{[L]}$, caracterized by the quantum numbers λ, is assumed to be a smooth enough function of r that $\partial B_\lambda / \partial r$ and $\partial^2 B_\lambda / \partial r^2$ can be neglected. The Adiabatic Basis $\{ B_\lambda^{[L]}(\Omega,r) \}$ is the symmetrical normalized combination

$$B_\lambda^{[L]}(\Omega,r) = H_{[L]}(\vec{x}) \sum_{i<j\leq A} P_\lambda (2r_{ij}^2/r^2 - 1, \, r) \quad (15)$$

where $P_\lambda(z,r)$ is a solution of (11) where $\partial^2/\partial r^2$ is removed and where the eigenpotential $U_\lambda(r)$ is substituted for E. The eigenvectors $B_\lambda^{[L]}(\Omega,r)$ are normalized for any r to

$$\int_{(r=1)} B_\lambda^*(\Omega,r)B_{\lambda'}(\Omega,r)d\Omega = \delta_{\lambda\lambda'}$$

The radial wave $u_\lambda(r)$ is a solution of

$$(-\frac{\hbar^2}{m}\frac{d^2}{dr^2} + U_\lambda(r) - E)u_\lambda(r) = 0 \qquad (16)$$

The eigenfunction $P_\lambda(z,r)$ can be obtained either by solving (11) for the single variable z or by using a Potential Polynomial expansion

$$P_\lambda(z,r) = \sum_K P_K^{[L]}(z) b_K^\lambda(r)$$

where $b_K^\lambda(r)$ are solution of the system of linear equations obtained from (5) by removing d^2/dr^2 and by substituting $U_\lambda(r)$ and $b_K^\lambda(r)$ for E and $u_K(r)$.

The Adiabatic Basis has a remarquable property : when $r \to \infty$ the derivatives with respect to r of the B_λ describing the various channels occuring in the many-body systems are cancelled, the states are fully decoupled and the limit of $U_\lambda(r)$ is the total energy available in the channel described where the various clusters are at rest.

In order to illustrate this important property we choose to solve the Positronium-Ion in which two electrons and one positron interact together[6]. The wave function is the sum of three components :

$$\Psi(\vec{x}) = \psi_e(r_{12},r) + \psi_p(r_{23},r) + \psi_p(r_{31},r)$$

where (1) and (2) refer to the electrons and (3) to the positron. The equations for ψ_e and ψ_p are

$$(T-E)\psi_e(r_{12},r) = -1/r_{12} \Psi(\vec{x})$$

$$(T-E)\psi_p(r_{23},r) = 1/r_{23} \Psi(\vec{x}) \quad \text{in atomic units}$$

and a similar equation for the pair (3,1).

The partial waves are written in terms of the Adiabatic Basis

$$\psi_\alpha(r_{ij},r) = B_n^\alpha(r_{ij},r)u_n^\alpha(r) \qquad \text{where } \alpha = p \quad \text{or} \quad e$$

The Adiabatic equations are solved by a potential harmonic expansion of B_m^α

followed by the solution of linear equations providing the eigenpotential $\mho_m(r)$. Up to 70 potential harmonics have been used and the lowest $\mho_m(r)$ are drawn in fig.1 in atomic units for r running from zero up to 80 Bohr radii.

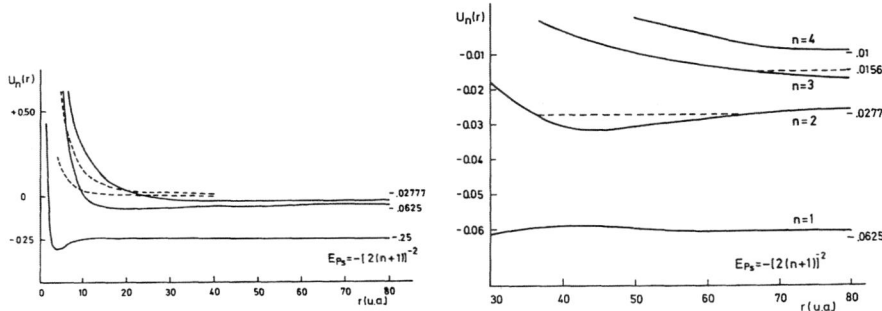

The asymptotic bahaviour of $\mho_m(r)$ where m is the number of nodes of the Positronium in S states is clearly seen : each eigenpotential reach asymptotically one of the eigenenergies $-(2(n+1))^{-2}$ of the positronium excited states. From the effective potential $\mho_m(r)$ introduced in (16) the S phase shifts in elastic scattering of an electron on the positronium can be calculated.

References

1. Fabre de la Ripelle, M. : Phys. Letters 135B, 5 (1984).
2. Fabre de la Ripelle, M. : Ann. Phys. (N.Y.) 147, 281 (1983).
3. Fabre de la Ripelle, M. : C.R. Acad. Sci. Paris 299 Serie II, 839 (1984).
4. Noyes, H.P. : Three-body problem in nuclear and particle physics. Eds : J.S.C. McKee and P.M. Rolph (North-Holland, Amsterdam, 1970).
5. Fabre de la Ripelle, M.,Fiedeldey, H. : Phys. Letters 171B, 325 (1986).
6. Fabre de la Ripelle, M., Larsen, S.Y. : to be published.

THREE-BODY CALCULATIONS AT LOS ALAMOS

J. L. Friar
Theoretical Division, Los Alamos National Laboratory
Los Alamos, NM 87545 U.S.A.

1. Introduction

Approximately seven years ago the Los Alamos-Iowa collaboration[1] began its program of investigating the trinucleon systems by solving the Faddeev equations in configuration space. This work was motivated by four goals: (1) By working in configuration space, where intuition is greatest, investigate graphically those trinucleon properties which are determined by specific features of wave functions; (2) Produce benchmark calculations against which new techniques and numerical methods can be measured; (3) Investigate the effect of the Coulomb interaction between the two protons in ^{3}He and in the p-d system; (4) Systematically investigate the various trinucleon observables. Configuration space is particularly well-suited for investigating the Coulomb problem. The singularity and discontinuity problems associated with the Coulomb (momentum space) t-matrix are transformed into boundary condition problems in configuration space. One simply adds the Coulomb potential to the strong interaction. In order to produce accurate numerical solutions powerful techniques were adopted which have not frequently been used in nuclear physics. These spline methods together with collocation techniques combine the power of Gaussian quadrature procedures with the flexibility and strength of finite element approaches to solving partial differential equations. Further details are given by G. L. Payne in his contribution to this workshop[2]. The union of these methods allows one to calculate wavefunctions at the same qualitative level of accuracy as the eigenvalues. Observables can therefore be calculated with considerable confidence.

2. The Faddeev Approach

2.1 Overview

The Faddeev approach[3] to solving the Schrödinger equation involves rather different formal procedures than other methods, such as the hyperspherical expansions or the Rayleigh-Ritz variational technique. The latter methods expand the wavefunction in some type of basis and solve directly for the expansion parameters of that basis. The Faddeev method, on the other hand, expands the potential as a (nonterminating) series, each term of which acts only in a single nucleon-nucleon partial wave (e.g., $^1S_0$). Each "interacting pair" of particles in a given partial wave, when coupled to partial waves of the remaining (noninteracting) "spectator", leads to one or two three-nucleon partial waves, or "channels". The infinite series of potential terms must be truncated in performing a numerical calculation, but the angular momentum barrier suppresses the contribution of higher partial waves at low energies and in the bound state.

2.2. Two-body Force Results

Including all nucleon-nucleon partial waves with total angular momentum $j \leq 4$ (34 channels) typically produces a converged answer at the level of 10 keV in the bound state, with the $j=4$ waves producing only 20 keV out of 58 MeV for the RSC potential[4]. This result, first achieved recently by the Los Alamos-Iowa group, improves the previous standards of 18 channels ($j \leq 2$) and 5 channels (positive parity, $j \leq 1$). Adding 29 channels to the latter case for the RSC potential enhances the binding by 320 keV. Commonly used potential models, such as Reid Soft Core (RSC)[5], Argonne V_{14} (AV14)[6], and Super Soft Core (C) (SSCC)[7], produce binding energies of [7.35, 7.67, 7.53] MeV, respectively. These are underbound by approximately 1 MeV. However, it has been known for several decades[8] that if the tensor force is weakened, the tendency is to increase the binding. The reason is that the deuteron (with one T=0 pair) is even more sensitive to the tensor force than the triton (with $\frac{3}{2}$ T = 0 pairs), where typically more than half the potential energy derives from that force. Weakening the tensor force requires an enhanced central force in order that the deuteron maintain the same binding energy. This drives up the triton binding energy because of the bigger deuteron lever arm with respect to the tensor force. Recent calculations[9-10] with unpublished versions of the Bonn potential[11], whose deuteron D-state probability is less than 5 percent, give 8.0-8.3 MeV binding. A major difficulty in pinning down the tensor force is the poor quality of some of the nucleon-nucleon data below 100 MeV, which allows considerable flexibility in the nucleon-nucleon forces.

Current experiments at Karlsruhe[12] and planned experiments at SIN are expected to help remedy this problem[13].

2.3 Three-Nucleon Forces

In addition to forces which depend only on the coordinates of two nucleons, there are other forces, three-nucleon forces, which depend in a nontrivial way on the simultaneous coordinates of the three nucleons[14]. Such forces are expected to be much weaker overall than the two-body potentials; order of magnitude estimates give 1 MeV potential energy (out of a total of 50 MeV). The archetype of three-body forces (TBF) in atomic, nuclear, and classical physics is the mutual "distortion" force caused by the change of shape of composite objects which interact with each other. The classical example is the earth's tidal force on a satellite, caused by the moon's attraction on the oceans. The atomic example is the long-range Axilrod-Teller force[15] from three mutually distorted atoms. The prototype long-range nuclear TBF arises from Δ-excitation of a nucleon by a pion emitted from a second nucleon, which pion is absorbed by a third nucleon after the virtual Δ decays. Many other mesonic processes also contribute. A major technical problem with these forces is that certain components depend on the way the nucleons in the nucleus are taken off-shell, which demands consistency between the two- and three-nucleon force models[16].

2.4 Three-Nucleon Force Results

Although it is trivial in principle to incorporate a TBF in bound state calculations, it is somewhat more complicated in practice, because the partial-wave series converges more slowly. Nevertheless, a number of calculations have been carried out at Los Alamos and elsewhere. Four TBF models have been used: Tucson-Melbourne (TM)[17], Brazilian (BR)[18], Urbana-Argonne (UA)[19], and Hajduk-Sauer (HS)[20]. The latter model has Δ-components in the wavefunction and only a single set of calculations exists. The TBF effect in this model is implicit; the net increase in binding is 0.3 MeV. Calculations with the other models[21] produce much more binding, in general, but have a disturbing sensitivity to the assumed short-range behavior, usually mediated by a pion-nucleon form factor. The range of this phenomenological cutoff is poorly known. The "standard" version increases the binding by roughly 1.5 MeV for the TM and BR models, and is insensitive to the two-body potential model used. However, if one assumes other plausible cutoffs[21] much more or much less binding accrual is found.

3. Observables

3.1 Scaling

Many combinations of two- and three-body force models have been solved at Los Alamos and yield a wide range of binding energies, E_B, most of which are not close to the physical triton energy. In those cases are the values of various calculated physical observables to be taken seriously? An example is the rms radius, which is very sensitive to the asymptotic portion of the wavefunction $\left[\sim\exp(-\kappa\rho)/\rho^{5/2}, \text{ where } \kappa \sim E_B^{1/2}\right]$. If one naively assumes that this asymptotic form holds everywhere, the radius should scale as $E_B^{-1/2}$. This is indicated in Fig. 1, where the individual points result from calculations at Los Alamos[22] of the (point nucleon) rms trinucleon charge radii, plotted versus the corresponding binding energies for diverse models including a Coulomb force in ^{3}He. The small spread of these "theoretical data" indicates scaling behavior with the binding energy, and the experimental data from a recent Saclay analysis[23] are in good agreement with the simple fits.

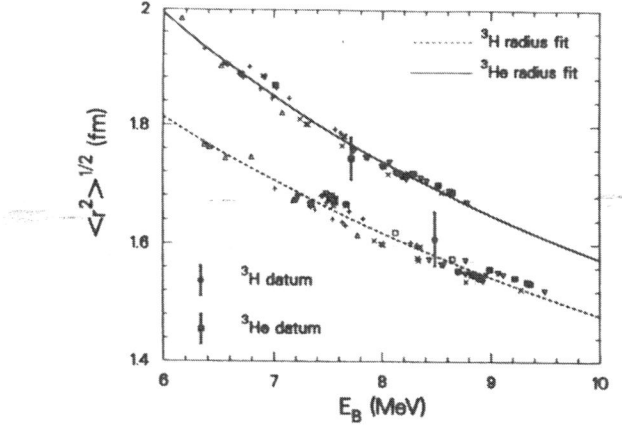

Fig. 1. Calculated trinucleon rms radii together with data.

The isoscalar, or mass, combination of radii does indeed vary as $E_B^{-1/2}$, while that combination sensitive to the ^{3}He-^{3}H density difference scales more nearly as E_B^{-1}. The latter density component is largely determined by the overlap of S- and S'-states, and the latter state decreases rapidly with increased binding.

A second example of scaling is the ^{3}He Coulomb energy, E_c, calculated in perturbation theory with a dipole nucleon form factor. The scaling plot is depicted in Fig. 2. At the physical triton binding energy one finds $E_c = 652\pm2$keV, which is slightly higher than the value produced by the hyperspherical approximation, which works well[22], used in conjunction with experimental charge form factor data. This difference can be traced to the apparent inability of the impulse approximation to reproduce all features of the trinucleon charge form factors. Both results are significantly lower than the observed 764 keV binding energy difference. Clearly, in these two examples of scaling any individual calculated point is virtually useless. Only the set of calculations provides substantial insight.

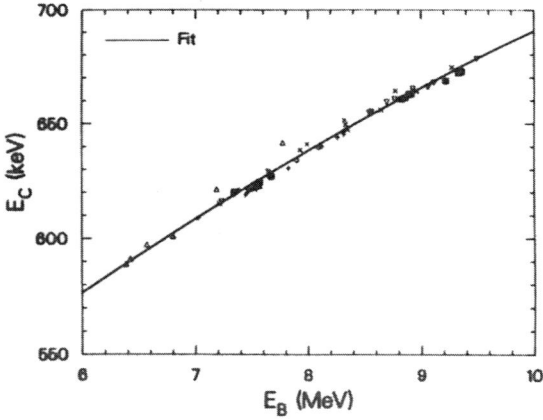

Fig. 2. Calculated trinucleon Coulomb energies.

3.2 Charge densities and form factors.

Interest in the trinucleon form factors, $F_{ch}(q^2)$, has been augmented by the recent tritium experiment at Saclay[23] and the nearly completed one at Bates[24]. The data are plotted in Figs. 3 and 4 versus squared momentum transfer, together with impulse approximation calculations from Los Alamos for 34 channels[25], the RSC two-nucleon potential model, and with various (or no) TBF models. The data in the region of the diffraction minima are in obvious disagreement with the calculations. All of the TBF results are similar, and correspond roughly to the same binding energy.

The shift in the curves when a TBF is added can be understood in terms of a simple schematic model, if we assume that the primary change results from the binding energy difference. Defining $\lambda = (E_B/E_B^0)^{1/2}$, where E_B and E_B^0 are the binding energies with and without a TBF, the asymptotic wavefunction displayed earlier can be written in the form: $\Psi(\lambda\vec{x}, \lambda\vec{y})\lambda^3$, where $\vec{x}$ and $\vec{y}$ are any combination of the two vectors necessary to specify a three-body system. Assuming that this form holds for the entire wavefunction changes the charge density $\rho_{ch}(r)$ to $\lambda^3\rho_{ch}(\lambda r)$, and the charge form factor, $F_{ch}(q)$, to $F_{ch}(q/\lambda)$. That is, the momentum transfer scale "stretches", preserving shapes and values of diffraction maxima, but moving positions of

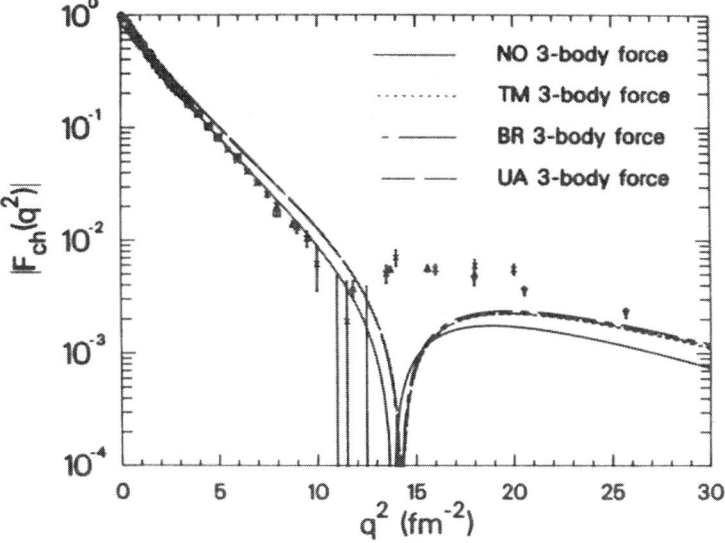

Fig. 3. Calculated [3]He charge form factors together with data.

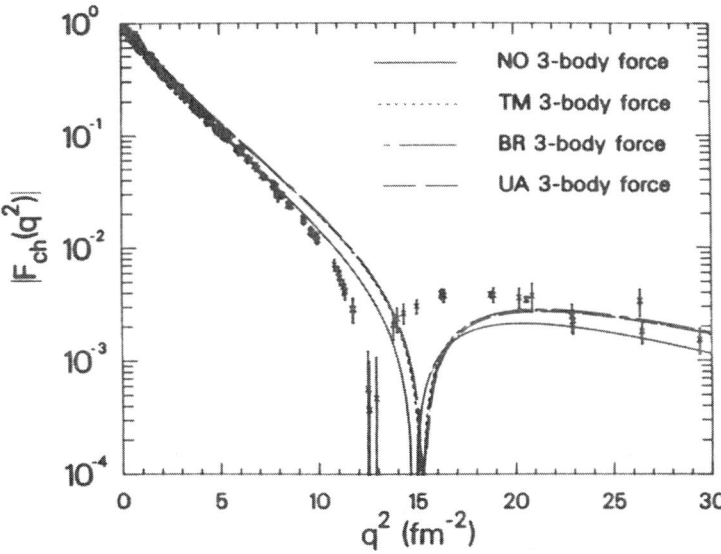

Fig. 4. Calculated ^{3}H charge form factors together with data.

minima outward. This model, which is only quantitative at quite low momentum transfers, accounts for the shift in the diffraction minima, but not
the increase in the height of the diffraction maxima.

The latter effect is outside the province of the stretching model, and
results from the strong angular dependence of three-body forces, which can
depress the charge density near the origin, in addition to the simple con-
traction of the density predicted by the stretching model. The ^{3}He charge
density corresponding to the Los Alamos-Iowa point-nucleon charge form factor shown earlier is depicted in Fig. 5, together with the quasi-
experimental data[26] obtained by removing the nucleon charge form factors

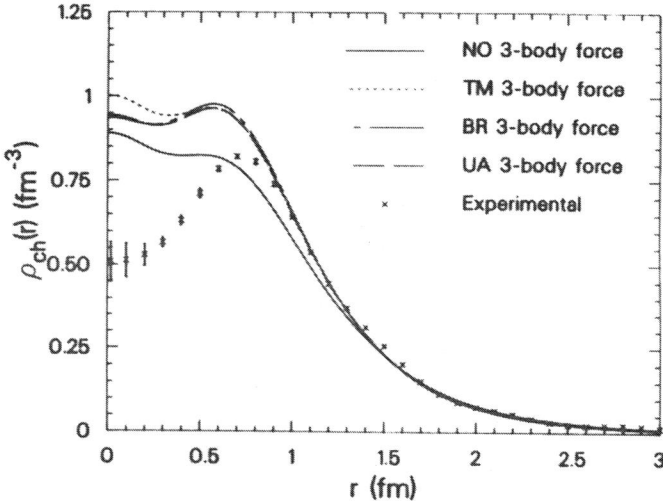

Fig. 5. Calculated ^{3}He charge densities and quasi-experimental data.

from the experimental nuclear form factor in an approximate way. The
dramatic "hole", which is not reproduced by the calculations, actually is
"small" in the sense that only about one percent of the ^{3}He charge would
fill it. The addition of a TBF does depress the central density somewhat,
relative to the surface density, but the effect is too small. A similar
change is observed in ^{3}H.

If the impulse approximation appears to fail at large momentum transfers, what mechanism would alleviate the problems? Figures 3 and 4 show
that any addition to F_{ch} of a small negative component which vanishes at
$q^2 = 0$ and large momentum transfers would tend to solve all of the
problems. Meson exchange currents have this shape property, and have occasionally been proposed as a solution. There is no question that they
help, if they are introduced in certain approximations. The difficulty is

that the pion-exchange part of the charge operator is a relativistic correction[27], is model dependent, and is also formalism dependent. The latter difficulty can lead to ambiguous matrix elements unless one also calculates the wave functions for those matrix elements in the same formalism (i.e., including relativistic corrections). This is not possible within the class of common nonrelativistic potential models, and at this stage calculations of exchange currents to ρ_{ch} must be regarded as *ad hoc*.

3.3 N-d scattering lengths.

The nucleon-deuteron system has two spin configurations: doublet and quartet. The latter configuration is primarily sensitive to the deuteron binding energy and is therefore not particularly interesting. The former has long been known to show scaling behavior for the n-d case (i.e., the Phillips line[28]). Recently, the Los Alamos-Iowa collaboration[29] showed that the p-d doublet scattering lengths behave similarly (Fig. 6), although the shape of the p-d and n-d curves is clearly different. Complicating the situation is a controversy over the definition of the p-d scattering

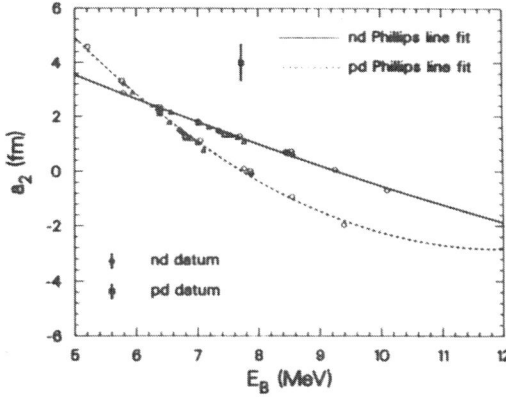

Fig. 6. Calculated Nd doublet scattering lengths together with data.

length, and a gross discrepancy between theory and experiment in that case. The theoretical problem has been resolved[30], but the problem with experiment remains. The n-d datum is in excellent agreement with the fit to the various calculations with and without three-body forces.

4. Conclusions.

The Los Alamos-Iowa group has recently extended the bound state Faddeev calculations to 34 channels, or j≤4. Most two-nucleon models have insufficient binding, although commonly used three-nucleon forces can sub-stantially increase this. Regrettably the latter forces are extremely sensitive to the assumed short-range behavior, which makes the binding ac-crual problematical. Many trinucleon observables display scaling behavior with the corresponding binding energy, and fits to existing calculations may provide the best theoretical estimates. The rms radii are in good agreement with recent experiments, while the charge form factors in impulse approximation are not. The extrapolated n-d doublet scattering length agrees well with experiment, while the p-d case does not.

Acknowledgement

This work was performed under the auspices of the U.S. Department of Energy.

References

1. Payne, G. L., Friar, J. L., Gibson, B. F., Afnan, I. R. , Phys. Rev. C22, 823 (1980).
2. Payne, G. L., contribution to this workshop.
3. Faddeev, L. D., Zh. Eksp. Teor. Fiz. 39, 1429 (1960).
4. Chen, C. R., Payne, G. L., Friar, J. L., Gibson, B. F., Phys. Rev. C31, 2266 (1985)
5. Reid, R. J. Jr., Ann. Phys. (N.Y) 50, 411 (1968).
6. Wiringa, R. B., Smith, R. A., Ainsworth, T. A., Phys. Rev. C29, 1207 (1984).
7. deTourreil, R., Sprung, D. W. L., Nucl. Phys. A201, 193 (1973).
8. Delves, L. M., in "Few Body Problems, Light Nuclei, and Nuclear Interactions", G. Paic and I. Slaus, eds., (Gordon and Breach, N.Y., 1968), p. 153.
9 Sauer, P. U., in "Proceedings of the International Symposium on the Three-Body Force in the Three-Body System", (Springer, Berlin, 1986).
10. Sasakawa, T., in Proceedings of "Eleventh International Conference on Few Body Systems in Particle and Nuclear Physics", Tokyo-Sendai (1986).
11. Machleidt, R., Holinde, K., Elster, Ch., Phys. Rep. (to appear).
12. Klages, H., in Proceedings of "Eleventh International Conference on Few Body Systems in Particle and Nuclear Physics", Tokyo-Sendai (1986).
13. Pickar, M. A., Sick, I., private communications.
14. Friar, J. L., Gibson, B. F., Payne, G. L., Ann. Rev. Nucl. Part. Sci. 34, 403 (1984).

15. Axilrod, B. M., Teller, E., J. Chem. Phys. 11, 299 (1943).

16. Friar, J. L., Coon, S. A., Phys. Rev. C34 [to appear].

17. Coon, S. A., *et al.*, Nucl. Phys. A317, 242 (1979).

18. Coelho, H. T., Das, T. K., Robilotta, M. R., Phys. Rev. C28, 1812 (1983).

19. Carlson, J., Pandharipande, V. R., Wiringa, R. B., Nucl. Phys. A401, 59 (1983).

20. Hajduk, C., Sauer, P. U., Streuve, W., Nucl. Phys. A405, 581 (1983).

21. Chen, C. R., Payne, G. L., Friar, J. L., Gibson, B. F., Phys. Rev. C33, 1740 (1986).

22. Friar, J. L., Gibson, B. F., Chen, C. R., Payne, G. L., Phys. Lett. 161B, 241 (1985).

23. Juster, F.-P., *et al.*, Phys. Rev. Lett. 55, 2261 (1985); Frois, B., and Martino, J., private communication.

24. Beck, D., private communication.

25. Friar, J. L., Gibson, B. F., Payne, G. L., Chen, C. R. , Phys. Rev. C (to appear).

26. Sick, I., Lecture Notes in Physics 87, 236 (Springer, Berlin, 1978).

27. Friar, J. L., Ann. Phys. (N.Y.) 104, 380 (1977).

28. Phillips, A. C., Rep. Prog. Phys. 40, 905 (1977).

29. Chen, C. R., Payne, G. L., Friar, J. L., Gibson, B. F., Phys.Rev. C 33, 401 (1986).

30. Chandler, C., in Proceedings of "Eleventh International Conference on Few Body Systems in Particle and Nuclear Physics", Tokyo-Sendai, (1986).

FEW BODY CALCULATIONS AT SENDAI

Tatuya Sasakawa

Department of Physics, Tohoku University

Sendai 980, Japan

From the triton calculations for various number of channels and various potentials without and with a three-nucleon potential, we conclude that the calculated D/S ratio of the asymptotic normalization constants, the charge radius, the momentum distribution are linearly correlated with the calculated binding energy. From this relation, we can deduce the values of these quantities, if we use the experimental binding energy. The effect of the three-body force is about 10% except for the Bonn potential.

1. INTRODUCTION

This is a report of the recent Sendai results for triton calculations, done by S. Ishikawa, A. Fukunaga, K. Soutome and T. Uchiyama.

In few-nucleon systems, we have a posssibility of solving the Schröxdinger equation for a given potential. It is so far limited to triton. However, once the triton wave function is obtained, we can anticipate to proceed to four and more nucleon systems. This fact makes few nucleon physics more and more importatnt as a solid testing ground for nuclear physics as well as a bridge between nuclear and particle physics.

Of course, we are aware that quarks and gluons are the building-stones of a nucleon and a boson. At this moment, however, the description of the triton in terms of quarks and gluons is almost impossible. Therefore, as a first step, we proceed a conservative approach based on the potential models without quarks. This approach seems not entirely pessimistic. Fig.1 shows the EMC effect for ^{3}He obtained from the triton wave function

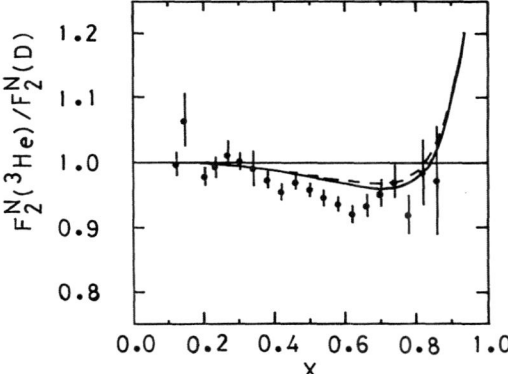

Fig.1   The EMC effect of ^{3}He from the 18-channel calculation of
RSC without (dashed line) and with (solid line) TM.

for the Reid soft core potential (RSC)[1] without (dashed line) and with
(solid line) the Tucson-Melbourne (TM) three-nucleon potential [2]. In
this figure, the experimental results for ^{4}He are shown. We urge that the
experiment for ^{3}He be done. We expect that the agreement between the
experimental and theoretical results will then be better, and the result
might show that the potential picture is enough to calculate the EMC effect, which was once thought to be a realization of the quark-gluon picture of nucleus.

Before we talk about the results, we have to explain how the word "
channel" is used in the triton calculations. The total spin of the triton
is 1/2. Under this constraint, the triton consists of infinite number of
partial wave states for the interacting pair and the spectator. The first
34 states were given in another article [3]. If we perform the triton calculation taking first 5(18,26,34) states into account, we say that we have
performed the 5(18,26,34)-channel calculations. The 18(26,34)-channel calculations corresponds to take the angular momentum of interacting pair to
be $J \le 2(3,4)$. The Hannover [4], Los Alamos-Iowa [5] and Sendai [3,6]
groups have solved the 34-channel Faddeev equation with a three-body
force; Hannover with the $\pi\Delta\pi$ force, Sendai with TM, Los Alamos and Iowa
with TM, Brazil (BR) and Urbana(UR). See, Table 1.

Fig.2 shows the convergence of the triton binding energy B_3 and the D
state probability with increasing number of channels. For the convergence
of other quantities, see [6]. It is clear that the 5-channel calculation
which was a main stream at a time of the Few Body X conference (1983) is
not enough.

Table 1. 34-channel calculations. Λ is the cut-off mass for πNN monople form factor.

3NP	Hannover[4] πΔπ	Los Alamos-Iowa[5]			Sendai[3,6] TM	
		TM 800	BR 800	UR 800	700	800
RSC [1]	7.38		7.35			
PARIS[7]					7.64	
TRS [8]					7.55	
AV [9]			7.67		7.68	
RSC + 3NP	7.7	8.86	8.89	8.70		
PARIS+3NP					8.32	9.18
TRS + 3NP					8.47	9.71
AV + 3NP		9.36	9.22	8.99	8.42	9.29
Exp	8.48					

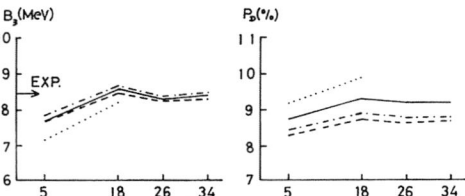

Fig. 2 Convergence of B_3 and P_D for 2NP + TM(Λ = 700MeV). 2NP means the two-nucleon potential. The dotted, solid, dashed and dash-dotted lines show the results of RSC, AV, PARIS and TRS taken for 2NP. The abscissa indicate the number of channels.

Besides RSC, AV, PRIS and TRS potentials, we also calculated the BONN potential. Since this potential yields very different values for various physical quantities, we discuss it in Sec. 3.

This talk consists of two parts: (1) A definitive conclusion about the triton properties and (2) subjects that need further study.

Using a computer SX1 installed recently at Tohoku University and the method of continued fractions, the Faddeev equation for the 34-channel triton with TM is solved in only one minutes.

2. TRITON PROPERTIES

Having calculated the triton properties for RSC, PARIS, TRS, AV and
BONN potentials without or with TM three-nucleon force, we conclude that
the triton physical quantities are linearly correlated to the calculated
binding energy without regard to the potential or the number of channels
[6,10]. In other words, any potential model that reproduce the triton
binding energy correctly can reproduce the triton physical quantities.
Also utilizing the linear relationship, we can predict the value of the
triton physical quantities, if we use the experimental value of the triton
binding energy. Let us look at the quantities from which the above conclusion was deduced.

(1) The asymptotic normalization constants [10]

Fig.3 shows the D/S ratio of the asymptotic normalization constants
vs. the triton binding energy, together with the new ETH experimental results, which was reported by W. Grüebler in Few-Body XI; 0.0432 $\pm$ 0.002.
The agreement with our predicted value [10], 0.0432 $\pm$ 0.0015, is remarkable.
Here we note that all potentials that we utilized have the one-pion tail.
A family of potentials with the same tail, but different from one-pion,
forms a different linear relationship. However, a potential whose tail is
different from one-pion is physically meaningless, and we should not draw
any conclusion from it.

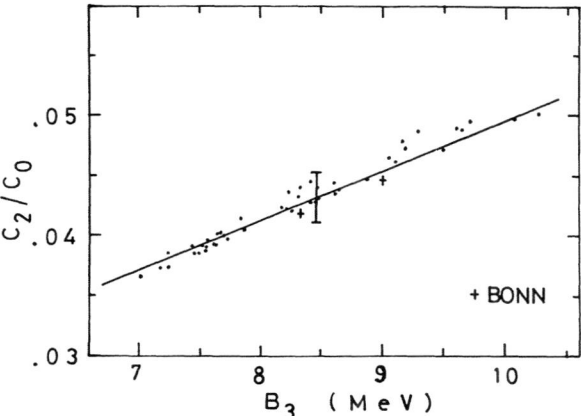

Fig. 3 All calculated results for the following 48 cases
plotted in the B_3-C_2/C_0 plane: RSC5,18; [RSC + TM(Λ)]5,18
for Λ = 700 and 800 MeV; 2NP5,18,26,34; [2NP + TM(Λ)]5,18,
26,34 for 2NP = AV, PARIS, TRS and Λ= 700 and 800 MeV; [R
SC + TM(1000)]5; [AV + TM(1000)]5; [AV + TM(1000) + (W_o =
9600, r_c = 0.50)]5; [AV + TM(1000) + (W_o = 28800, r_c =0.
45)]5; BONN34 and [BONN + TM(700)]34. For W_o and r_c, see
[10]. The new ETH value reported by W. Grüebler is shown.

(2) Momentum distribution [6]

We calculated the momentum distribution $\rho(q)$. The effect of the three-nucleon potential is seen by comparing the solid curve for AV34 and the dashed curve for [AV + TM(700)]34 in Fig.4. However, the difference between these curves is smaller than the experimental error for small q and the cross section is already very small for a large q. Therefore, it might not be able to deduce the effect of a three-nucleon force from an experimental quantity related to the momentum distribution.

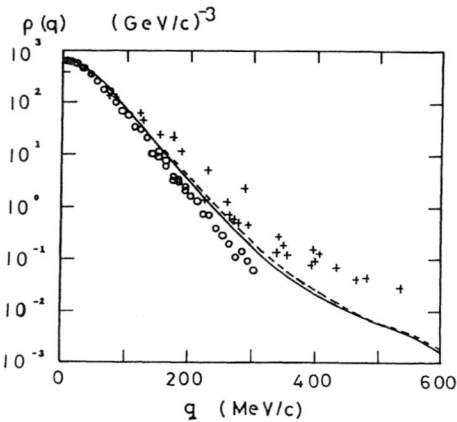

Fig.4 The momentum distribution $\rho(q)$ calculated for AV34 (solid line) and [AV + TM(700)]34 (dashed line) under PWI A. Results from (e,e') by Jans et al.(1982) are denoted by circles, while those from (p,pd) by Epstein et al.(1985) are shown by crosses.

The momentum distribution is deduced from the cross section on the assumption of PWIA. Therefore, the difference between (e,e') and (p,pd) experimental results might be due to the difference in the validity of PWIA. At low momentum such as q = 5 MeV/c, the (e,e') experiments fit our calculated results very well as shown in Fig.5.

(3) Charge radius [6]

From the calculated charge form factor F(q), we can obtain the charge radius by the formula $F(q) = 1 - r_{ch}q^2 + 0(q^4)$. Fig.6 shows the charge radius for the triton calculated from various potential models and the number of channels. The calculated values are compared with the experimental results. Again, we see that a model that yeilds the correct binding energy of triton reproduces the experimental charge radius.

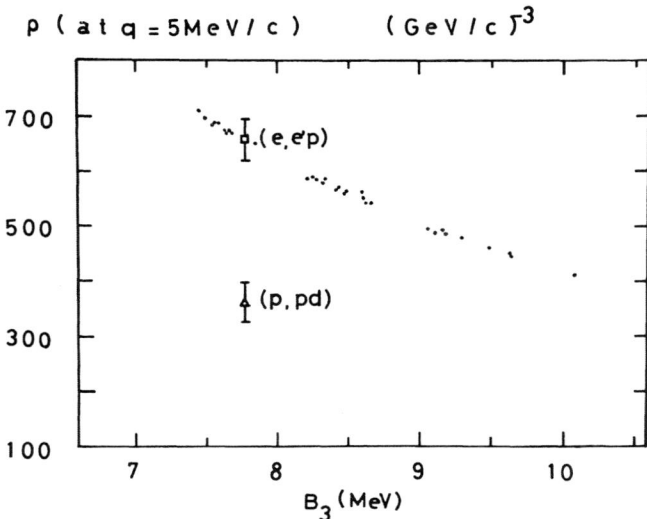

Fig. 5 The correlation between calculated $\rho(q)$ at $q = 5$MeV/c and B_3. The experimental values are placed at 7.72 MeV, the binding energy of ^{3}He, because the experiments were done for ^{3}He.

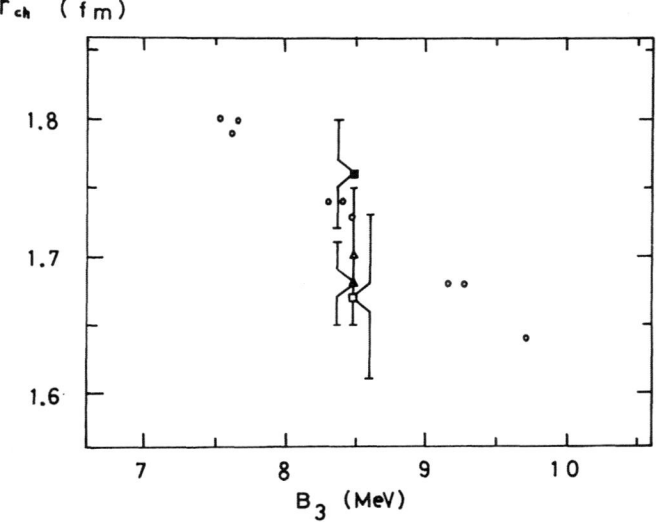

Fig.6 The calculated charge radius r_{ch} plotted against the corresponding B_3.

3. SUBJECTS THAT NEED FURTHER TUDY

(1) Binding energy

The results for AV, PARIS and TRS show a marked difference from the BONN results, especially for the binding energy and the D-state probability. In Table 2, we list the 34-channel values for various quantities with and without TM. The cut-off mass of the πNN form factor is taken to be 700 MeV. From this Table, we see that (1) AV, PARIS and TRS yield at most 7.7MeV for the binding energy and as a result, we need a three-nucleon force to fill up the discrepancy between experimental and theoretical values of about 10%, whereas BONN yields almost the correct binding energy without a three-nucleon potential, and (2) the P_D of BONN is much smaller than that of other potentials.

Table 2. Triton quantities for various potentials.

Potential	B_3		P_S		$P_{S'}$	
	2NP34	(2NP+TM)34	2NP34	(2NP+TM)34	2NP34	(2NP+TM)34
AV	7.68	8.42	89.85	89.71	1.12	0.94
PARIS	7.64	8.32	90.13	90.09	1.30	1.12
TRS	7.55	8.47	90.06	90.01	1.28	1.04
BONN	8.33	9.01	92.11	92.23	1.04	0.91

Potential	P_D		C_2/C_0		E_c	
	2NP34	(2NP+TM)34	2NP34	(2NP+TM)34	2NP34	(2NP+TM)34
AV	8.96	9.23	0.0402	0.0444	0.65	0.68
PARIS	8.50	8.67	0.0392	0.0431	0.66	0.68
TRS	8.60	8.81	0.0390	0.0439	0.66	0.69
BONN	6.81	6.79	0.0418	0.0446	0.697	0.718

Potential	$r_{ch}(^3H)$		$r_{ch}(^3He)$	
	2NP34	(2NP+TM)34	2NP34	(2NP+TM)34
AV	1.80	1.74	2.00	1.93
PARIS	1.79	1.68	2.00	1.93
TRS	1.80	1.64	2.01	1.91
BONN	1.73	1.69	1.91	1.86

Although we can not say much about the question as to whether the BONN potential is a desirable one or not, unless we make calculations on the Coulomb energy difference, the (e,e') scattering etc., we make some remarks about the BONN potential.

The BONN potential that we utilized here is the "nonrelativistic r-space version"[11]. This version was obtained in the following manner: First, the two nucleon phase shifts and deuteron properties are calculated field theoretically, especially including the two- and multi-boson ex-

exchanges. At this stage, no non-existing mesons are used. At the next step, the two- and multi-boson exchanges are represented by a scalar iso-scalar σ-meson exchange, whereby the relativistic nature of the propagator is kept. This is the"q-space relativistic version". Finally, in the r-space version, the static approximation is introduced and the coupling constants (c.c.) are adjusted a little to fit the experimental results. Their c.c. are larger than c.c. used by other authors, e.g., Brown-Jackson (BJ) [12]. In Table 3, $g_{\alpha}^2/4\pi$ are listed.

	BONN	BJ
π	14.6	14.3
ρ	0.95	0.53
ω	20.0	4.77
δ	3.7064	
η	3.0	
σ	8.0568	6.0

Table 3. The coupling constants $g_{\alpha}^2/4\pi$ for the r-space version of the BONN potential compared with those given by [12].

The increase of c.c. for σ makes the central potential more attractive, for ω does the short range central potential repulsive, and for ρ does the tensor potential a little bit repulsive. In sum, the BONN potential is more attractive than other potentials for the S-wave and more repulsive for the D-wave. This makes the binding energy of the triton larger, and the D-state probability smaller.

We have not calculated the charge form factor and the momentum distribution for the BONN potential as yet. In general, the major contribution to these quantities for large momentum comes from the triton D-state. Therefore, we may expect that the calculated charge form factor will agree less with the experimental results, if we utilize the BONN potential. On the other hand, the large attractive effect of this potential may push up the charge form factor and the momentum distribution at large momentum. Therefore, the BONN potential will not manifest itself in these quantities.

(2) Electron inelastic scattering from [3]He and y-scaling law [13]

In the inelastic electron scattering calculation, inclusion of the TM force improves the agreement with experiment as shown in Fig.7. However, comparison of the calculated and the experimental results suggests a need for further improvement by taking account of the distortion of the outgoing nucleon and further contributions from large momentum components, e.g. , due to the ρ- and ω-exchange three-nucleon potential.

Once it was said that the 6 quark-effect is seen at large |y|. However, this argument was based on a bad wave function. Fig.7 does not suggest any need for quarks.

(3) Coulomb energy difference [6]

We calculated the first-order Coulomb energy. Table 4 shows that the

112

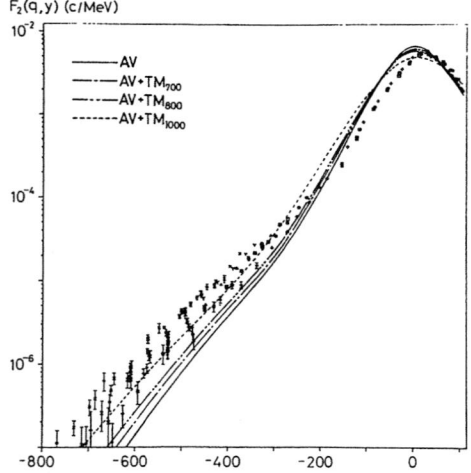

Fig.7 The scaling function $F_2(q,y)$ for the electron inelastic scattering: k_e = 14.7 GeV/c, q = 1.5GeV.

Table 4. Calculated Coulomb energy difference [6]

2NP		+TM(Λ=700)	+TM(Λ=800)
AV	0.65	0.68	0.70
PARIS	0.66	0.68	0.71
TRS	0.66	0.69	0.73

three-nucleon force makes the calculated value of the Coulomb energy larger, and encourage us to obtain the Coulomb energy difference of 0.763.

(4) Electromagnetic form factors [14]

The charge and magnetic form factors were calculated from the wave

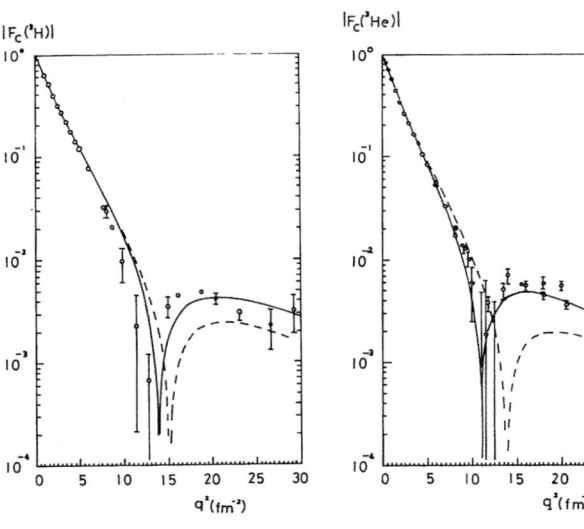

function for (RSC + TM(Λ=700MeV))18 . The dashed line shows the impulse approximation with this wave function. The solid line includes the meson exchange currents with the relativistic effect: π(ps)-, ρ-, ω-pair cur-

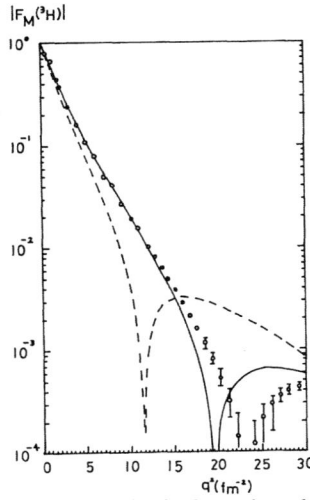 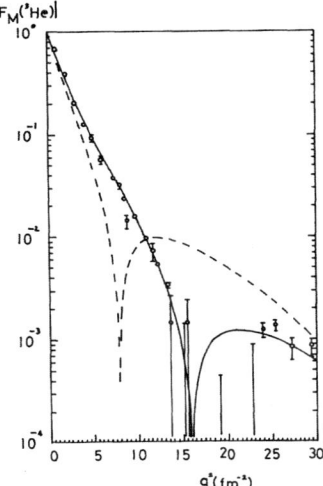

rents, pionic $\pi\pi\gamma$, $\rho\pi\gamma$, $\omega\pi\gamma$, $\pi N\Delta$ currents, and the retardation current that is the difference of recoil and renormalization currents, appeared at the third order Foldy-Wouthysen transformation. The experimental points are taken from unpublished Saclay data, prior to the later publication [15]. In this calculation, the fit to the triton is not so good as in ³He. We might have to perform a consistent calculation for the potential and exchange currents.

References

1. Reid, R. V.: Ann. Phys. (N.Y.) 50, 411 (1986)

2. Coon, S. A., Scadron, M. D., McNamee, P.C., Barrett, B. R., Blatt, D. W. E., McKeller, B. H.J.: Nucl. Phys. A317, 242 (1979); Coon, S. A., Glöckle, W.: Phys. Rev. C23, 1970 (1981)

3. Sasakwa, T., Ishikawa, S.: Few-Body Systems 1, 3 (1986)

4. Hajduk, Ch., Sauer, P.: Nucl. Phys. A369, 321 (1981); Hajduk, Ch.,Sauer, P., Strueve, W.: Nucl. Phys. A405, 581 (1983); Sauer, P.: Nuovo Cimento 76, 309 (1983)

5. Chen, C. R., Payne, G. L., Friar, J. L., Gibson, B. F.: Phys. Rev. Lett. 55, 374 (1985); Phys. Rev. 33, 1740 (1986).

6. Ishikawa, S., Sasakawa, T.: Few-Body Systems 1, 143 (1986)

7. Lacombe, M., Loiseau, B., Richard, J.M., Vinh Mau, R., Cote, J., Pires, P., de Tourreil, R.: Phys. Rev. 21, (1980) 861

8. de Tourreil, R.,Rouben, B., Sprung, D. W. L.:Nucl. Phys. A242, 445(1975)

9. Wiringa, R. B., Smith, R. A., Ainsworth, T. A.: Phys. Rev. C29, 1207 (1 984)

10. Ishikawa, S., Sasakawa, T.: Phys. Rev. Lett. 56, 317 (1986)

11. Machleidt, R.: Lecture note at the Los Alamos Workshop, June 1985, TRI-PP-85-68.

12. Brown, G. E., Jackson, A.D.: The Nucleon-Nucleon Interaction. Amsterdam 1979.

13. Soutome, T.: Master thesis at Tohoku University.

14. Fukunaga, A.: Master thesis at Tohoku University.

15. Juster, F. P., et al., Phys. Rev. Lett. 55, 2261 (1985)

HIGH ORDER PERTURBATION THEORY WITH THREE-BODY FORCES
IN THE FADDEEV SCHEME

A. Bömelburg, Institut für Theoretische Physik II

Ruhr Universität Bochum, 4630 Bochum, West Germany

Realistic two-body potentials, i.e. potentials that describe the two-nucleon observables quantitatively correct are not adequate to reproduce the fundamental observables of e.g. the A=3 nucleus like the binding energy. There are several theoretical approaches to overcome this discrepancy : On the "classical" level which we will adopt one may ask if meson-theoretical three-body forces can explain it. On the other hand one can extend the Hilbert space to take into account non-nucleonic degrees of freedom. On this level one may first regard the inclusion of Δ's and π's or even more pretentiously treat the nucleons as composite particles built up by quarks.

Since the inclusion of three-body potentials in realistic three-nucleon-calculations involves severe technical problems, it seems necessary to use different methods to check the results. In order to produce benchmarks we employ the widely used Reid- and Paris two-nucleon potentials /1,2/ together with the Tucson-Melbourne two-pion-exchange three-nucleon potential (2πE-3NP) /3/.

Our perturbational treatment of three-body potentials is based on the observation that the energetic effects of the 2πE-3NP are small compared to the expectation value of relistic two-body forces. Therefore we extended the standard Faddeev equations to include the dynamical effects of a three-body potential to arbitrary order in our perturbational expansion. /4/

Starting from the Schrödinger equation in integral form

$$(1) \qquad \Psi = G_0 \sum_{i=1}^{3} (V_i + W_i) \Psi$$

with the free resolvent G_0

(2) $\qquad G_0 = (E - H_0)^{-1}$

the pair interactions $V_i = V_{jk} \ (j, k \neq i)$ and the decomposition of the three-potential W

(3) $\qquad W = \sum_{i=1}^{3} W_i$

which is obvious for the $2\pi E$-3NP we used.

Then we split up the full wavefunction Ψ into 4 parts

(4) $\qquad \Psi = G_0 \sum_{i=1}^{3} V_i \Psi + G_0 W \Psi \equiv \sum_{i=1}^{3} \Psi_i + \Psi_4$

and make use of the Lippmann-Schwinger equations for the scattering by V_i resp. W :

(5a) $\qquad t_i = V_i + V_i G_0 t_i$

(5b) $\qquad T_4 = W + W G_0 T_4$

Now we proceed in the usual way to set up integral equations for the Faddeev components Ψ_i and Ψ_4, ending with the coupled set

(6a) $\qquad \Psi_i = G_0 t_i (P \Psi_i + \Psi_4)$

(6b) $\qquad \Psi_4 = G_0 T_4 (1 + P) \Psi_i$

where the permutation operator P in the subsystem 1 is given by

(7) $\qquad P = P_{12} P_{23} + P_{13} P_{23}$

As we are treating identical particles we restrict our attention to one subsystem and leave the index i behind. The insertion of equations (6a) and (6b) then gives :

(8) $\qquad \Psi = G_0 t \left(P + G_0 T_4 (1 + P) \right) \Psi$.

Now we expand the Faddeev component Ψ around the unperturbed solution $\Psi^{(0)}$ (W=0) corresponding to the energy $E^{(0)}$

116

$$(9) \qquad \psi^{(0)}(\epsilon^{(0)}) = G_o(\epsilon^{(0)}) \, t(\epsilon^{(0)}) \, P \, \psi^{(0)}(\epsilon^{(0)}) \; ,$$

define the following abbreviations

$$(10a) \qquad \psi^{(0)} \equiv \psi^{(0)}(\epsilon^{(0)})$$

$$(10b) \qquad G_o \equiv G_o(\epsilon^{(0)})$$

$$(10c) \qquad t_o \equiv t_o(\epsilon^{(0)})$$

and write for our perturbational ansatz

$$(11a) \qquad E = \epsilon^{(0)} + \Delta E$$

$$(11b) \qquad \psi = \psi(E) = \psi^{(0)} + \psi'$$

$$(11c) \qquad \psi' = \sum_{m \geqslant 1} \psi^{(m)}$$

After expanding the energy dependent terms in equ. (8) i.e. $G_o(E^{(0)} + \Delta E)$, $t(E^{(0)} + \Delta E)$ and $T_4(E^{(0)} + \Delta E)$ around $E^{(0)}$, we end up with the following equations, which are correct up to N-th order in W and ΔE :

$$(12a) \qquad \sum_{m=0}^{N} \psi^{(m)} = G_o \, t_o \, P \sum_{m=0}^{N} \psi^{(m)} + \varepsilon \sum_{m=0}^{N-1} \psi^{(m)}$$

$$+ \; G_o t_o \, G_o \sum_{m=0}^{N-1} \sum_{n=0}^{m} \omega^n \, W \, (1+P) \, \psi^{(m-n)}$$

$$(12b) \qquad \Psi^{(N)} = (1+P) \sum_{m=0}^{N} \psi^{(m)} + G_o \sum_{m=0}^{N-1} \sum_{n=0}^{m} \omega^n \, W \, (1+P) \, \psi^{(m-n)}$$

$$(12c) \qquad \Delta E^{(N)} = \langle \psi^{(0)} | W | \Psi^{(N-1)} \rangle \; / \; \langle \psi^{(0)} | \Psi^{(N-1)} \rangle$$

with the normalisation condition

$$(13) \qquad \langle \psi^{(0)} | \Psi^{(N)} \rangle = 1$$

and

$$(14a) \qquad \varepsilon \equiv - \Delta E \left(G_o + G_o t_o G_o \right)$$

$$(14b) \qquad \omega \equiv (W - \Delta E) \, G_o$$

It should be noticed that the kernel K ($K = G_o\, t_o\, P$) of our integral equations for the perturbed part of ψ is the same as for the unperturbed problem. Furthermore the inhomogenities can be calculated from the terms of lower orders by simple quadratures.

In addition to the physical eigenstate of the unperturbed problem

$$(15) \qquad \lambda_o\, \psi^{(o)} = K\, \psi^{(o)} \qquad , \qquad \lambda_o = 1$$

there exists an unphysical eigenstate $\psi^{(-)}$ which stems from the repulsive core of the two-nucleon potential

$$(16) \qquad \lambda^{(-)}\, \psi^{(-)} = K\, \psi^{(-)} \qquad , \qquad \lambda^{(-)} < -1 \quad .$$

Since we want to solve our equations iteratively, which is mandatory because of the very large dimension of the discretised integral kernel, which is of the order of 5000, we transform our inhomogeneous equation

$$(17) \qquad \psi' = \phi + K\, \psi'$$

into

$$(18a) \qquad \phi' = \frac{1}{1 - \lambda^{(-)}}\, \phi$$

$$(18b) \qquad K' = \frac{1}{1 - \lambda^{(-)}}\, \left(K - \lambda^{(-)}\, \mathbb{1} \right)$$

$$(18c) \qquad \psi' = \phi' + K'\, \psi'$$

Now it can be shown by simple estimates that the Neumann series of (18c) is convergent if the irrelevant part of ψ' proportional to $\psi^{(o)}$ is projected out. We introduce the projection operator Λ

$$(19) \qquad \Lambda = \mathbb{1} - \frac{|\psi^{(o)}\rangle\langle\tilde{\psi}^{(o)}|}{\langle\tilde{\psi}^{(o)}\,|\,\psi^{(o)}\rangle}$$

where $\tilde{\psi}^{(o)}$ is the lefthanded eigenfunction to K

$$(20) \qquad \tilde{\psi}^{(o)} = \tilde{\psi}^{(o)}\, K$$

which obviously has the desired properties. Finally we solve

$$(21) \qquad \Lambda\, \psi' = \phi' + K\Lambda\, \psi' \qquad , \qquad \Lambda\, \psi' = \psi'_\perp$$

and determine the admixture of $\psi^{(o)}$ to $\psi^{(1)}$ setting

$$(21) \qquad \psi^{(1)} = \psi^{(1)}_{\perp} + \alpha \, \psi^{(o)}$$

where α is fixed by the normalisation condition (13).

All of our calculations are based on a decomposition of the operators and wavefunctions into three-body partial waves, where all 18 states with the total angular momentum of the subsystem $j \leqslant 2$ are taken into account. The numbers given in the following tables represent the following : $E^{(i)}$ is the energy shift in i-th order, E_i the summation of these effects up to i-th order and $\sum^{(i)}$ the resulting triton binding energy.

Table 1 Triton binding energy in perturbation theory for the Reid softcore potential

i	$E^{(i)}$/ MeV	E_i/ MeV	$\sum^{(i)}$/ MeV
o	–	–	-7.24
1	-0.99	-0.99	-8.23
2	-0.69	-1.68	-8.92
3	-0.10	-1.78	-9.02
4	-0.06	-1.84	-9.08
5	-0.01	-1.84	-9.08

Table 2 Triton binding energy in perturbation theory for the Paris potential

i	$E^{(i)}$/ MeV	E_i/ MeV	$\sum^{(i)}$/ MeV
o	–	–	-7.33
1	-0.72	-0.72	-8.05
2	-0.93	-1.66	-8.99
3	-0.07	-1.73	-9.06
4	-0.13	-1.86	-9.19
5	-0.00	-1.86	-9.19

A closer look at the table reveals that the main contributions to the energy shift come from the first 2(3) orders, in fact they make up about 90 (97) percent of the total effect.

We summarize our results stating that the $2\pi E$-3NP yields a substantial additional binding energy in the triton, now overbinding the triton by an amount of about 0.6 ... 0.7 MeV. This reduces the discrepancy between experiment and theory, when only two-nucleon potentials are used by roughly a factor of 2. A recent nuclear matter calculation with π-ρ - and ρ- ρ - 3NP's has shown a repulsive effect of these forces. /5/ Therefore one can hope that the inclusion of these potentials in a realistic three-body calculation could reproduce the binding energy of the triton on this "classical" level.

References

1. Reid, R. : Ann. Phys. (N.Y.) 50, 411 (1968)
2. Lacombe, M., Richard, M., Vinh Mau, R., Côté, J., Pirès, P., de Tourreil, R. : Phys. Rev. C21, 861 (1980)
3. Coon, S.A., Scadron, M.D., Mc Namee, P.C., Barrett, B.R., Blatt, D.W.E., Mc Kellar, B.H.J. : Nucl.Phys. A317, 242 (1979)
4. Bömelburg, A. : Phys. Rev. C34, 14 (1986)
5. Ellis, R.G., Coon, S.A., Mc Kellar, B.H.J. : Nucl.Phys. A438, 631 (1985)

FEW-BODY CALCULATIONS AT SAPPORO

Yoshinori Akaishi

Department of Physics, Hokkaido University, Sapporo 060, Japan

ATMS is the abbreviation of "Amalgamation of Two-body correlations into Multiple Scattering process" , and is a method for treating few-body systems with realistic interactions. The accuracy of the method is investigated for ^{3}H with the Reid soft core (V_8) potential. By the use of ATMS, the properties of ^{4}He including momentum distributions, the possible existence of s-shell Σ-hypernuclei and the structure of the (dtµ) molecule are discussed as topics in few-body problems.

1. Introduction

In order to investigate structures of realistic four-body systems the present author and others (Sapporo group) proposed the ATMS method twelve years ago [1]. The four-nucleon system is of fundamental interest to nuclear physics: The ^{4}He nucleus is the lightest system to exhibit the basic nuclear structure property of excited states and is the most important cluster unit in nuclei. ATMS makes it possible to treat the four-nucleon system as well as the three-nucleon system on the basis of realistic nuclear interactions.

The ATMS method originates from the nuclear matter theory of Brueckner et al. [2] which established the essential role of the independent-pair correlation in infinite nuclear matter. Nagata and the present author [3] found that the two-body correlation dominates even in three- and four-nucleon systems as well as in nuclear matter. Thus, the idea of using two-body correlation functions is combined with the multiple scattering theory of Watson [4] into ATMS: The realistic wave function of the few-nucleon system is constructed by amalgamating two-body correlation functions into the multiple scattering process.

ATMS is reformulated by Tanaka on the basis of the generalized wave-matrix theory [5]. This general formulation suggests that ATMS may be applicable not only to the few-nucleon system but also to various few-body systems. In fact, hypernuclei and few-atom molecules are successfully treated by Shinmura [6], Kurihara [7], Nakaichi-Maeda [8] and others. ATMS possesses wider applicability than considered earlier.

2. Accuracy of ATMS

ATMS constructs the wave function of a system in the following form;

$$\Psi = F \Phi , \qquad (1)$$

where Φ is a reference (model) wave function and F is a correlation function. The wave function Ψ is proved to be an exact solution of the original Schrödinger equation, when F is the multiple scattering operator $\hat{F}$

$$\hat{F} = 1 + \sum_{(ij)} \frac{Q}{e} g_{ij} \hat{F}_{ij} ,$$

$$\hat{F}_{ij} = 1 + \sum_{(k\ell)}' \frac{Q}{e} g_{k\ell} \hat{F}_{k\ell} \qquad (2)$$

with the appropriately defined propagator (Q/e) and reaction matrix g [9]. The ATMS correlation function F is obtained by replacing $(Q/e)g$'s with a few kind of two-body correlation functions: The better F can be produced systematically by increasing the number of the kind.

The accuracy of ATMS is investigated by Morita for the case of realistic three- and four-nucleon systems. The two-body correlation functions are determined by the Euler-Lagrange equation derived from the variational principle;

$$\delta [< \Psi H \Psi > - \lambda <\Psi|\Psi>] = 0. \qquad (3)$$

The NN potentials employed are the Reid soft core V_8 model (RSCV$_8$) used in Variational Monte-Carlo (VMC) calculation [10] and the Reid soft core $^1S_0$,

$^3S_1$–$^3D_1$ potential (RSC5) used in Faddeev 5-channel calculations [11].

The results are shown in Table 1. In the case of RSCV$_8$, ATMS gives the upper-bound energy lower by about 0.27 MeV than the result of VMC.

Table 1. Energies (MeV) of ^{3}H with RSC potentials.

NN potential	RSCV$_8$		RSC5	
Method	ATMS	VMC	ATMS	Faddeev
Total energy	−7.13	−6.86±0.08	−7.02	−7.023
Kinetic energy	49.76	47.81	49.17	
Potential energy	−56.89	−54.68	−56.19	
Central	−14.15		−13.85	
Tensor	−43.75		−43.45	
LS	1.01	0.95	1.11	
P(S) (%)	89.75		89.63	88.91
P(S') (%)	1.47		1.71	1.67
P(D) (%)	8.79		8.66	9.34
Mass rms radius (fm)	1.81		1.83	
Charge rms rad. (fm)	1.68		1.70	1.70

In the case of RSC5, several Faddeev calculations give consistent results for ^{3}H. Except for the D-state probability to which the energy is rather insensitive, ATMS reproduces almost completely the Faddeev result. These comparisons demonstrate the high accuracy of ATMS.

It should be stressed that this ATMS method can be extended straightforwardly to the four-nucleon system. Table 2 shows an preliminary

Table 2. Energies (MeV) of ^{4}He with the RSCV$_8$ potential.

Method	ATMS	VMC	Exp.
Total energy	−21.8	−22.9±0.5	−28.3
Kinetic energy	102.9	106.9	
Potential energy	−124.7	−129.8	
Central	−37.6		
Tensor	−89.1		
LS	2.0	4.05	
P(S) (%)	88.98		
P(S') (%)	0.17		
P(D) (%)	10.86		
Rms radius (fm)	1.54		1.47

ATMS result for the RSCV$_8$ potential, where the number of two-body correlation functions introduced is small compared to the three-nucleon case. The final calculation will be soon accomplished.

For the Malfliet-Tjon central potential, the Green Function Monte-Carlo (GFMC) method gives the energy of -31.3±0.2 MeV [12], which is accepted as the exact one though its error is not small enough. The result of VMC is -31.19±0.05 MeV [13]. ATMS gives the upper bound of -31.36 MeV and the lower bound of -32.8 MeV. The two bounds enable us to estimate the exact energy with a much smaller error than the GFMC's [9].

In conclusion, we can say that ATMS is a powerful method to solve few-body systems with high accuracy. It should be mentioned that the form of Eq. (1) is a successful description of physical contents of the realistic few-body system. The correlated variational harmonic-oscillator (CVHO) method by Ciofi degli Atti and Simula [14] expresses the wave function in the same form. In CVHO, the correlation function F is of simple Jastrow-type but Φ is expanded on a large enough harmonic-oscillator basis. In ATMS, though Φ is taken to be simple, the correlation F is dynamically determined. One could find a more practical method in between.

3. Nuclear Short-Range Correlation and Spectral Function of ^{4}He

The ^{4}He results obtained from ATMS calculations are summarized in Ref. [9]: They are ground- and excited-state energies by Sakai, charge form factor by Katayama, effects of three-body force by Sato and so on. Here, we show momentum distributions in ^{4}He and discuss how to discover the short-range correlation inside the nucleus.

The momentum distributions of a single nucleon and an NN cluster in ^{4}He are calculated with the ATMS wave function obtained with RSCV$_8$ and are parametrized into two-range Gaussian form;

$$W(q) = C \left[\exp(-Bq^2) + s \exp(-Bq^2/t)\right] , \qquad (4)$$

where the second term represents the high-momentum component. The results are given in Table 3, where SN, TNC and TNR denote single nucleon, two-nucleon center-of-mass and two-nucleon relative momentum distributions, and q's are the momenta conjugate to $\vec{r}-\vec{R}_{cm}$, $(\vec{r}_1+\vec{r}_2-\vec{r}_3-\vec{r}_4)/4$ and $\vec{r}_1-\vec{r}_2$. In the case of SN the high-momentum component manifests itself above 2 fm^{-1}. Half of the high-momentum component comes from the short-range correlation and half from the D-wave correlation caused by the tensor force. The latter reflects long-range behavior of the wave function and conceals the detail of the short-range correlation.

Table 3. Parametrized momentum distributions in ^{4}He.

	B	β (fm^{-2})	t	s
w^{SN}	$3/(4\beta)$	0.42	12	0.00286
w^{TNC}	$1/(4\beta)$	0.42	12	0.00286
w^{TNR}	$2/\beta$	0.42	9	0.00634

In order to know what really happens when nucleons come close together, we have to remove the obstacle of the D-wave correlation from the observed high-momentum component. This can be done by a coincidence experiment ^{4}He(e, e'p): From the spectral function of ^{4}He we can get crucial information on the nuclear short-range correlation.

The spectral function $S(\vec{k}, E)$ is a probability of finding a nucleon with a momentum $\vec{k}$ and an energy E in the nucleus [15] and is calculated with the ground-state wave function Ψ_4 of ^{4}He and the three-nucleon wave function Ψ_3^f in the final state as

$$S(\vec{k}, E) = \sum_f | \tilde{\phi}_{fi}(\vec{k})|^2 \delta (E - [E_f^{3N} - E_0^{4N}]),$$

$$\tilde{\phi}_{fi}(\vec{k}) = (2\pi)^{-3/2} \int d\vec{R} \exp(-i\vec{k}\vec{R}) \phi_{fi}(\vec{R}) ,$$

$$\phi_{fi}(\vec{R}) = \sqrt{4} \int d\vec{\xi}_1 \int d\vec{\xi}_2 \ \Psi_3^{f*}(\vec{\xi}_1, \vec{\xi}_2) \Psi_4(\vec{\xi}_1, \vec{\xi}_2, \vec{R}). \tag{5}$$

We define the "ground" and "excited" spectral functions by

$$S^{gr}(\vec{k}) = \int_{E_{min}}^{E_{min}+\Delta} S(\vec{k}, E) \, dE, \qquad S^{ex}(\vec{k}) = \int_{E_{min}+\Delta}^{\infty} S(\vec{k}, E) \, dE, \tag{6}$$

which are related to the momentum distribution as

$$S^{gr}(\vec{k}) + S^{ex}(\vec{k}) = 4 \left(\frac{4}{3}\right)^3 w^{SN}(\frac{4}{3}\vec{k}). \tag{7}$$

Figure 1 shows the result. The NN potentials used are the RSCV$_8$ and super soft-core (SSC) [16] potentials. In the region of a small k the ground S^{gr} exhausts more than 90 % of the total, while S^{ex} amounts to 99% at around k = 2.4 fm^{-1}. S^{gr} has a dip at k = 2.2 fm^{-1} and a second maximum at k = 2.7 fm^{-1}. An interesting fact is that this peak comes only from the short-range correlation without any mask from the D-wave correlation and

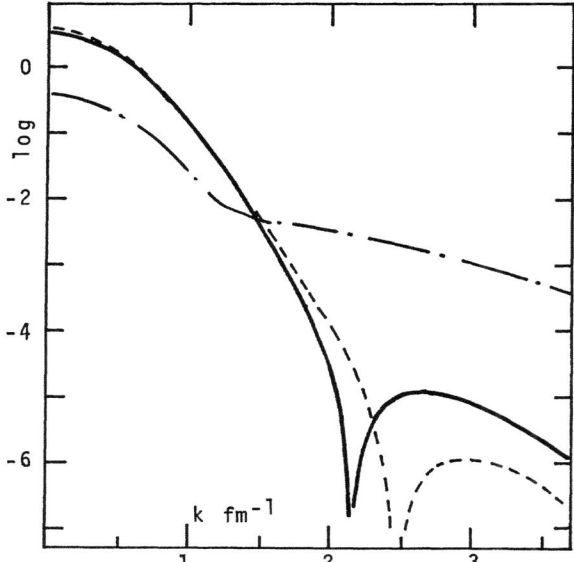

Fig. 1. The spectral function of ^{4}He. The ground one is shown by the solid line for RSC and by the dotted line for SSC. The dot-dashed line denotes the excited one.

its height depends on the nuclear potential model as seen from the comparison between RSC and SSC. We should notice a difference between the situations of ^{3}He and ^{4}He. In the case of ^{3}He the D-wave correlation can contribute to the ground spectral function, but in the case of ^{4}He it can not do because the spins of p and ^{3}H and the L=2 angular momentum can not couple to the spin 0 of ^{4}He. This is the reason why the short-range correlation can be seen not in ^{3}He but in ^{4}He.

Thus, the "ground" y-scaling function [17] of ^{4}He calculated by

$$F^{gr}(y) = 2\pi \int_{|y|}^{\infty} S^{gr}(\vec{k}) \, k \, dk \tag{8}$$

is a good quantity to see the short-range correlation in the nucleus.

4. Possible Existence of Light Σ-Hypernuclei

Kurihara et al.[7] have shown that the effective Λ-^{4}He potential has a central repulsion coming from the strong repulsive core of NΛ interaction and the compactness of the alpha particle, as seen in Fig. 2. This central repulsion strongly affects the Λ-density distribution and, therefore, the pionic decay rate of ^{5}He, and brings about an overbinding problem of $^9_\Lambda$Be. The problem would be discussed elsewhere.

126

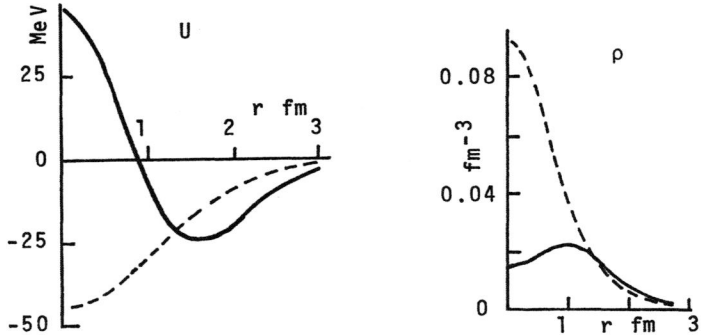

Fig. 2. The Λ-^{4}He potential and the Λ-density distribution.
The dotted lines are conventional ones.

Here, we discuss the possible existence of s-shell Σ-hyper nuclei
investigated by Harada et al., which enable us to investigate the "narrow-
width" problem for Σ-hypernuclei. A central NΣ potential of two-range
Gaussian type is constructed so as to reproduce low-energy experimental
data and Nijmegen phase shifts: (r in fm)

$$v = 3000 \exp(-(r/0.40)^2) + V_L \exp(-(r/1.00)^2) \qquad (9)$$

with V_L = -200, 0, -21, -207.4 MeV for (isospin, spin) = (1/2,1), (1/2,0),
(3/2,1), (3/2,0) states. The potential is strongly state-dependent and the
system with much (1/2,1) and (3/2,0) interactions has a chance to bound.

In the (NNNΣ) four-body systems, we can find the bound states. They
are listed in Table 4.

Table 4. Possible bound states of (NNNΣ).

T_z	S	$-B(\Sigma)$	Produced by
-1/2	0	-1.3 MeV	^{4}He(K$^-$, π^0)
1/2	0	-3.3 MeV	^{4}He(K$^-$, π^-)

The binding energy $B(\Sigma)$ is measured from ^{3}H $+ \Sigma^0$ for T_z = -1/2 and from ^{3}H
$+ \Sigma^+$ for T_z = 1/2. The mixing of the total isospin T is calculated
variationally: It is found that the bound systems are in almost pure T=1/2
states and the mixing of T=3/2 state is only less than 1 %. The NΣ - NΛ
conversion width can be investigated in detail for these bound hypernuclei.

The first observation of the bound Σ-hypernucleus, the alpha particle
with strangeness, is now planned at KEK by Yamazaki et al. It would be
fruitful to develope the few-body calculation to hypernuclear problems.

5. Structure of (dtμ) Molecule

Recently, the muonic molecule (dtμ) has aroused considerable attention in relation to the realization of a useful muon-catalyzed fusion [18]. It is interesting to see how well ATMS works for this Coulomb three-body system. We solve the multiple scattering equation (2) by introducing three kinds of two-body functions $\widetilde{f}$, $\overline{f}$ and f for each pair and obtain

$$
\begin{aligned}
\Psi = \; & \widetilde{f}_{31}(r_{31}) \, \overline{f}_{23}(r_{23}) \, f_{12}(r_{12}) + \widetilde{f}_{23}(r_{23}) \, \overline{f}_{31}(r_{31}) \, f_{12}(r_{12}) \\
& + \widetilde{f}_{12}(r_{12}) \, \overline{f}_{31}(r_{31}) \, f_{23}(r_{23}) + \widetilde{f}_{31}(r_{31}) \, \overline{f}_{12}(r_{12}) \, f_{23}(r_{23}) \\
& + \widetilde{f}_{23}(r_{23}) \, \overline{f}_{12}(r_{12}) \, f_{31}(r_{31}) + \widetilde{f}_{12}(r_{12}) \, \overline{f}_{23}(r_{23}) \, f_{31}(r_{31}) \\
& - \widetilde{f}_{31}(r_{31}) \, \widetilde{f}_{23}(r_{23}) \, [f_{12}(r_{12}) + 2 \, \overline{f}_{12}(r_{12})] \\
& - \widetilde{f}_{12}(r_{12}) \, \widetilde{f}_{31}(r_{31}) \, [f_{23}(r_{23}) + 2 \, \overline{f}_{23}(r_{23})] \\
& - \widetilde{f}_{23}(r_{23}) \, \widetilde{f}_{12}(r_{12}) \, [f_{31}(r_{31}) + 2 \, \overline{f}_{31}(r_{31})] \\
& + 4 \, \widetilde{f}_{31}(r_{31}) \, \widetilde{f}_{23}(r_{23}) \, \widetilde{f}_{12}(r_{12}),
\end{aligned}
\tag{10}
$$

which we call 2nd ATMS wave function. This 2nd ATMS reduces to 1st ATMS if $\widetilde{f}_{ij} = \overline{f}_{ij}$, and to 0th ATMS if $\widetilde{f}_{ij} = \overline{f}_{ij} = f_{ij}$. The 0th ATMS, the simplest case of ATMS, is nothing but a Jastrow-type wave function.

The ground-state energies of (dtμ) measured from the (tμ)+d threshold are −220.84 eV, −312.64 eV and −319.18 eV for 0th ATMS, 1st ATMS and 2nd ATMS, respectively. The 1st ATMS is superior to the 0th ATMS. The 2nd ATMS attains a still higher accuracy by the better treatment of the multiple scattering process. The 2nd ATMS result of −319.18 eV can satisfactorily be compared to the best value −319.15 eV of Vinitskii et al.[19] obtained from an very elaborate adiabatic calculation. Thus, the ATMS method can treat with high accuracy the Coulomb few-body system.

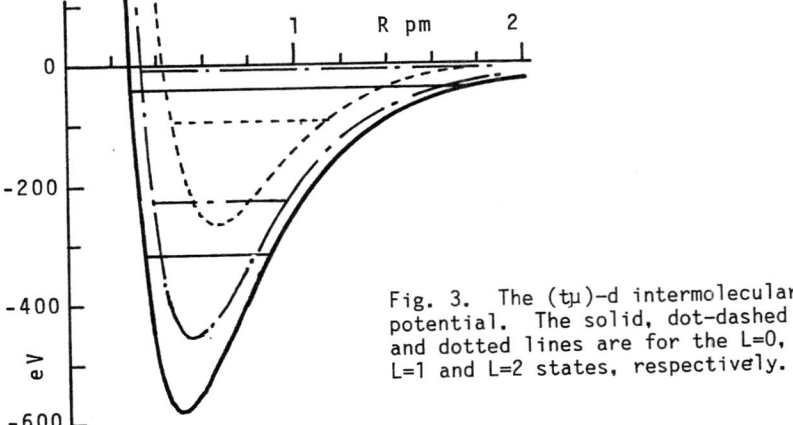

Fig. 3. The (tμ)−d intermolecular potential. The solid, dot-dashed and dotted lines are for the L=0, L=1 and L=2 states, respectively.

128

By the use of the ATMS wave function, we can derive the intermolecular potential between ($t\mu$) and d shown in Figure 3. The L=0 potential can be simulated by a Morse-potential form;

$$U(R) = D \left[\exp(-2a(R-R_0)) - 2 \exp(-a(R-R_0))\right] \qquad (11)$$

with D = 580.3 eV, R_0 = 0.509 pm and a = 3.0136 pm^{-1}.

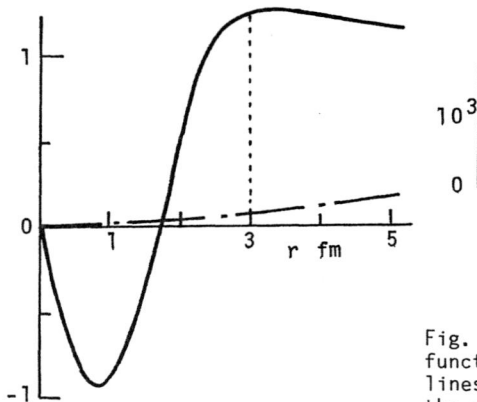

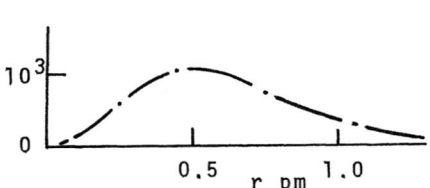

Fig. 4. The d-t relative wave function. The solid and dot-dashed lines are the cases with and without the nuclear potential.

The fusion cross section of ($dt\mu$) is proportional to a probability that d and t come closer to the range of nuclear interaction. Figure 4 shows the d-t relative wave function. The nuclear interaction between d and t has a near-threshold resonance at +66 KeV. This nuclear resonance strongly couples with the ($dt\mu$) molecular states. The solid line is the case including the nuclear potential: The wave function is greatly enhanced in the nuclear interaction range of about 3 fm, and the fusion rate becomes more than 200 times as large as that of the only Coulomb-interaction case denoted by the dot-dashed line.

It would be interesting to note a "doubly resonant" feature that the ($dt\mu$) molecular states of size ~pm resonantly couples with D_2 molecular states [18] of ~Å and with the d-t nuclear state of ~fm. ATMS can treat a variety of few-body systems of current interest.

The author thanks Prof. H. Tanaka, Dr. T. Katayama, Dr. S. Nakaichi-Maeda, Dr. S. Shinmura, Dr. Y. Kurihara, Mr. H. Morita, Mr. M. Teshigawara, Mr. S. Tadokoro and Mr. T. Harada for discussions and collaborations.

References

1. Akaishi,Y., Sakai,M., Hiura,J., Tanaka,H.: Prog. Theor. Phys. Suppl.
   No. 56, 6 (1974).
2. Brueckner,K.A., Levinson,C.A.: Phys. Rev. 97, 1344 (1955).
   Gomes,L.C., Walecka,J.D., Weisskopf,V.F.: Ann. of Phys. 3, 241 (1958).
3. Akaishi,Y., Nagata,S.: Prog. Theor. Phys. 48, 133 (1972).
4. Watson,K.M.: Phys. Rev. 89, 575 (1953).
5. Tanaka,H.: ATMS Method and Its Application. International Symposium
   in Nanning, 1985.
6. Shinmura,S., Akaishi,Y., Tanaka,H.: Prog. Theor. Phys. 71, 546 (1984).
7. Kurihara,Y., Akaishi,Y., Tanaka,H.: Phys. Rev. C31, 971 (1985).
8. Nakaichi-Maeda,S., Lim,T.K., Akaishi,Y., Tanaka,H.: J. Chem. Phys. 71,
   4430 (1979).
9. Akaishi,Y., International Rev. Nucl. Phys. Vol.4. Singapore:
   World Scientific 1986.
10. Lomnitz-Adler,J., Pandharipande,V.R., Smith,R.A.: Nucl. Phys. A361, 399
    (1981).
11. Hajduk,Ch., Sauer,P.U.: Nucl. Phys. A369, 321 (1981).
    Ishikawa,S., Sasakawa,T., Sawada,T., Ueda,T.: Phys. Rev. Lett. 53, 1877
    (1984).
    Chen,C.R., Payne,G.L., Friar,J.L., Gibson,B.F.: Phys. Rev. C31, 2266
    (1985).
12. Zabolitzky,J.G., Kalos,M.H.: Nucl. Phys. A356, 114 (1981).
13. Carlson,J., Pandharipande,V.R.: Nucl. Phys. A371, 301 (1981).
14. Ciofi degli Atti,C., Simula,S.: Nuovo Cim. 41, 101 (1984).
15. Ciofi degli Atti,C.: Prog. Part. Nucl. Phys. 3, 163 (1980).
16. de Tourreil,R., Rouben,B., Sprung,D.W.L.: Nucl. Phys. A242, 445 (1975).
17. Ciofi degli Atti,C., Pace,E., Salmè,G.: Phys. Lett. 127B, 303 (1983).
18. Bracci,L., Fiorentini,G.: Phys. Rep. 86, 169 (1982).
19. Vinitskii,S.I., Melezhik,V.S., Ponomarev,L.I., Puzynin,I.V.,
    Puzynina,T.P., Somov,L.N., Truskova,N.F.:
    Soviet Phys. JETP 52, 353 (1980).

VARIATIONAL MONTE CARLO CALCULATIONS OF FEW-BODY NUCLEI

R. B. Wiringa
Physics Division, Argonne National Laboratory
Argonne, IL 60439-4843 USA

Abstract

The variational Monte Carlo method is described. Results for the binding energies, density distributions, momentum distributions, and static longitudinal structure functions of the ^{3}H, ^{3}He, and ^{4}He ground states, and for the energies of the low-lying scattering states in ^{4}He are presented.

The variational Monte Carlo (VMC) method has been used for a wide variety of studies in both few- and many-body systems. These include calculations of the ground-state binding energy and density of infinite liquid atomic ^{3}He and ^{4}He [1-3] and of liquid drops containing up to 240 atoms of ^{3}He or ^{4}He [4,5]. This note will review the application of VMC methods to nuclear few-body systems [6-13]. New results using the 34-channel Faddeev wave function of the Los Alamos-Iowa group [14-15] confirm the binding energy and density distributions reported by them, and the first momentum distribution and longitudinal structure function calculations with these wave functions are presented.

In the VMC method a variational wave function $\psi_v(\alpha)$ is constructed (with parameters denoted by α) and a Rayleigh-Ritz upper bound to the ground-state energy,

$$E_v(\alpha) = \frac{\langle \psi_v(\alpha) | H | \psi_v(\alpha) \rangle}{\langle \psi_v(\alpha) | \psi_v(\alpha) \rangle} \geq E_0 \tag{1}$$

is evaluated. The Metropolis Monte Carlo algorithm [16] is used for the 3A-dimensional integration. The parameters α are varied until the lowest upper bound is found; the resulting E_v is taken as the energy of the system, and other properties are computed using ψ_v.

The VMC method has several advantages. In particular, it has wide applicability to problems in both nuclear and condensed matter physics. The general structure for good candidate ψ_v is well established both from the VMC studies of atomic helium systems and from variational hypernetted chain (HNC) studies of nuclear matter [7,11,17,18]. The Monte Carlo integration procedure can provide arbitrary accuracy (although at the cost of a quadratic increase in computer time) and no partial-wave expansion of the Hamiltonian, or any other operator, is required.

The chief disadvantage is that the wave function is variational. Better wave functions can generally be found when more exact methods can be brought to bear, e.g., the recent Faddeev calculations for ^{3}H [14,15,19,20] or the Green's function Monte Carlo (GFMC) calculations of atomic helium drops [4]. Further, while variational methods are economical in computer usage for a single evaluation of the energy, a search in the parameter space α must be performed, requiring many energy evaluations to determine the best ψ_v. This determination is made more difficult by the statistical error of the Monte Carlo integration procedure.

In nuclear systems we wish to take the expectation value of a many-body Hamiltonian of the form

$$H = \sum_i \frac{\hbar^2}{2m} \nabla_i^2 + \sum_{i<j} v_{ij} + \sum_{i<j<k} V_{ijk} . \tag{2}$$

The variational wave function for ^{3}H and ^{4}He is written as [6]

$$\psi_v = \{S \prod_{i<j} F_{ij}\} \phi , \tag{3}$$

$$V_{ij} = f_c(r_{ij}) [1 + \sum_{k\neq i,j} f_3(r_{ij}, r_{ik}, r_{jk}) \{u_\sigma(r_{ij}) \sigma_i \cdot \sigma_j$$

$$+ u_{t\tau}(r_{ij}) S_{ij} \tau_i \cdot \tau_j\}] . \tag{4}$$

Here S is a symmetrization operator, f_c provides both short-range correlation and long-range confinement, f_3 is a three-body correlation that acts only on the state-dependent correlations u_σ and $u_{t\tau}$, and ϕ is an anti-symmetrized spin-isospin state. The f_c, u_σ, and $u_{t\tau}$ are obtained from the solution of coupled Euler-Lagrange equations which minimize the two-body cluster energy, subject to boundary conditions that constrain their asymptotic long-range behavior in an appropriate manner for pd and ppn breakup in ^{3}He and for pt and dd breakup in ^{4}He [11]. In our calculations ψ_v is stored as a vector with $(2^A A!/(A-Z)!Z!)$ components, each of which specifies the spin and isospin of each particle.

Alternatively a Faddeev wave function, ψ_F, or any other suitable wave function can be used in the Monte Carlo integrations [9]. In fact, analysis of the 5-channel ψ_F led to a significant improvement in the long-range boundary conditions used for the ψ_v of Eq. (3). We are now studying the 34-channel ψ_F for possible further improvements in the variational ansatz that can be applied to ^{4}He and larger nuclei.

The Metropolis Monte Carlo algorithm [1,16] provides a way for drawing points $\{R_i\}$ in 3A-dimensional space from a normalized probability distribution $P(R) = |\psi(R)|^2 / \int |\psi(R)|^2 dR$. The central limit theorem then implies that for any quantity $f(R)$,

$$\lim_{N \to \infty} \frac{1}{N} \sum_i f(R_i) = \frac{\int |\psi(R)|^2 f(R) dR}{\int |\psi(R)|^2 dR} , \tag{5}$$

and in particular,

$$E_v = \lim_{N \to \infty} \frac{1}{N} \psi(R_i) H\psi(R_i) / |\psi(R_i)|^2 . \tag{6}$$

The binding energy results for ^{3}H using several Hamiltonians and wave functions are shown in Table 1. The v_{ij} used is the Argonne v_{14}(AV14) model [21] and V_{ijk} is either the Urbana VII (U7) model [11] or Tucson-Melbourne (TM) model [22]. Three different ψ_F have been used, which were obtained by solving for AV14, AV14 + U7, and AV14 + TM Hamiltonians, respectively. These expectation values all agree with the values reported by the Los Alamos-Iowa group [14,15]. Also shown are the results obtained by using ψ_v of the form of Eq. (3) for both ^{3}H and ^{4}He. These results are from larger Monte Carlo samplings than (but are consistent with) previously reported values [11].

Table 1. Binding energies in MeV for ^{3}H and ^{4}He for different H and ψ. Numbers in parentheses are one-standard-deviation Monte Carlo sampliing errors in the last digit.

		AV14	AV14+U7	AV14+TM
^{3}H	ψ_F(AV14)	7.68(2)	8.71(4)	8.41(4)
	ψ_F(AV14+U7)	7.27(4)	9.04(3)	8.56(3)
	ψ_F(AV14+TM)	6.50(4)	8.24(4)	9.332(15)
	ψ_F(AV14+U7)	7.05(4)	8.28(4)	7.35(6)
^{4}He	ψ_v(AV14+U7)	21.5(3)	27.4(4)	23.9(4)

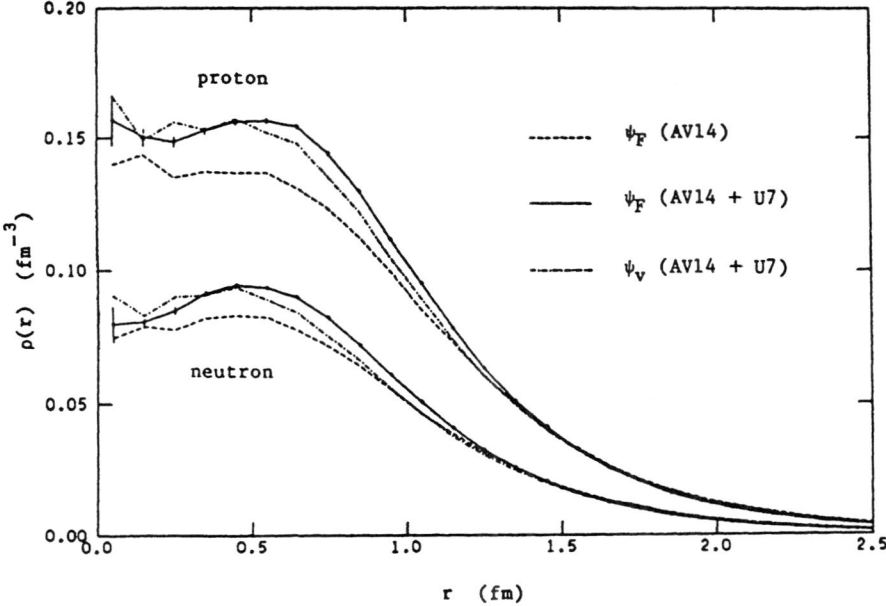

Figure 1. Density distributions in ³He for several ψ.

The current ψ_V gives 0.7 to 0.8 MeV less binding in ³H for AV14 + U7 than the 34-channel ψ_F. If this deficiency of the variational ansatz scales with the total potential energy, then the true ⁴He binding energy for AV14 + U7 might be ≈ 29.2 MeV; if it scales as $\langle V_{ijk} \rangle$ the binding energy might be ≈ 30.5 MeV.

The proton and neutron density distributions for several of these wave functions as shown in Fig. 1. The density for AV14 + U7 shows a small central depression, in agreement with the results reported by the Los Alamos-Iowa group.

Nucleon momentum distributions can be evaluated by sampling the quantity

$$N_{tz}(k) = \frac{\int d\mathbf{r}_1' \, \psi_V(\mathbf{R}) \, \exp[i\mathbf{k} \cdot (\mathbf{r}_1' - \mathbf{r}_1)] \, P_{tz}(1) \, \psi_V(\mathbf{R}')}{\left| \psi_V(\mathbf{R}) \right|^2} \tag{7}$$

where $\mathbf{R} = (\mathbf{r}_1, \mathbf{r}_2, \mathbf{r}_3)$, $\mathbf{R}' = (\mathbf{r}_1', \mathbf{r}_2, \mathbf{r}_3)$, and P_{tz} is an isospin projection operator. The proton momentum distributions in ³He calculated for several wave functions are shown in Fig. 2, along with (e,e'p) data analyzed in plane wave impulse approximation (PWIA) [23]. The $N_p(k=0)$ intercept scales as E^{-3} and ψ_V comes closest to experiment because its binding energy is

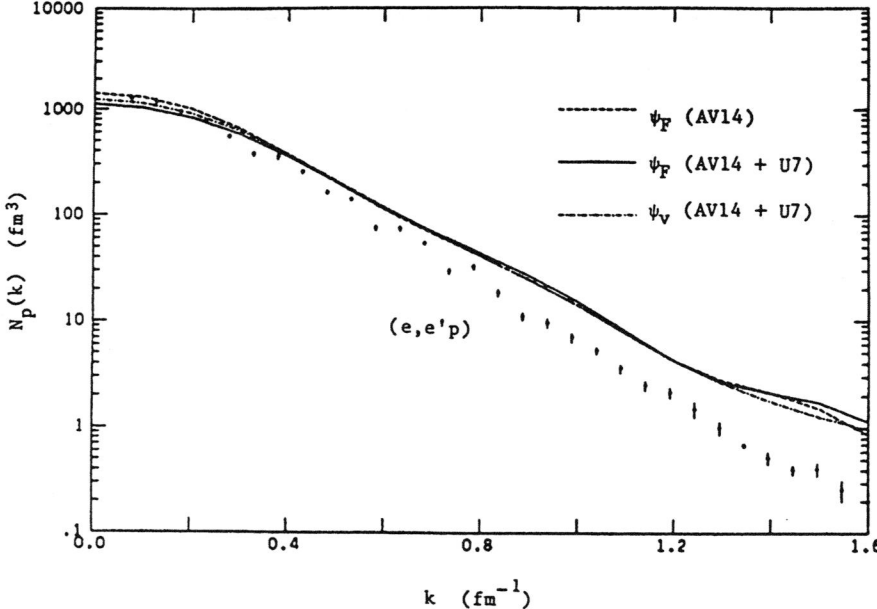

Fig. 2. Proton momentum distributions in ^{3}He for several ψ.

closer than any of the ψ_F (even though it is not the best ψ_V for H). As k increases all the wave functions give an $N_p(k)$ increasingly above the data. Since these high-k components arise when the nucleons are close together, final state interactions may be expected to play an important role and the PWIA analysis will not be adequate. Calculations in the VMC framework are currently under way to study the final state interactions [24].

Two-body breakup amplitudes can also be calculated giving $N_{dp}(k)$ in ^{3}He and $N_{tp}(k)$ and $N_{dd}(k)$ in ^{4}He. The $N_{dp}(k)$ exhibit the same behavior as $N_p(k)$, with the k = 0 value well predicted, but an increasing disparity with increasing k compared to the PWIA analysis of available (e,e'p) data [11]. The D_2 values, which are related to the asymptotic ratio of D-state to S-state wave functions in the d + p and d + d amplitudes, are also computed with ψ_V and give −0.24 and −0.16, respectively, for AV14 + U7.

Other quantities of interest are the dynamic and static longitudinal structure functions $S_L(k,\omega)$ and $S_L(k)$,

$$S_L(k,\omega) = 1/Z \sum_I \langle 0 | \rho^\dagger_{L,k} | I \rangle \langle I | \rho_{L,k} | 0 \rangle \; \delta(\omega - \omega_I) \; , \tag{8}$$

$$\rho_{L,k} = \sum_i \exp(i\mathbf{k}\cdot\mathbf{r}_i) \; 1/2 \; \{(1+\tau_3(i)) - \mu_n[k^2/4(k^2+m^2)] \; (1-\tau_3(i))\} \; , \qquad (9)$$

$$S_L(k) = \int_{0^+}^{\infty} S_L(k,\omega) \; d\omega \; . \qquad (10)$$

In Eq. (8) the sum is over all states $|I\rangle$, $|0\rangle$ is the ground state, and ω_I is the energy of state I with respect to the ground state. The sum in Eq. (9) is over all particles, $1/2(1\pm\tau_3(i))$ are proton and neutron projection operators, and the μ_n term comes from electron–neutron scattering. The lower limit of the integral in Eq. (10) is taken for convenience as 0^+ to eliminate the elastic contribution. $S_L(k)$ can be evaluated in the light nuclei by using closure and subtracting the elastic part explicitly:

$$S_L(k) = 1/Z \; [\langle 0|\rho_{L,k}^{\dagger} \; \rho_{L,k}|0\rangle - |\langle 0_R|\rho_{L,k}|0\rangle|^2] \; . \qquad (11)$$

Here the recoiling ground state $|0_R\rangle = \exp(1/A\Sigma\mathbf{k}\cdot\mathbf{r}_i\rangle|0\rangle$, and the expectation values in Eq. (11) are easily evaluated by Monte Carlo integration. The results of such calculations using ψ_v for ^{3}H, ^{3}He, and ^{4}He are shown in Fig. 3, along with experimental data on ^{3}He [25]. The calculations agree with experiment very well; substitution of ψ_F does not significantly alter the results. The calculation of $S_L(k,\omega)$ is more involved, but is currently underway [24].

The VMC method has also been used to study the excited resonance states in ^{4}He [8]. This is done with a modified R-matrix method which converts the scattering problem into a bound-state problem with special boundary conditions. The variational wave function for the excited states is taken as:

$$\psi_v(J^\pi,T,n) = (S \prod_{i<J} F_{ij}) \; \phi(J^\pi,T,n) \; , \qquad (12)$$

$$\phi(J^\pi,T,n) = [\sum_i \xi_{n\ell}(\mathbf{R}_i) \; 0_i(J^\pi,T)]\phi \; . \qquad (13)$$

Here the correlation operator of Eq. (4) is now applied to an uncorrelated excited state $\phi(J^\pi,T,n)$ which is constructed from a sum over single particles i of a radial wave function $\xi_{n\ell}(\mathbf{R}_i \equiv A[\mathbf{r}_i-\mathbf{R}_{c.m.}]/[A-1])$ with nodal and angular quantum numbers n and ℓ, and an operator $0_i(J^\pi,T)$ of angular momentum J, parity π, and isospin T. For ^{4}He the lowest excited states are $(J^\pi,T) = (0^+,0)$, $(0^-,0)$, and $(2^-,0)$; the corresponding (n,ℓ) values are $(1,0)$, $(0,1)$, and $(0,1)$, and the 0_i are 1, $\sigma_i\cdot\mathbf{R}_i$, and $3\sigma_{3i}\cdot\mathbf{R}_{3i}-\sigma_i\cdot\mathbf{R}_i$, respectively. The radial dependence of $\xi_{n\ell}$ is controlled by a few variational parameters, while F_{ij} is taken from the ground-state calculation.

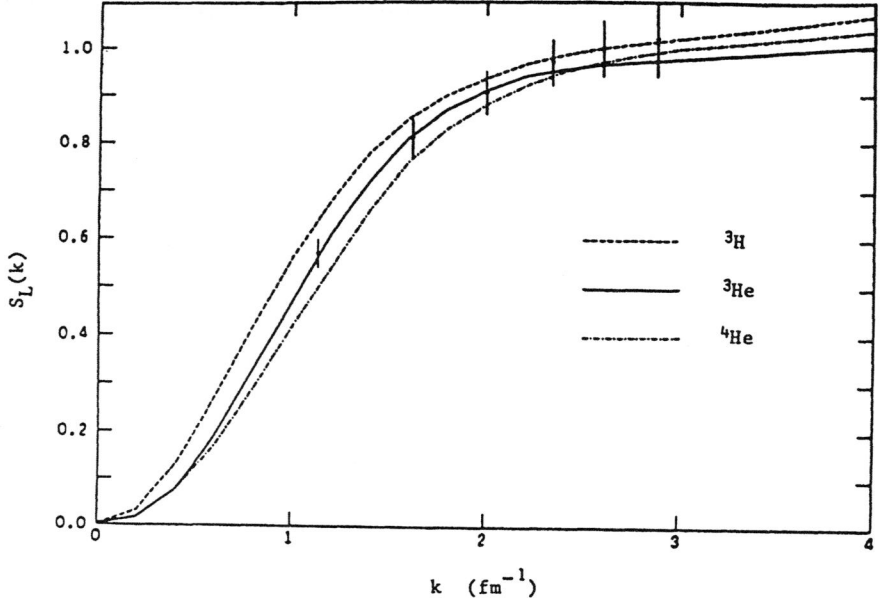

Figure 3. Static longitudinal structure function $S_L(k)$.

The excited-state energy $E(J^\pi,T,n)$ is the expectation value of Eq. (1) with $\psi_v(J^\pi,T,n)$ substituted for ψ_v. If we ignore the Coulomb force, the $(0^+,0)$ state is weakly bound and can be treated as a regular variational problem. Because the (J^π,T) quantum numbers are the same as the ground state, $E(0^+,0,1)$ is minimized subject to the orthogonality condition $\langle\psi_v(0^+,0,1)|\psi_v\rangle = 0$. For the negative parity states we calculate the phase shift δ_ℓ,

$$\tan(\delta_\ell) = j_\ell(kR_n)/n_\ell(kR_n) , \tag{14}$$

$$k = (2\mu E_{rel}/\hbar^2)^{1/2} , \tag{15}$$

$$E_{rel} = E(J^\pi,T,n) - E(^3H) . \tag{16}$$

Here E_{rel} is the relative energy of a proton scattering off ^3H. It is obtained by picking a nodal boundary R_n where $\xi_{n\ell}(R_n) = 0$, and minimizing $E(J^\pi,T,n)$ inside this boundary. The calculation is performed for several values of R_n, thus mapping out $\delta(E)$. Both the 0^- and 2^- states show resonant behavior, and the energy at which $\partial\delta(E)/\partial E$ is a maximum is identified as the energy of the resonance.

In the calculation for ^{4}He, the states come out in the correct order, with the 0^+ state at -8.6 ± 0.4 MeV, and the 0^- and 2^- states at -6.5 ± 0.5 MeV and -4.0 ± 0.5 MeV, which may be compared to Coulomb corrected experimental values of -8.5, -7.4, and -6.4, respectively. The V_{ijk} contributes significantly to the excitation by lowering the ground-state energy much more than the resonance states, as one would expect. More recently, this method has been applied by the Courant group to the five-body problem of α-neutron scattering [13]. They calculate both the $J = 1/2$ and $3/2$ scattering states using a semirealistic v_{ij} and V_{ijk} and obtain most of the observed spin-orbit splitting without recourse to relativistic effects.

Future work with the VMC method will involve: 1) improving the ansatz for constructing ψ_v, 2) extending the calculations to more properties of interest, 3) extending the calculations to larger nuclei, and 4) further refining the nuclear many-body Hamiltonian. The improvements in ψ_v include the introduction of L·S correlations and better three-body correlations. The 34-channel ψ_F sets a goal for performance in ^{3}H, and any improvements we can make there should carry over to the larger systems. The study of final state interactions in (e,e'p) scattering and of $S_L(k,\omega)$ are examples of additional properties beyond the ground-state binding energy that can be calculated [24]. Meson-exchange currents are another area that should be investigated with VMC.

The extension to larger systems has already begun with an impressive first calculation of ^{16}O by the Courant group [10]. Using the semirealistic Reid v_6 interaction (containing tensor, but no spin-orbit forces) they obtain a binding energy per particle of -6 ± 1 MeV. The calculation is computationally very time consuming, and further improvement in the algorithm is necessary, as well as the use of more realistic interactions. Nevertheless, with the continuing rapid increase in computer power, such microscopic calculations may become commonplace in a few years. Realistic studies of smaller systems like ^{6}Li and ^{8}He are probably feasible now.

The long term goal of all these studies is the improvement of our understanding of the nuclear many-body Hamiltonian. The VMC method will continue to be a quantitatively valuable tool in the study of possible interactions, e.g., in the study of alternative V_{ijk} models [12]. Future work may include the study of v_{ij} models with explicit Δ components, as a possible alternative to V_{ijk} [19,21] or the study of semirelativistic H where the consequences of using $\sqrt{p^2+m^2}$ for the kinetic energy may be explored.

138

The work reported here was done in collaboration with many people in the Urbana-Argonne and Los Alamos-Iowa groups. I particularly wish to thank G. L. Payne for providing the 34-channel ψ_F and R. Schiavilla for the calculations of $S_L(k)$. I also wish to thank J. L. Friar, V. R. Pandharipande, and S. C. Pieper for many useful discussions. This work was supported by the U.S. Department of Energy, Nuclear Physics Division, under contract W-31-109-ENG-38.

References

[1] McMillan, W. L., Phys. Rev. A 138, 442 (1965).

[2] Schmidt, K., Kalos, M. H., Lee, M. A., and Chester, G. V., Phys. Rev. Lett. 45 573 (1980).

[3] Schmidt, K. E., Lee, M. A., Kalos, M. H., and Chester, G. V., Phys. Rev. Lett. 47, 807 (1981).

[4] Pandharipande, V. R., Zabolitzky, J. G., Pieper, S. C., Wiringa, R. B., and Helmbrecht, U., Phys. Rev. Lett. 50, 1676 (1983).

[5] Pandharipande, V. R., Pieper, S. C., and Wiringa, R. B., Phys. Rev. B 34, 4571 (1986).

[6] Lomnitz-Adler, J., Pandharipande, V. R., and Smith, R. A., Nucl. Phys. A361, 399 (1981).

[7] Carlson, J., Pandharipande, V. R., and Wiringa, R. B., Nucl. Phys. A401, 59 (1983).

[8] Carlson, J., Pandharipande, V. R., and Wiringa, R. B., Nucl. Phys. A424, 47 (1984).

[9] Wiringa, R. B., Friar, J. L., Gibson, B. F., Payne, G. L., and Chen, C. R., Phys. Lett. 143B, 273 (1984).

[10] Carlson, J., and Kalos, M. H., Phys. Rev. C 32, 2105 (1985).

[11] Schiavilla, R., Pandharipande, V. R., and Wiringa, R. B., Nucl. Phys. A449, 219 (1986).

[12] Keister, B. D., and Wiringa, R. B., Phys. Lett. 173B, 5 (1986).

[13] Carlson, J., Schmidt, K. E., and Kalos, M. H., in Condensed Matter Theories, Vol. 1, edited by F. B. Malik (Plenum, New York 1986).

[14] Chen, C. R., Payne, G. L., Friar, J. L., and Gibson, B. F., Phys. Rev. Lett. 55, 374 (1985).

[15] Chen, C. R., Payne, G. L., Friar, J. L., and Gibson, B. F., Phys. Rev. C 33, 1740 (1986).

[16] Metropolis, N., Rosenbluth, A. W., Rosenbluth, M. N., Teller, A. M., and Teller, E., J. Chem. Phys. 21, 1087 (1953).

[17] Pandharipande, V. R., and Wiringa, R. B., Rev. Mod. Phys. 51, 821 (1979).

[18] Day, B. D., and Wiringa, R. B., Phys. Rev. C 32, 1057 (1985).

[19] Hadjuk, Ch., Sauer, P. U., and Strueve, W., Nucl. Phys. A405, 581 (1983).

[20] Ishikawa, S., and Sasakawa, T., Few-body Sys. 1, 143 (1986).

[21] Wiringa, R. B., Smith, R. A., and Ainsworth, T. L., Phys. Rev. C 29, 1207 (1984).

[22] Coon, S. A., Scadron, M. D., McNamee, P. C., Barrett, B. R., Blatt, D. W. E., and McKellar, B. H. J., Nucl. Phys. A317, 242 (1979).

[23] Jans, E., et al., Phys. Rev. Lett. 49 974 (1982).

[24] Schiavilla, R., private communication.

[25] Marchand, C., et al., Phys. Lett. 153B, 29 (1985).

HYPERSPHERICAL CALCULATIONS FOR FOUR-NUCLEON SYSTEMS

J.L. Ballot

Division de Physique Théorique, Institut de Physique Nucléaire,
91406 Orsay Cedex, France

Abstract We develop hyperspherical calculations on the bound states of four-nucleon systems and particularly. the fondamental level and the first 0^+ excited states. With neglect of the Coulomb effect, we analyze the convergence of the optimal subset expansion for the binding energies calculated for central or realistic potentials.

1. Outline of the method

In recent years the ground-state of the four-nucleon system has been calculated by many authors using different techniques based on : The Yakubovsky-Faddeev integral equations [1-8], the coupled cluster method [9,10], the ATMS [12,13], Montecarlo [14] methods and the hyperspherical expansion [15-19,33-35] Most of these methods have given successful results in the approach to the three nucleon physics.

It is an interesting problem to understand the structure of the four-nucleon system because its spectrum is more rich than the two and three-body one. Indeed besides the ground-level and the first 0^+ excited state there are many excited unstable levels with different spins and parities which have been observed.

In this talk we discuss the four-nucleon problems calculated with non relativistic Schrödinger equations with pair-wise forces in the framework of the hyperspherical formalism.

The ground-state of the ^{4}He nucleus has total angular momentum J=0, total isospin T=0 and even π parity. The Schrödinger equation for such four-nucleon system is written as :

$$\left\{ -\frac{\hbar^2}{m} \sum_{i=1}^{4} \nabla_{x_i}^2 + \sum_{i,j>i} V(\vec{r_{ij}}) - E \right\} \Psi(\vec{x_1},\vec{x_2},\vec{x_3},\vec{x_4}) = 0 \quad (1)$$

where $\vec{x}_i$ represents the coordinate of the nucleon i.

The wave function of the four-nucleon is totally antisymmetric under particle permutation and can be written as products of elements in configuration spin and isospin states. Each of these elements can be constructed in such a way to belong to a specific representation of the permutation group S_4. The representations of this group can be symmetric [S], antisymmetric [A] or have a mixed symmetry [M] under the exchange of any pair of particles. The representations [S] and [A] are one-dimensional, whereas the last has two elements, M^+ and M^-, which are respectively even or odd under the exchange of nucleons 1 and 2 which we take to be the pair of reference.

We denote the isospin-spin state by $\Gamma^{SS_z}_{TT_z}(R)$ where S and S_z are the total and third component of spin, T and T_z are the corresponding isospin components and R indicates the type of permutation symmetry. The various states needed in the study of four-nucleon are

$$\Gamma^{00}_{00}(S) \equiv S(T,S) = \frac{1}{\sqrt{2}}\left[|+\rangle_T|+\rangle_S + |-\rangle_T|-\rangle_S\right], \quad \Gamma^{00}_{00}(A) \equiv A(T,S) = \frac{1}{\sqrt{2}}\left[|+\rangle_T|-\rangle_S - |-\rangle_T|+\rangle_S\right]$$

$$\Gamma^{00}_{00}(M^+) \equiv S'(T,S) = \frac{1}{\sqrt{2}}\left[|+\rangle_T|+\rangle_S - |-\rangle_T|-\rangle_S\right], \quad \Gamma^{00}_{00}(M^-) \equiv A'(T,S) = \frac{1}{\sqrt{2}}\left[|+\rangle_T|-\rangle_S + |-\rangle_T|+\rangle_S\right]$$

(2)

where the labels T and S indicate isospin and spin states. In both cases we have

$$|-\rangle = \left|\left(\left[0\tfrac{1}{2}\right]\tfrac{1}{2}\tfrac{1}{2}\right)00\right\rangle, \quad |+\rangle = \left|\left(\left[1\tfrac{1}{2}\right]\tfrac{1}{2}\tfrac{1}{2}\right)00\right\rangle, \quad |0\rangle = \left|\left(\left[1\tfrac{1}{2}\right]\tfrac{3}{2}\tfrac{1}{2}\right)20\right\rangle$$

in using the usual notation for the coupling of spin of four particles in the sequence $|[(S_{12}S_3)S_{123}S_4]S_{1234}\rangle$. The configuration part of the wave function is denoted by $\phi^{LL_z}_S(R)$ where L is the total angular momentum that can be 0,1,2 L_z is its third component, R indicates the representation of the permutation group S_4 and δ is a label for possible degenerate states.

The various components of the full four-nucleon wave function are obtained by coupling these tensors $\phi^{LL_z}_S(R)$ to the isospin-spin expressions (2) in such a way as to produce antisymmetric states with total angular momentum J=0. We denote the coupled states by $[^SL]^{JM}_{TT_z}$, where L is S, P or D depending on whether the orbital momentum is 0,1 or 2. The rules for reducing products of representation of the permutation group S_4 yield eight combinations among them we keep only the following S and D states.

$$[^0S]^{00}_{00} = \phi^{00}(S)\,\Gamma^{00}_{00}(A)$$

$$\left[S' \right]_{00}^{00} = \phi^{00}(M^+) \Gamma_{00}^{00}(M^-) - \phi^{00}(M^-) \Gamma_{00}^{00}(M^+) \tag{3}$$

$$\left[D \right]_{00}^{00} = \sum_{\lambda} \langle 2\lambda 2 \lambda | 00 \rangle \left[\phi^{2\lambda}(M^+) \Gamma_{00}^{20}(M^-) - \phi^{2\lambda}(M^-) \Gamma_{00}^{20}(M^+) \right]$$

These equations are convenient parametrizations of the various components of the total wave function. All the dynamical informations are contained in the special wave function $\phi_S^{LL}{}_{z}$ (R) which can be calculated by solving the Schrödinger equation in the framework of the hyperspherical harmonic (H.H) method.

We eliminate the center of mass in using the Jacobi coordinates

$$\vec{\xi}_1 = \sqrt{3/2}\left[\vec{x}_4 - (\vec{x}_1 + \vec{x}_2 + \vec{x}_3)/3 \right] = 2\sqrt{2/3} \,(\vec{x}_4 - \vec{X})$$

$$\vec{\xi}_2 = 2/\sqrt{3}\left[\vec{x}_3 - (\vec{x}_1 + \vec{x}_2)/2 \right] \tag{4}$$

$$\vec{\xi}_3 = \vec{x}_2 - \vec{x}_1 \qquad\qquad \vec{X} = (\sum_{i=1}^{4} \vec{x}_i)/4$$

$$\xi_1 = |\vec{\xi}_1| = \xi \sin\phi_3 \sin\phi_2$$

$$\xi_2 = |\vec{\xi}_2| = \xi \sin\phi_3 \cos\phi_2$$

$$\xi_3 = |\vec{\xi}_3| = \xi \cos\phi_3 \tag{5}$$

where ξ and ϕ_ℓ are respectively the hyper radius and the hyperangles given by

$$\xi^2 = \xi_1^2 + \xi_2^2 + \xi_3^2 \qquad tg\,\phi_\ell = \sqrt{\sum_{j=1}^{\ell-1} \xi_j^2} \,/ \xi_\ell \tag{6}$$

with this change of variables the kinetic energy operator must be written as

$$T = -\frac{\hbar^2}{m}\left[\frac{\partial^2}{\partial\xi^2} + \frac{8}{\xi}\frac{\partial}{\partial\xi} + \frac{\mathcal{L}^2(\Omega)}{\xi^2} \right] \tag{7}$$

where $\mathcal{L}^2$ is the "grand orbital" operator depending of height angles

$$\Omega \equiv \left[\hat{\vec{\xi}}_1, \hat{\vec{\xi}}_2, \hat{\vec{\xi}}_3, \phi_2, \phi_3 \right]$$

The hyperspherical harmonics are the eigenfunctions of $\mathcal{L}^2(\Omega)$ operator,
$[\mathcal{L}^2 + L(L+7)]\, \mathcal{Y}_{[L]}(\Omega) = 0$ and can be written as

$$\mathcal{Y}_{[L]}(\Omega) = Y_{\ell_1}^{m_1}(\hat{\xi}_1)\, Y_{\ell_2}^{m_2}(\hat{\xi}_2)\, Y_{\ell_3}^{m_3}(\hat{\xi}_3)\, {}^{[2]}P_{L_2}^{\ell_2 \ell_1}(\phi_2)\, {}^{[3]}P_{L}^{\ell_3 L_2}(\phi_3) \quad (8)$$

where
$$ {}^{[2]}P_{L_2}^{\ell_2 \ell_1}(\phi_2) = D_{L_2}^{\ell_2 \ell_3}\, [\cos\phi_2]^{\ell_2}[\sin\phi_2]^{\ell_1}\, P_{\frac{L_2-\ell_1-\ell_2}{2}}^{\ell_1+\frac{1}{2},\,\ell_2+\frac{1}{2}}(\cos 2\phi_2) \quad (9)$$

$$ {}^{[3]}P_{L}^{\ell_3 L_2}(\phi_3) = C_{L}^{\ell_3 L_2}\, [\cos\phi_3]^{\ell_3}[\sin\phi_3]^{L_2}\, P_{\frac{L-\ell_3-L_2}{2}}^{L_2+2,\,\ell_3+\frac{1}{2}}(\cos 2\phi_3) \quad (10)$$

$$ D_{L_2}^{\ell_2 \ell_1} = \left[\frac{2(L_2+2)\,\Gamma(L_2/2+\ell_1/2+\ell_2/2+2)\,\Gamma(L_2/2-\ell_1/2-\ell_2/2+1)}{\Gamma(L_2/2+\ell_1/2-\ell_2/2+3/2)\,\Gamma(L_2/2+\ell_2/2-\ell_1/2+3/2)} \right]^{\frac{1}{2}} \quad (11)$$

$$ C_{L}^{\ell_3 L_2} = \left[\frac{2(L+7/2)\,\Gamma(L/2-L_2/2-\ell_3/2+1)\,\Gamma(L/2+L_2/2+\ell_3/2+7/2)}{\Gamma(L/2+L_2/2-\ell_3/2+3)\,\Gamma(L/2-L_2/2+\ell_3/2+3/2)} \right]^{\frac{1}{2}} \quad (12)$$

the $P_n^{\alpha,\beta}$ is the usual Jacobi polynomial.

We use the kinematic rotation vector [30] to identify a pair of particles by two angles Ψ_2 and Ψ_3. Let $\vec{z}$ be any linear combination of the coordinates $\vec{\xi}_i$:

$$ \vec{z} = \sum_{i=1}^{3} a_i \vec{\xi}_i = c \sum_{i=1}^{3} \frac{a_i}{c}\, \vec{\xi}_i \quad , \quad c^2 = \sum_{i=1}^{3} a_i^2 \quad . \quad (13)$$

We define the kinematic rotation vector $\vec{U}(\Psi) = \vec{\xi}_1 \sin\Psi_2 \sin\Psi_2$
$+\ \vec{\xi}_2 \sin\Psi_3 \cos\Psi_2 + \vec{\xi}_3 \cos\Psi_3$ (14) and $\vec{z} = c\,\vec{U}(\Psi)$ with
$\cos\Psi_i = a_i / \{\sum_{j=1}^{i} a_j^2\}^{1/2}$. With our previous choice of coordinates $\vec{\xi}_i$
we obtain the following correspondance :

$$\vec{x}_2 - \vec{x}_1 = \vec{\xi}_3 \qquad\qquad \Psi_2 = 0 \quad \Psi_3 = 0$$

$$\vec{x}_3 - \vec{x}_2 = \frac{\sqrt{3}}{2}\vec{\xi}_2 - \frac{1}{2}\vec{\xi}_3 \qquad \Psi_2 = 0 \quad \Psi_3 = 2\pi/3$$

$$\vec{x}_1 - \vec{x}_3 = \frac{\sqrt{3}}{2}\vec{\xi}_2 + \frac{1}{2}\vec{\xi}_3 \qquad \Psi_2 = 0 \quad \Psi_3 = -2\pi/3$$

$$\vec{x_4} - \vec{x_1} = \sqrt{\frac{2}{3}}\; \vec{\xi_1} + \frac{1}{2\sqrt{3}}\vec{\xi_2} + \frac{1}{2}\vec{\xi_3} \qquad \Psi_2 = \alpha \quad \Psi_3 = \pi/3$$

$$\vec{x_4} - \vec{x_2} = \sqrt{\frac{2}{3}}\; \vec{\xi_1} + \frac{1}{2\sqrt{3}}\vec{\xi_2} - \frac{1}{2}\vec{\xi_3} \qquad \Psi_2 = \alpha \quad \Psi_3 = 2\pi/3$$

$$\vec{x_4} - \vec{x_3} = \sqrt{\frac{2}{3}}\; \vec{\xi_1} - \frac{1}{\sqrt{3}}\vec{\xi_2} \qquad\qquad \Psi_2 = \beta \quad \Psi_3 = \pi/2 \tag{15}$$

$$\cos\alpha = 1/3 \qquad \sin\alpha = 2\sqrt{2}/3$$

$$\cos\beta = -1\sqrt{3} \qquad \sin\beta = \sqrt{2/3}$$

In using these set of angles (Ψ_2, Ψ_3) we are able to construct spacial wave functions $\Phi^{00}(S)$, $\Phi^{00}(M^+)$, $\Phi^{00}(M^-)$, $\Phi^{2\lambda}(M^+)$ and $\Phi^{2\lambda}(M^-)$ which have definite symmetries under the exchange of nucleons. This is done by acting three symmetrization operators $\Sigma_0, \Sigma_+, \Sigma_-$ on the polynomial product $^{[2]}P^{\ell_2\ell_1}_{L_2}(\Psi_2)$ $\times\; ^{[3]}P^{\ell_3 L_2}_{2K}(\Psi_3)$ and which generate irreducible representations of S_4, totally symmetric for $\mathcal{E} = 0$ and mixed symmetry for $\mathcal{E} = \pm$ [31].

$$^{\mathcal{E}}G^{\ell_2\ell_1\ell_3 L_2}_{L_2\, 2K} = \sum_{\mathcal{E}}\; ^{[2]}P^{\ell_2\ell_1}_{L_2}(\Psi_2)\; ^{[3]}P^{\ell_3 L_2}_{2K}(\Psi_3) \tag{16}$$

The ^{4}He wave function is then expanded on the optimals subsets $^{\mathcal{E}}P_{2K}(\Omega)$ and $^{\mathcal{E}}D_{2K+2}(\Omega)$ as follows:

$$\xi^4\, \Psi(\xi\Omega) = \sum_{K=0}^{\infty}\, ^0P_{2K}(\Omega)\, A(T,S)\, \mathcal{U}^0_{2K}(\xi)$$

$$+ \sum_{K=1}^{\infty} \frac{1}{\sqrt{12}}\left[^+P_{2K}(\Omega) A'(T,S) + \bar{P}_{2K}(\Omega) S'(T,S)\right] \mathcal{U}^H_{2K}(\xi)$$

$$+ \sum_{K=0}^{\infty}\left[^+D_{2K+2}(\Omega)|-\rangle_T - \bar{D}_{2K+2}(\Omega)|+\rangle_T\right] \mathcal{U}^D_{2K}(\xi) \tag{17}$$

with $^{\mathcal{E}}P_{2K}(\Omega) = {}^{\mathcal{E}}N_{2K}\sum_{[2K]} \hat{\ell_1}\hat{\ell_2}\hat{\ell_3}\,\Delta(\ell_1\ell_2\ell_3) \begin{pmatrix}\ell_1 & \ell_2 & \ell_3 \\ m_1 & m_2 & m_3\end{pmatrix} Y^{m_1}_{\ell_1}(\hat{\xi_1}) Y^{m_2}_{\ell_2}(\hat{\xi_2})$

$$Y^{m_3}_{\ell_3}(\hat{\xi_3})\; ^{\mathcal{E}}G^{\ell_2\ell_1\ell_3 L_2}_{L_2\, 2K}\; ^{[2]}P^{\ell_2\ell_1}_{L_2}(\phi_2)\; ^{[3]}P^{\ell_3 L_2}_{2K}(\phi_3) \cdot \tag{18}$$

and $^{\mathcal{E}}D_{2K+2}(\Omega) = {}^DN_{2K+2}\sum_{\alpha\mu\,[2K+2]} (-1)^{\mu+m_3}\hat{\ell_1}\hat{\ell_2}\hat{\ell_3}\,\hat{\alpha^2}\,\Delta(\ell_1\ell_2\alpha)\,\Delta(\alpha\ell_3 2)$

$$Y^{m_1}_{\ell_1}(\hat{\xi_1}) Y^{m_2}_{\ell_2}(\hat{\xi_2}) Y^{m_3}_{\ell_3}(\hat{\xi_3})\; ^{\mathcal{E}}G^{\ell_2\ell_1\ell_3 L_2}_{L_2(2K+2)}\; ^{[2]}P^{\ell_2\ell_1}_{L_2}(\phi_2)\; ^{[3]}P^{\ell_3 L_2}_{2K+2}(\phi_3)$$

$$\begin{pmatrix}\ell_1 & \ell_2 & \alpha \\ m_1 & m_2 & \beta\end{pmatrix}\begin{pmatrix}\alpha & \ell_3 & 2 \\ -\beta & m_3 & \gamma\end{pmatrix} |([1\tfrac{1}{2}]\tfrac{3}{2}\tfrac{1}{2})2 - \mu\rangle_S \cdot \tag{19}$$

where $\widehat{\ell_i} = \sqrt{2\ell_i + 1}$ and $\Delta(\alpha\beta\gamma)$ stands for the 3J symbol $\begin{pmatrix} \alpha & \beta & \gamma \\ 0 & 0 & 0 \end{pmatrix}$.
In expanding the total wave function $\Psi(^4He)$ on the optimal subsets (18,19) the Schrödinger equation is transformed into an infinite set of second-order coupled differential equations :

$$\left\{ -\frac{\hbar^2}{m}\left[\frac{d^2}{d\xi^2} - \frac{(2K+4)(2K+3)}{\xi^2} \right] - E \right\} u_{2K+\ell}^{R}(\xi)$$

$$+ 6\sum_{\substack{KK'K'' \\ \ell'\ell''R'\gamma}} (-1)^{K''\gamma} \mathcal{U}_{2K''}^{\gamma \quad (9,\ell')}(\xi) \begin{bmatrix} R & \gamma & R' \\ 2K+\ell & 2K'+\ell'' & 2K'+\ell' \end{bmatrix} u_{2K'+\ell'}^{R'}(\xi)$$

$$\tag{20}$$

with two kinds of multipoles $\mathcal{U}_{2K''}^{\gamma \quad (9,\ell)}(\xi)$ with respect to the component γ of the interaction. For the tensor component the multipole

$$^T\mathcal{U}_{2K''}^{(9,2)}(\xi) = 2\frac{\Gamma(K''+1)\Gamma(11/2)}{\Gamma(K''+3)\Gamma(5/2)} \int_0^1 du\, u^4(1-u^2) \, ^T\mathcal{U}(u\xi) \, P_{K''}^{5/2,2}(1-2u^2)$$

$$\tag{21}$$

is expressed in term of Jacobi polynomials and for all the other components

$$^c\mathcal{U}_{2K''}^{(9,0)}(\xi) = (-1)^{K''}\frac{\Gamma(K''+1)\Gamma(K''+3/2)}{\Gamma(K''+3)\Gamma(K''+7/2)} \int_0^1 du\, u(1-u^2)^2 \, ^c\mathcal{U}(u\xi) \, C_{2K''+1}^{5/2}(u)$$

$$\tag{22}$$

the multipole is expressed in terms of Gegenbauer polynomials. The symbols $\begin{bmatrix} R & \gamma & R' \\ 2K+\ell & 2K''+\ell'' & 2K'+\ell' \end{bmatrix}$ represent the hyperspherical coefficients [31].

2. Results and discussion

The energy of the ground-state of the four-nucleon is obtained by integrating the N first coupled equations of the infinite set (20) for Volkov and Afnan-Tang S3 potentials. The final results are listed in tables 1 and 2 respectively, where we give the binding energy for the coupled equations (CE) and also for different adiabatic approximation equations [19] which give, for the eigenvalues of the system, a lower-bound with the extreme adiabatic approximation (EAA) and an upper-bound with the uncoupled adiabatic approximation [UAA]. For a fixed grand orbital quantum number (L=2K) the upper bound E[UAA] is found to be very closed to the exact energy E[CE]. In fact we have the approximate relation :

$E(CE) \simeq E[UAA] + 0.2$ (E(EAA) - E(UAA)) within an accuracy of $1^\circ/_{\circ\circ}$. If for the Volkov potential our results are as good as variational ones, for the S3 interaction our binding energy is 1.9 % weaker than variational ones. Then we have to analyze the contribution of non potential harmonics which exist in the four-nucleon from K=3. In tables 1 and 2 we give also the kinetic energy versus K and the potential energy for a wave K in taking into account all the coupling with the others. We see that the major part of the potential energy becomes from the coupling with the first K=0 wave. The convergence of the energy in terms of the quantum number K is determined by the shape of the interaction [31]. For Gaussian interaction [Volkov,S3] the increase of the binding energy converges exponentially in terms of K . The rule $\Delta E(K) = E(k) - E(K+2) = \exp[-AK-B]$ can be used to obtain an extrapolated energy (K$\rightarrow\infty$). For the four realistic potentials [GPDT,SSCB,SSCC,URBANA] tables 3 to 6 show the convergence of the lower bound (EAA) and of the upper bound (UAA) of the four-nucleon ground-state energies calculated without Coulomb force. For L=22 the convergence is not yet reached. We have to calculate all the different kinds of the hyperspherical coefficients up to K=15 (L=30) to get it. For the three super soft core potentials the ground-state energy E(UAA) is found to be -20.92 MeV for SSCC, -22.75 MeV for SSCB and -27.33 for GPDT. These results are in agreement with those obtained by Fomin and Efros [33] for GPDT and SSCC interactions. For the potential SSCB we are in agreement only with the version (PH) of their calculations. A first excited 0^+ state for these three potentials is found to be :

GPDT $E0^+[UAA]$=-5.48 MeV, P(S)=94.16, P(S')=1.89, P(D)=3.94,R_{mat}=3.57 fm
SSCB $E0^+[UAA]$=-1.98 MeV, P(S)=95.13, P(S')=.46, P(D)=4.41, R_{mat}=3.82 fm
SSCC $E0^+[UAA]$=-1.04 MeV, P(S)=95.75, P(S')=.19, P(D)=4.05, R_{mat}=4.11 fm.
The four-nucleon ground-state energy calculated with the Urbana potential [27] is : E(UAA) = 18.88 MeV. Here also the convergence is not yet reached

for $L = 22$. Note that for this potential negative eigenvalue is found from $L=8$, then the convergence should be slower, but we can expect to get it for $L=30$. For the ARGONNE potential [36], negative eigenvalue starts at $L=10$ with $E(UAA) = -0.655$ MeV for $L=22$ we have obtained : $E[UAA] = -15.65$ MeV. The calculation of the ^{4}He charge form factor with the different realistic wave functions exhibits the same defects than those of the ^{3}H form factor. The diffraction minimum is at about $q^2 = 13$ fm^{-2} and the second peak too low around $q^2 = 18$ fm^{-2}.

Conclusion

The hyperspherical method is successfully applied to four-nucleon calculation. Even if the convergence has to be improved in integrating coupled equations for $L=2K > 22$, the different binding energies obtained for realistic potentials in using an expansion of the wave function on the potential basis are promising. The use of the different adiabatic approximations has shown their efficiency to calculate the excited states of the four-nucleon. A deficiency of several MeV 6 to 8.5 following the interactions shows that as for the trinucleon the realistic two body forces are not realistic for the four-nucleon.

REFERENCES

1. Narodetsky, I.M.: Nucl. Phys. A221, 191 (1974)

2. Gibson, B.F., Lehman, D.R.: Phys. Rev. C14, 685 (1976)

3. Sofianos, S., Fiedelday, H., McGurk, N.J.: Phys. Lett. 68B, 117 (1977); Nucl. Phys. A294, 49 (1978)

4. Perne, R., Sandhas, W.: Phys. Rev. Lett. 39, 788 (1977)

5. Kröger, H., Sandhas, W.: Phys. Rev. Lett. 40, 834 (1978)

6. Narodetsky, J.M.: preprint 1979

7. Tjon, J.A.: Phys. Lett. 56B, 217 (1975)

8. Tjon, J.A.: Phys. Rev. Lett. 40, 1239 (1978)

9. Zabolitzky, J.G.: Nucl. Phys. A228, 272 (1974)

10. Kümmel, H., Lührmann, K.M., Zabolitzky, J.G.: Phys. Rev. 36C, 11 (1978)

11. Zabolitzky, J.G.: Phys. Lett. **100B**, 5 (1981)

12. Akaishi, Y., Sakai, M., Hiura, J., Tanaka, H.: Prog. Theo. Phys. **56**, 7 (1975)

13. Sakai, M., Shimodaya, I., Akaishi, Y., Hiura, J., Tanaka, H.: Prog. Theo. Phys. **56**, 32 (1975)

14. Carlson, J., Pandharipande, V.R., Wiringa, R.B.: Nucl. Phys. **A401**, 59 (1983)

15. Demin, V.F., Pokrovsky, Y.E., Efros, V.D.: Phys. Lett. **44B**, 227 (1973)

16. Ballot, J.L., Beiner, M., Fabre de la Ripelle, M.: The nuclear many body problem, eds. Calogero, F., Ciofi Degli Atti, C., Vol. 1, Editrice Compositori, Bologna, 1973 p. 565

17. Ballot, J.L.: Z. Phys. **A302**, 347 (1981)

18. Ballot, J.L.: Phys. Lett. **127B**, 399 (1983)

19. Ballot, J.L., Fabre de la Ripelle, M., Levinger, J.S.: Phys. Rev. **C26**, 2301 (1982)

20. Afnan, I.R., Tang, Y.C.: Phys. Rev. **175**, 1337 (1968)

21. Volkov, A.B.: Nucl. Phys. **74**, 33 (1965)

22. Fantoni, S., Panattoni, L., Rosati, S.: Nuov. Cim. **67A**, 80 (1970)

23. Malfliet, R.A., Tjon, J.A.: Nucl. Phys. **A127**, 161 (1969)

24. Gogny, D., Pirès, P., De Tourreil, R.: Phys. Lett. **B32**, 591 (1970)

25. Merkuriev, S.P., Yakoviev, S.L., Gignoux, C.: Nucl. Phys. **A431**, 125 (1984)

26. De Tourreil, R., Sprung, D.W.L.: Nucl. Phys. **A201**, 193 (1973)

27. Lagaris, I.E., Pandharipande, V.R.: Nucl. Phys. **A359**, 331 (1981)

28. Goldhammer, P.: Phys. Rev. **C29**, 1444 (1984)

29. Tjon, J.A.: Phys. Rev. Lett. **40**, 1239 (1978)

30. Ballot, J.L., Fabre de la Ripelle, M.: Ann. Phys. (NY) **127**, 62 (1980)

31. Ballot, J.L.: to be published

32. Schneider, T.K.: Phys. Lett. **40B**, 439 (1972)

33. Fomin, B.A., Efros, V.D.: Sov. J. Nucl. Phys. **31**, 748 (1980); Phys. Lett. **98B**, 389 (1981); Sov. J. Nucl. Phys. **34**, 327 (1981)

34. Fomin, B.A.: Sov. J. Nucl. Phys. **33**, 612 (1981)

35. Fomin, B.A., Efros, V.D.: Sov. J. Nucl. Phys. **34**, 485 (1981).

VOLKOV Ref.[21]						
K	-E[EAA]	-E[UAA]	-E[CE]	E(CE)	T_{Kin}	$-V_{Pot}(K)$
0			28.578		48.345	78.495
2	29.505	29.187	29.283	.705	.778	.926
3	30.057	26.698	29.805	.521	.564	.618
4	30.410	30.042	30.150	.345	.365	.3947
5	30.528	30.155	30.263	.113	.116	.1253
6	30.620	30.238	30.349	.085	.087	.0932
7	30.649	30.264	30.375	.026	.0259	.0277
8	30.665	30.276	30.389	.0139	.0139	.0147
9	30.672	30.282	30.394	.0056	.0056	.0059
10	30.675	30.285	30.397	.0026	.0025	.0027
11	30.6771	30.2857	30.3981	.0001	.001	.001
12	30.6775	30.2862	30.3986	.00005	.0006	.0005
Fantoni et al. [22]			30.317		50.307	80.705
Afnan-Tang [20]			29.947			
Mercuriev [25]			30.2			

Table 1 - Convergence of the ^{4}He ground-state in terms of the grand orbital L=2K for Volkov interaction, without Coulomb forces for each adiabatic approximation EAA, UAA and Coupled equations (CE). The kinetic energy and the potential energy (see text) are given in terms of K. All the energies are in MeV.

S3 AFNAN-TANG Ref.[20]						
K	-E[EAA]	-E[UAA]	-E[CE]	E(CE)	T_{Kin}	$-V_{Pot}(K)$
0			7.193		48.773	73.83
2	10.768	10.236	10.331	3.138	2.959	3.236
3	15.701	15.109	15.219	4.888	3.902	4.149
4	21.500	20.835	20.967	5.747	4.380	4.61
5	23.479	22.763	22.903	1.936	1.566	1.642
6	25.318	24.454	24.624	1.721	1.522	1.591
7	25.989	25.070	25.248	.624	.567	.591
8	26.413	25.441	25.627	.379	.356	.370
9	26.624	25.619	25.810	.1827	.173	.179
10	26.738	25.713	25.908	.0977	.0926	.096
11	26.793	25.758	25.953	.0458	.0435	.045
12	26.825	25.785	25.980	.0273	.0261	.027
13	26.8418	25.7976	25.9937	.0129	.0122	.012
14	26.8510	25.8055	26.0012	.0075	.0074	.0075
15	26.8559	25.8095	26.0053	.0040	.0038	.004
Fantoni Ref.[22]			26.47		64.386	90.392
Afnan-Tang [20]			27.286			
Mercuriev [25]			25.5			

Table 2 - Convergence of the ^{4}He ground-state energy in terms of the grand orbital L=2K for S3 interaction. All the energies are in MeV.

URBANA Ref.[27]							
L=2K	NS.NS'.ND	-E[EAA]	-E[UAA]	P(S)	P(S')	P(D)	E[UAA]
4	2. 2. 1	-	-	-	-	-	-
6	3. 3. 2	-	-	-	-	-	-
8	4. 4. 3	4.151	3.873	92.94	.14	6.91	
10	5. 5. 4	9.032	8.6595	92.44	.12	7.44	4.786
12	6. 6. 5	13.962	13.422	92.18	.14	7.68	4.762
14	7. 7. 6	16.183	15.545	92.15	.18	7.67	2.123
16	8. 8. 7	17.856	17.121	92.08	.22	7.69	1.576
18	9. 9. 8	18.837	18.032	92.02	.25	7.72	.911
20	10. 10. 9	19.425	18.576	92.06	.26	7.68	.543
22	11. 10. 10	19.762	18.887	92.09	.26	7.65	.311

Table 3 - Convergence of the ^{4}He ground-state binding energies in terms of the grand orbital (L=2K) for Urbana [27] interaction, without Coulomb forces, for each approximation EAA, UAA. The percentages of the S,S' and D components of the wave functions are those of the [UAA] solution. All the energies are in MeV.

G.P.D.T. Ref.[24]							
L=2K	NS. NS' ND	-E[EAA]	-E[UAA]	P(S)	P(S')	P(D)	E[UAA]
4	2. 2. 1	17.915	17.539	97.37	.06	2.56	-
6	3. 3. 2	23.313	22.832	95.04	.107	4.84	5.29
8	4. 4. 3	25.801	25.284	94.45	.116	5.43	2.45
10	5. 5. 4	26.864	26.290	94.19	.16	5.64	1.00
12	6. 6. 5	27.548	26.905	94.05	.197	5.746	.615
14	7. 7. 6	27.797	27.123	93.99	.22	5.78	.218
16	8. 8. 7	27.943	27.246	93.95	.24	5.80	.122
18	9. 9. 8	28.002	27.294	93.95	.25	5.80	.048
20	10. 10. 9	28.032	27.219	93.95	.25	5.80	.025
22	11. 10. 10	28.048	27.332	93.95	.25	5.80	.013

Table 4 - Convergence of the ^{4}He ground-state binding energies in terms of the grand orbital [L=2K] for G.P.D.T. [24] interaction, without Coulomb forces, for each approximation EAA, UAA. The percentages of the S,S' and D components of the wave functions are those of the [UAA] solution. All the energies are in MeV.

S S C.C Ref.[26]								
L=2K	NS.	NS'.	ND	-E[EAA]	-E[UAA]	P(S)	P(S')	P(D)	E[UAA]
4	2.	2.	1	-	-	-	-	-	-
6	3.	3.	2	5.265	4.979	93.53	.134	6.33	
8	4.	4.	3	12.321	11.922	91.54	.029	8.42	6.94
10	5.	5.	4	16.086	15.580	91.04	.047	8.90	3.66
12	6.	6.	5	18.927	18.267	90.77	.114	9.11	2.69
14	7.	7.	6	20.195	19.452	90.67	.17	9.15	1.185
16	8.	8.	7	21.027	20.209	90.58	.23	9.18	.756
18	9.	9.	8	21.467	20.600	90.55	.276	9.167	.391
20	10.	10.	9	21.706	20.817	90.59	.26	9.14	.216
22	11.	10.	10	21.827	20.924	90.62	.26	9.12	.108

Table 5 - Convergence of the ^{4}He ground-state binding energies in terms of the grand orbital [L=2K] for SSCC [26] interaction, without Coulomb forces, for each approximation EAA, UAA. The percentages of the S,S' and D components of the wave functions are those of the [UAA] solution. All the energies are in MeV.

S S C.B Ref.[26]								
L=2K	NS.	NS'	ND	-E[EAA]	-E[UAA]	P(S)	P(S')	P(D)	E[UAA]
4	2.	2.	1	.726	.576	98.01	.23	1.75	
6	3.	3.	2	8.164	7.810	93.38	.08	6.53	7.23
8	4.	4.	3	15.703	15.219	91.89	.034	8.07	7.41
10	5.	5.	4	19.122	18.514	91.70	.086	8.21	3.29
12	6.	6.	5	21.572	20.796	91.67	.173	8.16	2.28
14	7.	7.	6	22.551	21.693	91.68	.233	8.09	.89
16	8.	8.	7	23.179	22.252	91.68	.29	8.03	.56
18	9.	9.	8	23.496	22.587	91.69	.32	7.98	.27
20	10.	10.	9	23.669	22.683	91.73	.31	7.96	.156
22	11.	10.	10	23.755	22.759	91.76	.3	7.94	.076

Table 6 - Convergence of the ^{4}He ground-state binding energies in terms of the grand orbital [L=2K] for SSSB [26] interaction, without Coulomb forces, for each adiabatic approximation EAA, UAA. The percentages of S, S' and D components of the wave functions are those of [UAA] solution. All the energies are in MeV.

DISCUSSION ON REALISTIC CALCULATIONS FOR THREE- AND FOUR-NUCLEON SYSTEMS

Peter U. Sauer
Theoretical Physics, University Hannover
3000 Hannover, Germany

1. Motivation for the Calculations

In this session various groups presented their recent calculations of the three- and four-nucleon ground states. The calculations employ realistic interactions, but use different computational techniques. The presentation was followed by a discussion on the motivation and goals for the calculations. This note tries to reflect the spirit of the discussion. Admittedly, it also reflects the bias of the discussion leader.

The calculations test our microscopic understanding of the nucleus. Since they are of profound importance for the structure of nuclear theory in general, technical reliability is indispensible. Thus, a discussion of numerical accuracy can never be avoided. Discrepancies between the results of various groups still exist for the ^{3}H binding energy, e.g., certainly in case of the Paris potential [1,2] and maybe in case of the Bonn potentials also, and they exist for the ^{3}He and ^{3}H electromagnetic (e.m.) form factors [1,3,4]. Since in future many calculations will not be done simultaneously by several groups, an honest and independent error analysis has to be attempted for each individual calculation, supplementing and replacing cross-checks between groups.

We would like to learn from these calculations what a nucleus really is. Is the description of the nucleus by simple hadronic degrees of freedom, i.e., in terms of nucleons, some well-known isobars and mesons, sufficient? Or does the compositeness of nucleons manifest itself by important multi-quark-gluon contributions already in low-energy and intermediate-energy ob-

servables? This is the relevant question which, however, none of the presented calculations has addressed itself to explicitly. The restraint is a
healthy one. Simpler questions should be answered first: We are still pre-
occupied with establishing the failure of the traditional picture of the
nucleus in terms of nucleons only and with patching up this picture by corrections. In this context three problem areas reqiure our attention. They
are discussed in turn.

2. Conceptual Problems

2.1 The Three-Nucleon Force

The three-nucleon force is a fact of life. It exists, since nucleons
are composite. One may still debate its most economical representation by
an instantaneous potential [5] or by the explicit treatment of some non-
nucleonic degrees of freedom [6] . But the phenomenon "three-nucleon force"
exists. Its size is nonnegligible, since all realistic two-nucleon potentials underbind the few-nucleon bound states - 1 MeV for ^3H and 6 MeV for
^4He are characteristic numbers. Two-nucleon potentials are theoretically
not determined at small relative distances and the phenomenology chosen
there may differ dramatically [7] . Nevertheless, the spread in computed
binding energies is much smaller than the discrepancies with the experimental values. Thus, the importance of the three-nucleon force for few-
nucleon systems is well established. (Two-nucleon potentials with an extremely low deuteron D-state probability may require a comparatively weak
three-nucleon force [1] , but these potentials are suspected to be unrealistic in other respect, e.g., for e.m. phenomena).

We have to define our expectations with respect to the three-nucleon
force. In the few-nucleon bound states, do we test the predetermined three-
nucleon force or do we adjust its largely phenomenological form to the experimental binding energies? We remember, the two-nucleon force is never
really tested, but always calibrated in the two-nucleon system. We had hoped
to do better for the three-nucleon force in the three-nucleon system: Short-
ranged two-nucleon correlations are thought to keep nucleons so effectively
apart that only the outer and understood part of the three-nucleon force [5]
matters. The recent calculations of the three-nucleon bound states [1,8]
disprove this expectation.

The approach of the Hannover group to the three-nucleon force [6] still
follows alternative one. According to Fig. 1 it constructs a force model
with explicit Δ-isobar and pion degrees of freedom. The force model is

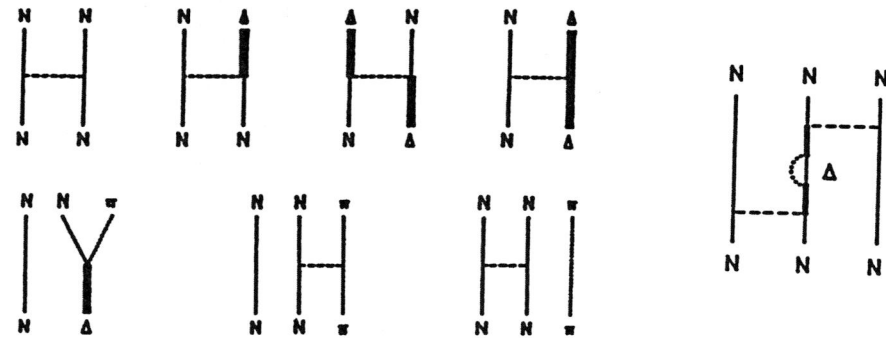

Fig. 1

Building blocks (left) of the force model with Δ-isobar and pion degrees of freedom. The force model, discussed in Ref. [9], yields a three-nucleon force (right) in the nuclear medium besides other Δ-isobar effects.

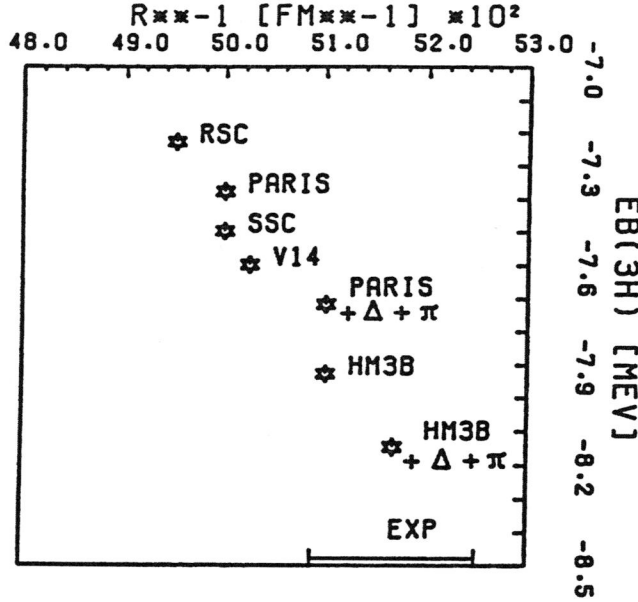

Fig. 2

Scaling property of binding energy and nuclear size. The observation of Ref. [1,8] is also made in the context of alternative one [6] for the three-nucleon force from which this figure is taken. The ³H binding energy is plotted versus the inverse of the ³He charge radius.

tested in two-nucleon scattering above pion threshold [9] and determines a contribution to the three-nucleon force, consistent with the two-nucleon potential and testable in few-nucleon systems. The approach relates physics processes conceptually far apart by the same hamiltonian, e.g., pion-deuteron scattering and the three-nucleon force. This unifying aspect is the beauty of the approach. However, only the part of the three-nucleon force due to the mechanism of pion production and absorption through the Δ-isobar is provided. Thus, the resulting three-nucleon force is necessarily incomplete. This fact is the weakness of the approach.

The other groups [1,8,10] have adopted alternative two. They adjust parameters in the three-nucleon force to account for the ³H binding energy. The ³H binding energy becomes a data point for the fit of the hadronic hamiltonian, whose real test then lies beyond the three-nucleon bound states. This procedure appears limited, but has produced some remarkable and useful results:

(a) Long-range properties of the three-nucleon bound states scale with the binding energy. Binding energy and the nuclear size, e.g., charge radii and asymptotic normalization, come simultaneously into agreement with experiment [1,8]. An example is given in Fig. 2. The Coulomb energy also scales, but does not account for the full ³He-³H binding-energy difference. The violation of charge symmetry cannot be cured by a charge-symmetric three-nucleon force.

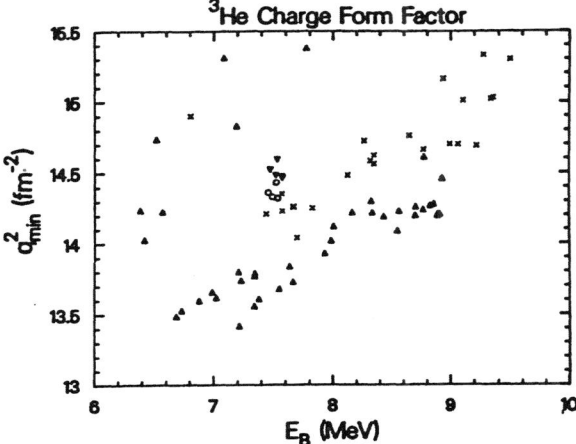

Fig. 3

Position of the first diffraction minimum of the ³He charge form factor plotted versus the corresponding ³H binding energy for different combinations of two-nucleon and three-nucleon force models. The figure is taken from Ref. [11].

(b) Short-range properties of the three-nucleon bound states do not scale with the binding energy at all. They yield a more complicated dependence on the hadronic hamiltonian. An example is given in Fig. 3.

Thus, also an adjustable three-nucleon force does not resolve all puzzles encountered in the three- and four-nucleon bound states.

With respect to the three-nucleon force we have to define our long-range goals. Is a combination of the two approaches [5,6], presently persued complementarily, a fruitful compromise?

2.2 Relativistic Corrections

Relativistic corrections are as much a fact of life as the three-nucleon force is. They are also sizeable enough to worry about them. The ^{3}He and ^{3}H charge form factors provide inescapable evidence for their importance in few-nucleon ground state properties.

In nonrelativistic approximation the charge operator is assumed to be of one-body nature. In order that the nonrelativistic one-body charge operator accounts for the experimental ^{3}He charge form factor, the ^{3}He point-proton distribution has to have a deep central depression [12]. No theoretical calculation with two- and three-nucleon forces is able to yield that. Though the three-nucleon force polarizes the point-proton distribution, compared to the result with two-nucleon forces only, as shown in Fig. 4, the achieved improvement [11,13] of the charge form factors in the region of the secondary maxima is minor. Thus, three-nucleon forces consistent with the ^{3}H binding energy yield no cure for the existing discrepancies between theory and experiment in the three-nucleon charge form factors. This fact is illustrated in Fig. 5.

The only presently known way for a somehow quantitative description of the experimental data is the use of two-body operators which are of relativistic order [1,3,4,14]. These corrections have a dramatic size in the region of the secondary maxima according to Fig. 5. Due to their size a consistent treatment of hadronic and e.m. corrections is required. Such a consistent approach is well known [15]. It has not been implemented in the present round of calculations [1,3,4], which threrefore remain unsatisfactory with respect to e.m. form factors. In this light it is encouraging to observe that — despite missing consistency and despite observation (b) of Sect. 2.1 — the used operator corrections yield rather similar results for different potential models [14]. Thus, the conceptual inadequacies of the present form factor calculations may not be too consequential after all.

Relativistic calculations are at present the domain of a specialist elite [16]. More and realistic calculations are urgently needed.

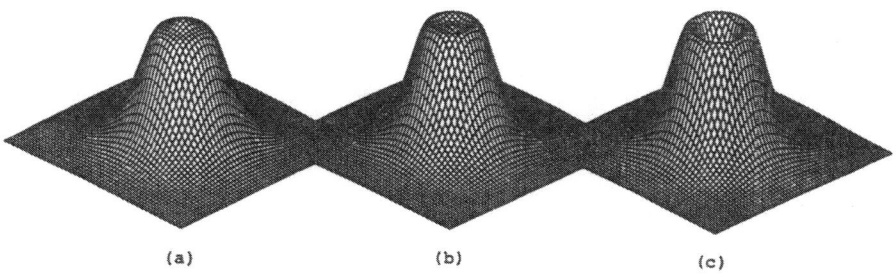

(a)   (b)   (c)

Fig. 4

³He point-proton densities. The densities without (a) and with (b) three-nucleon force are compared with the one required (c) when one-nucleon charge operators are assumed to account for the experimental data. The result is taken from Ref. [13]. It is consistent, even in detail, with the result of Ref. [11].

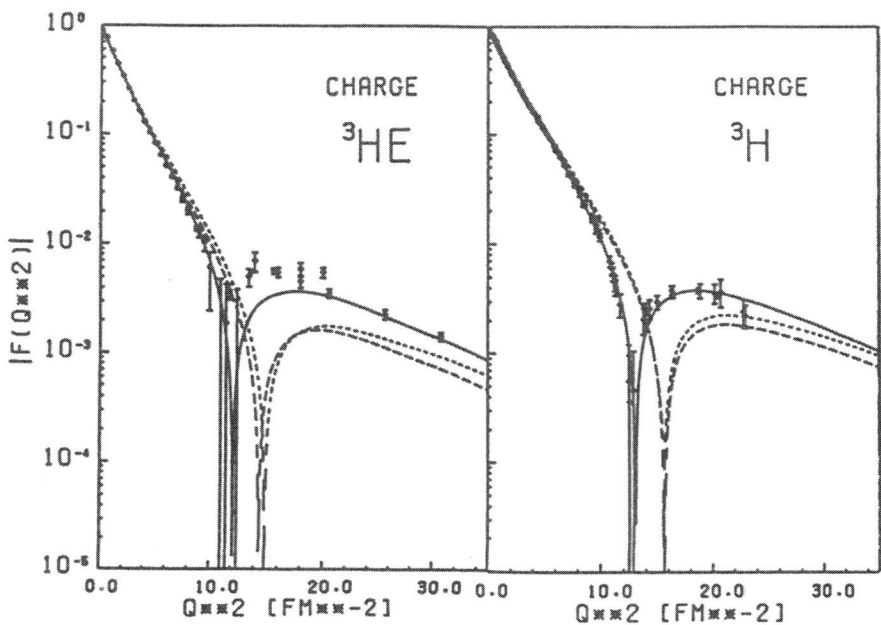

Fig. 5

³He and ³H charge form factors. The dashed (dotted) curves refer to a calculation without (with) three-nucleon force. The solid curves include the corrections of relativistic order. The results are taken from Ref. [14]. They are consistent with Ref. [11].

2.3 Corrections beyond Simple Hadronic Degrees of Freedom

One may argue that

(i)   the absence of sizeable dibaryon effects in two-nucleon scattering
      at intermediate energies [9] and

(ii) the successful description of the EMC effect in ^{4}He according to
      Fig. 6 with traditional nuclear constituents [1,17]

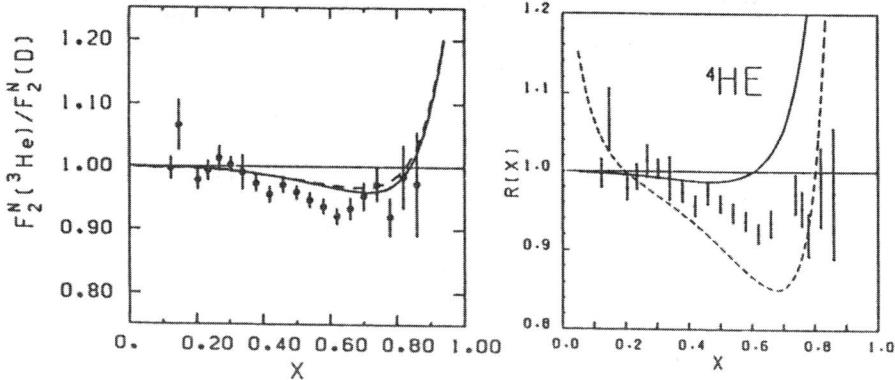

Fig. 6

Ratio of inelastic structure functions versus Bjorken scaling variable. The
isoscalar combination of the ^{3}He and ^{3}H inelastic structure functions is
compared to He data. Whereas the result (left) of Ref. [1] employs nucleonic degrees of freedom only with two- and three-nucleon forces, the result
(right) of Ref. [17] also requires Δ-isobar and meson degrees of freedom.
In the latter case the dashed curve refers to the full calculation, the
solid one to the nucleonic contribution.

makes the worry about nonhadronic degrees of freedom obsolete. Indeed the
picture of the nucleus in terms of nucleons, some well-known isobars and
mesons is so successful that it should not be abandoned entirely. However,
the size of possible corrections due to exotic multi-quark-gluon structures
has to be explored. The corrections require a treatment, consistent and complete with respect to hadronic and e.m. interactions. This is attempted in
Ref. [18,19], but lacking in the hybrid quark-hadron approach of Ref. [20].
The realistic calculations of this session are still a long way away from
going beyond simple hadronic degrees of freedom.

3. Conclusion

The calculations of three- and four-nucleon bound states deserve the
best numerical techniques available to us. They will only make the expected
impact on nuclear theory in general, if their conceptual goals are well
thought of and clearly defined.

References

1. Sasakawa, T.: The Sendai Trinucleon Calculations, Talk at this Conference

2. Hajduk, Ch., Sauer, P.U.: Nucl. Phys. A369, 321 (1981)

3. Hadjimichael, E., Bornais, R., Goulard, B.: Phys. Rev. C27, 831 (1983)
   Lina, J.M., Goulard, B.: Phys. Rev. C34, 714 (1986)

4. Strueve, W., Hajduk, Ch., Sauer, P.U., Theis, W.: ^{3}He and ^{3}H Electromagnetic Form Factors, Hannover Preprint ITP-UH 2/86

5. McKellar, B.H.J.: Springer Lecture Notes in Physics 260, 7 (1986)

6. Sauer, P.U.: Springer Lecture Notes in Physics 260, 107 (1986)

7. Sauer, P.U.: AIP Conference Proceedings 41, 195 (1978)

8. Friar, J.L.: The Los Alamos-Iowa Trinucleon Calculations, Talk at this Conference

9. Sauer, P.U.: Nucleon-Nucleon INteraction above Pion Threshold, Talk at this Conference

10. Wiringa, R.B.: Monte Carlo Three- and Four-Nucleon Bound State Calculations, Talk at this Conference
    Pandharipande, V.R.: Springer Lecture Notes in Physics 260, 59 (1986)

11. Friar, J.L., Gibson, B.F., Payne, G.L., Chen, C.R.: Los Alamos Preprint LA-UR-86-1991

12. Sick, I.: Springer Lecture Notes in Physics 86, 299 (1978)

13. Hajduk, Ch., Sauer, P.U., Strueve, W.: Nucl. Phys. A405, 581 (1983)

14. Sauer, P.U.: Electromangetic Form Factors of the Three-Nucleon Bound States, Talk Presented at Few Body XI Conference, Tokyo/Sendai 1986, Hannover Preprint ITP-UH 7/86

15. Friar, J.L.: Phys. Rev. C12, 2127 (1975)
    Gari, M., Hyuga, H.: Z. Physik A277, 291 (1976)

16. Tjon, J.A.: Relativistic Calculations in Few Body Systems, Talk at this Conference;
    Gross, F.: Relativistic Equations, Talk at this Conference

17. Oelfke, U., Sauer, P.U.: Czech. J. Phys. B36, 926 (1986) and Contribution to Few Body XI, Tokyo/Sendai 1986

18. Bakker, B.L.G.: The Nucleon-Nucleon Interaction and the Structure in Few-Nucleon Systems within the Quark Compound Bag Model, Talk at this Conference

19. Sauer, P.U., Wiese, U.-J.: Springer Lecture Notes in Physics 197, 326 (1984)

20. Kisslinger, L.S.: Springer Lecture Notes in Physics 260, 432 (1986) and References there

THE NUCLEON-DEUTERON INTERACTION

Ivo Šlaus

The Rugjer Bošković Institute, P.O.B. 1016,
41000 Zagreb, Croatia, Yugoslavia

Abstract

We discuss the present status of experimental and theoretical studies of nucleon-deuteron interaction. Predictions of three nucleon models are compared with the data and implications of discrepancies are analyzed.

1. The Present Status of the Experimental Research

The progress in scientific research technologies during the last decade has been remarkable: **i)** Beam intensities of polarized H^+ and D^+ are now in the mA and of polarized H^- and D^- in the μA region. Maximum polarized ion beam intensities have steadily increased from 1965 on with a doubling time of less than 2 years [1]. Further progress may come from the controlled fusion program, since it has been realized that significant advantages stem from the use of polarized fuel [2]. High intensity polarized neutron beams are also available. **ii)** Several polarized proton and deuteron targets are in use [1]. E.g. deuteron vector polarization of -0.5 and tensor polarization of 0.2 have been obtained in an ND_3 target at 200 mK and 3.5 . Jet targets with polarized atom densities of 10^{13} cm^{-3} have been studied. Such targets can be made of pure hydrogen, the sign of polarization can be easily reversed with kHz rate and pure vector or tensor deuteron polarization can be achieved [1].
iii) Multidetector systems have become now almost a standard setup. Several types of polarimeters (based on elastic scattering from ^{4}He, ^{12}C and ^{28}Si for proton and on the reaction ^{3}He(d,p)^{4}He for deuteron polari-

meters) have been developed which have enabled high precision measurement of spin observables [1].

Therefore, it is natural that an impressive amount of very high accuracy nucleon-deuteron (Nd) data has recently been acquired [3]. Here are some highlights: **a)** Scattering matrix M describing Nd elastic scattering is a 6x6 matrix. Symmetries reduce these 36 amplitudes to 12 independent amplitudes, i.e. a complete determination of M requires measurements of 23 independent observables. Two extensive studies of proton-deuteron (pd) scattering have been done: at 10 MeV 23 observables have been measured and 19 are independent "Fig. 1",

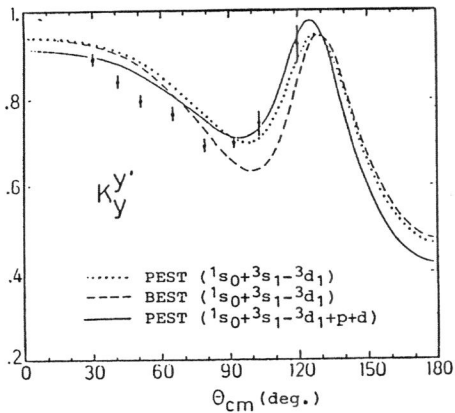

Fig. 1: The proton-proton polarization transfer coefficient data [4] and calculations with different nucleon-nucleon potentials [5]

and at 800 MeV a complete set of 24 observables has been measured [6] enabling thus for the first time to determine M from the data.

b) Significant improvements in the accuracy of spin observable measurements have been achieved. Uncertainties in analyzing power measurements in the pd elastic scattering at E_d=10 MeV are ±0.0003 [7] "Fig. 2" and the accuracy of A_y in neutron-deuteron (nd) elastic scattering are ±0.002 [8]. High precision nd data are now available up to 50 MeV [9] "Fig. 3".

c) Measurements of Nd breakup cross section and spin observables have been performed in a kinematically overdetermined mode with typical accuracies for pd of 2-4% and ±0.02 - 0.04 [10,11] and for nd ±0.04 [12]and in a kinematically underdetermined mode a typical accuracy of A_y for

162

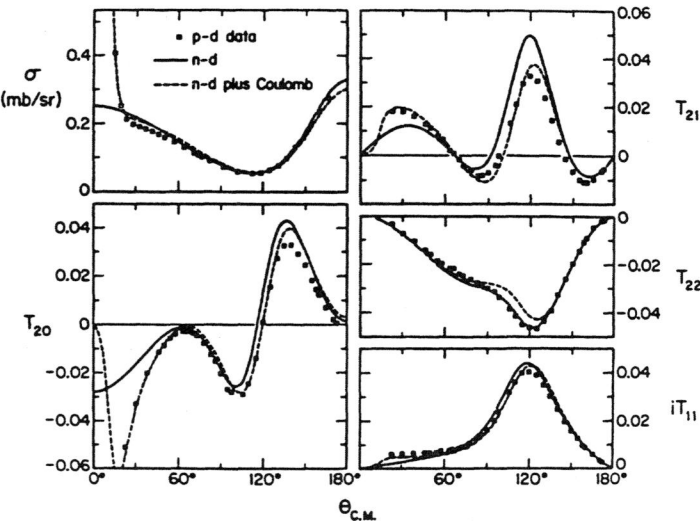

Fig. 2 : Cross section and analyzing power data at
$E_d = 10$ MeV and PEST(^1S_o, tensor)-Doleschall(PD) calculation with and without Coulomb correction [7]

$D(\vec{n}, np(^1S_o))n$ is $\pm 0.01 - 0.02$. **d)** Total cross section, angular distribution
and $A_y(\theta)$ for the nd capture have been measured [13,14] up to $E_n = 25$ MeV
with accuracies of 3-10 %, 4 % and ± 0.03, respectively and results agree
with those of the triton photodisintegration. **e)** Extensive measurements
of cross section and spin observables in the πd elastic and inelastic
processes at several energies up to 350 MeV have been recently done [15].

2. The Present Status of Theoretical Research

The Nd scattering calculations are now being done using as inputs
a variety of nucleon-nucleon (NN) forces ranging from rather simple S wave
separable to sophisticated realistic potentials: local [16,17] (de Tourreil-
Rouben-Sprung, RSC), separable [18] (Graz II, Doleschall's B and R) and
a separable representation of meson theory based potentials [19] (Paris,
Bonn). We will name reasonable representations of Paris and Bonn potentials obtained using EST method [20] PEST-N and BEST-N (N is the rank of

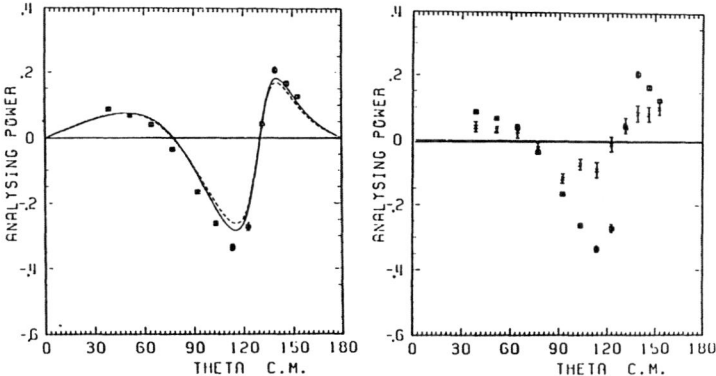

Fig. 3 : Left: elastic n⃗d A_y data at 30 MeV and calculations using Graz II (solid) and PEST 4 (dashed) potentials. Right: A_y for the reaction D(n⃗,n)d* (crosses) compared with elastic scattering (squares) [10]

the potentials). Comparison between on-shell and off-shell Paris and PEST-N shows that convergence (in the sense that there is no change in the results when rank is increased from N to N+1) has been so far achieved only for $^1S_0$ PEST-3 (PEST-N should be tested in several calculation to assure that convergence is achieved [21]). Therefore, we are still not using the best available NN potentials in the three nucleon models (3NM) and present conclusions on more exotic aspects such as the three nucleon force (3NF) and quark degrees of freedom can at best be only indications. Even for the nd system calculations assume charge symmetry and use pp forces. The calculations for the pd system are being done neglecting the Coulomb force or treating it only approximately [18]. Calculations of pd phase parameters from threshold to 50 MeV done in the framework of the method of Alt et al [22] show that they differ from nd phase parameters: the difference is largest at 10-20 MeV, but even at 50 MeV it is not negligible. Predictions of two approximate treatments of Coulomb forces [18,23] disagree with those of Ref. 22. Thus, we have no solid ground for believing the results of these approximations. On the other hand, the calculation of Ref.[22] has also resorted to approximation: the two body Coulomb amplitude has been replaced by its Born approximation. Kröger has developed a different exact approach to include

Coulomb forces and has applied it to low energy pd breakup data [24].
From Refs. 22 and 24 we conclude that Coulomb forces appreciably influence 3NM predictions at energies below 50 MeV. Also, approximate treatments of Coulomb forces modify 3NM predictions "Figs. 2 and 4". It
has been pointed out that Coulomb forces influence even the quasifree
scattering (QFS) region [25]. It is important to emphasize that so far
there is no Nd calculation which includes realistic nuclear and Coulomb
forces. Only one calculation includes 3NF but for the NN force uses
an S wave [26].

The status of Nd calculations has to be compared to that of the
bound state, where up to 34 channel calculations have been done using
several realistic NN forces and the 3NF aspects have been approached in
two ways including: a) explicitly the 3NF or b) isobar components.

The πNN system has been treated in terms of models based on baryon
and meson degrees of freedom: the relativistic three body Faddeev theory
[27] and the coupled NN-πNN theory [28]. Several questions still remain
controversial, the treatment of the real absorption of pions the P_{11}
amplitude and its separation into pole and non-pole pieces. Recently,
coupled NN-πNN theory has been extended [29] to include N and Δ on an
equal footing.

3. The Comparison of Theoretical Models with the Data

Elastic and inelastic Nd cross section data are quite well reproduced by 3NM even using only S wave potentials. Inclusion of higher partial waves globally improves the fit "Fig. 4", but in specific kinematic
conditions 3NM with less realistic forces sometime give better fits. In
addition, just like experiments, calculations too contain their numerical
uncertainties. At this stage we are no longer interested in qualitative
features and precise quantitative studies demand high accuracies of both
theory and experiment. High precision data compared to good theories
have h lped solve some long standing controversies. New nd elastic σ(θ)
data and calculations with PEST-4 S waves are in excellent agreement at
backward angles up to 50 MeV and there is little sensitivity to off shell
and higher partial waves [9]. In general, 3NM with realistic NN forces
and approximately including Coulomb forces give a remarkably good fit to
the pd data. Again, however, high quality data demonstrate discrepancies:
10 and 14.1 MeV σ(θ) and A_y(θ) elastic pd data [31] disagree with predictions of such calculations.

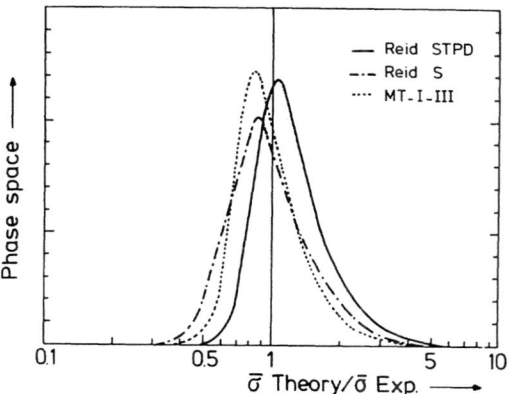

Fig. 4 : Discrepancy spectrum for three 3NM calculations with different NN forces compared with D(p,2p)n 50 MeV data [30]

The relativistic three body Faddeev theory provides a good fit to πd elastic and breakup processes [27]. A significant exception is a backward angle σ(θ). Present Faddeev calculations do not include NΔ interactions. A phenomenological ΔN ⟶ ΔN interaction is added in the intermediate state of πd elastic scattering to the Faddeev amplitude and its parameters are varied to fit σ(θ) and iT_{11}(θ) data at 142, 180 and 256 MeV. Good fit is obtained for all data except for the backward angle σ(θ) at 256 MeV which shows how strongly limited is the choice of ΔN interaction parameters [32].

3.1. Three Nucleon Force Effects in Nd Interactions

The 3NF has a complex spin-isospin structure and the triton binding energy and ^{3}He charge form factor suggest that 3NF might be attractive in isosceles and repulsive in collinear configurations [33]. Nd interaction, particularly the breakup process offers a rich field of different observables which provide information on 3NF effects. Since these effects in general amount to only few percent, it is necessary to choose observables where the dominant ordinary physics is somehow suppressed and where both experiment and calculation can be performed with a high degree of accuracy [34]. At this stage nd studies are favoured though one should keep in mind the possibility that charge symmetry violations may be present.

Evidence for 3NF effects have been recently discussed [35]. We will concentrate here only on Nd observables: **i)** Comparison between PEST predictions and elastic scattering A_y data at 10-14 MeV suggests a discrepancy around 120°. That can be caused by 3NF since A_y (120°) is very sensitive to ^3P_i which couple to $N\Delta$ waves (compare triton energy calculations). Similarly, it has been observed at higher energies that the fit becomes worse with the better potential (off energy properties of Graz II are inferior to those of PEST as judged by electron-deuteron t_{20} measurements [36], and Graz II gives a better fit to A_y, "Fig. 3"). It is necessary to investigate the sensitivity of A_y for the reaction $D(\vec{n},n')$ $d_0^*(^1S_0)$ on 3NF effects since high accuracy data are now available. At present there are no other nd spin observable data and it would be necessary to contemplate polarized deuteron target experiments. An interesting insight can be provided [37] by studying $D(\vec{d},pd)n$ and $D(\vec{d},nd)p$ $N\vec{d}$QFS, since successful four body calculations have now been [38] done. **ii)** Collinear and star configurations have been studied in the $D(p,2p)n$ reaction at several energies [10,30,39], but there is no reliable calculations with the Coulomb force "Fig. 5". At present one can only conclude that spin observables do not reveal any appreciable signature of 3NF effects. Three $D(n,np)n$ measurements investigating these configurations are in progress: a 20 detector system (4 out of plane) to measure cross sections at these and other configurations in a kinematically overdetermined mode at 13 MeV [40], a similar study at 10.3 MeV [41], and a kinematically underdetermined measurement ($\vec{p}_n$ and E_p are measured) of A_y [12]. **iii)** Cross section data of the reaction H(d,2p)n at 0.275 MeV c.m. energy above threshold have been used to place an upper limit of 30% on 3NF effects [42]. **iv)** The NN final state interaction (FSI) cross section is influenced by 3NF [43] and consequently the values for scattering lengths extracted from Nd breakup experiments are in general modified by the 3NF. Two different calculations show that 3NF affects NN FSI enhancement differently for knockon and pickup $D(n,p)nn$ processes [26,43]: due to the 3NF scattering length a_{nn} extracted from a knockon process is more negative and from a pickup process is less negative than a_{nn} for free nn scattering. Since a_{nn} is a magnifying glass of the force, the 3NF effect is relatively large 1-2 fm and its value is strongly dependent

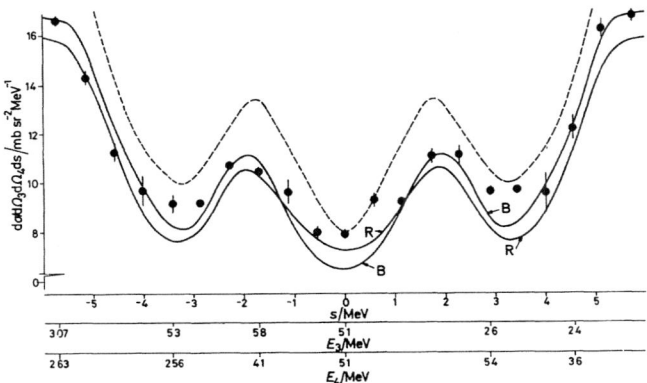

Fig. 5 : Cross section for the reaction H($\vec{d}$,2p)n around col-
linearity (s=0) at $\theta_3 = \theta_4 = 33.4°$, $E_d = 16$ MeV compared with calculations using Doleschall R and B potentials with (dashed)
and without Coulomb corrections

on the NN and 3NF used. The value of a_{np} extracted from Nd breakup should
be also influenced by 3NF. There are several measurements of angular distributions of np FSI, [44] but only one [45] provides explicit values of
extracted a_{np}. Their results seem to confirm those found for a_{nn} [43].
The calculation of the 3NF effect on the extracted a_{nn} shows that it decreases for knockon process as incident energy increases. A recent
study of the knockon process at 63 MeV confirms this result and it adds
an additional value $a_{nn} = -18.8 \pm 1.0$ fm [46] which together with that
from the reaction D(π^-,γ)2n yields $a_{nn} = -18.51 \pm 0.42$ fm. The difference
between a_{nn} and a_{pp} is in agreement with the calculation [47] : 1.2 fm
and with the ^{3}H - ^{3}He binding energy difference. It should be emphasized
that this provides a measure of class III charge symmetry breaking forces
while np analyzing powers measure class IV forces [48]. Considerations of
3NF implies the need to include other small effects, e.g. magnetic interaction in the nnp system [49]. **v)** Study of the radiative capture
D(n,γ)T provides a sensitive test of mixed symmetry S' wave function, of
meson exchange current (MEC) contributions and of 3NF because the

contribution due to the dominant S wave vanishes. The contribution of MEC is essential and the 3NF brings theoretical predictions in better agreement with data [50]. It is not clear why a non-relativistic E1 + E2 calculation fails to fit the measured asymmetry in the D(n,γ)T angular distribution [13]. **vi)** Certain regions of the three nucleon phase space in the reaction ^{3}He(γ,pp)n will be enhanced by a 3NF. A calculation with NN forces predicts that a yield through a quasi-diproton process is 100 times smaller than through a quasideuteron. Measurement of the ratio of portions in the Dalitz plot that correspond to these processes provides informations on 3NF [51].

The Nd observables are mainly determined by NN forces and thus, evidence for 3NF effects critically depends on NN phase parameters. Unfortunately, they are not always adequately known. Some Nd observables are more sensitive to specific aspects of the NN force than NN observables and this makes the three nucleon system an important laboratory for studying NN forces.

3.2. Possible Evidence for Quark Degrees of Freedom

The backward Nd elastic scattering has been since 1953 associated with one nucleon exchange mechanism [52]. High energy data led to models which assumed exchanges of one (N* 1688) or several isobars, triangle diagrams and recently contribution of tribaryon resonances which can be derived in the model of stretched rotating quark bags [53]. Excitation functions of $\sigma(180^\circ)$ and $T_{20}(180^\circ)$ are still not well understood [54].

The deuteron is dominantly an np system. Quark-quark interactions generate admixtures of the color-polarized states and effects which are beyond the πNΔ description. A study of spectator momentum distribution in Nd breakup induced by 3.3 GeV/c deuterons show that impulse approximation, double scattering and FSI do not fit the data, while the inclusion of model six-quark corrections "Fig. 6" provides a better fit [55].

Acknowledgemnt

I express my thanks to Drs. R. van Dantzig, H. Klages, and G. Rauprich for sending me their results.

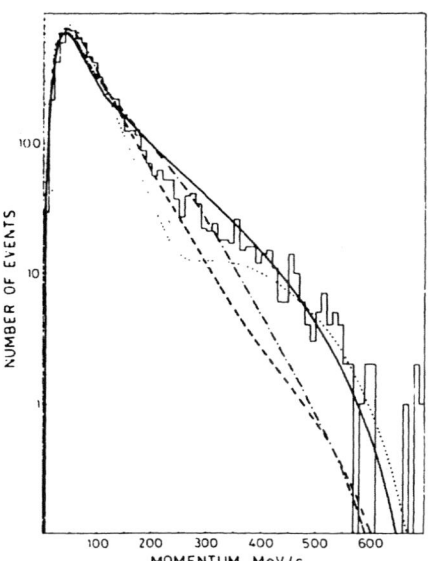

Fig. 6 : Momentum distribution of
neutron spectators for |t|<0.4 (GeV/c)².
Curves: impulse approximations (IA)
using Paris potential (dashed), IA+FSI
(dashed-dotted), IA+ 6 quark corrections
(dotted), IA+FSI+double scattering + 6
quark corrections (solid) [55].

References

1. Grüebler, W.: Suppl. J. Phys. Soc. Japan 55, 435 (1986); ibid
   Clegg, T. B.: p. 535.

2. Knize, R. J.: Suppl. J. Phys. Soc. Japan 55, 412 (1986).

3. Tilley, D. R., Hasan, H. H., Weller, H. R.: to be published (1986).

4. Sperisen, F., Grüebler, W., König, V., Schmelzbach, P. A., Elsner,
   K., Jenny, B., Schweizer, C., Ulbricht, J.: Nucl. Phys. A422, 81(1984).

5. Koike, Y.: Suppl. J. Phys. Soc. Japan 55, 272 (1986).

6. Sperisen, F.: Suppl. J. Phys. Soc. Japan 55, 852 (1986); Aas, B. et
   al, ibid, p. 854.

7. Intl. Symp. Three Body Force (Washington 1986), contributed papers; Sowinski, J., Pun Casavant, D. D., Knutson, L. D.: to be published.

8. Tornow, W., Byrd, R. C., Howell, C. R., Pedroni, R. S., Walter, R. L.: Phys. Rev. C27, 2439 (1983).

9. Klages, H.: Few Body XI, Sendai (1986).

10. Brown, R.E., et al: to be published.

11. Karus, M., Buballa, M., Helten, J., Oswald, H., Rauprich, G., Paetz gen Schiek, H.: Phys. Rev. C31, 1112 (1985).

12. Howell, C. R., et al : to be published.

13. Mitev, G., Colby, P., Roberson, N. R., Weller, H. R., Tilley, D. R.: Phys. Rev. C34, 389 (1986).

14. Mösner, J., Möller, K., Pilz, W., Schmidt, G., Stiehler, T.: Few Body Systems 1, 83 (1986).

15. Gyles, W., Boschitz, E. T., Garcilazo, H., List, W., Mathie, E. L., Ottermann, C. R., Smith, G. R., Tacik, R., Johnson, R. R.: Phys. Rev. C33, 583 and 595 (1986), and references therein.

16. Takemiya, T.: Prog. Theor. Phys. 74, 301 (1985).

17. Stolk, C., Tjon, J. A.: Nucl. Phys. A319, 1 (1979).

18. Doleschall, P., Grüebler, W., König, V., Schulzbach, P. A., Sperisen, F., Jenny, B.: Nucl. Phys. A380, 72 (1982).

19. Koike, Y., Plesas, W., Zankel, H.: Phys. Rev. C32, 1796 (1985); Koike, Y., Taniguchi, T.: Few Body Systems 1, 13 (1986).

20. Ernst, D. J., Shakin, C. M., Thaler, R. M.: Phys. Rev. C8, 46 (1973).

21. Lampl, H., Weigel, M. K.: Phys. Rev. C33, 1834 (1986).

22. Alt, E. O., Sandhas, W., Ziegelmann, H.: Nucl. Phys. A445, 429 (1985).

23. Haftel, M. I., Zankel, H.: Phys. Rev. C24, 1322 (1981).

24. Kröger, H., Nachabe, A. M., Slobodrian, R. J.: Phys. Rev. C33, 1208 (1986) and references therein.

25. Kok, L. P., van Haeringen, H.: Phys. Rev. Lett. 46, 1257 (1981); Bajzer, Ž.: to be published.

26. Meier, W., Glöckle, W.: Phys. Lett. 138B, 329 (1984).

27. Garcilazo, H.: Phys. Rev. Lett. 53, 652 (1984), preprint (1986).

28. Locker, M. P., Sainio, M. E., Švarc, A.: Advances in Nucl. Phys. (to be published).

29. Afnan, I. R., Blankleider, B.: Phys. Rev. C32, 2006 (1985).

30. Blommestijn, G. J. F., Van Dantzig, R., Haitsma, Y., Mooy, R. B. M., Šlaus, I.: Nucl. Phys. A365, 202 (1981), and to be published.

31. Rauprich, G., Buballa, M., Hähn, H.J., Karus, M., Koike, Y., Laumann, B., Niessen, P., Paetz gen. Schieck, H.: Suppl. J. Phys. Soc. Japan 55, 858 (1986).

32. Andrade, S. C. B., Ferreira, E., Dosch, H. G.: Phys. Rev. $\underline{C34}$, 226 (1986).

33. Fabre de la Ripelle, M.: C. R. Acad. Sci. $\underline{B288}$, 325 (1979).

34. Šlaus, I.: Intl. Symp. Three Body Force (Washington 1986), Contributed papers.

35. Šlaus, I.:Few Body Methods and Their Application. World Sci. Publ., to be published.

36. Schultze, M. E., Beck, D., Farkhondeh, M., Gilard, S., Goloskie, R., Holt, R. J., Kowalski, S., Laszewski, R. M., Leitch, M. J., Moses, J. D., Redwine, R. P., Saylor, D. P., Specht, J. R., Stephenson, E. J., Stephenson, K., Turchinetz, W., Zeideman, B.: Phys. Rev. Lett. $\underline{52}$, 597 (1984).

37. Howell, C. R.: Priv. Com.

38. Mdlalose, T. E., Fiedeldey, H., Sandhas, W.: to be published.

39. Barinack, B., Börker, G., Kamke, D., Lekkas, P., Stephan, M. : Intl. Symp. Three Body Force (Washington 1986), contributed papers.

40. Böttcher, J., et al. (Tornow, W.: Priv Com.).

41. Bodek, K., Kamke, D., Krug, J., Obermanns, S., Stephan, M.: Intl. Symp. Three Body Force (Washington, 1986), contributed papers.

42. Slobodrian, R. J.: Few Body Problems in Physics. (Eds.: Faddeev, L.D., Kopaleishvili, T.I.), World Sci. Publ., 1985, p. 304.

43. Šlaus, I., Akaishi, Y., Tanaka, H.: Phys. Rev. Lett. $\underline{48}$, 1993 (1982) and to be published.

44. Doornbos, J., Krijgsman, W., van der Wey, R., Jonker, C. C.: Nucl. Phys. $\underline{A297}$, 412 (1978), and references therein.

45. Brückmann, H., et al: Nucl. Phys. $\underline{A157}$, 209 (1970).

46. Koori, N.: Priv. Com.

47. Coon, S. A., Scadron, M. D.: Phys. Rev. $\underline{C26}$, 562 and 2402 (1982).

48. Abegg, R., et al: Phys. Rev. Lett. $\underline{56}$, 2571 (1986).

49. Slobodrian, R. J.: Phys. Lett. $\underline{135B}$, 17 (1984).

50. Torre, J., Goulard, B.: Phys. Rev. $\underline{C28}$, 529 (1983).

51. O'Connell, J. S.: Intl. Symp. Three Body Force (Washington, 1986), Contributed papers.

52. Christian, R. S., Gammel, J. L.: Phys. Rev. $\underline{91}$, 100 (1953).

53. Kondratyuk, L.A., Lev, F. M., Schevehenko, L. V.: Phys. Lett. $\underline{100B}$, 448 (1981).

54. Arvieux, J.: J. Phys. Soc. Japan $\underline{55}$, 304 (1986).

55. Deloff, A., Siemiarczuk, T.: Nucl. Phys. $\underline{A449}$, 603 (1986).

PREDICTIONS OF MESON-THEORETICAL NUCLEON-NUCLEON INTERACTIONS
FOR NUCLEON-DEUTERON SCATTERING OBSERVABLES

W. Plessas[*]
Institute for Nuclear Physics
KFA Jülich
D-5170 Jülich, FRG

J. Haidenbauer and Y. Koike
Research Center for Nuclear Physics
Osaka University
Ibaraki, Osaka 567, Japan

Abstract

We report on our investigations of nucleon-deuteron scattering with meson-theoretical nucleon-nucleon interactions. In this paper we consider the Paris and Bonn potentials which are introduced into momentum-space Faddeev calculations via separable expansions. Results are presented for neutron-deuteron integrated and differential cross sections below $E_n = 20$ MeV and for spin-transfer coefficients of the reactions $^2H(\vec{N},\vec{N})^2H$ as well as $^2H(\vec{N},\vec{d})N$ at $E_N = 10$ MeV.

1. Introduction

At present meson-theoretical potentials provide for the best description of the nucleon-nucleon (N-N) interaction at non-relativistic energies [1]. While they have already been employed in calculations of the trinucleon bound states [2-5], there was in the past no feasible way of using them also in 3-N scattering calculations. Only recently, by the advent of

*   On leave from Institute for Theoretical Physics, University of Graz,
    Universitätsplatz 5, A-8010 Graz, Austria

refined separable expansions, has it become possible to introduce such models as the Paris [6] or Bonn [7] potentials in a reliable manner into Faddeev calculations of 3-N scattering. Thereby these forces can be subject to an additional test in the 3-N system. This is important for judging on their off-shell behaviour [8-10] which is not constrained from considering the N-N system alone.

With respect to the trinucleon bound states it has become clear that 2-N meson-exchange forces alone cannot explain neither the binding energies nor the electromagnetic form factors [2,3,11]. There is still the possibility that the inclusion of an appropriate 3-N force can remedy the situation [12]. For 3-N scattering no such evidences exist as yet. Due to the lack of investigations with realistic N-N forces no places with manifest effects from 3-N forces have been located so far and no calculations with 3-N forces have been performed in the scattering domain except at 3-N threshold [13-15]. It is therefore of great interest to see, how genuine meson-exchange forces perform in 3-N scattering and, in more general terms, whether or not meson-exchange dynamics can account for the whole 2-N and 3-N phenomenology at low and moderate energies.

We have pursued the aim of making any 2-N interaction of arbitrary form amenable to a treatment in momentum-space Faddeev calculations of the 3-N system. In the course of our investigations it turned out that a reliable method of introducing all relevant features of a given N-N model into such calculations consisted in the technique of separable expansions. In particular a method suggested by Ernst, Shakin, and Thaler (EST) [16] lent itself to great use, since it allowed to approximate both the on-shell and off-shell behaviour of the underlying interaction by a separable form. Still the resulting separable potentials are not too complicated as to be applicable in present-day computer codes for the 3-N (scattering) system.

2. Separable representations of the Bonn and Paris potentials

We have so far constructed separable representations of the Paris potential [17] and of a one-boson-exchange version of the Bonn potential derived in the framework of the Blankenbecler-Sugar equation [18]. In both cases we employed the EST method. The so-called PEST and BEST parametrizations are presented in refs. [19] and [20], where also a detailed discussion is given of their properties in the 2-N system. In the present report we are also going to employ more recent separable expansions of the Paris potential which were considered in ref. [21]. They are characterized by a new analytical representation of the (originally numerical) EST form factors, namely via Gegenbauer polynomials instead of rational functions.

The EST method uses the wave functions or alternatively half-off-shell transition matrices of the given interaction V to construct the form factors of the separable expansion $\tilde{V}$. Thus it allows $\tilde{V}$ to reproduce the on-shell as well as half-off-shell properties of V at certain fixed energies of the 2-N system. If N energy points, preferably from the spectrum of V, are chosen, one ends up with a rank-N expansion (in a certain two-body partial-wave state). In practice it turns out that for the N-N potentials mentioned above a rank up to 3 is required in each partial wave (i.e., up to 6 in coupled partial-wave states) so as to achieve convergence at the levels of the 2-N as well as 3-N systems [21].

3. Results for elastic N-d scattering

3.1. Comparison of the Paris and Bonn potentials by means of elastic n-d differential cross sections

As an example, how two different meson-exchange models of the N-N interaction compare for 3-N scattering, we consider the elastic neutron-deuteron (n-d) differential cross section between E_n = 8 and 12 MeV (Fig. 1). For this purpose we employ for both cases, the Paris and Bonn potentials, separable approximations of rank 3 in $^1S_0$ and of rank 4 in $^3S_1$-$^3D_1$ [19,20]. The effect of higher N-N partial waves is here taken into account via phenomenological separable potentials developed by Doleschall [22]. This makes sense, because the n-d differential cross section at these low energies is predominantly governed by the low N-N partial waves, especially by the $^3S_1$-$^3D_1$ state [23]. It is also only this latter state, where some obvious differences between the Paris and Bonn potentials occur [24].

In general, a satisfactory reproduction of the angular dependence of the n-d cross section is achieved in both cases. The BEST results are somewhat higher than the ones of PEST at forward and backward angles, while they are lower at the cross-section minimum. This is likely to be caused by the different potential behaviour in $^3S_1$-$^3D_1$. However, at present it is hard to tell, whether it is an on-shell or off-shell effect. Further investigations are needed to disentangle effects from different proper-

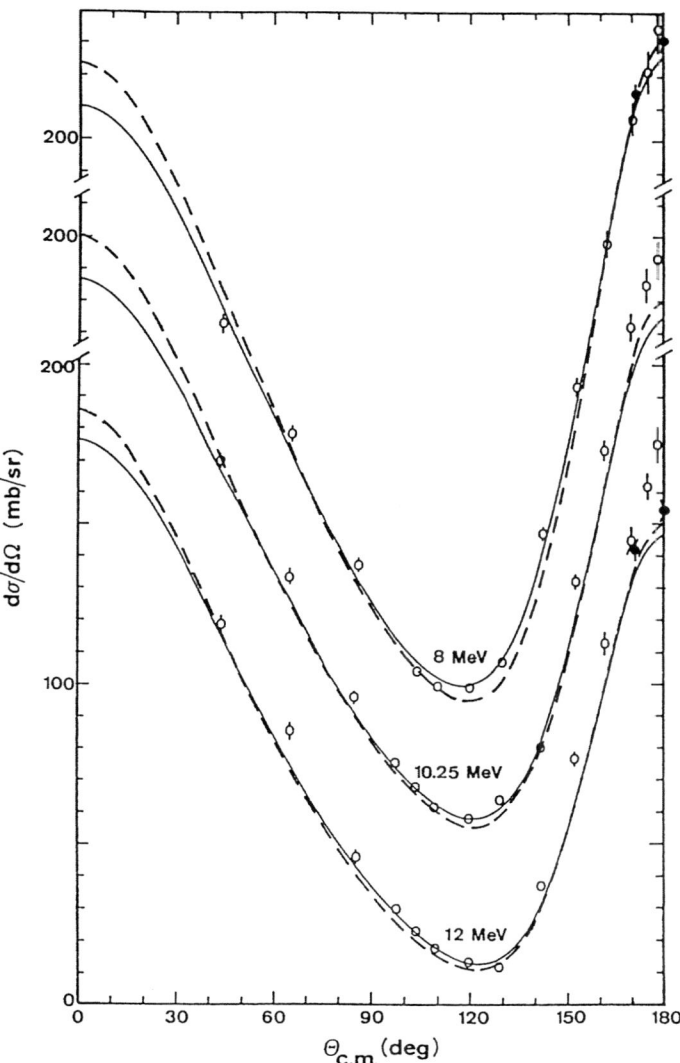

Fig. 1. Differential cross section for n-d elastic scattering at E_n = 8, 10.25, and 12 MeV with PEST3($^1S_0$)+PEST4($^3S_1$-$^3D_1$)+Doleschall(P+D) (solid lines) and BEST3($^1S_0$)+BEST4($^3S_1$-$^3D_1$)+Doleschall(P+D) (dashed lines). Open circles are experimental data of the Karlsruhe group [25] and full circles of the Uppsala group [26].

ties such as low-energy parameters, D-state probability, asymptotic wave-function normalization etc.

3.2. Paris-potential predictions for low-energy N-d observables

For the case of the Paris potential we already succeeded in applying the presently most refined separable representations (of higher rank) [19,21] in all N-N partial waves. With regard to $^1S_0$ and $^3S_1$-$^3D_1$ detailed convergence tests at the 3-N level have shown that corresponding rank-3 and rank-6 expansions are adequate and also sufficient for the present N-d results [21]. The same can be expected from the rank-2 approximations in the higher uncoupled partial waves as well as the rank-3 and rank-4 approximations in $^3P_2$-$^3F_2$ and $^3D_3$-$^3G_3$, respectively. This is especially true for the n-d cross sections and the spin-transfer coefficients shown below, since they receive only a small influence from higher N-N partial waves.

Here we give only results for n-d scattering, i.e. without considering Coulomb effects, because we cannot yet treat them in a reliable manner and the situation is still quite controversial in this respect [27]. For the n-d cross sections we can compare to corresponding experimental data.

Table 1. Paris-potential predictions for the n-d total cross section as calculated with the separable representations PEST3-G($^1S_0$)+PEST6-G($^3S_1$-$^3D_1$)+PEST2($^1P_1$,$^3P_0$,$^3P_1$,$^1D_2$,$^3D_2$)+PEST3($^3P_2$-$^3F_2$)+PEST4($^3D_3$-$^3G_3$) from refs. [19,21] (see also ref. [29]) in comparison to experimental data.

E_n (MeV)	σ_{calc} (mb)	σ_{exp} (mb)	
		ref. [30]	ref. [25]
6	1469	1478±16[*]	1471±20
7	1330	1296±10[*]	1337±10
8	1225	1207±13	1224±10
9	1129	1118±10[*]	1128±10
10	1047	1055±10	
10.25	1028		1038±10
12	912	913±13	923±10
14	807		824±10
18	650		666±7
20	587	584±10[*]	603±6

[*] These experiments are at slightly different energies $\Delta E_n \approx \pm 0.1$ MeV.

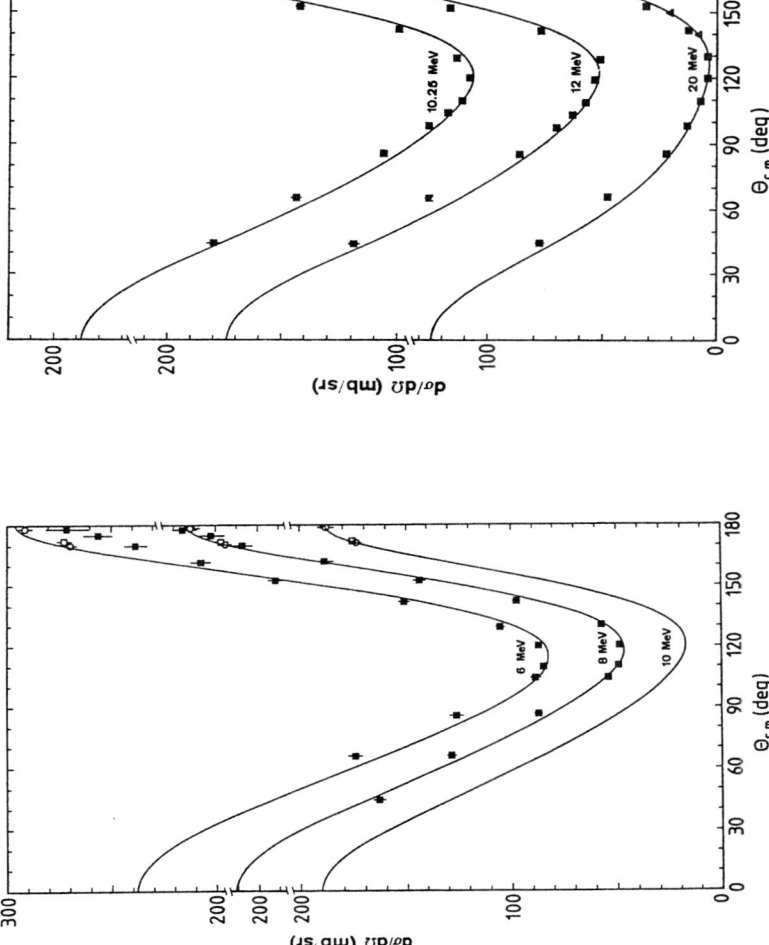

Fig. 2. Paris-potential predictions for the n-d differential cross section as calculated with the separable representations specified in Table 1. Open circles and squares are Uppsala data [26], full squares are Karlsruhe data [25], and triangles recent measurements of Hofmann [31].

178

For the spin transfers we can make the comparison only to p-d data, what may be justified, however, since in these cases Coulomb effects are expected to be small [8,28].

The predictions we obtain for the Paris potential from our presently most advanced Faddeev calculations [29] are given in Table 1 and Figs. 2-4. A remarkably good agreement is achieved in all cases. We can take this as a first indication that the behaviour of the Paris potential might be quite reasonable especially in the $^1S_0$ and $^3S_1$-$^3D_1$ partial waves both on-shell and off-shell. Above all the description of the backward differential cross section, which has been a long-standing problem for purely phenomenological forces [23], now seems to be brought about by the meson-exchange models (Figs. 1 and 2). Also the spin observables in Figs. 3 and 4

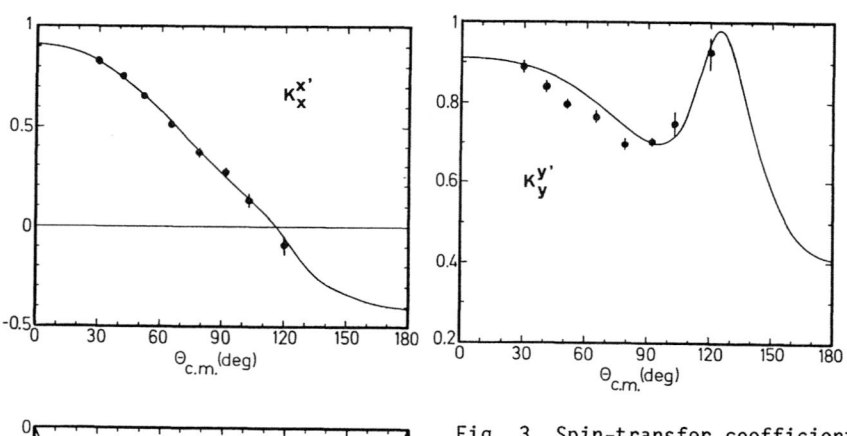

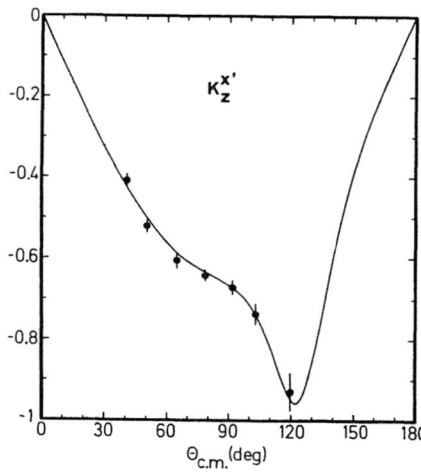

Fig. 3. Spin-transfer coefficients of the reaction $^2H(\vec{p},\vec{p})^2H$ at $E_p =$ = 10 MeV for the Paris potential as calculated with the separable representations specified in Table 1 without Coulomb corrections. Experimental data are from ref. [28].

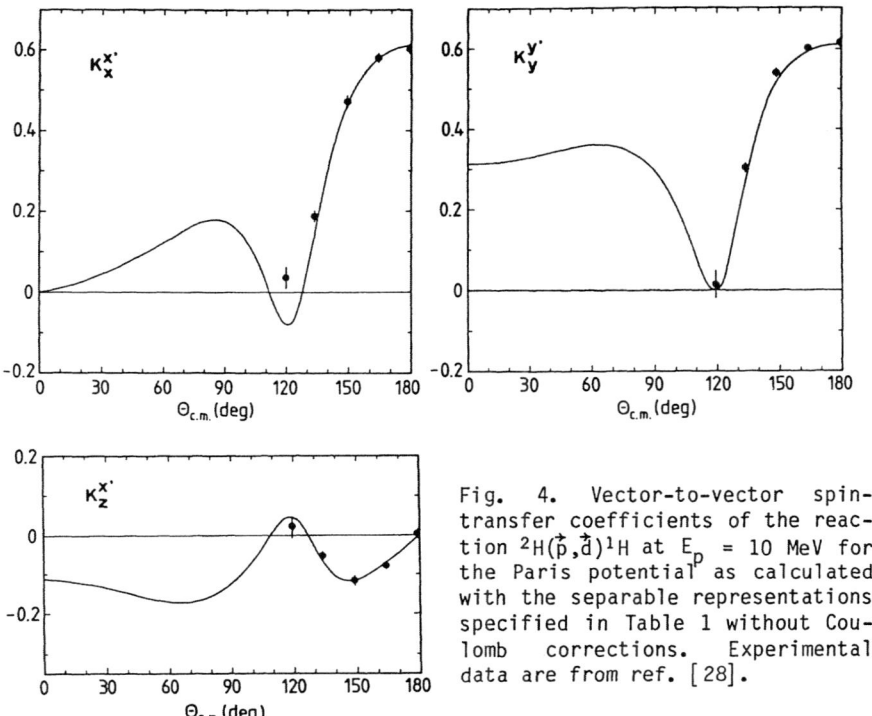

Fig. 4. Vector-to-vector spin-transfer coefficients of the reaction $^2H(\vec{p},\vec{d})^1H$ at E_p = 10 MeV for the Paris potential as calculated with the separable representations specified in Table 1 without Coulomb corrections. Experimental data are from ref. [28].

provide stringent tests for specific details of the N-N interaction [8,28]. It was found before that they rule out unrealistic off-shell behaviours of the N-N interaction [9]. The present agreement between experiment and theory provides evidence that meson-exchange dynamics constitutes a reliable concept for the N-N interaction in the domain probed by these reactions.

In view of the results achieved so far one must not forget, however, that several other questions remain to be answered. For instance, we have not yet dealt with the problem of deuteron vector and tensor polarizations, since there we have the tricky complications with Coulomb effects and more vigorous effects from higher N-N partial waves. In these respects further substantial efforts are required in the solution of the 3-N equations in order to derive reliable results.

Acknowledgment

This work was supported by Fonds zur Förderung der Wissenschaftlichen Forschung in Österreich, project 5733, and the Japanese Ministry of Education via a Monbusho scholarship.

References

1. See, e.g., de Swart, J.J.: these proceedings.

2. Hajduk, C. and Sauer, P.U.: Nucl. Phys. A369, 321 (1981);
   Strueve, W., Hajduk, C., and Sauer, P.U.: ibid. A405, 620 (1983).

3. Hadjimichael, E., Bornais, R., and Goulard, B.: Phys. Rev. Lett. 48, 583 (1982);
   Hadjimichael, E., Goulard, B., and Bornais, R.: Phys. Rev. C27, 831 (1983).

4. Ishikawa, S. et al.: Phys. Rev. Lett. 53, 1877 (1984).

5. Bömelburg, A.: Phys. Rev. C34, 14 (1986).

6. Cottingham, W.N. et al.: Phys. Rev. D8, 800 (1973);
   Vinh Mau, R.: in: Mesons and Nuclei I (Rho, M. and Wilkinson, D., eds.), p. 151. Amsterdam-New York-Oxford: North-Holland 1979.

7. Machleidt, R.: in: Quarks and Nuclear Structure (Bleuler, K., ed.), p. 352. Heidelberg-Berlin: Springer 1984;
   Machleidt, R., Holinde, K., and Elster, C.: Phys. Rep., to appear.

8. Zankel, H., Plessas, W., and Haidenbauer, J.: Phys. Rev. C28, 538 (1983);
   Zankel, H. and Plessas W.: Z. Phys. A317, 45 (1984).

9. Loiseau, B. et al.: Phys. Rev. C32, 2165 (1985).

10. Haidenbauer, J. et al.: Proceedings of the Workshop on Few-Body Approaches to Nuclear Reactions, Tokyo 1986 (Sawada, T. et al., eds.). Singapore: World Scientific, to appear.

11. Sasakawa, T. and Ishikawa, S.: Few-Body Systems 1, 3 (1986);
    Ishikawa, S. and Sasakawa, T.: ibid. 143 (1986).

12. See, e.g., the corresponding review resp. rapporteur talks at the 11th Int. Conf. on Few-Body Systems in Particle and Nuclear Physics, Tokyo-Sendai 1986. Nucl. Phys. A, to appear.

13. Torre, J., Benayoun, J.J., and Chauvin, J.: Z. Phys. A300, 319 (1981).

14. Delfino, A. and Glöckle, W.: Phys. Rev. C30, 376 (1984).

15. Chen, C.R. et al.: Phys. Rev. C33, 401 (1986).

16. Ernst, D.J., Shakin, C.M., and Thaler, R.M.: Phys. Rev. C8, 46 (1973).

17. Lacombe, M. et al.: Phys. Rev. C21, 861 (1980).

18. Machleidt, R. and Holinde, K.: in: Few-Body Problems in Physics (Zeitnitz, B., ed.), p. 79. Amsterdam-New York-Oxford: North-Holland 1984.

19. Haidenbauer, J. and Plessas, W.: Phys. Rev. C30, 1822 (1984); ibid. C32, 1424 (1985).

20. Haidenbauer, J., Koike, Y., and Plessas, W.: Phys. Rev. C33, 439 (1986).

21. Haidenbauer, J. and Koike, Y.: Phys. Rev. C34, 1187 (1986).

22. Correll, F.D. et al.: Phys. Rev. C23, 960 (1981).

23. Koike, Y. and Taniguchi, Y.: Few-Body Systems 1, 13 (1986).

24. Plessas, W., Schwarz, K., and Mathelitsch, L.: in: Perspectives in Nuclear Physics at Intermediate Energies (Boffi, S. et al., eds.), p. 53. Singapore: World Scientific 1984.

25. Schwarz, P. et al.: Nucl. Phys. A398, 1 (1983).

26. Janson, G. et al.: in: Few-Body Problems in Physics (Zeitnitz, B., ed.), p. 529. Amsterdam-New York-Oxford: North Holland 1984; Janson, G.: private communication.

27. Kok, L.P.: in: Few-Body Problems in Physics (Faddeev, L.D. and Kopaleishvili, T.I., eds.), p. 252. Singapore: World Scientific 1985.

28. Sperisen, F. et al.: Phys. Lett. 102B, 9 (1981); ibid. 110B, 103 (1982).

29. Koike, Y., Haidenbauer, J., and Plessas, W.: preprint 1986.

30. Davies, J.C. and Barschall, J.J.: Phys. Rev. C3, 1798 (1971).

31. Hofmann, K.: Thesis (Univ. Karlsruhe), 1985.

COULOMB MODIFIED FADDEEV CALCULATIONS OF LOW-ENERGY P-D OBSERVABLES *

G.H. Berthold, A. Stadler and H. Zankel

Institut für Theoretische Physik, Universität Graz

A-8010 Graz, Austria

Abstract

A separable representation in the S-waves of the Paris potential was used in an exact momentum space p-d calculation of differential cross sections and some second order polarization observables below breakup threshold. The influence of the three-body doublet S-state was studied in these observables and significant sensitivity was found mainly in the spin correlation parameter $C_{y'y}$. The lack of experimental information in this observable prevents a clarification of the inconsistency between model and phase shift analysis doublet S-state effective range function at this stage. An important step to establish exact model results of first order p-d polarizations that are experimentally available has now been made by calculating ^{3}He binding energies where the S- and P-waves and the $^3S_1 - ^3D_1$ state of the Paris-like potential have been taken into account. Our results compare favourably to results obtained with the Reid potential.

It is now already 10 years ago that the first Faddeev type proton – deuteron (p-d) scattering calculation [1] which took into account exactly the Coulomb interacion between the two protons was published. At that time the simplest nucleon – nucleon (N-N) interaction was taken, namely a separable one term Yamaguchi potential, and assumed to act in N-N S-states only. As a conse-

* This work was supported in part by Fonds zur Förderung der wissenschaftlichen Forschung in Österreich, project 5797.

quence it was possible to calculate just the spin averaged differential cross section, i.e. just one of the p–d observables has been measured at that time. Today further and more precisely measured first and second order p–d polarization data are available, but p–d calculations have not been pushed much further ever since [2] remaining almost at it's status of 1976.

There has been lately, however, some development on the theoretical side in connection with low–energy parameters, in particular the p–d scattering lengths. Configuration space calculations employing the same N–N potentials were performed by two different groups each of which used a different method for obtaining the scattering lengths. One (the Los Alamos – Iowa group) calculated at elastic threshold [3] whereas the other (Leningrad group) used low energy extrapolation [4] to extract the scattering lengths. Unfortunately, the calculated scattering lengths are at variance both in doublet and quartet scattering. Again the Leningrad group has taken N–N S–wave potentials solely, though of local type. The Los Alamos – Iowa group, on the other hand, has also used the tensor force in the $^3S_1 - ^3D_1$ state. In course of the dispute roused by the contradictory results the problem of Coulomb polarization was rediscovered [5] and used [6] as an argument against the Los Alamos – Iowa results. The contraversy seems to have come to an end now, since our own calculation [7] which used low energy extrapolation of p–d phase shifts derived from a momentum space calculation clearly supports the Los Alamos – Iowa results. Having subtracted three–body Coulomb polarization we were able to show for a set of different N–N potentials (including $^3S_1 - ^3D_1$ tensor force within an angular momentum truncation approximation) p–d Phillips line relations and quartet scattering lengths compatible with the one previously found by the Los Alamos – Iowa group.

The striking point of our result is the disagreement with the low energy parameters as derived from phase shift analysis [8]. In particular, our doublet effective range function displays a totally different behaviour than the one given by data analysis. To make sure that the pole–type behaviour of the p–d doublet effective range function is not an artifact of the assumed model we have started to calculate with a more realistic potential low energy p–d observables to be compared with available data.

In the first place we had to check differential cross sections because they are readily available within our present p–d calculation and are the observables mainly governing the phase shift analysis below $E_p = 1$ MeV. With a separable representation of the Paris potential we were able to demonstrate [7] reasonable agreement with data between $E_p = 0.4$ and 1.0 MeV while the pole–type behaviour in the doublet effective range function still persists. Given this result it seems to be adequate to search also for aspects on the side of the experimental analysis that could help to resolve the discrepancy.

One rather trivial aspect is certainly that no data below E_p=0.4 MeV are at hand to be included in the phase shift analysis and it is exactly that energy domain below 0.4 MeV where the theoretical pole-type behaviour sets in. Another aspect is that the differential cross section is not very sensitive to doublet scattering and hence phase shift analysis relying on such data would not sense details of the doublet contribution. To give a more quantitative picture of that aspect we have calculated the doublet contribution to the differential cross section and the result is shown in Fig. 1.

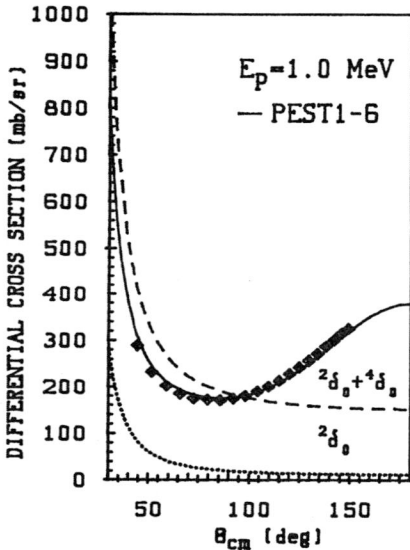

Figure 1. Differential cross section: solid line represents the p-d result obtained with N-N S-state potential PEST1-6; the dotted line is the contribution of the doublet S-state phase shift ($^2\delta_0$) and the dashed line is the sum of doublet and quartet S-state ($^2\delta_0 +^4\delta_0$). Squares are the experimental data ([11]).

Just with N-N S-waves and also when tensor force in the $^3S_1 - ^3D_1$ state is taken within the angular momentum truncation approximation [9] first order p-d polarization observables cannot be obtained, but some second order polarizations like the spin correlation parameter $C_{y'y}$ or the spin transfer parameter $K_{y}y'$ are already largely determined by the S-wave part of the N-N interaction. Consequently we have also looked for the sensitivity of these observables to doublet contributions and have found even less doublet dependence for $K_y{}^{y'}$ but a quite significant influence of doublet scattering in the case of $C_{y'y}$ (Figs. 2a and 2b). Unfortunately, measurements of these observables at energies as low as 1 MeV or less are presumably beyond present day experimental techniques and we may not be able to pin down the doublet problem.

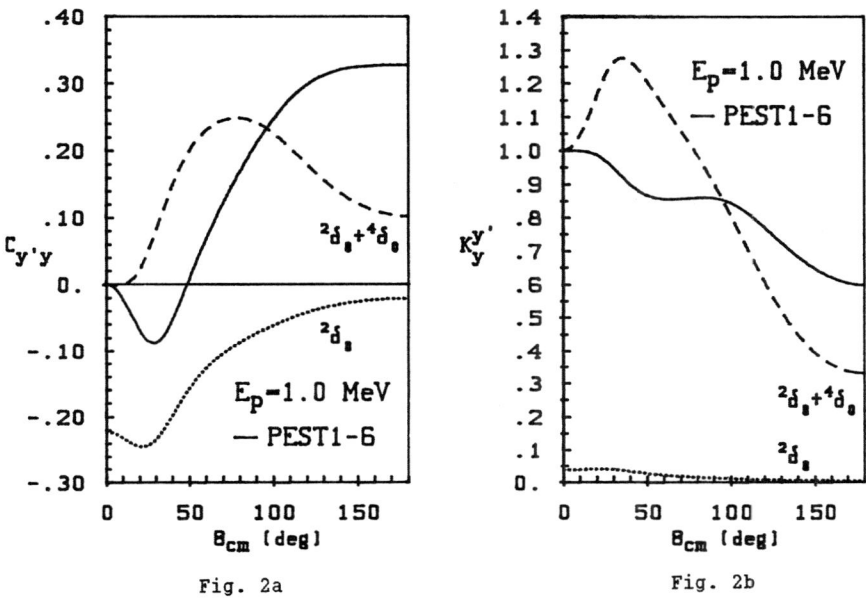

Fig. 2a

Fig. 2b

Figure 2. Spin correlation parameter $C_{y'y}$ (Fig. 2a) and spin transfer parameter $K_y{}^{y'}$ (Fig. 2b): same description as in Fig. 1.

In view of this difficult situation for improving experimental information at very low energies it is important to improve matters on the theoretical side, namely to investigate whether calculated p-d observables other than the one shown here do agree with existing data at lower energies. If we were able to prove that our calculation is compatible with the measured tensor polarization T_{20} at $E_p=1.5$ MeV [10] and with the proton analysing power A_y measured at $E_p=0.8$ to 1.0 MeV [11] while simultaneously being in line with differential cross sections at $E_p=0.4$ MeV we could speak of a realistic model calculation and the pole-type behaviour would have to be taken as a necessary feature of the doublet S-state effective range function which may only be confirmed by more sophisticated experiments.

However, such a procedure demands nothing less than performing an exact p-d calculation that also includes higher N-N partial waves. The fact that the p–d calculations have not yet reached this level within 10 years from their original introduction may indicate the grade of difficulty involved in such a calculation. Nevertheless we have started to push our p–d calculations in this direction, because we are convinced that the calculation of first order polarization below breakup threshold is feasible for us. The first step in this project is the calculation of the ³He binding energy. This calculation itself does not yet provide new information, although it is, to our knowledge, the first one yielding the ³He binding energy for a Paris–like potential including $^3S_1-^3D_1$ tensor force and P–waves. The real importance of said calculation within the context of our project lies in the solution of a main portion of the problems, like the effective three–body potential, which are present in the scattering below breakup threshold. The extension to scattering energies will mainly effect the requirement for computer time which is already quite big at this stage reflecting the complexity of the calculation even at the level of the bound state. Another reason for performing the bound state calculation is that one can compare the resulting Coulomb energy with results from other calculations, mainly perturbative [12] ones, thus being provided with a test case to check, at least, part of the aimed for p–d calculation.

To faciliate the discussion on the various levels of sophistication of our bound state calculation and their comparison with results from other calculations we have listed in Table 1 the three–body channels that couple in the Faddeev equations for $J^\pi=1/2^+$ when N–N S– and P–waves plus the $^3S_1-^3D_1$ tensor force are included. Here the possibility of the isospin $I_\alpha=3/2$ was neglected consistently. We have, however, included the $I_\alpha=3/2$ contribution in one calculation of the ³He binding energy when $^1S_0$ and $^3S_1-^3D_1$ N–N interaction was taken only, i.e. in a eight channel calculation (which would be a 6– channel calculation if N–N charge independence would have been assumed) and have found practically no influence of these states. This finding proofs the validity of the assumption of Payne et al. [13] that the $I_\alpha=3/2$ channels do not contribute to the binding energy.

Just for the purpose of illustration we also show the homogenious version of the p–d Faddeev equations which read

$$T_{\beta,a}(q_\beta,q_\alpha;E) = 4\pi \sum_\gamma \int_0^\infty dq_\gamma q_\gamma^2\, V_{\beta,\gamma}(q_\beta,q_\gamma;E)\, G_0^\gamma(E-\tfrac{3}{4}q_\gamma^2)\, T_{\gamma,\alpha}(q_\gamma,q_\alpha;E) \qquad (1)$$

In order to increase the numerical stability of our results we have introduced a subtraction method similar to the one used in [9].

Table 1. The three particle states in the channel spin and isospin coupling scheme.

2p.-state	α	$(\sigma_\alpha, l_\alpha)\bar{j}_\alpha$	(S_α, L_α)	J	$\bar{I}_\alpha$	$\bar{I}_{\alpha 3}$	I_α
$^1S_0(np)$	1	(0,0) 0	(1/2,0)	1/2	1	0	1/2
$^1S_0(pp)$	1	(0,0) 0	(1/2,0)	1/2	1	1	1/2
$^3S_1(np)$	2	(1,0) 1	(1/2,0)	1/2	0	0	1/2
$^3D_1(np)$	2	(1,2) 1	(1/2,0)	1/2	0	0	1/2
$^3S_1(np)$	3	(1,0) 1	(3/2,2)	1/2	0	0	1/2
$^3D_1(np)$	3	(1,2) 1	(3/2,2)	1/2	0	0	1/2
$^1P_1(np)$	4	(0,1) 1	(1/2,1)	1/2	0	0	1/2
$^1P_1(np)$	5	(0,1) 1	(3/2,1)	1/2	0	0	1/2
$^3P_0(np)$	6	(1,1) 0	(1/2,1)	1/2	1	0	1/2
$^3P_0(pp)$	6	(1,1) 0	(1/2,1)	1/2	1	1	1/2
$^3P_1(np)$	7	(1,1) 1	(1/2,1)	1/2	1	0	1/2
$^3P_1(pp)$	7	(1,1) 1	(1/2,1)	1/2	1	1	1/2
$^3P_1(np)$	8	(1,1) 1	(3/2,1)	1/2	1	0	1/2
$^3P_1(pp)$	8	(1,1) 1	(3/2,1)	1/2	1	1	1/2
$^3P_2(np)$	9	(1,1) 2	(3/2,1)	1/2	1	0	1/2
$^3P_2(pp)$	9	(1,1) 2	(3/2,1)	1/2	1	1	1/2
$^3P_2(np)$	10	(1,1) 2	(5/2,3)	1/2	1	0	1/2
$^3P_2(pp)$	10	(1,1) 2	(5/2,3)	1/2	1	1	1/2

α ... effective channel in the integral equation (1)

σ_α ... 2 particle spin

l_α ... 2 particle angular momentum

$\bar{j}_\alpha$ ... total angular momentum of 2 particles

S_α ... channel spin

L_α ... angular momentum of third particle

J ... total angular momentum of 3 particles

$\bar{I}_\alpha$ ... 2 particle isospin

$\bar{I}_{\alpha 3}$ ... z-component of 2 particle isospin

I_α ... total isospin of 3 particles

In presenting the binding energies we start with calculational results obtained with a set of Yamaguchi potentials that are identical in the $^1S_0$-state but differ in the $^3S_1 - {}^3D_1$ state by their D-state probability [9] From Table 2 the scaling behaviour of the Coulomb energy can be read off. Also we would like to mention that the angular momentum truncation used in a previous paper [9], where only the first 3 channels listed in Table 1 are taken into account for the effective potential, is a rather good approximation, not only for ^3H as we had shown before, but also for ^3He. Another possibility to reduce the number of the coupled integral equations is the neglection of the $^3S_1 - {}^3D_1$ states where the third particle is in a relative d-state (4 channels), but this truncation gives an even worse result for the ^3He compared with the 3-channel calculation.

Table 2. Trinucleon binding energies with $^3S_1-^3D_1$ potentials of varying D-state probability obtained from 3, 4 and 6 channel calculations.

NN-pot.	E (^{3}H) [MeV]			E (^{3}He) [MeV]			E_C [MeV]		
	3 ch.	4 ch.	6 ch.	3 ch.	4 ch.	6 ch.	3 ch.	4 ch.	6 ch.
2.15%	-9.88	-9.60	-9.80	-9.01	-8.75	-8.94	0.87	0.85	0.86
4%	-9.11	-8.62	-9.00	-8.29	-7.82	-8.19	0.82	0.80	0.81
5.5%	-8.49	-7.94	-8.43	-7.71	-7.17	-7.64	0.78	0.77	0.79
7%	-7.94	-7.39	-7.94	-7.19	-6.65	-7.17	0.75	0.74	0.77
8.98%	-7.37	-6.85	-7.44	-6.67	-6.14	-6.71	0.70	0.71	0.73

Finally in Table 3 we present our results employing a separable representation of the Paris potential in the $^1S_0$, $^3S_1-^3D_1$ and all P-states. The most elaborated calculation is therefore the 18-channel calculation and the according Coulomb energy is the first one obtained with the Paris (or a Paris-like) potential at this level. Neglecting all P-waves yields the results denoted by 6-channel (5-channel when one uses a charge independent interaction). A calculation using the Paris potential [14] is given in Table 4, which one should compare with our 6-channel result. But since to our knowledge no ^{3}He or Coulomb energy calculations with the Paris potential have been performed we have also included results obtained with the Reid potential [12], from where we can get also information about the influence of P-waves on the Coulomb energy.

Table 3. Trinucleon binding energies with a PEST1 potential in $^1S_0$, $^3S_1-^3D_1$ and P-states from 3, 4, 6, 8 and 18 channel calculations.

		18 ch.	8 ch.	6 ch.	4 ch.	3 ch.
E (^{3}H)	[MeV]	-7.46	-7.44	-7.44	-6.87	-7.06
E (^{3}He)	[MeV]	(-6.84)	-6.83	-6.83	-6.27	-6.44
E_C	[MeV]	(0.62)	0.61	0.61	0.60	0.62

Considering the difference in the binding energy between the Paris and the Reid potential our results seem to be compatible with the results of Friar et al. [12]. The difference between 6- and 4-channel results (5- and 3-channel in the Reid case) compares favourably. The magnitude of the contribution of the P-waves which is given by comparing our 6 and 18 channel calculation is about the same as for the Reid soft core potential. This can be seen from Table 4 looking at the results RSC18 and RSC9. It should be mentioned that the difference of RSC18 and RSC9 yields the influence of the odd partial waves for $J_\alpha \leqslant 2$ and that the calculation with the RSC-potential assumes charge independence thus explaining the different number of channels.

Table 4. Trinucleon binding energies with the Paris potential ([14]) and the Reid potential ([12]).

| | | RSC18 | RSC9 | RSC5 | RSC3 | Paris |
		(+ P-waves)			($^1S_0$, $^3S_1-^3D_1$)	
E (^{3}H)	[MeV]	-7.23	-7.21	-7.02	-6.38	-7.48
E (^{3}He)	[MeV]			-6.39	-5.77	
E_c	[MeV]	0.62	0.62	0.63	0.61	

In summary we have demonstrated that an exact calculation of p-d polarization observables is feasible since the first step in this project has been proven successfully, namely to obtain the ^{3}He binding energy using a N-N interaction where besides the $^1S_0$ and $^3S_1 - ^3D_1$ states also all P-states are taken into account.

References

1. Alt, E.O., Sandhas, W., Zankel, H., Ziegelmann, H.: Phys. Rev. Lett. 37, 1537 (1976)

2. Chandler, C.: in Proceedings of the Eleventh International Conference on Few Body Systems in Particle and Nuclear Physics, Tokyo and Sendai, 1986

3. Friar, J.L., Gibson, B.F., Payne, G.L.: Phys. Rev. C28, 983 (1983); Friar, J.L., Gibson, B.F., Payne, G.L., Chen, C.R.: Phys. Rev. C30, 1121 (1981)

4. Kvitsinskii, A.A.: Pis'ma Zk. Eskp. Fiz. 36, 375 (1982); Merkuriev, S.P.: in Few Body Problems in Physics, edited by Faddeev, L.D., and Kopaleishvilli, T.I., Singapore, 1985, p.269

5. Kok, L.P.: in Few Body Problems in Physics, edited by Faddeev, L.D., and Kopaleishvilli, T.I., Singapore, 1985, p.252

6. Kvitsinskii, A.A., Merkuriev, S.P.: Yad. Fiz. 41, 647 (1985); Sov. J. Nucl. Phys. 41, 3 (1985)

7. Berthold, G.H., Zankel, H.: Phys. Rev. C (in press)

8. Huttel, E., Arnold, W., Baumgart, H., Berg, H., Clausnitzer, G.: Nucl. Phys. A406, 443 (1983)

9. Berthold, G.H., Zankel, H., Mathelitsch, L., Garcilazo, H.: Nuov. Cim. 93A, 89 (1986)

10. Gruebler, W.: private communication

11. Huttel, E., Arnold, W., Berg, H., Krause, H.H., Ulbricht, J., Clausnitzer, G.: Nucl. Phys. A406, 435 (1983)

12. Chen, C.R., Payne, G.L., Friar, J.L., Gibson, B.F.: Phys. Rev. C31, 2266 (1985); Sasakawa, T., Okuno, H., Sawada, T.: Phys. Rev. C23, 905 (1981)

13. Payne, G.L., Friar, J.L., Gibson, B.F.: Phys. Rev. C22, 832 (1980)

14. Sasakawa, T., Ishikawa, S.: Few-Body Sys. 1, 3 (1986)

AN ALGEBRAIC APPROACH TO MULTICHANNEL SCATTERING
BY NEGATIVE-ENERGY WEINBERG STATES

G.Cattapan, G.Pisent, L.Canton
Dipartimento di Fisica dell'Università
Istituto Nazionale di Fisica Nucleare, Sezione di Padova
v. F. Marzolo 8, 35131 Padova, Italy

Abstract.- The single and multichannel scattering problem for elastic and i-
nelastic collisions between two composite particles is solved by resorting
to basis sets of negative-energy Sturmian functions referring to auxiliary
potentials which allow an easy solution of the Sturmian eigenvalue equation.
By means of these functions the original coupling potentials are approximated with finite-rank interactions. The coupled-channel equations can be then
reduced to sets of algebraic linear equations which can be solved by standard matrix-inversion techniques. For illustrative purposes, our formalism
is applied to the elastic neutron-alpha scattering. The results of these
preliminary calculations are encouraging and show a fast convergence of the
method in the whole energy region we have considered ($0 \div 20$ MeV).

1. Introduction

The search for fast and convenient methods for solving large sets of
coupled differential or integro-differential equations is a subject of con-
siderable importance both in nuclear |1| and in atomic and molecular physics
|2|. In particular, one is interested in viable ways of reducing the given
(integro) differential equations to algebraic equations which can be solved
by standard matrix-inversion techniques. To this end, an interesting possibility is represented by the employment of Sturmian functions (Weinberg states). These functions, which have a rather long history in the physics literature, have been proposed as a useful tool for treating the conventional
multichannel scattering problem in the pioneering papers by the Heidelberg

school (see ref. |3| and references quoted therein). Recently, particularly interesting results have been obtained by Rawitscher |4| . The central idea of this approach is the introduction of complete sets of Sturmian functions, built up on suitable auxiliary potentials which allow an easy solution of the Sturmian eigenvalue problem. By resorting to these Sturmian basis sets, Rawitscher is able to get a full algebrization of the multichannel scattering problem, inasmuch as both the full wave function and the relevant scattering matrix elements can be evaluated through algebraic manipulations. In this approach the Weinberg states are evaluated at the physical (positive) energy, and satisfy non-trivial boundary conditions.

Here we develop an alterantive approach to the multichannel scattering problem. As in |4| we also introduce auxiliary potentials; however we employ *negative-energy* Sturmian functions, to provide the basis sets, in terms of which all relevant quantities may be represented. By exploiting the completeness property of the negative-energy Weinberg states, we replace the complex interactions appearing in the multichannel scattering problem by finite-rank potentials, which can reproduce the original interactions to an arbitrary degree of accuracy. By transforming the coupled differential equations into a set of coupled integral equations, it is then straightforward to reduce the original scattering problem to the solution of a set of linear algebraic equations. Our general formalism will be outlined in the next section; preliminary calculations in the single-channel case are briefly presented in sect. 3.

2. General formalism

We consider the elastic and inelastic scattering between two composite particles. As is well-known, such a scattering problem can be often reduced to the solution of a set of coupled radial second-order differential equations of the form

$$(H_{oc}-E_c)f_c(r)=- \sum_{c'=1}^{N} V_{cc'}(r)f_{c'}(r) \tag{1}$$

for given values of total angular momentum and parity. Here E_c is the usual channel energy, the quantities $V_{cc'}(r)$ can be evaluated in terms of the interaction between the two colliding fragments and of the corresponding internal bound-state wave-functions, and the solution $f_c(r)$ represents the radial scattering wave function in channel c for a given asymptotic initial configuration. The second-order differential operator H_{oc} contains, in addition to the usual centrifugal term, a local distorting potential $V_o(r)$

such as, for instance, the monopole long-range Coulomb interaction in nucleus-nucleus scattering.

To solve eq. (1), supplemented by the usual scattering boundary conditions, we first consider the Sturmian eigenvalue problems

$$\langle \overline{G}_{oc}(B_c), \overline{V}_c \phi_{cm} \rangle (r) = \eta_{cm} \phi_{cm}(r) \tag{2}$$

where $\langle \overline{G}_{oc}(B_c), \overline{V}_c \phi_{cm} \rangle (r)$ is a short-hand notation for $\int_0^\infty \overline{G}_{oc}(r,r';B_c) \overline{V}_c(r') \times \times \phi_{cm}(r') dr'$. In (2) the quantities $\overline{G}_{oc}(r,r';B_c)$ are the free partial-wave Green's functions, associated to the differential operators $\overline{H}_{oc} \equiv H_{oc} - V_o$, evaluated at the negative energies B_c, and the quantities $\overline{V}_c(r)$ are auxilia ry local hermitian potentials which, if necessary, can be chosen in a different way in correspondence to the various channels c (c=1,...,N). The nega tive-energy Sturmian functions $\phi_{cm}(r)$ are normalized according to

$$\int_0^\infty \phi_{cm}(r) \overline{V}_c(r) \phi_{cm'}(r) dr \equiv \langle \phi_{cm}, \overline{V}_c \phi_{cm'} \rangle = \eta_{cm} \delta_{mm'}, \tag{3}$$

with c=1,...,N. We can now resort to (3) as well as to the completeness pro perty of the Sturmian functions |5| to write

$$V_{cc'}(r') \delta(r-r') = \sum_{m=1}^\infty \overline{V}_c(r) \phi_{cm}(r) \eta_{cm}^{-1} \phi_{cm}(r') V_{cc'}(r'). \tag{4}$$

By truncating the sum in this identity to a finite m=M one approximates the local channel-channel coupling potentials $V_{cc'}(r')$ by the non-local M-rank interactions

$$V_{cc'}^{(M)}(r,r') = \sum_{m=1}^M \overline{V}_c(r) \phi_{cm}(r) \eta_{cm}^{-1} \phi_{cm}(r') V_{cc'}(r'). \tag{5}$$

The use of (4) and (5) needs some specification. If, for instance, finite--range potentials $\overline{V}_c(r)$ are employed, the completeness relation for the Weinberg states holds only within the range of these potentials, so that the approximation (5) has to be used in the interaction region only. However, as we hsall see, this, together with the appropriate boundary conditions for the $f_c(r)$'s, is enough to guarantee a good reproduction of the scattering data.

To get an approximate solution of (1) in correspondence to the model potentials (5) we first transform these differential equations into a set of coupled integral equations, which we symbolically write as

$$f_c = f_o \delta_{c1} + \sum_{c'} G_{oc} V_{cc'} f_{c'}. \tag{6}$$

Here G_{oc} is the distorted resolvent for the operator E_c-H_{oc}, and f_0 satisfies the unperturbed equation $(H_{01}-E_1)f_0=0$. We have denoted the elastic channel with $c=1$. Eqs. (6), with the approximation (5) for the coupling potentials $V_{cc'}$, can be straightforwardly solved by algebraic manipulations |6|. Indeed, multiplying by $\phi_{c''}(r)V_{c''c}(r)$ from the left, integrating over r and summing over c one can transform (6) into a set of coupled algebraic equations for the quantities $X_{cm}=\sum_{c'}<\phi_{cm},V_{cc'}f_{c'}>$. These equations can be readily solved by standard matrix-inversion techniques to yield the approximate expression

$$f_c^{(M)}(r)=f_0(r)\delta_{c1}+\sum_{m=1}^{M}<G_{oc},\overline{V}_c\phi_{cm}>(r)\sum_{m'=1}^{M}\sum_{c'=1}^{N}M^{-1}_{cm,c'm'}<\phi_{c'm'},V_{c'1}f_0> \qquad (7)$$

for the radial wave functions $f_c(r)$. In this equation M^{-1} is a $(N\times M)^2$ matrix with

$$M_{cm,c'm'}=\eta_{cm}\delta_{cc'}\delta_{mm'}-<\phi_{cm},V_{cc'}<G_{oc'},\overline{V}_{c'}\phi_{c'm'}>(r)>. \qquad (8)$$

Eqs. (7) and (8) represent the main result of our formalism. The scattering matrix elements can be extracted from (7) by analysing the asymptotic behaviour of $f_c^{(M)}(r)$, the proper boundary conditions coming into play through the Green's functions $G_{oc}(r,r')$. Since there is a large arbitrary in the choice of the auxiliary potentials $\overline{V}_c(r)$ as well as of the negative energies B_c in (2), our formalism enjoys a large degree of flexibility; moreover, the free Green's functions $G_{oc}(B_c)$ occurring in (2), can be always assumed to act in S-wave only, independently of how many partial waves are involved in the considered scattering problem, because we only need the completeness property (4) to simulate the local coupling potentials $V_{cc'}(r)$ by means of finite-rank interactions. We finally observe that, if the free Green's functions $\overline{G}_{oc}$ are replaced by the distorted Green's functions G_{oc} in (2), evaluated at the physical positive energies E_c, our equations (7) and (8) straightforwardly reproduce the results of ref. |4| .

3. A numerical example: single-channel neutron-alpha scattering

To test our method on a physical example, we have chosen the purely elastic neutron-alpha scattering, in other words, we have considered a single-channel problem with no Coulomb interaction. In such a case our eqs. (7) and (8) reduce to

$$f_{j1}^{(M)}(r)=F_1(r)+\sum_{m=1}^{M}<G_{o1},\overline{V}\phi_m>(r)\sum_{m'=1}^{M}(M_{j1}^{-1})_{mm'}<\phi_{m'},V_{j1}F_1> \tag{9}$$

and

$$(M_{j1})_{mm'}=\eta_m\delta_{mm'}-<\phi_m,V_{j1}<G_{o1},\overline{V}\phi_{m'}>(r)>, \tag{10}$$

respectively, for each (j,1) partial-wave state. Here $F_1(r)$ is the usual Riccati-Bessel function, $G_{o1}(r,r')$ represents the partial-wave free Green's function, and with $V_{j1}(r)$ we have denoted the real Woods-Saxon potential describing the neutron-alpha interaction, namely

$$V_{j1}(r)\equiv-V(1+\exp(\{r-R\}/a))^{-1}+\left(\frac{\hbar}{m_\pi c}\right)^2 V_s\vec{L}\cdot\vec{\sigma}\frac{1}{r}\frac{d}{dr}(1+\exp(\{r-R_s\}/a_s))^{-1}. \tag{11}$$

The parameters for this potential have been taken from ref. |7|.

By taking into account the explicit expression for $G_{o1}(r,r')$, it is straightforward to get the asymptotic behaviour of the radial wave functions $f_{j1}^{(M)}(r)$. One can then show |6| that the elastic scattering matrix S_{j1} can be written in the form

$$S_{j1}=\frac{\det M_{j1}^\star}{\det M_{j1}}=\exp(2i\delta_{j1}). \tag{12}$$

As for the auxiliary potential $\overline{V}(r)$, we have chosen a square-well whose range b is assumed sufficiently larger than the range R of $V_{j1}(r)$. It turned out that the depth of $\overline{V}$ does not appear in the final expressions, and is therefore irrelevant.

The convergence of the method has been tested both with respect to the range parameter b and with respect to the reference negative energy B. As for b, two conflicting requirements have to be satisfied. It cannot be too small, because this would imply too severe a truncation for the original potential $V_{j1}(r)$; at the same time, it cannot be so large to waste the convergence with respect to M. On the ground of our numerical calculations, we have assumed b=1.5R. As far as B is concerned it turned out that this parameter is practically uneffective on the results. We have assumed $b\sqrt{2m|B|/\hbar^2}=1$.

The convergence with respect to M has been checked by comparing the approximate phase shifts calculated through eq. (12) with the phase shifts obtained from the exact potential (11). An analysis up to E=20 MeV and 1=2 has shown that convergence is already obtained with M=3 (see fig. 1). By assuming M=5 we have been able to reproduce the optical phase shifts within 0.1 degree in the whole energy region we have considered. Fig. 2 shows the

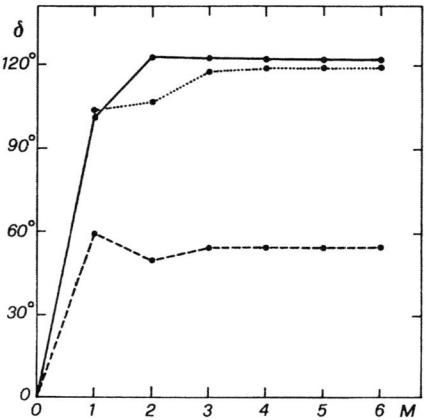

Fig. 1 - Convergence of the separable expansion (5) with respect to M for the $S_{1/2}$ (full line), $P_{1/2}$ (dashed line) and $P_{3/2}$ (dotted line) phase shifts at E=6 MeV.

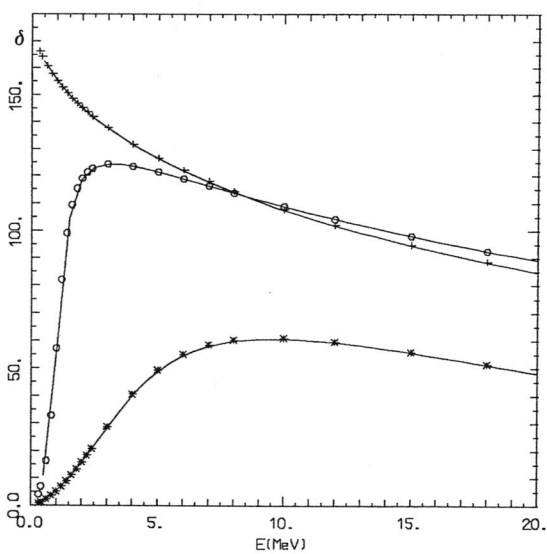

Fig. 2 - The $S_{1/2}$, $P_{1/2}$ and $P_{3/2}$ approximate phase shifts as functions of the neutron energy E for M=5. Crosses, asterisks and circles represent the "exact" corresponding values.

$S_{1/2}$, $P_{1/2}$ and $P_{3/2}$ phase shifts as functions of the neutron energy E for $0<E<20$ MeV. The curves represent the approximate phase shifts evaluated by means of eq. (12), whereas crosses, asterisks and circles exhibit the corresponding values calculated from the original local potential (11).

In conclusion, the scattering data referring to the elastic neutron-alpha scattering can be reproduced by means of our method to a high degree of accuracy with a few terms in the separable expansion (5). It is worth stressing that this expansion does not provide a mere separable approximation, phase equivalent to the original interaction at given energies; on the contrary, it yields a finite-rank potential able to reproduce the original local potential faithfully, at least as M goes to infinity. It is therefore expected that, provided M is chosen sufficiently high, the model potential (5) can successfully simulate given local interactions both on- and off-the-energy-shell. On the ground of these considerations, it will be interesting to apply our method to realistic nucleon-nucleon interactions. At the same time, if the quick convergence with respect to M will be confirmed in the multichannel case, our method will provide an efficient way of solution of the coupled-channel problem.

References

1. Satchler, G.R.: Direct Nuclear Reactions. Oxford-New York: Oxford University Press, 1983.
2. Schneider, B.I., Collins, L.A.: Phys. Rev. A33, 2970, 2982 (1986).
3. Mahaux, C., Weidenmüller, H.A.: Shell-Model Approach to Nuclear Reactions. Amsterdam-London: North-Holland, 1969;
   Romo, W.J.: Nucl. Phys. A191, 65 (1972).
4. Rawitscher, G.H.: Phys. Rev. C25, 2196 (1982).
   Rawitscher, G.H., Delic, G.: Phys. Rev. C29, 747, 1153 (1984).
5. Newton, R.G.: Scattering Theory of Waves and Particles. New York-Heidelberg-Berlin: Springer, 1982.
6. Pisent, G., Canton, L.: Nuovo Cim. 91A, 33 (1986).
7. Satchler, G.R., Owen, L.W., Elwin, A.J., Morgan, G.L., Walter, R.L.: Nucl. Phys. A112, 1 (1968).

FADDEEV-YAKUBOVSKY CALCULATION OF 4-ALPHA PARTICLE SYSTEM WITH REALISTIC ALPHA-ALPHA INTERACTIONS

S. Oryu[*], T. Nishino and H. Kamada

Department of Physics, Faculty of Science and Technology,
Science University of Tokyo, Noda, Chiba 278, Japan

ABSTRACT

The low-lying energy levels of ^{12}C and ^{16}O nuclei are calculated by using the Faddeev and Yakubovsky equations based on the 3- and 4-alpha models, respectively. We used two off-shell different alpha-alpha potentials which represent the Pauli exclusion property successfully. The first one is the KNFT potential which is given by Fujiwara and Tamagaki to remove the Coulomb effects from the OCM equivalent Kukulin-Neudatchin potential. The second one is the FBOM equivalent UIM potential, in which both the totally and partly Pauli forbidden states are excluded, thoroughly. In the 3-alpha calculation, almost all of the important low-lying energy levels are well reproduced with both potentials except that some spurious states appear with the KNFT potential. It is found that not only the Coulomb force but also the three-cluster force are very important obtaining the correct order of the energy spectra as well as a good binding energy for the ground state of the ^{12}C nucleus, in which we proposed a simple form of the three-cluster force. The sub-amplitudes for [3+1] as $^{12}C + \alpha$ system, and also for [2+2] as $^{8}Be + ^{8}Be$ system, are represented by using rank-1 and rank-2 Hilbert-Schmidt expansion methods, respectively. The convergences of the energy spectra for the low-lying 0_1^+, 0_2^+ and 0_3^+ states in the ^{16}O nucleus are investigated for several higher partial waves.

* oral presentation

I. INTRODUCTION

The three-nucleon system has been investigated by many authors based on the Faddeev formalism[1]. As compared to this, only few calculations have been carried out within this framework for the three-composite particle systems[2]. One of the reasones is due to the fact that the interactions between composite particles systems cannot be chosen as a simple separable potential. Another one is probably due to the existence of the N-body Faddeev formalism which should be solved rigorously. On the other hand, it is well known that the alpha-alpha interaction is well represented in the resonating group method (RGM)[3], in which the optical potential is introduced based on the microscopic treatment in statical ways with the Pauli principle. Generally, those potential have not only the local term but also the non-local one which originates from the nucleon exchanges. Our motivation is that such a simple equation with optical potentials could be performed with some statical approximations of the N-body equations, together with several properties of the nucleon forces. The three-alpha model of ^{12}C nucleus is well reproduced from 12-nucleon system based on the RGM treatment with three times of the alpha-alpha interaction for three different partitions and the residual three-alpha-cluster force[4]. Therefore, such a three-body system could be calculated rigorously with those cluster-cluster potentials based on the Faddeev equation. Here, such a potential has two Pauli forbidden states for the S-wave, and one for the D-wave which are from the exchange interaction kernel of the RGM. Those unphysical states can be removed in the orthogonality condition model (OCM)[5]. Recently, we have proposed a multi-rank alpha-alpha separable potential which is on- and off-shell equivalent with the fish-bone-optical model (FBOM) potential[6]. In this model, the Pauli exclusion principle is treated thoroughly. Our potential is constructed by applying the unitary interpolation method, called UIM, to the FBOM. The UIM potential is of rank 4 for the S-wave, of rank 3 for the D-wave, and of rank 2 for the G-wave, respectively.[7] The two of the ranks in the S-wave and one in the D-wave were used to eliminate the above mentioned Pauli forbidden states. The remaining terms are for the partly Pauli forbidden states and

serve to represent the off-shell structure.   Therefore, our
potential is a generalization of the OCM potential and, also of
the rank-1 KNFT potential which is modified by Fujiwara and
Tamagaki from the Kukulin-Neudatchin potential.[8,9]
   In the three-alpha model of the ^{12}C nucleus, energy spectra
were calculated by using the Faddeev equation with both the UIM
and the KNFT potentials without three-cluster-force.[9,10]   It was
demonstrated that our UIM alpha-alpha potential reproduces
almost all the important low-lying levels of ^{12}C nucleus
successfully.[10]   No spurious states occur in the UIM case.
   In this report, we would like to review some important
results with the three-cluster-force.[11]   On the extension from
^{12}C nucleus, the ^{16}O nucleus is not only a simple 4-cluster
model but also very interesting nucleus because of the magic
number.   Since it was found that the Faddeev formalism is a
useful tool not only in the well developed cluster region near
the breakup threshold, and also in the investigation of the
ground state of such a closed shell nucleus, then the energy
spectra of the ^{16}O nucleus in the 4-cluster model could be
well described by using the Faddeev-Yakubovsky (FY) equation.[12]
Therefore, these nucleus are very useful to probe the relation
between the cluster model and the shell model of the nucleus.
Furthermore, it is expected that one could find some origin of
the nuclear saturation property.[4]

II.   TWO- AND THREE-BODY SUB-SYSTEMS

   1)   Two-body sub-amplitudes
        The two-body separable t-matrix is given by the OCM
equivalent rank-one separable Kukulin-Neudatchin (KN) potential
and also the Fujiwara-Tamagaki (FT) potential.   Here, the formfactors are chosen such that the inner oscillator of the overlapping region between alpha-alpha interaction is reproduced.
The phase shift of alpha-alpha scattering is reproduced by the
KN potential but it seems to be inconsistent with the alpha-
alpha binding energy.   On the other hand the FT potential was
introduced to remove the Coulomb effects from the KN potential.
Hereafter, we will use the FT potential as the KNFT one.
Moreover, the UIM potential uses rank 4, rank 3 and rank 2
separable forms to reproduce the S-wave, D-wave and G-wave
potentials, respectively:

$$t(k,k';z) = \sum_{ij} g_i(k) \tau_{ij}(z) g_j(k').$$

2) [3+1] and [2+2] sub-amplitudes

The [3+1] sub-amplitudes $X_{jj'}(p,p';z)$ satisfy the Amado-Lovelace-Mitra (ALM) equation.[13] Recently, we obtained almost all important low-lying energy spectra of ^{12}C nucleus by using the Faddeev equation with the UIM potential, but without the 3-cluster force.[10] We found that some shell-like states 0_1^+, 2_1^+ and 4_1^+ gain very large binding energies, however, the well-developped cluster-like states 0_2^+ and 2_2^+, give less binding energies. We adopted a phenomenological 3-cluster force as follows:[14]

$$V = V_0 \exp[-a(r_{12}^2 + r_{23}^2 + r_{31}^2).$$

After a simple calculation with the Coulomb interaction, the 3-cluster force which has two parameters, can reproduce the correct order of the energy spectra as well as a good binding energy for the ground state, in which the 3-cluster range parameter a is very sensitive to the order of energy levels, after considering the Coulomb corrections and fitting the potential depth V_0. The figure illustrates the energy levels vs the range parameters. The good fitting parameters are given by $V_0 = 824.3$ MeV and $a = 0.2$ fm^{-2}, respectively.[11]

On the other hand, [2+2] sub-amplitudes $Y_{jj'}(k,k';z)$ are given by the Lippmann-Schwinger type integral equations. These sub-amplitudes are approximated by the multi-rank Hilbert-Schmidt expansion. That is,

$$X_{jj'}(p,p';z) = \sum_n x_j^n(p;z) H_n(z) x_{j'}^n(p';z),$$

and

$$Y_{jj'}(k,k';z) = \sum_n y_j^n(k;z) G_n(z) y_{j'}^n(k';z),$$

where these form-factors $x_j^n(p;z)$ and $y_j^n(k;z)$ are determined by eigenvalue equations, and the propagators are given by the related functions with those form-factors.

III. 4-BODY FADDEEV-YAKUBOVSKY EQUATION

1) Integral equation

In order to obtain the ground state energy of the ^{16}O nucleus based on the 4-alpha model, the Faddeev-Yakubovsky equation is given by[15]

$$\alpha_n(q;z) = \sum_{n'} \int_0^\infty K_{nn'}(q,q';z) H_{n'}(z-2q'^2/3m) \alpha_{n'}(q';z) \frac{q'^2 dq'}{2\pi^2}$$

where the part of the kernel is defined by

$$K_{nn'}(q,q';z) = E_{nn'}(q,q';z)$$
$$+ \sum_{n''} \int_0^\infty 2 F_{nn''}(q,q'';z) G_{n''}(z-\frac{q''^2}{2m}) F_{n''n'}(q'',q';z) \frac{q''^2 dq''}{2\pi^2}$$

with

$$E_{nn'}(q,q';z) = \frac{1}{2} \sum_{jj'} \int_{-1}^1 dx\, x_j^n (P[q,q',x];z-2q^2/3m)$$

$$\tau_{jj'}(z-\frac{3q^2}{4m} - \frac{3q'^2}{4m} - \frac{qq'}{2m}x) x_{j'}^{n'}(P[q',q,x];z-2q'^2/3m),$$

and

$$F_{nn'}(q,q';z) = \frac{1}{2} \sum_{jj'} \int_{-1}^1 dx\, x_j^n (Q[q,q',x];z-2q^2/3m)$$

$$\tau_{jj'}(z - \frac{q^2}{m} - \frac{3q'^2}{4m} - \frac{qq'}{m}x) y_{j'}^{n'}(R[q,q',x];z-q'^2/2m),$$

where some characteristic momenta used in the above equation are defined as follows,

$$P[q,q',x] = (q^2/9 + q'^2 + 2qq'x/3)^{1/2},$$

$$Q[q,q',x] = (4q^2/9 + q'^2 + 4qq'x/3)^{1/2},$$

and

$$R[q,q',x] = (q^2 + q'^2/4 + qq'x)^{1/2}.$$

2) Recoupling coefficients

The state channels are specified by particle spins s_i, 2-body orbital angular momenta L_{ij}, 3-body orbital angular momenta $L_{ij,k}$, between the center of mass of the two particle subsystems and the respective spectator particles, and 4-body orbital angular momenta L_1 between the center of mass of the three-particle subsystems and the respective spectator particles, in the [3+1] channel, respectively. Moreover, in the [2+2] channel, additional orbital angular momenta $L_{ij,kl}$ are defined between two subsystems (ij) and (kl). Generally, the state vectors N_1 are defined by

$$N_1 = [(s_i + s_j) + L_{ij} = J_{ij}, \quad (J_{ij} + s_k) + L_{ij,k} = J_{ij,k},$$
$$(J_{ij,k} + s_1) = K_1, \quad K_1 + L_1 = J].$$

Since the spin of the alpha-particle is zero, it is sufficient to specify N_1 by the orbital angular momenta sets $(L_{ij}, L_{ij,k}, L_1)$, and $(L_{ij}, L_{kl}, L_{ij,kl})$ for the fixed value $J = 0^+$. In the kernel of the FY equation, the constituents $E_{nn'}(q,q';z)$ and $F_{nn'}(q,q';z)$ are replaced by $E_{nn'}(N_i,N_j;q,q';z)$ and $F_{nn'}(N_i,N_j;q,q';z)$, respectively. Recoupling coefficients which remain in the kernels, are $(L_{ij}, L_{ij,k}, L_1) = (0,0,0), (0,1,1), (0,2,2), \ldots$, and also $(L_{ij}, L_{kl}, L_{ij,kl}) = (0,0,0), (0,2,2), (0,4,4), \ldots$, and so on. It is very interesting that one of the alpha-alpha sub-systems is only S-wave. In other words, the ^{12}C nucleus which contains the full energy spectra is not pointed to in the ^{16}O ground state at all. Therefore, 4-alpha particles almost overlap in the ^{16}O nucleus region, and makes the shell-like structure in which 4-cluster force is very important as well as the 3-cluster force in the ^{12}C nucleus by means of the Pauli principle.

Table 1

[2+2] $L_{ij,kl}$ / [3+1] $L_{ij,k}$ ($L_{ijk,l}$)	0	0,2	0,2,4	Experiments
0	-5.1 -11.0	-5.3 -11.4	-5.6 -11.4	
0, 2	-5.4 -5.7 -12.6			
0, 1, 2	-5.1 -5.5 -13.3			
0, 2, 4	-5.6 -6.5 -12.7			-2.39 -8.39 -14.44

Table 2

[2+2] $L_{ij,kl}$ / [3+1] $L_{ij,k}$ ($L_{ijk,l}$)	0	0,2	0,2,4	Experiments
0	-6.3 -11.5	-6.0 -11.8	-6.3 -11.8	
0, 2	-5.7 -6.3 -13.0			
0, 1, 2	-5.6 -6.4 -14.2			
0, 2, 4	-5.8 -7.1 -13.0			-2.39 -8.39 -14.44

Table 3

[3+1] $L_{ij,k}$ ($L_{ijk,l}$)	0	0,2	0,1,2	0,2,4
[2+2] $L_{ij,kl}$ = 0	-4.78	-4.80	-4.88	-4.81

IV. RESULTS AND DISCUSSIONS

Calculated 4-alpha binding energies (MeV) are shown in the Table 1 for the HSE rank-one method in [3+1] and [2+2] sub-amplitudes. And also the HSE rank-two calculations are given in the Table 2, together with the experimental data.

It becomes clear that the importance of the higher waves and the many ranks, however, those results show rather good convergence except for the case $L_{ij,k} = 0$ only.
The higher partial wave effects of [2+2] amplitudes are less than 0.5 MeV, and those of [3+1] amplitudes are almost similar to the [2+2] cases for 0_2^+ and 0_3^+ states.
Table 3 is a preliminary results for the UIM potential.
We need more computer space to fill up full tables, however, these results give rather good fitting to the experimental values even for such a full overlaping region of the four alpha clusters in the ground state of ^{16}O nucleus.

REFERENCES

1) L. D. Faddeev, JETP 12, 1014 (1961).
   J. Haidenbauer, Y. Koike and W. Plessas, Phys. Rev. C33, 439 (1986).
2) B. Charnomordic, C. Fayard and G. H. Lamot, Phys. Rev. C15, 864 (1977).
   Y. Matsui, Phys. Rev. C22, 2591 (1980).
   Y. Koike, Nucl. Phys. A337, 23 (1980).
   K. Miyagawa, Y. Koike, T. Ueda, T. Sawada and S. Takagi, Prog. Theor. Phys. 74, 1264 (1985).
   K. Hahn and E. W. Schmid, Phys. Rev. C31, 325 (1985).
   E. W. Schmid, Nucl. Phys. A416, 347c. E. W. Schmid, M. Orlowski and Bao Cheng-guang, Z. Phys. A308, 237 (1982).
3) K. Wildermuth and Y. C. Tang, A Unified Theory of the Nucleus, in Clustering Phenomena in Nuclei (Vieweg, Braunschweig,1977) vol 1.
4) G. Spitz, H. Klar and E. W. Schmid, Z. Phys. A 322, 49 (1985).
   S. Saito, Y. Akaishi, H. Klar, S. Nakaichi-Maeda, E. W. Schmid, and G. Spitz, 320, 399 (1985).
   K. Hahn and E. W. Schmid and P. Doleschall, Phys. Rev. C31, 325 (1985).
5) S. Saito, Prog. Theor. Phys. 40, 893 (1968); 41, 705.
6) E. W. Schmid, Z. Phys. A297, 105 (1980).
7) R. Kircher and E. W. Schmid, Z. Phys. A299, 241 (1981).
   R. Kircher, H. Kamada and S. Oryu, Prog. Theor. Phys. 73, 1442 (1985).
8) V. I. Kukulin and V. G. Neudatchin, Nucl. Phys. A157, 609 (1970).
9) Y. Fujiwara and R. Tamagaki, Prog. Theor. Phys. 56, 1503 (1976); R. Tamagaki and Y. Fujiwara, Prog. Theor. Phys. Suppl. 61, 229 (1977).
10) S. Oryu and H. Kamada, Prog. Theor. Phys. 76-6 (1986)
11) H. Kamada and S. Oryu, Proceedings of Few-Body XI edited by T. Sasakawa, K. Nishimura, S. Oryu and S. Ishikawa, (North-Holland Pub. 1986).
12) O. A. Yakubovsky, Sov. J. Nucl. Phys. 5 (1967), 937.
13) A. W. Thomas, ed. Modern Three-Hadron Physics (Springer Verlag 1977).
14) O. Portilho and S. A. Coon, Z. Phys. A290, 93 (1978).
15) S. Nakaichi-Maeda, Y. Akaishi and H. Tanaka, Prog. Theor. Phys. 64, 1315 (1980); Phys. Rev. A26, 32 (1982).

ALPHA-DEUTERON BREAK-UP CALCULATIONS WITH COULOMB INTERACTION

P. Doleschall [†]

Central Research Institute for Physics, H-1525 Budapest, Pf 49, Hungary

C. Chandler

University of New Mexico, Albuquerque, U.S.A.

M. Bruno, F. Cannata, M. D'Agostino, M.L. Fiandri

Dipartimento di Fisica dell'Università and INFN, Bologna, Italy

Abstract: Isospin breaking in the $\alpha + d \longrightarrow \alpha+p+n$ reaction has been studied experimentally by searching for the production of the d* singlet (T=1) state. Such production manifests itself as narrow F.S.I. peak in the breakup for $E_{np}=0$. Theoretically the treatment of $\alpha +d \longrightarrow \alpha +p+n$ has been done within the framework of Faddeev equations. The "trivial" background which provides isospin breaking is given by the αp Coulomb potential which allows then the coupling to the singlet ^1S_o np interaction. We study here in detail how much of the experimental effects can be accounted by such a mechanism.

The α-d break-up process of the low energy provide us with a good opportunity to investigate the importance of the Coulomb interaction in the low energy nuclear reactions, and it may be a good test of the approximation schemes which incorporate the Coulomb interaction into the Faddeev calculations.

If the α-nucleon interaction respects isospin symmetry the n-p ^1S_o interaction may play a role in the α-d system only because of the presence of the Coulomb interaction in the α-p subsystem. In turn the ^1S_o n-p interaction manifest itself with the appearence of the

† supported by INFN – Laboratori Nazionali di Legnaro, Italy

n-p FSI peaks in the α-d break-up process. A large part of the experimental work [1] was devoted to search for these peaks at low energies where the effect of the Coulomb interaction is expected to be stronger. A difficulty at low energies comes from the fact that the n-p FSI regions and the ^{5}He $P_{3/2}$ resonance regions overlap. On the other side at higher energies where more clean n-p FSI regions exist the n-p FSI peaks are less prominent [2], and their identification requires a spin dependent analysis [3].

The first Faddeev calculations, performed by Koike [4], did not include any Coulomb interaction; Coulomb effects were only includeded as final state corrections. This approach produced very good results in some break-up configurations [1], but failed to give an acceptable description of the experimental data in other configurations, in particular at lower energies [1]. There are indeed prominent peaks at E_α^{Lab}=11.3 MeV in the α+d --→ α+p+n reaction where the n-p FSI and ^{5}He regions overlap and this type of calculation is far from predicting the measured peaks correctly [1], the main reason beeing the absence of the Coulomb and consequently of the ^1S_o interaction.

A full inclusion of the Coulomb interaction is out of question [5], because of the complexity [5]. Phenomenologically, because of the long range of the Coulomb interaction, a positive barrier of 0.4 MeV height can differentiate α-p from α-n potential around 6 fm where the nuclear part is already vanishing. This barrier may influence the break-up process by suppressing the kinematical region of small α-p relative energy. In order to describe this physics we assumed a cut-off function which leaves the Coulomb interaction nearly unchanged between the 0-10 fm region, and suppresses it between 10-20 fm. The cut-off function was chosen as $e^{-(\alpha r)^n}$, with α 0.07 and 0.08 fm, and n=4 and 6 respectively.

For the nuclear part of the α-d interaction we selected only the $S_{1/2}$, $P_{1/2}$, $P_{3/2}$ partial wave components. Basically local interations in form of a sum of Gaussian functions was fit to the 0-20 MeV α-n phase shifts. Two types of $S_{1/2}$ interaction were constructed:

A) a pure local repulsive one;

B) a local attractive one plus a very strong separable repulsive one.

For the Faddeev calculations a separable expansion of these potentials was made using the EST [6] method. Instead of using positive energy mesh points, in most cases negative values were chosen.

The α-p interaction have the following structure:

1) the α-n and the cut-off Coulomb potentials were added and a separable expansion was performed globally;

2) a rank-3 separable expansion of the cut-off Coulomb potential was added to the existing separable form of the α-n interaction; in this way by construction the nuclear part of the α-n and α-p potentials are exactly identical.

For the n-p subsystem, except for one calculation, we used the rank-1 Haidenbauer-Plessas [6] approximation of the Paris potential, and we included only the $^3S_1$-$^3D_1$ and ^1S_o states.

Six calculations were performed at E_α^{Lab}=11.3 MeV. In a first Koike--type calculation we checked the reliability of our α-n interactions and we basically reproduced the results obtained by Koike. Another calculation was devoted to check the sensitivity to a rank-4 $^3S_1$-$^3D_1$ and a rank-3 ^1S_o force. We found that at least at E_α^{Lab}=11.3 MeV the rank-1 interactions give pratically the same results. Two different calculations were performed to check the sensitivity of the results to the details of the $P_{1/2}$ and $P_{3/2}$ α-n interactions; we compared the results obtained with rank-1 and rank-2 approximations of the local P-wave α-n interactions. The results indicate that sometimes the difference can reach 10% and therefore we conclude that at least the rank-2 interaction is needed. We also checked the insensitivity to the different ways of including the cut-off Coulomb interactions. Finally we noted a sensitivity to the α-n $S_{1/2}$ interaction, especially at large laboratory proton angle measurements. In Fig. 1 and 2 are shown together with the experimental results some results of the calculations. The solid lines show the calculations performed with rank-1 $^3S_1$-$^3D_1$ and ^1S_o interactions, with rank-2 α-n interactions in the $S_{1/2}$, $P_{1/2}$, $P_{3/2}$ channels, with rank-4 $S_{1/2}$ and $P_{1/2}$ and rank-3 $P_{3/2}$ α-p interactions. The $S_{1/2}$ interaction was a pure repulsive interaction (A) and we used the method 1) to include the Coulomb interaction. The dashed

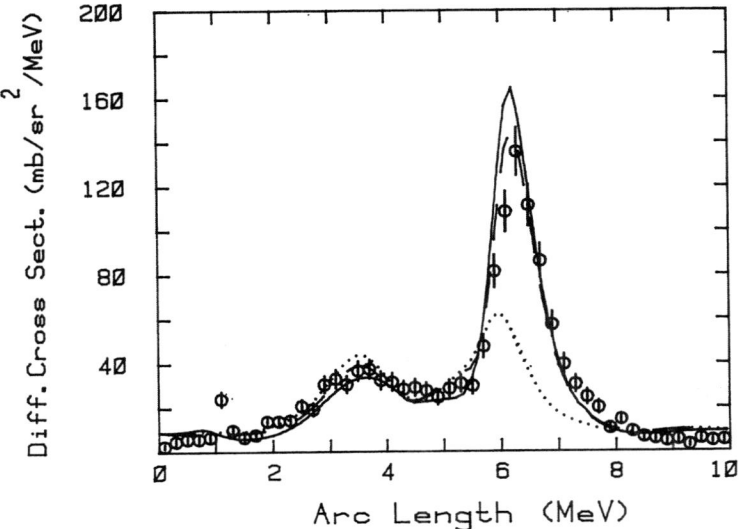

Fig. 1 - Experimental cross sections for alpha incident energy 11.3 MeV, alpha angle
15° and proton angle 14.8° (from Ref. 1), together with theoretical predictions
as described in the text.

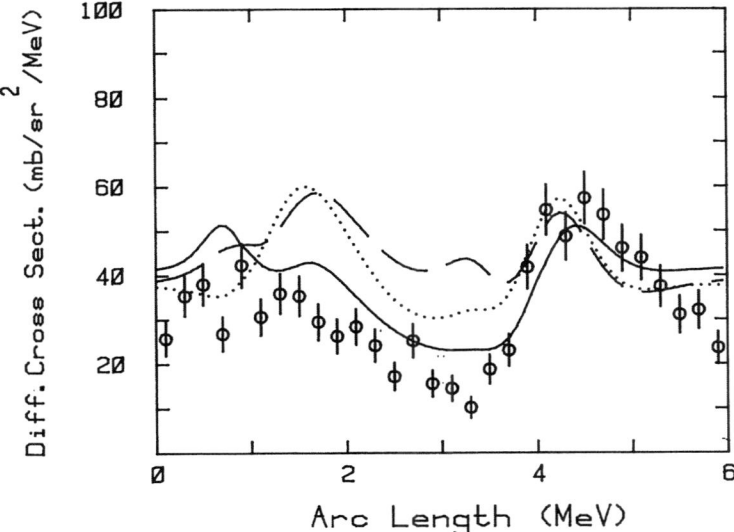

Fig. 2 - Experimental cross sections for alpha incident energy 11.3 MeV, alpha angle
17° and proton angle 35° (from Ref. 7), together with theoretical predictions
as described in the text.

lines show the results of the calculation with the rank-1 $^3S_1 - ^3D_1$ and ^1S_o interaction, with rank-2 α-n interactions and rank-5 α-p interactions. The $S_{1/2}$ interaction was of the type B and we used the method 2) to include the Coulomb interaction. The dotted lines show the results of the last calculation without the inclusion of the n-p ^1S_o interaction.

In conclusion:

1) the inclusion of the Coulomb interaction and consequently the ^1S_o interaction is very important at low energy. In particular one notes a large improvement on the earlier calculation generating spurious peak (see Fig. 1d of Ref. 7) in the n-p FSI region.

2) There is a sensitivity to the nuclear α-N interaction; a still puzzling result is that the $S_{1/2}$ α-n interaction which is qualitatively near to RGM α-n interaction [8] obtained by the orthogonality condition model seems not to agree with measurements at large proton angle.

References

1) M. Bruno, F. Cannata, M. D'Agostino, M.L. Fiandri, M. Frisoni, G. Vannini and M. Lombardi, Nucl. Phys. A386, 269 (1982).

   I. Koersner, L. Glantz, A. Johansson, B. Sundqvist, H. Nakamura and H. Noya, Nucl. Phys. A286, 431 (1977).

2) T. Rausch, H. Zell, D. Wallenwein and W. Von Witsch, Nucl. Phys. A222, 429 (1974).

3) C. Werntz and F. Cannata, Phys. Rev. C24, 349 (1981).

4) Y. Koike, Prog. Theor. Phys. 59, 87 (1978); Nucl. Phys. A301, 411 (1978); Nucl. Phys. A337, 23 (1980).

5) E.O. Alt - this workshop.

6) J. Haidenbauer and W. Plessas, Phys. Rev. C27, 63 (1984).

7) M. Bruno, F. Cannata, M. D'Agostino, M.L. Fiandri, H. Oswald, P. Niessen, J. Schulte-Vebbing, H. Paetz gen. Schieck, P. Doleschall and M. Lombardi - to be published.

8) K. Hahn, E.W. Schmid, P. Doleschall, Phys. Rev. C31, 325 (1985).

SYMMETRY BREAKING IN NUCLEAR REACTIONS:
ASYMMETRY IN α+d → ³He + ³H

M.Bruno[°], F.Cannata[°], M.D'Agostino[°], M.Frisoni[°], H.Herman[°], B.Vuaridel[+], V.König[+],

W.Grüebler[+], K.Elsener[+], P.A.Schmelzbach[+], J.Ulbricht[+], Ch.Forstner[+], M.Bittcher[+],

D.Singy[+], H.M.Hofmann[++]

° Dipartimento di Fisica dell'Università and INFN, Bologna, Italy

+ Institut for Medium Energy Physics, ETH, CH-8093 Zürich, Switzerland

++ Institut für Theoretische Physik, Universität Erlangen-Nürnberg, West Germany

The general idea of an invariance property characterizing a quantum mechanical state is familiar. For example take a particle (nucleon) bound in a potential well and consider an external probe transferring momentum $\vec{q}$ to the bound particle. One argues that due to the "symmetry" $\vec{p}_i \rightarrow -\vec{p}_i$ (this symmetry of course holds if there is rotational invariance in three dimensional space, central potential) the two configurations

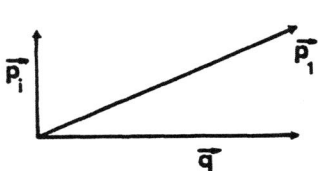 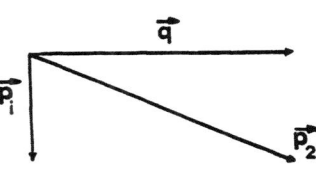

Fig. 1 - Nucleon knock out with initial nucleon momentum $\vec{p}_i$ (a) and $-\vec{p}_i$ (b)

for the ejected nucleon p_1 and p_2 would occur with equal probability.

The argument however goes wrong if the probe-nucleon interaction has a spin-flip term and the potential has a spin-orbit component. Roughly speaking the probe can polarize the nucleon normal to the scattering plane and the polarization is subsequently analyzed by the spin-orbit potential acting on the nucleon in the scattering state [1]. An asymmetry is thus generated like in a double scattering experiment. In this case clearly it is the spin-orbit potential which breaks rotational invariance in three dimensional space. For a detailed discussion see Ref. 1. In a very similar way one can discuss isospin invariance properties of the wave function.

Let us consider the reaction [from now on we work in the center of mass frame]

$$\alpha + d \rightarrow {}^3\text{He} + {}^3\text{H}$$

If isospin is a good symmetry the initial and as a consequence also the final state are pure T=0 states. ^{3}He and ^{3}H are assumed to belong to a T=½ isospin multiplet being the T_3=+½ and T_3= -½ components respectively. We easily recognize that the wave function involving space and spin must be symmetric for interchange of ^{3}He and ^{3}H if we want the full wave function to be antisymmetric [i.e. ^{3}He and ^{3}H are treated us undistinguishable particles: generalized Pauli principle].

Let us first consider a simple dynamics for which both spin and isospin are conserved. Then the final state like the initial state has S=1 and consequently the space part of the wave function is purely symmetric. As a consequence the counting rate of tritons at angle θ and of ^{3}He at the same angle is the same and this implies (we are working in the center of mass) symmetry for θ → π - θ. In terms of partial waves it means that only even partial waves are allowed. Non conservation of spin i.e. S=0 and S=1 mixtures in the final state don't change the symmetry since S=0 and S=1 states are orthogonal and as a consequence even and odd waves do not interfere. In turn if an asymmetry between θ and π - θ is experimentally observed it implies an interference between even and odd partial waves i.e. isospin breaking of some kind [2].

To illustrate the general discussion let us consider in particular the mechanisms:

Fig. 2 - One nucleon exchange mechanism

The symmetry we have discussed before follows if for the vertices both equa
lities hold

$$_3\alpha \rightarrow {}^3He + n = {}_3\alpha \rightarrow {}^3H + p$$
$$^3He \rightarrow d + p = {}^3H \rightarrow d + n$$

One can test the last relation by other reactions e.g. $\pi^{\pm}$ on ${}^3He$ and ${}^3H$

by comparing

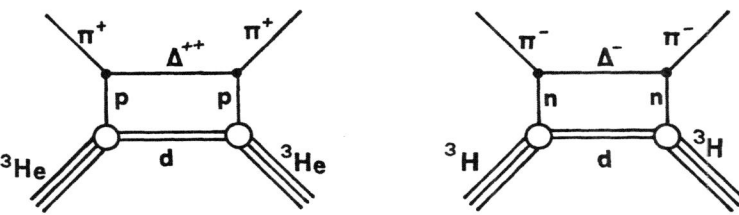

Fig. 3 - Pion scattering in impulse approximation

The results of the experimental data [3] seem to suggest [4] that the di
stribution of protons in ${}^3He$ is more extended in space than the distribution of neutrons in ${}^3H$.

The real interest of genuine isospin breaking effects (not induced by

Coulomb) comes from the fact that we know from QCD based phenomenology [5]

that there is a genuine isospin breaking effect induced from the underlying

quark degrees of freedom $[m_u \neq m_d]$. It has been proposed e.g. that the observed isospin breaking constrains the constituent quark·masses [6].

Symmetry breaking may lead to very far reaching consequences. An example of a very interesting SU (3) flavour mixing is provided by the η,η' mesons which belong respectively to a SU(3) octet and singlet. It is the SU(3) singlet which can couple to purely gluonic states but if there is a breaking of SU(3) flavour also the η can have a gluonic admixture. Such effects seem indeed to occur in the η production via G(1590) meson [7].

$$\pi^- \; p \; \rightarrow \; G \; \; n$$
$$\lfloor\!\longrightarrow \; \eta \; \eta \; (\eta')$$

This leads possibly to completely new mechanisms for absorption (or production) of η in nuclei [8].

After this general discussion let us now come back to a more detailed discussion of the asymmetry in the reaction

$$\alpha \; + \; d \; \rightarrow \; {}^3He \; + {}^3H \tag{1}$$

This reaction was experimentally investigated by a ETH-Zürich and INFN-Bologna collaboration [9]. The experiment has been carried out at the SIN injector cyclotron, using a polarized deuteron beam and a gaseous Helium target. The full angular distribution has been obtained measuring in the forward region both ^{3}He and tritons, since the last ones correspond to ^{3}He at $(\pi - \theta)$ cm angle. The particle discrimination was obtained with telescopes of three surface barrier silicon detectors, the first two to identify ^{3}He and the second and the third one to identify tritons. Four telescopes have been used at the same time at two different angles on the left and the right side of the beam. This technique minimizes errors due to solid angle uncertainties, allowing for a more reliable determination of asymmetries.

To investigate reaction (1) theoretically we have attempted a microscopic multichannel and multistructure calculation in the frame of the refined resonating group method [10]. Eight fragmentations of ^{6}Li were considered, as shown in Table 1.

Table 1

Fragmentation	Threshold energy (MeV)		
	exp	present calc.	Ref.11
^{4}He + d	0	0	0
^{5}He + p	3.12	4.57	4.97
^{5}Li + n	4.19	5.40	5.89
^{5}He* + p	6.43±2.1	6.26	10.44
^{5}Li* + n	8.02±2.1	7.08	11.36
^{3}He + ^{3}H	14.32	14.10	14.09
^{5}He** + p	19.88	21.12	21.31
^{5}Li** + n	20.85	22.08	22.36
^{6}Li(gr.st.)	-1.47	-1.51	-0.9

Fragments with A $\leqslant$ 4 were assumed to have no internal structure, while those with A=5 were treated as alpha plus nucleon with relative angular momentum ℓ=1 (ground and first excited states) or as ^{3}He (^{3}H) plus deuteron with
no relative angular momentum (second excited states). In Table 1 the threshold energies obtained, relative to α+d threshold, are compared with the
experimental results and with the results of previous calculations [11].
It appears that the agreement with experimental values has been improved.

For the reaction (1) one expects that the gross feature of cross sections and analyzing powers should be given in terms of positive parity
channels. [11] Therefore we have limited ourselves only to these channels to
have a first guess of theoretical predictions. In this limit the Barshay-
Temmer theorem [2] is respected since negative parity channels are relevant
for the asymmetry [11].

Results of calculations together with experimental data are shown in
Figs. 4÷7 for two different energies.

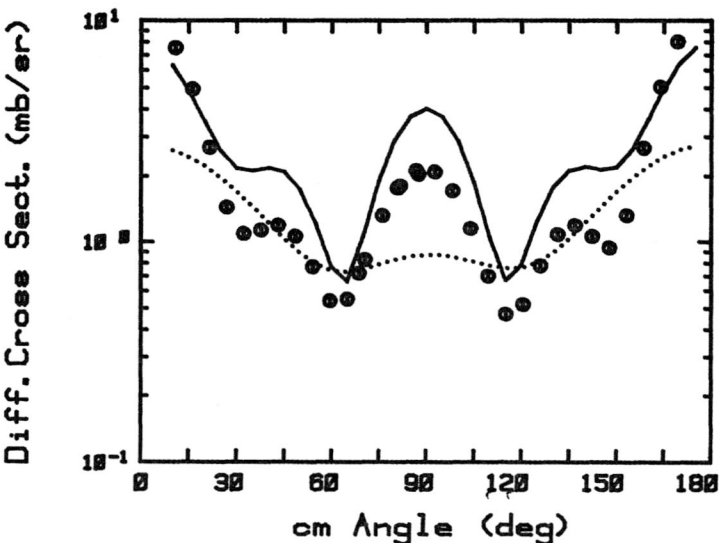

Fig. 4 — Differential cross section for ^{4}He(d, ^{3}H)^{3}He at 23.4 MeV c.m. energy. Theoretical predictions including all fragmentations (solid line) and only ^{4}He+d together with ^{3}H+ ^{3}He fragmentations (dotted line) were calculated at 23 MeV c.m. energy.

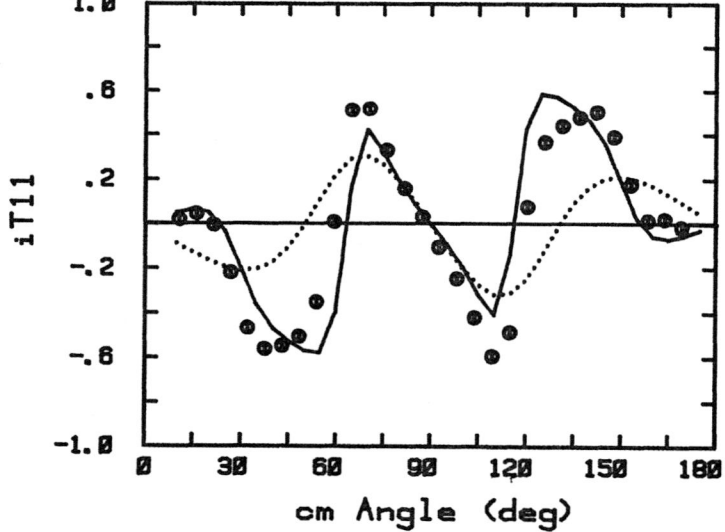

Fig. 5 — The same as Fig. 4 but for vector analysing power.

In Figs. 4 and 5 calculations including only α+d and ^{3}He+^{3}H fragmentations are presented (dotted lines). From the results it appears that other fragmentations besides the incoming and outgoing ones are needed.

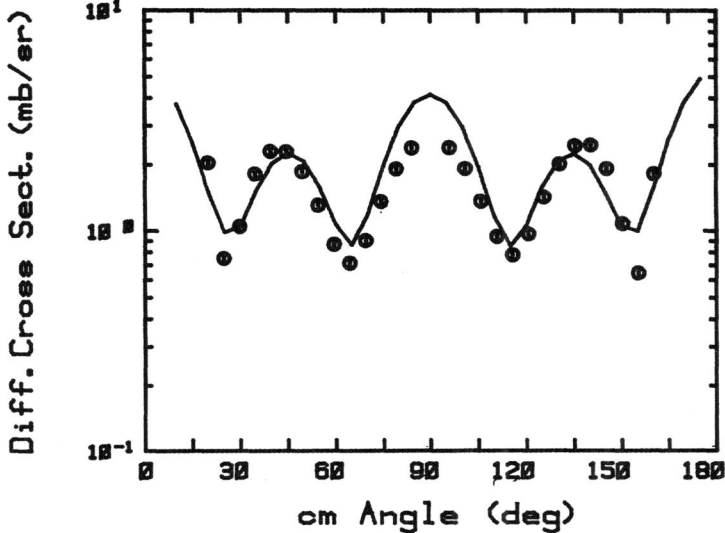

Fig. 6 — Differential cross section for ^{4}He(d,^{3}H)^{3}He reaction at 29.9 MeV c.m. energy. Solid line represents calculations at 30 MeV c.m. energy.

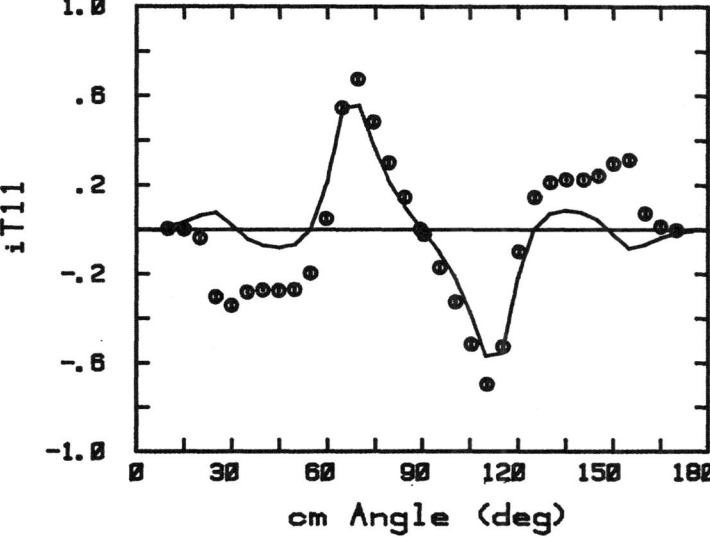

Fig. 7 — The same as Fig. 6 but for vector analysing power.

It can be seen that calculations, including all the structures shown in Ta
ble 1, account.ng only for positive parity channels, generate the correct
shape of the angular distributions. Also energy variations are reasonably
predicted. On the other side at the quantitative level some improvements
are still needed and we have therefore not yet attempted to include negative parity channels in order to calculate asymmetries.

Experimental asymmetries of cross sections defined as

$$W(\theta) = [\frac{d\sigma}{d\Omega}(\theta) - \frac{d\sigma}{d\Omega}(\pi - \theta)] / [\frac{d\sigma}{d\Omega}(\theta) + \frac{d\sigma}{d\Omega}(\pi - \theta)]$$

are shown in Fig.8 for two different energies.

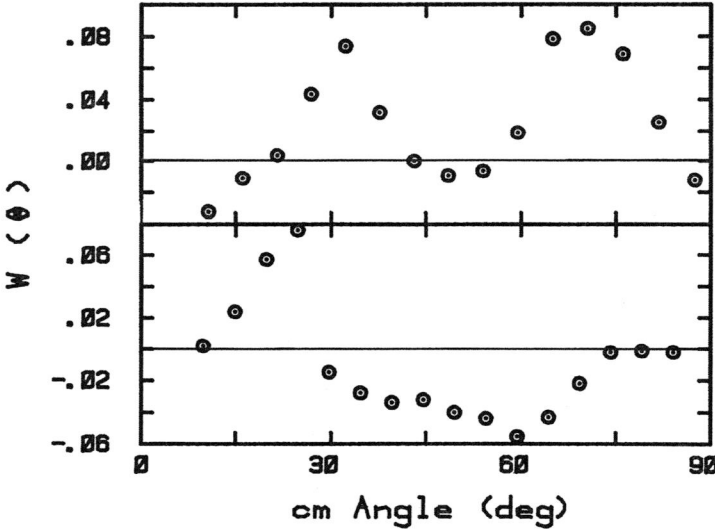

Fig. 8 - Angular distribution of differential cross section asymmetry measured at 23.4 MeV
c.m. energy (upper) and at 29.9 MeV (lower).

The asymmetries of the vector analyzing powers, defined as

$$D(\theta) = iT_{11}(\theta) + iT_{11}(\pi - \theta)$$

are shown in Fig.8 for the same energies.

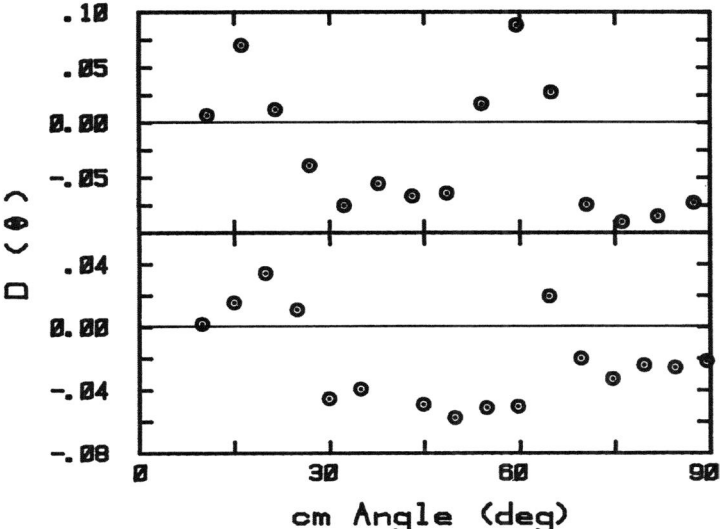

Fig. 9 - Angular distribution of vector analyzing power asymmetry measured at 23.4 MeV c.m. energy (upper) and at 29.9 c.m. energy (lower)

The asymmetry reaches about 8%, showing that, even at these relatively high energies, the Barshay-Temmer theorem is not strictly respected. The energy dependence of the asymmetry is rather smooth, as far as the shape and the absolute value are concerned. A similar behaviour was found also for the asymmetry of vector analyzing powers.

In order to speculate about the possible physical effects contributing to the asymmetries more accurate theoretical reproduction of observables is needed. To this end calculations are still in progress.

References

1) F.Cannata,J.P.Dedonder,J.R.Gillespie: "Spin orbit forces in knockout reactions" preprint.

2) S.Barshay and G.M. Temmer: Phys. Rev. Lett. 12,728 (1964).

3) B.M.K. Nefkens et al: Phys.Rev.Lett. 52,735(1984).

4) C.Werntz,F.Cannata:"Charge symmetry in the A=3 Nuclei",preprint.

5) W-Y.P.Hwang:"Quark Interchange and isospin symmetry violations in Nucleon Nucleon scattering at low energies" Phys.Rev.Lett. to be published.

6) S.Godfrey and N.Isgut:Phys.Rev.D32 189 (1985) and D34, 899 (1986).

7) F.Binon et al: Sov.J.Nucl.Phys. 39, 526 (1984).

   S.S.Gershtein et al.:Sov.J.Nucl.Phys. 39, 156 (1984).

   See however: S.A.Coon, B.H.J.Mc Kellar,M.D.Scadron to be published in Phys.Rev. D.

8) F.Cannata: "Color degrees of freedom in nuclei" Proc. of Winter School, Folgaria (1986), ed. by T.Bressani (north Holland Publ.Comp.).

9) B.Vuaridel, V.König, W.Grüebler, K. Elsener,P.A.Schmelzbach,J.Ulbricht, Ch. Forstner,M.Bittcher,D.Singy, M.Bruno, F.Cannata,M.D'Agostino,I.Borbely-Proceedings Sixth International Symposium Polarization Phenomena in Nuclear Physics, Osaka, 1985 - J.Phys.Soc. Jpn.Suppl. 55, 874(1985); B.Vuaridel et al. -contributed paper to Workshop on Three body Force in the Three-Nucleon System, Washington, 1986;

   B.Vuaridel et al.- contributed paper to XI International Conference on few body systems in particle and nuclear physics, Sendai, Japan, August 1986.

10) H.M. Hofmann et al. Nucl.Phys. A410, 208 (1983).

   H.H.Hackenbroich, Symp. on present status and novel developments in the nuclear many-body problem, Rom-1973.

11) W.Schütte, thesis, Köln 1977.

Three-Body Calculation of ^{9}Be *

E. Cravo and A.C.Fonseca

Centro de Física Nuclear, Av.Gama Pinto 2
1699 Lisbon, PORTUGAL

Since both ^{5}He and ^{8}Be are unbound nuclei and ^{8}Be shows large α-reduced width near the Wigner limit, one expects to be able to describe the low lying states of ^{9}Be through a three-body model made up of two structure reless α-particles and a neutron n interacting by pair wise potentials. Although our aim is ultimately to include realistic α-α interactions that take into account the effects of the Pauli principle between the α's together with the repulsive Coulomb force, at this stage we report the results of a less ambitious calculation.

In the present work we neglect the Coulomb force between the α's and take for n-α and α-α interactions one-or two-term separable potentials with form factor $f_\ell(k)$ of the type

$$f_\ell(k) = k^\ell / (k^2+\beta^2)^{\ell+1}$$

where ℓ is the pair angular momentum quantum number. For the n-α interaction we choose one term spin and angular momentum dependent potentials in channels $s^{\frac{1}{2}}$, $p^{\frac{1}{2}}$ and $p^{\frac{3}{2}}$ whose parameters are fitted to low energy n-α phase shifts and whose values are given in ref. [1]. Since there is some ambiguity on how to choose the $s^{\frac{1}{2}}$ potential we take here, for simplicity, a repulsive interaction. In the future we plan to use the other option proposed by Saito [2] and Kukulin et al [3] which involves an attractive $s^{\frac{1}{2}}$ interaction with the bound state pole removed to infinity. Both methods reflect different ways to take into account the Pauli principle between the neutron and the α-particle in a potential model. Likewise for the α-α interaction we take a separable interaction in channels s,d and g. The s-wave potential given below involves two terms, one attractive and one repulsive,

$$V_o = \lambda_o^a |f_o^a><f_o^a| + \lambda_o^r |f_o^r><f_o^r|$$

---

*Paper submitted in absentia

while the d-and g-waves only a single attractive term. Since in the presen
te work we neglect the Coulomb force the parameters are fitted to the α-α
binding energy and low energy phase shifts obtained with the Ali-Bodmer[4]
potential alone. The parameters are given in Table I and the corresponding
fit in Fig.1.

Table 1 - Parameters of the α-α interaction

λ_o^a (fm^{-3})	-33.0
β_o^a (fm^{-1})	1.4
λ_o^r (fm^{-3})	19.3
β_o^r (fm^{-1})	1.0
λ_2^a (fm^{-7})	-4.92×10^2
β_2^a (fm^{-1})	1.55
λ_4^a (fm^{-11})	-9.83×10^6
β_4^a (fm^{-1})	2.47

The results of our calculation for the ground state of ^{9}Be are shown in Ta
ble II. The interactions, are included one by one and the effect of each
term carefully studied. We find that the $s^{\frac{1}{2}}$ n-α interaction changes the
binding energy of the α + n + α system by ≈1.60 MeV while the $p^{\frac{1}{2}}$ only contri
butes ≈0.10 MeV to the binding energy. We also find that the d-wave α-α
contribution is larger than the s-wave while the g-wave contribution is the
smallest of them all. Furthermore the importance of the $s^{\frac{1}{2}}$ n-α interaction

Table 2 - Binding energies from present calculation

INTERACTIONS		
n - α	α - α	Binding Energy ^{9}Be (MeV)
$p^{\frac{3}{2}}$	none	-3.57
$p^{\frac{1}{2}};p^{\frac{3}{2}}$	none	-3.74
all	none	-0.10
all	s	-1.72
all	s;d	-5.48
all	all	-5.72
$p^{\frac{3}{2}}$	all	-7.19
$p^{\frac{1}{2}};p^{\frac{3}{2}}$	all	-7.28
$p^{\frac{3}{2}}$	s	-5.00
$p^{\frac{1}{2}};p^{\frac{3}{2}}$	s	-5.25

is reduced when the d-wave α-α interaction is included.

Finally we looked for excited states of α + n + α in channels J^p equals $1/2^+$ and $5/2^-$ and found none. Work is in progress to include other n-α and α-α potentials as well as Coulomb force.

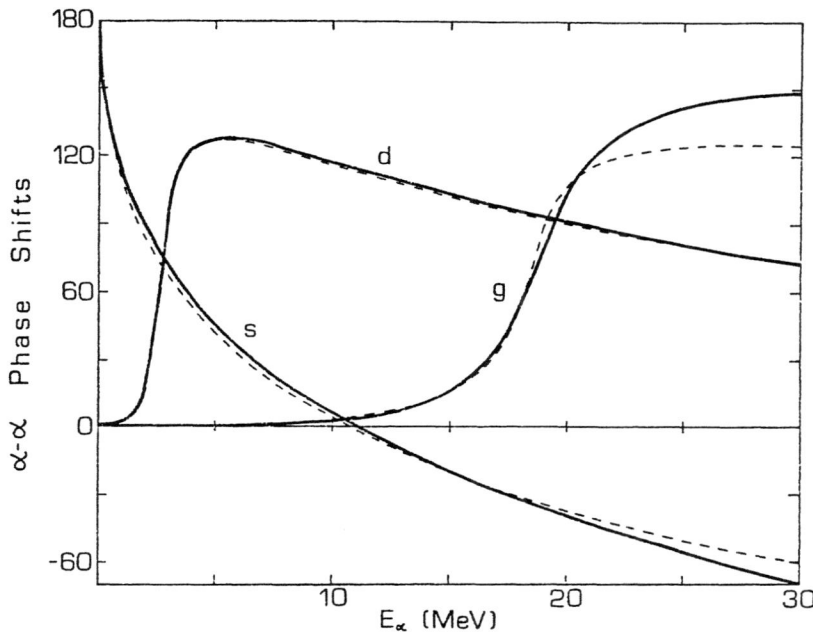

Fig.1 Phase-shifts for α-α in s-, d- and g- channels. Solid line corresponds to Ali-Bodmer potential without Coulomb and das̲ hed line to the separable potential of our choice.

1. A.C.Fonseca, J.Revai and A.Matveenko, Nucl.Phys. A326,182(1979)

2. S.Saito, Prog.Theor.Phys. 41, 705(1969)

3. V.I.Kukulin et.al., Nucl.Phys. A245, 429(1975);
   Yad. Fiz. 24, 298(1976)

4. S.Ali and A.R.Bodmer, Nucl.Phys. 80, 99(1966)

POLARIZATION EFFECTS IN PHOTO- AND ELECTRODISINTEGRATION OF THE DEUTERON[*]

H. Arenhövel and W. Leidemann
Institut für Kernphysik
Johannes Gutenberg-Universität
D-6500 Mainz
Federal Republic of Germany

Abstract

Polarization observables in deuteron photodisintegration like photon and target asymmetries and one- and two nucleon polarization of the final state are studied with respect to their sensitivity to subnuclear degrees of freedom. Furthermore it is shown that the electric form factor of the neutron has a sizeable influence on the polarization asymmetry in $d(\vec{e},e'n)p$ with polarized electrons. Uncertainties arising from potential model dependence are discussed.

1. Introduction

The search for effects from subnuclear degrees of freedom in nuclei has focussed the attention on polarization observables which contain in principle much richer information on the dynamics of the system under study than attainable without projectile and target polarization and without polarization analysis of the final state. Because these polarization observables contain interference terms of the various amplitudes in different combinations and thus can be more sensitive to small contributions of subnuclear degrees of freedom.

[*]Supported by Deutsche Forschungsgemeinschaft (SFB 201)

In view of the recent technical improvements [1-3] for polarized beams and targets it appears therefore timely to study polarization observables in deuteron disintegration by photons and electrons - a process of fundamental importance in nuclear physics - in order to see what kind of information is buried in the various observables, in particular, what we can learn about the role of subnuclear degrees of freedom like meson and isobar degrees of freedom.

2. Polarization observables for photodisintegration

Polarization phenomena are most easily described by a density matrix for the initial state of the incoming photon and the deuteron

$$\rho = \rho^\gamma \times \rho^d \qquad (1)$$

in terms of which the differential cross section is given by

$$d\sigma/d\Omega = \text{Tr}(T\rho T^+), \qquad (2)$$

where T denotes the reaction matrix for deuteronphotodisintegration. Correspondingly one has for the polarization of one outgoing nucleon "i"

$$\vec{P}(i)d\sigma/d\Omega = \text{Tr}(\vec{\sigma}(i)T\rho T^+) \qquad (3)$$

or of both outgoing nucleons

$$P_{x_i x_j}(1,2)d\sigma/d\Omega = \text{Tr}(\sigma_{x_i}(1)\sigma_{x_j}(2)T\rho T^+). \qquad (4)$$

We will not discuss the photon and deuteron density matrices in detail but only remind the reader, that the photon polarization can be described uniquely by two parameters P_C^γ and P_ℓ^γ describing the degree of circular and linear polarization, respectively and the direction of linear polarization. The deuteron density matrix is in general characterized by eight parameters. However, for present day sources of polarized deuterons two parameters, P_1^d and P_2^d, i.e., vector and tensor polarization with respect to a given orientation axis $\hat{d}$ are sufficient. They are given in terms of the probabilities p_m, the probabilities for finding a deuteron with spin projection m on $\hat{d}$

$$P_1^d = P_{10}^d = \sqrt{3/2} \, (p_1 - p_{-1}) \qquad (5)$$

$$P_2^d = P_{20}^d = (p_1 + p_{-1} - 2p_0)/\sqrt{2} \qquad (6)$$

Various methods for polarizing deuterons are described in refs.[1-3].

In the cm-frame the reaction matrix may be given in the form

$$T_{sm\ \lambda m} = e^{i(\lambda+m_d)\phi}\ t_{sm\ \lambda m}\ (\theta).\qquad(7)$$

Here the initial state is characterized by the incoming photon polarization λ and the deuteron spin projection m_d with respect to the incoming photon momentum $\vec{\omega}$. We associate a frame of reference with z-axis parallel to $\vec{\omega}$ and x-axis in the direction of maximal linear photon polarization. The final state is characterized by the p-n relativ momentum $\vec{k}$ having the spherical angles (θ,ϕ) with respect to the chosen frame of reference and by the total spin s and projection m_s on $\vec{k}$. For the description of nucleon polarization we introduce another frame of reference $(\bar{x},\bar{y},\bar{z})$ for the final state with $\bar{z}$-axis parallel to $\vec{k}$ and $\bar{y}$-axis parallel to $\vec{\omega}\times\vec{k}$.

Of the 24 amplitudes only 12 are independent if parity is conserved. Since one phase remains undetermined one needs 23 independent observables for a complete description of deuteronphotodisintegration. These will be provided by various polarization observables.

Without giving details, which can be found in ref.[4], one obtains for the differential cross section including photon and target asymmetries

$$d\sigma/d\Omega = d\sigma_o/d\Omega\ [1 - P_\ell^\gamma\ \Sigma(\theta)\cos2\phi\qquad(8)$$
$$+ P_1^d\ T_{11}(\theta)\sin(\phi-\phi_d)d_{10}^1(\theta_d)$$
$$+ P_2^d\ \Sigma_{M\geq0}\ T_{2M}(\theta)\cos M(\phi-\phi_d)d_{M0}^2(\theta_d)$$
$$- P_\ell^\gamma\ (P_1^d\Sigma_{M=-1,1}T_{1M}^\ell\sin\Psi_M\ d_{M0}^1(\theta_d)$$
$$+ P_2^d\ \Sigma_{M=-2,2}T_{2M}^\ell\cos\Psi_M d_{M0}^2(\theta_d))$$
$$+ P_c^\gamma\ (P_1^d\Sigma_{M\geq0}\ T_{1M}^C\cos M(\phi-\phi_d)d_{M0}^1(\theta_d)$$
$$+ P_2^d\ \Sigma_{M>0}\ T_{2M}^C\sin M(\phi-\phi_d)d_{M0}^2(\theta_d))]$$

where

$$\Psi_M = M(\phi-\phi_d)-2\phi.\qquad(9)$$

Furthermore $d_{M'M}^I$ denote the reduced rotation matrices. The unpolarized differential cross section and the target and photon asymmetries are given in terms of certain quadratic forms of the t-matrix elements (see ref.[4]).

Similarly one finds for the polarization of one outgoing nucleon including photon or target asymmetries

$$P_{x/z}\ d\sigma/d\Omega = (- P_\ell^\gamma\ P_{x/z}^\ell(\theta)\sin2\phi+P_c^\gamma P_{x/z}^C(\theta)\qquad(10)$$
$$+ P_1^d\Sigma_{M>0}\ P_{x/z}^{1M}(\theta)\cos M(\phi-\phi_d)d_{M0}^1(\theta_d)$$
$$+ P_2^d\Sigma_{M=1}\ P_{x/z}^{2M}(\theta)\sin M(\phi-\phi_d)d_{M0}^2(\theta_d))\ d\sigma/d\Omega,$$

$$P_y \, d\sigma/d\Omega = (P_y(\theta) - P_\ell^\gamma \, P_y^\ell(\theta)\cos 2\phi$$
$$+ P_1^d \, P_y^{11}(\theta)\sin(\phi-\phi_d)d_{10}^1(\theta_d)$$
$$+ P_2^d \Sigma_{M>0} P_y^{2M}(\theta)\cos M(\phi-\phi_d)d_{M0}^2(\theta_d)) \, d\sigma/d\Omega. \qquad (11)$$

Expressions for simultaneous photon and target asymmetries are given in ref.[4].

The vanishing of $P_{x/z}$ for unpolarized beam and target is a consequence of parity conservation because they are related to the expectation value of pseudoscalar quantities, whereas P_y is a scalar as given by $\langle\vec{\sigma}\cdot(\hat{\omega}\times\hat{k})\rangle$ with $\hat{\omega}$ and $\hat{k}$ being the unit vectors along photon momentum $\vec{\omega}$ and relative p-n momentum $\vec{k}$. For the same reason the double polarization of both outgoing nucleons $P_{x_i x_j}(1,2)$ vanishes for the combinations

$$(x_i x_j) = xy, \ yx, \ zy, yz. \qquad (12)$$

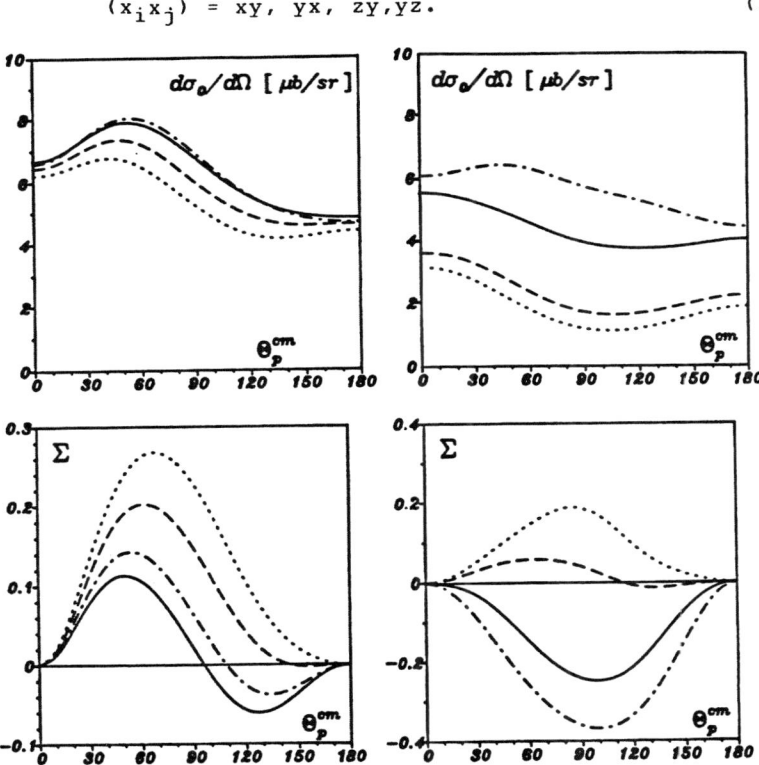

Fig.1 Differential cross section and photon asymmetry at 100 MeV (left) and 260 MEV (right). Normal (N) theory: dotted curves, N+MEC: dashed curves, N+MEC+IC(IA) dash-dot curves, N+MEC+IC(CC): full curves.

3. Results for photodisintegration

Now we will present some illustrative examples of what can be learned from polarization observables. These observables are calculated using a nonrelativistic description for the bound deuteron and the p-n continuum states. The waves functions are obtained by numerical integration of the Schrödinger equation using some realistic NN-potential, in the following the Reid soft core potential (RSC)[5]. In addition isobar degrees of freedom are included either in the impulse approximation (IA) or in a coupled channel approach (CC)[6]. The electromagnetic current contains besides the classical one-body operators also two-body contributions from π-meson exchange (MEC).

First we will discuss the differential observables. Fig.1 shows the differential cross section and its photon

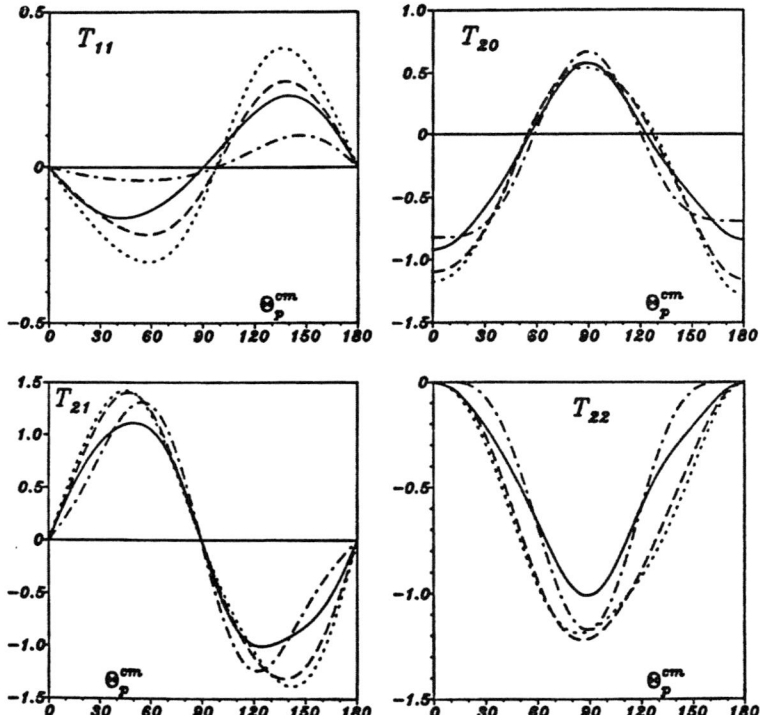

Fig.2 Target asymmetries of differential cross section at 260 MeV. Notation as in fig.1.

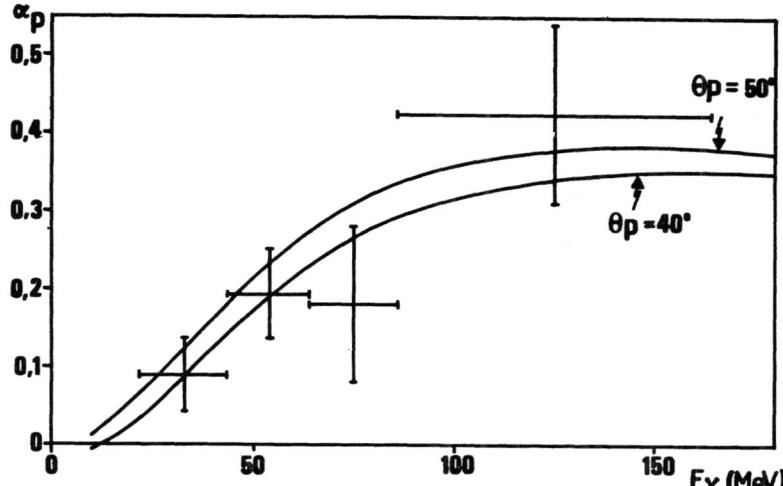

Fig.3 Experimental results for d(γ,p)n with tensor polarized
deuterons. α_p is proportional to T_{21} (see ref.[7]).
Theory: full curves for two proton angles.

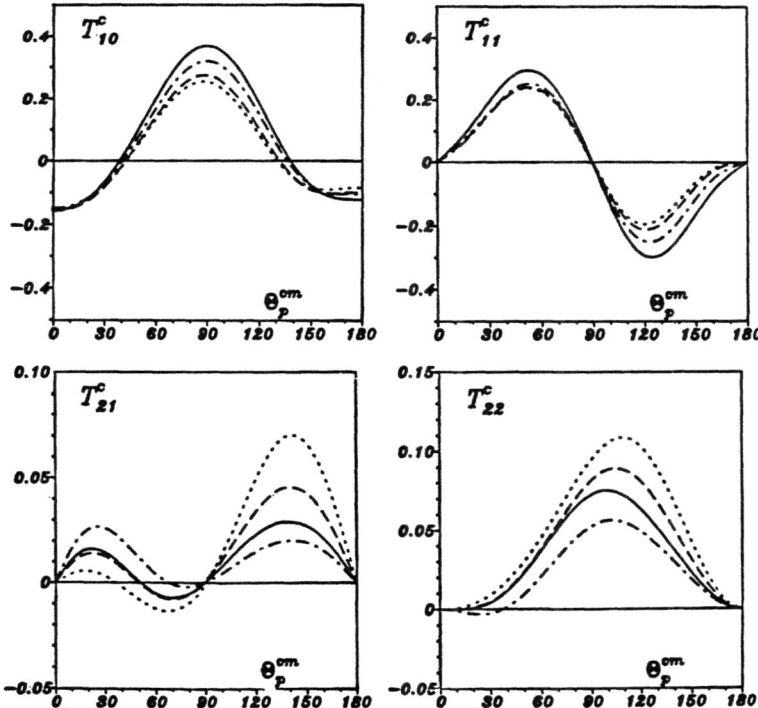

Fig.4 Circularly polarized photon and target asymmetries at
100 MeV. Notation as in fig.1.

asymmetry for E_γ=100 MeV and 260 MeV. At 100 MeV the photon asymmetry Σ is strongly affected by exchange effects, reducing Σ considerably and changing even the sign above 120°. The target asymmetries T_{IM} on the other hand show little influence at this energy. For this reason we show T_{IM} only at 260 MeV, where they are more sensitive to non-nucleonic degrees of freedom, in particular T_{11}. There is also a marked difference between Δ-IC in the impulse approximation and a coupled channel approach.

A recent experimental result[7] for T_{21} is shown in fig.3 with our calculation for the Reid potential including MEC and IC contributions, which however are unimportant at these energies as mentioned above. The agreement is quite satisfactory, but certainly more accurate measurements are necessary.

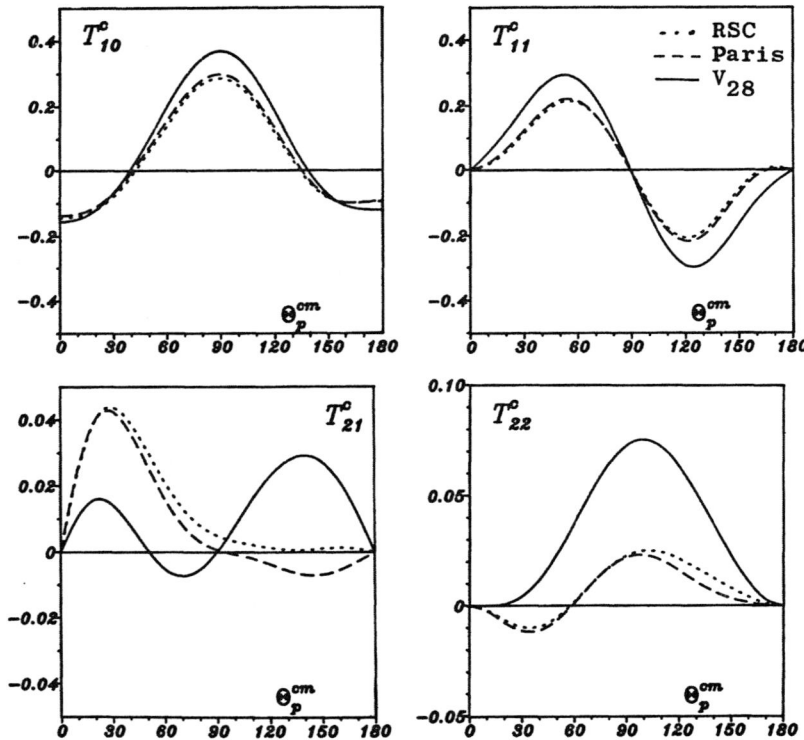

Fig.5   As fig.4 at 100 MeV for different potentials.

Double asymmetries T^C_{IM} for circularly polarized photons and polarized deuterons are displayed in figs.4 and 5. These observables are more influenced by subnuclear degrees of freedom. Already at 100 MeV T^C_{IM} show a significant change by MEC and IC, particularly strong for tensor polarized deuterons. At 100 MeV MEC and IC show equal importance and one notes again a significant difference for IC whether calculated in IA or in CC.

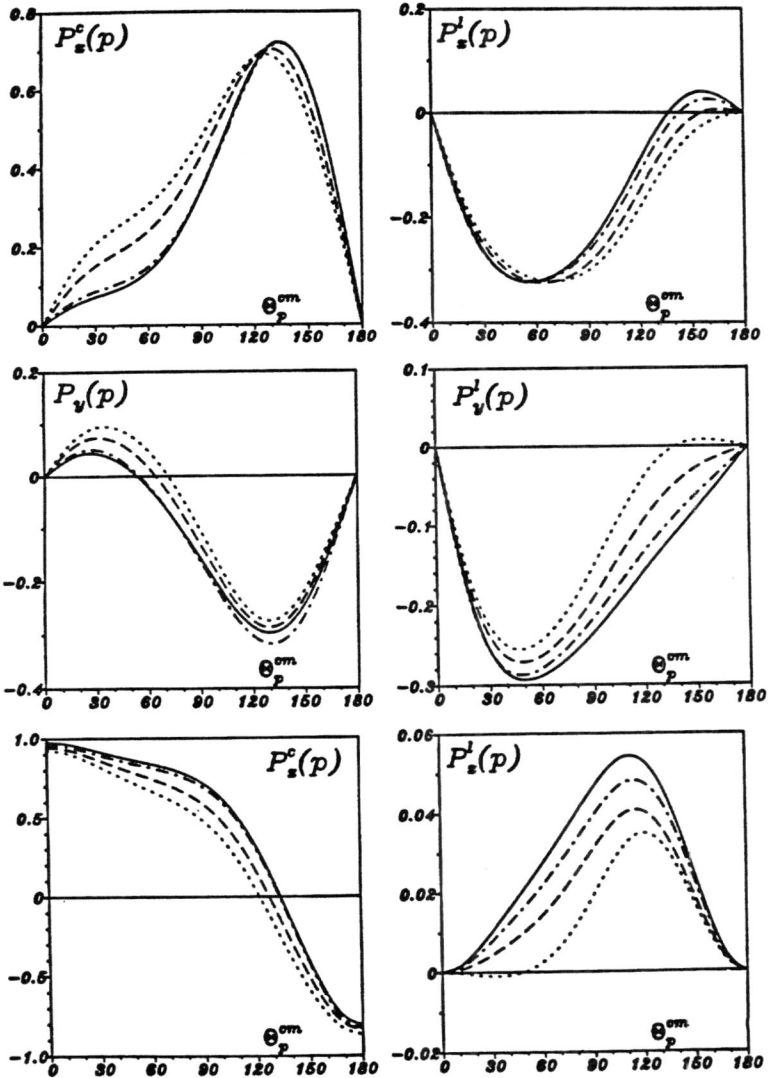

Fig.6   Polarization of outgoing proton for polarized photons at 100 MeV. Notation as in fig.1.

232

The potential model dependence is shown in fig.5. Particular T_{21}^{C} and T_{22}^{C} exhibit quite a different behaviour if IC are calculated in a coupled channel approach.

As an example for single nucleon polarization observables we will discuss for an outgoing proton all photon asymmetries $P_x^{\ell,C}(p)$, shown in fig.6 and for the x-component all target asymmetries $P_x^{IM}(p)$ as shown in fig.7. At 100 MeV one notes in general a modest influence from MEC and IC on $P_x^{\ell/C}(p)$ except for $P_z^{\ell}(p)$, where these effects are more distinct (fig.6). Isobar contributions become more pronounced at 260 MeV for $P_{Y/z}^{\ell}(p)$ and $P_x^{C}(p)$, where the former two depend also on the dynamical treatment of the Δ. On the other hand the target asymmetries $P_x^{IM}(p)$ at 100 MeV do not change significantly if MEC and IC are included (fig.7) and also the potential model dependence is small. But again at 260 MeV one readily sees a strong influence from Δ degrees of freedom and its dynamics.

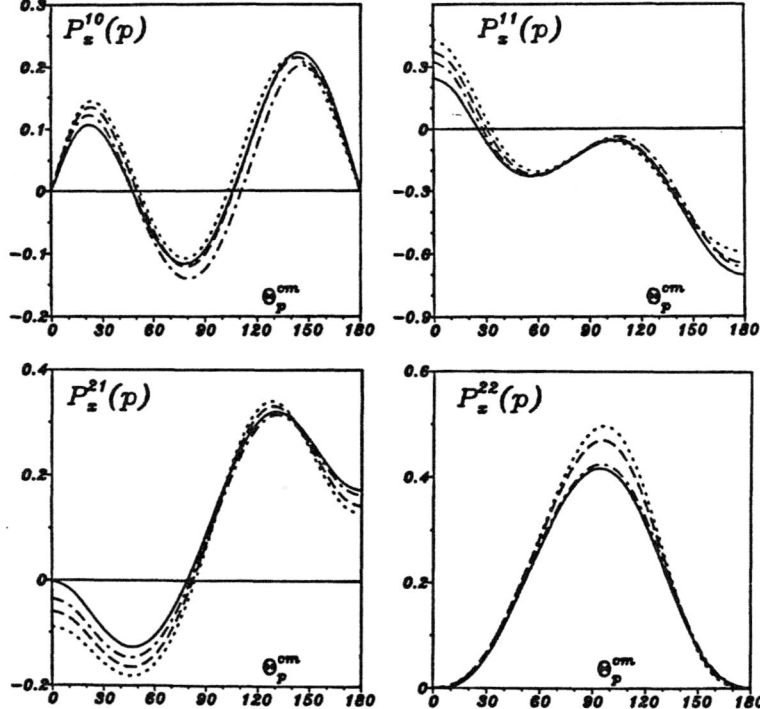

Fig.7 x-component of proton polarization for oriented deuterons at 100 MeV. Notation as in fig.1.

Finally we show all two nucleon polarization observables $P_{x_i x_j}$ for unpolarized photons and deuterons in fig.8. While at 100 MeV only two of them, P_{xx} and P_{yy}, are strongly

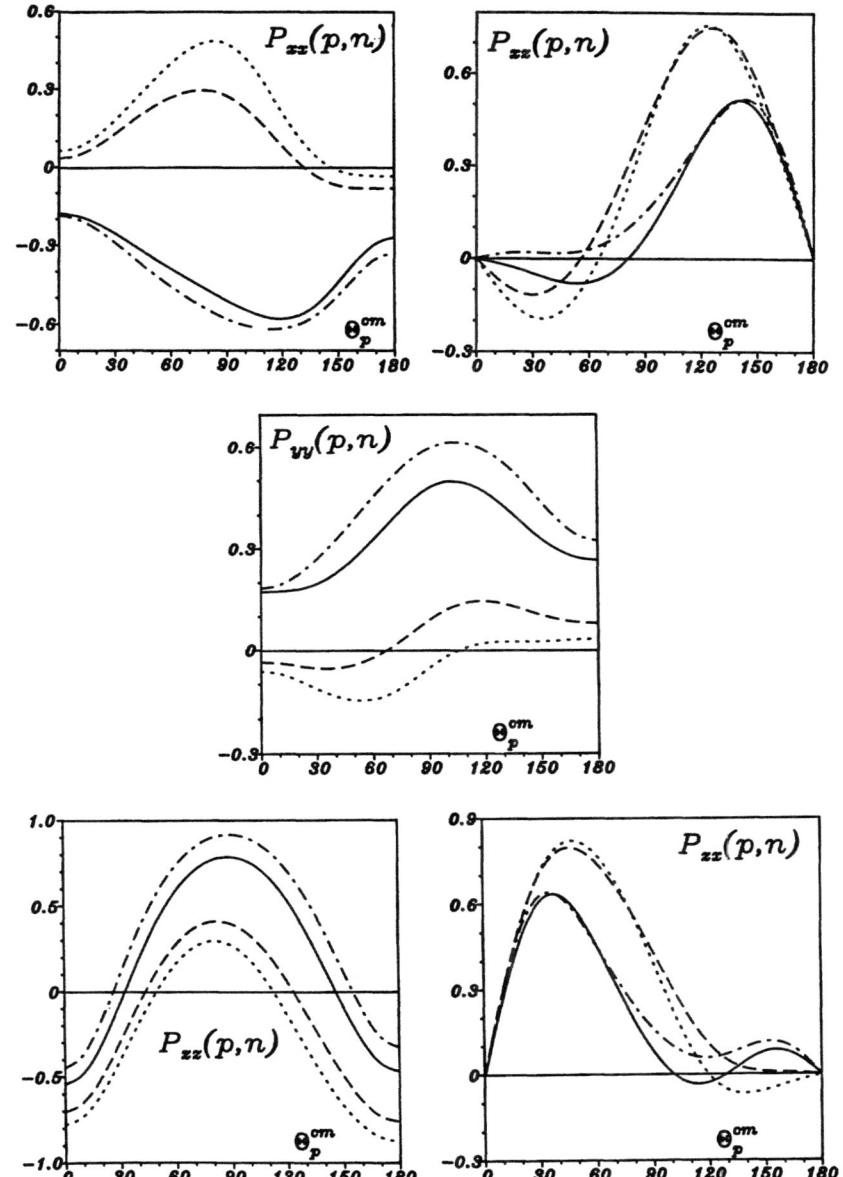

Fig.8 Polarization of both outgoing nucleons for unpolarized photons and deuterons at 260 MeV. Notation as in fig.1

affected by MEC and IC, this effect becomes quite dramatic at 260 MeV leading even to a sign change for P_{xx} and a drastic enhancement for P_{yy}. At this energy also the other double polarization observables in fig.8 show a clear sign of subnucleonic degrees of freedom.

4. Disintegration by polarized electrons

In this last section we will discuss the influence of the neutron electric form factor on the polarization asymmetry in $d(\vec{e},e'n)p$ in quasi-free scattering of longitudinally polarized electrons [8], where the neutron is detected in coincidence with the scattered electron. The differential cross section for $d(e,e'N)N$ is given by [9,10]

$$d^3\sigma/dk'd\Omega_e, d\Omega^{cm} = (\alpha/6\pi^2)(k_o'/k_o q_\nu^4)(\rho_L\ f_L\ +\ \rho_T\ f_T \qquad (13)$$
$$+\ \rho_{LT}\ f_{LT}\cos\phi\ +\ \rho_{TT}\ f_{TT}\cos2\phi$$
$$+\ h\ \rho_{LT}'\ f_{LT}'\sin\phi).$$

Here h describes the degree of longitudinal electron polarization. The dependence of $d\sigma$ on the electron polarization is governed by the fifth structure function f_{LT}', which is the imaginary part of the longitudinal transverse interference term. Therefore it vanishes due to Watson's final state theorem, if the final state interaction is neglected. The electron polarization asymmetry is directly proportional to f_{LT}'.

$$\Sigma_e = -\ \rho_{LT}'f_{LT}'\ /\ (\rho_L f_L + \rho_T f_T - \rho_{TT}f_{TT}) \qquad (14)$$

In fig.9 we show Σ_e for two kinematics in the quasi-free region calculated with the RSC potential as final state interaction. In order to find sizeable effects from the neutron electric form factor we had to choose high momentum transfers. In both cases the normal theory (N) without sub-nuclear degrees of freedom exhibits two characteristic minima in the forward and backward direction. While at backward angles Σ_e is very little influenced by subnuclear degrees of freedom one finds a strong reduction at forward angles, especially by MEC. At these angles also the neutron electric form factor (dipole fit of ref.[13]) shows a sizeable even though not very pronounced effect, which requires corresponding precision in the experiment. Since, however,

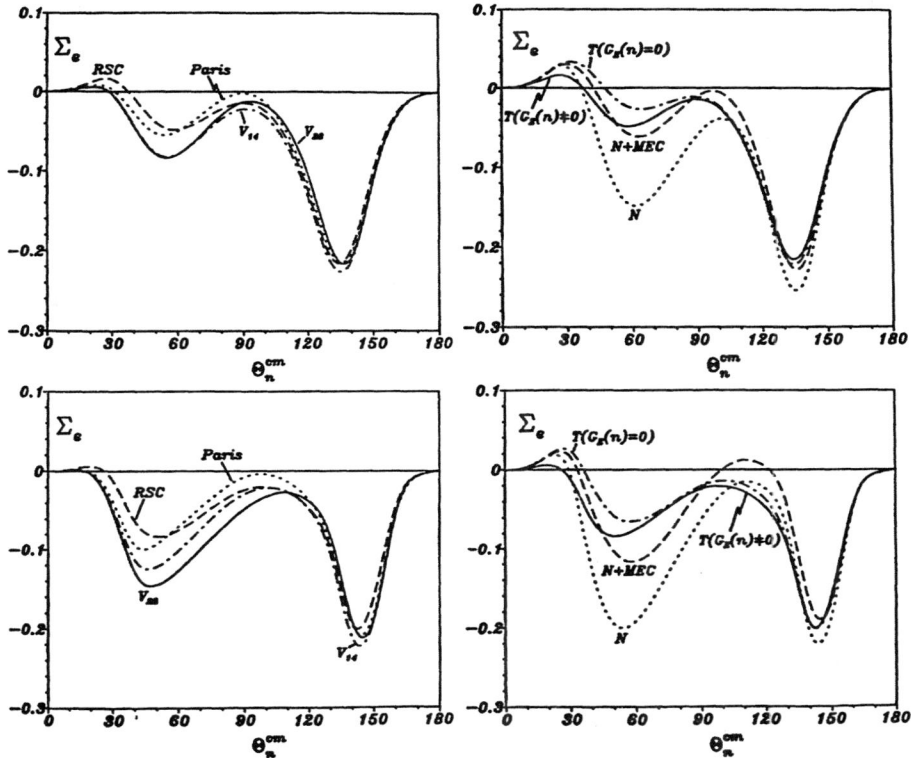

Fig.9 Separate contri-
butions to electron polarization asymmetry Σ_e for
$d(\vec{e},e'n)p$; top: $E_{np}=120$ MeV,
$q^2=12$ fm^{-2}, bottom: $E_{np}=200$
MeV, $q^2=20$ fm^{-2}. dotted:
normal contribution (N),
dashed: normal plus MEC,
dash-dot: T=N+MEC+IC, all
with $G_E(n)=0$, full: total
contribution with $G_E(n)=0$.

Fig.10 Electron polarization
asymmetry Σ_e for $d(\vec{e},e'n)p$
calculated with various
potential models (RSC[5],
Paris [11], Argonne V_{14} and
V_{28}[12], all with $G_E(n)=0$.
Same kinematics as in fig.9.

the extraction of the form factor needs theoretical input,
one has to see how model dependent these results are.

Fig.10 shows the dependence of Σ_e on various realistic
models for the final state interaction. One readily sees that
the different models introduce a variation of Σ_e, in
particular at forward angles, which has about the same size
than the effect of the neutron electric form factor. This
model dependence, which is even more pronounced at 200 MeV
excitation energy, seems to spoil the experimental
determination of $G_E(n)$. However, one should keep in mind that

at these rather high excitation energies one is near or beyond the limit of applicability of some of these potential models. Thus some of the model dependence might disappear if one admits only such potential models, which give a good NN scattering description at these high energies.

In view of the large influence of MEC (see fig.9) we have also studied the possible dependence on the exchange current model. To this end we have chosen the Paris potential using in one case only regularized π-MEC and in the other case the consistent gauge invariant exchange current for the Paris potential [14,15] and found a considerably smaller model dependence than the one introduced by the various potential models[8].

Thus we may conclude that the measurement of the electron polarization asymmetry Σ_e in the coincidence experiment d(e,e'n)p in the quasi-free region offers an interesting possibility for the determination of the electric form factor of the neutron. However, a very good knowledge of the NN interaction at these higher energies which we have considered is mandatory.

With this we will close this brief survey of polarization observables in deuteron photo- and electrodisintegration hoping to have convinced the reader that a wealth of interesting physics is buried in these polarization observables.

References

[1] Grüebler, W.,Proc. Workshop on Polarized Targets in Storage Rings,
     Argonne National Laboratory 1984, ANL-85-50, p.223

[2] Meyer, W., Lecture Notes in Physics, vol.234, p.413
     Proc. CEBAF 1985 Summer Workshop, p.237

[3] McNaughton, M.W., Proc. CEBAF 1985 Summer Workshop, p.271

[4] Arenhövel, H., Leidemann, W., to be published

[5] Reid, R.V.,Ann. Phys. 50, 411 (1968)

[6] Leidemann, W., Arenhövel, H., preprint MKPH-T-86-5, Nucl. Phys. (in print)

[7] Vesnovsky, D.K., et al., preprint 86-75, Novosibirsk

[8] Leidemann, W., Arenhövel, H., preprint MKPH-T-86-7, Phys.Lett. (in print)

[9] Donnelly, T.W., Proc. of CEBAF 1985 Summer Workshop,p.57

[10] Boffi, S., Giusti, C., Pacati, F.D., Nucl. Phys. A435 697 (1985)

[11] Lacombe, M. et al., Phys. Rev. C21,861 (1980)

[12] Wiringa, R.W., Smith, R.A., Ainsworth, T.L., Phys. Rev. C29, 1207 (1984)

[13] Galster, S. et al., Nucl. Phys. B32, 221 (1971)

[14] Riska, D.O., Phys. Scripta 31, 107 (1985); 31, 4471 (1985)

[15] Buchmann, A., Leidemann, W., Arenhövel, H., Nucl. Phys. A443, 726 (1985)

POLARIZED RADIATIVE CAPTURE EXPERIMENTS AND THE D-STATE OF ^{4}HE AND ^{3}HE

H. R. Weller

Duke University and Triangle Universities Nuclear Laboratory

Durham, North Carolina 27706

The tensor polarized capture reactions ^{2}H(d,γ)^{4}He and ^{1}H(d,γ)^{3}He are a sensitive means for measuring D-state effects in ^{4}He and ^{3}He. Data and recent calculations are considered. Values for the D-to-S state asymptotic ratio of -0.2 ± 0.05 for ^{4}He and 0.032 ± 0.014 for ^{3}He are deduced from the capture data. Results are compared to the results of other recent measurements of D-state effects in these systems. The effect of the D-state of ^{4}He on the low energy ^{2}H(d,γ)^{4}He cross-section is shown to be profound.

Previous experimental studies of the ^{2}H(d,γ)^{4}He reaction were limited to the use of unpolarized beams.[1-4] Differential cross sections were measured as a function of energy and angle over a broad range of energies. These data were interpreted as indicating that the d-d capture reaction proceeds primarily via E2 radiation in the low energy (E_d of 5-20 MeV) region. Furthermore, the angular distributions were consistent with D-wave capture forming a 2$^+$ (S=0) state followed by an E2 transition to the 0$^+$, L=0, S=0 ground state of ^{4}He. If we use the notation ^SL_J where L+S = J to label the initial and final states, this corresponds to the $^0D_2$ (E2)$\rightarrow$$^0S_0$ transition.

Dramatic evidence for the presence of strength in the ^{2}H(d,γ)^{4}He reaction other than that corresponding to E2 capture to the ground state of ^{4}He which has L=0, S=0, J=0 has been obtained. This evidence was discovered in measurements of the tensor analyzing power produced when a 10 MeV beam of tensor polarized deuterons was captured by a deuterium target.[5] The preponderance of E2 radiation in this low energy region is expected from the fact that the reaction involves the collision of two identical (T=0) bosons. This symmetry restricts the scattering states ^SL_J to those for which L+S is even. Of course this restriction could be broken by multistep processes such as deuteron break-up. However, within the framework of the one-step model of the reaction, the formation of a 1$^-$ state would require

L=S=1. This means that the strong (ΔS=0) E1 transition to the ground state (S=0) would be forbidden. Also, since, in the long wavelength limit (and neglecting the spin dependent part of the E1-operator) ΔT=0, E1 transitions are forbidden in self-conjugate nuclei, the T=0 states states of ^{4}He formed in the d+d capture channel should not decay via E1-transitions to the ground state. Isospin selection rules also give a considerable inhibition to ΔT=0 M1 transitions in ^{4}He. Therefore, we expect E2 radiation to dominate this reaction, although small amounts of M2, spin-flip E1 and M1 strength could occur.

If the ground state of ^{4}He contains a D-state admixture it would be expected to be present largely in the two-deuteron like configuration. As will be seen below, this representation of the ground state is not orthogonal to the single particle components and has been calculated as having a 67% overlap with the "physical" ground state. Since J = L+S, L=2 requires S=2 in the ground state of ^{4}He (the D-state). The E2 capture to this state should come primarily from 2^+ (S=2) continuum states. Since we can form a 2^+ S=2 state with two deuterons having L=0, 2 or 4, there will be three E2 transition matrix elements having S=2: $^2S_2$ (E2)$\rightarrow$$^2D_0$, $^2D_2$ (E2)$\rightarrow$$^2D_0$ and $^2G_2$ (E2)$\rightarrow$$^2D_0$. The observables of the ^{2}H($\vec{d},\gamma$)^{4}He reaction, when considering only E2 radiation, can therefore be written in terms of four-transition-matrix elements: the three above, plus the strong $^0D_2$ (E2)$\rightarrow$$^0S_0$ term. The expression for the angle integrated cross section, for example, is given by

$$\sigma_T = 4\,\pi A_0 = 4\,\pi\,(0.556)\,[\,|^0D_2|^2 + |^2S_2|^2 + |^2D_2|^2 + |^2G_2|^2\,]\,,$$

where we have used a shortened notation for the four E2-transition matrix elements.

The presence of these three S=2 transition-matrix-elements (TME's) (which are expected to involve the D-state of ^{4}He) will be difficult to detect in the cross section. Not only will they represent only a small contribution to σ_T, but, since states of different channel spin don't interfere in the cross section observables, the angular distribution coefficients for $\sigma(\theta)$ can be expected to show only small effects from these S=2 type terms. If we consider vector polarized deuterons, we find that the vector analyzing powers are also insensitive to these terms since S, S', and 1 must triangulate here. So with, for example, S=0 and S'=2, there will be no vector analyzing power. Therefore the interference term involving the large S=0 T-matrix element and the small S=2 matrix element is absent from this observable.

The tensor analyzing powers ($T_{2q}(\theta)$), however, arise from the product of an S=0 and an S'=2 type TME (S, S' and 2 triangulate here). In the context of the E2 model of this reaction, the expressions for these observables can be written in terms of three quantities A, B and C where,

$$A \equiv {}^0D_2{}^2S_2\cos\phi\,({}^0D_2 - {}^2S_2) \tag{1}$$
$$B \equiv {}^0D_2{}^2D_2\cos\phi\,({}^0D_2 - {}^2D_2)\,,\text{ and}$$
$$C \equiv {}^0D_2{}^2G_2\cos\phi\,({}^0D_2 - {}^2G_2)\,.$$

The expressions for the $T_{kq}(\theta)$ will, in general, involve two types of terms: products of two S=2 TME's and terms which are products of an S=2 and an S=0 type TME (A, B and C).

However, since the S=2 TME's are expected to be small compared to the S=0 one (the S=2 terms are proportional to the D-state strength in ^{4}He), the $(S=2)^2$ type terms should be of secondary importance. If only the first order terms are kept, the results are:

$$T_{20}(\theta) = + 0.497A - 0.594B + 0.797C, \tag{2}$$
$$T_{22}(\theta) = - 0.203A - 0.242B - 0.055C,$$
$$T_{21}(\theta) = (- 0.145A + 0.087B + 0.155C) \times (2.8 - 5.6 \cos^2\theta) / \sin 2\theta,$$

and $iT_{11}(\theta) = 0$,

where the TME's $^0D_2$, $^2S_2$, $^2D_2$, and $^2G_2$ are normalized such that $5/9 \sum (^SL_J)^2 = 1.0$. These expressions indicate that the assumption of pure E2 radiation leads to the result that, if second order terms (products of two S=2 TME's) are neglected, then $T_{20}(\theta)$ and $T_{22}(\theta)$ will be isotropic. Of course deviations from this isotropy should be expected near 0°, 90° and 180° where the cross section due to the $^0D_2$ TME goes to zero so that the second order terms (and other "small" contributions) become important.

The most extensive set of polarized deuteron capture data to date have been obtained for incident deuteron beams of 10 MeV.[5,6] The gas target cell used in both the TUNL and the Wisconsin work was such that the center of target beam energy was about 9.6 MeV. The combined data for $\sigma(\theta)$, $iT_{11}(\theta)$, $T_{20}(\theta)$ and $T_{22}(\theta)$ are shown in Figs. 1 and 2.

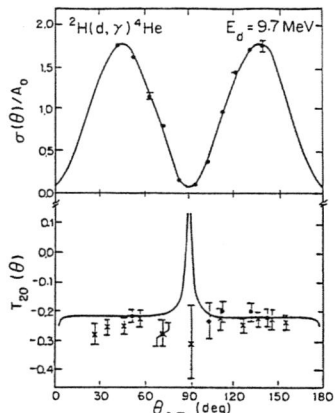

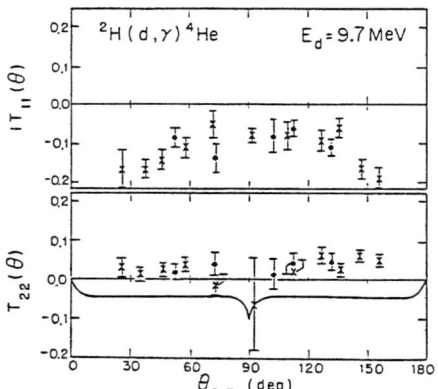

Fig. 1. $\sigma(\theta)$ and $T_{20}(\theta)$. TUNL (·); WISC (x). Curves are calculations. (See text).

Fig. 2. $T_{22}(\theta)$ and $iT_{11}(\theta)$ data as in Fig. 1. Curve for $T_{22}(\theta)$ is calculation described in text.

An examination of these results reveals the following impressions. First of all, $\sigma(\theta)$ is dominated by a $\sin^2 2\theta$ form, as expected for D-wave capture followed by an E2 transition to the $^0S_0$ ground state of ^{4}He. An inspection of the tensor analyzing powers reveals that $T_{20}(\theta)$ and $T_{22}(\theta)$ are remarkably isotropic, especially if the data point near 90° is neglected.

Finally, the large vector analyzing power, having the same sign at all measured angles, indicates the presence of non-E2 radiation. An evaluation of this will be considered below.

The first attempt to relate these experimentally determined quantities to the D-state of the α-particle was presented (along with the first data) in Ref. 5. In this rather simple calculation, a direct E2 mechanism and point deuterons were assumed. The ground-state wave function was constructed from two Woods-Saxon potentials which bound the two point deuterons by 23.84 MeV with L=0 (V_0=54.6 MeV, r_0=2.0 fm, a=1.0 fm), and L=2 (V_0=120.8 MeV, r_0=2.0 fm, a=1.0 fm), respectively. The continuum wave functions for each channel (S=0 and S=2) were generated from these same two potentials. Siegert's form of the E2 (spin-independent) operator was used to construct the four relevant E2 TME's. The results of this calculation, which includes the "second order" $(S=2)^2$ type terms, are shown in Fig. 1 for a two-deuteron ground state ^{4}He wavefunction, which contains a 5% L=2 admixture. This amount of D-state in ^{4}He gives a value of -0.22 for the constant value of $T_{20}(\theta)$, which is the value obtained when a constant is fit to the $T_{20}(\theta)$ data (-0.220 ± 0.014).[5]

One outstanding problem in the comparison of this simple model calculation with these data is in the vector analyzing power data $iT_{11}(\theta)$ (or $A_y(\theta)$, where $iT_{11}(\theta) = (\sqrt{3}/2)A_y(\theta)$). Pure E2 radiation gives rise to finite vector analyzing powers only through terms which are products of two S=2 amplitudes. This vector analyzing power in the pure E2 model is predicted to be very small (~0.01) and to have opposite signs for angles greater and less than 90°. In order to produce the observed result, radiation of opposite parity must be present. While either E1 or M2 are likely candidates, M2 seems preferred since E1 is especially difficult to occur in this reaction, the most likely contribution coming from the spin dependent isoscalar E1 operator (see below for an evaluation of this strength). In the M2 case, due to the restriction that L+S be even, the strength must originate from a continuum state having $^SL_J = {}^1P_2$ or $^1F_2$. The presence of finite b_1 and b_3 coefficients due to M2-E2 interference will, in fact, be essentially indistinguishable in the vector analyzing power from that due to E1-E2 interference. Both produce the result that $b_3 = 0.67 b_1$ if only the leading terms (interference with the dominant $^0D_2$ E2 strength) are considered.

In order to examine the data for the possible presence and the implications of M2 (E1) radiation, we took the results of the direct capture model for the four E2-TME's which gave the experimental T_{20} as a given. The two (complex) M2 matrix elements ($^1P_2$ and $^1F_2$) were then added and searched on to fit the $A_y(\theta)$ data. The data of TUNL[5] and WISC[6] at 9.6 MeV were included here. The resulting fit is shown in Fig. 3. The M2 contribution represented ~3% of the total integrated capture cross-section. We then used these M2 TME's and the direct capture model E2 TME's and calculated $\sigma(\theta)$ and $T_{20}(\theta)$. The results are shown in Fig. 4. It is evident that the M2 radiation which is able to account for the vector analyzing power has only a minor effect on the tensor analyzing power data $T_{20}(\theta)$. The addition of the 3% M2 strength only effects $T_{20}(\theta)$ in the region of 90° where, in fact, it

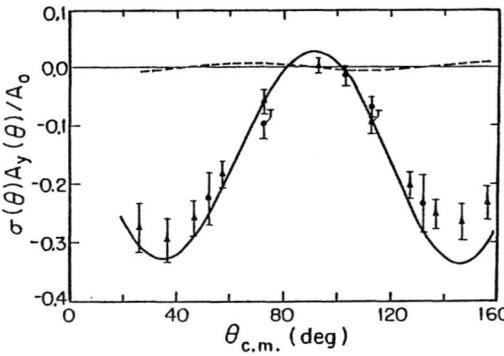

appears to improve the agreement between theory and experiment. The insensitivity of the tensor analyzing power to the presence of M2 strength is understood since the M2 contributions arise from S=1 terms. Such terms can interfere with the dominant $^0D_2$ (E2) strength to give

Fig. 3. TUNL ($\bullet$); WISC ($\blacktriangle$).
- - - Pure E2. —— E2 + M2 fit.

vector analyzing power, but must interfere with other S=1 or (small) S=2 terms to give rise to tensor analyzing power. Hence the extraction of D-state information on ^{4}He from the tensor analyzing powers observed in the ^{2}H(d,γ)^{4}He reaction appears to be quite insensitive to the M2 (E1) radiation which generates a sizeable vector analyzing power.

As mentioned earlier, one can consider either the spherical tensor analyzing powers $T_{kq}(\theta)$ or the cartesian tensor analyzing powers $A_{ij}(\theta)$. There is, in fact, some interesting physics which occurs depending on which representation is used. It is easily shown that the relationship of $A_{yy}(\theta)$ to the spherical tensor analyzing powers $T_{20}(\theta)$ and $T_{22}(\theta)$ is given by

$$A_{yy}(\theta) = -\frac{1}{\sqrt{2}} T_{20}(\theta) - \sqrt{3}\, T_{22}(\theta).$$

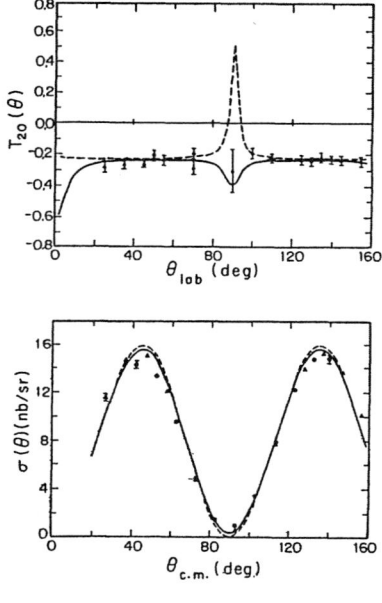

Fig. 4. Data: TUNL ($\cdot$), WISC ($\blacktriangle$).
Curves as in Fig. 3.

If we use equations (2) for $T_{20}(\theta)$ and $T_{22}(\theta)$ we find $A_{yy}(\theta) = 0.839\,B - 0.468\,C$. The quantity $A_{yy}(\theta)$ is independent of A, which represents the $^0D_2$ $^2S_2$ interference term. The $^2S_2$ TME, which corresponds to S-wave capture to the D-state of ^{4}He, is expected to be the most problematical S=2 TME. There are two reasons for this: First, this term could include a possible contribution from the deuteron D-state to the $^2S_2$ (E2)$\rightarrow$$^0S_0$ transition, not present in the other terms. Furthermore, the scattering phase shifts indicate that the S-partial-waves are highly distorted, whereas the D and G-waves are only weakly distorted. Thus, A_{yy} is insensitive to both the effects of the deuteron-D-state and the larger entrance-channel distortions.

There have been several attempts[7,8] to improve the theoretical analysis of this problem beyond that previously described. The contribution of Ref. 7 consisted of using the same bound state wavefunctions as those of Ref. 5 (Woods-Saxon wells), but replacing the continuum wavefunctions by ones which reproduced the scattering phase shifts. A more recent calculation which is an extension of this work will be described below.[8]

A significant advance in the theoretical development of this problem was made in Ref. 9. In this work variational wavefunctions having a pair-correlation operator containing central, tensor and spin correlations were constructed. A realistic Hamiltonian which included a three-nucleon interaction and which gave a reasonable value for the binding energy of ^{4}He was employed. The results of this work indicated that the probability that two T=0 pairs in ^{4}He be simultaneously in deuteron states is 0.67. Therefore it appears as though measuring the D-state of ^{4}He contained in the two deuteron like configurations is actually measuring the major part of the D-state in ^{4}He. The authors of Ref. 9 tabulate the wavefunctions of ^{4}He in momentum space for two realistic two-body interaction models: the Argonne and the Illinois interactions. These two models yield values for the ^{4}He D-state parameter D_2 of -0.16 and -0.24 fm^2, respectively. Tostevin[8] has recently used these wavefunctions in a calculation of the ^{2}H$(\vec{d},\gamma)^4$He reaction observables. In his work the wavefunctions were constructed from Woods-Saxon wells with geometries chosen to model the results of Ref. 9 as closely as possible, and depths adjusted to reproduce the d-d separation energy. The continuum wavefunctions for this calculation were obtained from separable potentials which were fitted to the one-channel resonating group model phase shifts of Chwieroth et al.[10] for the channels $(\ell\ s) = (0\ 2), (2\ 0), (2\ 2),$ and $(4\ 2)$. Two additional channel wavefunctions were needed. $\chi_{\ell s} = \chi_{31}$ was noted to be very weakly distorted and so $\chi_{31} = 4\pi j_3(k\rho)$ was assumed, while χ_{11} was calculated in an attractive spherical square well chosen to reproduce δ_{11}. The full operators, including the spin-dependent terms, for E1, M1, E2 and M2 multipoles in the long-wavelength approximation were employed.

The results of these calcualtions are compared with the experimental data for $A_y(\theta)$, $A_{yy}(\theta)$ and $T_{20}(\theta)$ in Figs. 5, 6 and 7. A reasonable description of $A_y(\theta)$ is obtained when either M2, E1 or both are included. Note that the E1 strength here arises from the isoscalar spin dependent part of the E1 operator. In the present two-deuteron model the isovector E1 strength is zero. Furthermore, the E2 + E1 and the E2 + M2 results are essentially identical in their angular dependence since F-wave capture is very small and no ^1P_J phase shift splitting is present. The ^{4}He wavefunction used here is that of the Argonne interaction.

The results for $A_{yy}(\theta)$ are also quite remarkable. Either two-body potential (Argonne

244

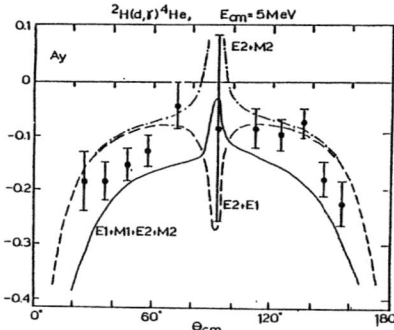

Fig. 5. WISC A_y data (Ref. 6) and Tostevin calculations (Ref. 8).

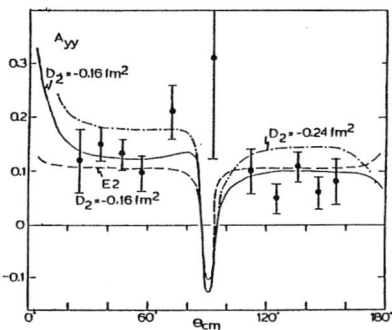

Fig. 6. A_{yy} data of Ref. 6. E2 only curve uses Argonne (D_2=-0.16). Other two curves have E1+M1+E2+M2.

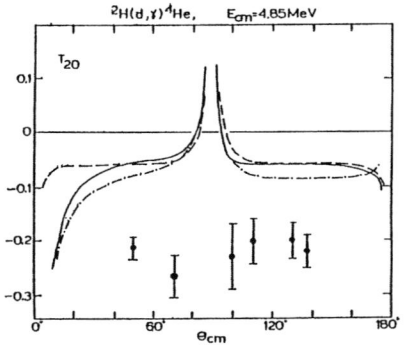

Fig. 7. TUNL data. Curves coded as in Fig.6.

with D_2 = -0.16 or Illinois with D_2 = -0.24) gives a good description of the data. The inclusion of multipoles other than E2 gives rise to a slight asymmetry in $A_{yy}(\theta)$ which appears to be present in the data. However, the E2 only calculation (dashed curve) gives most of the observed experimental result, as expected.

Finally, we see that the model fails to reproduce the $T_{20}(\theta)$ observable. This presumably reflects the fact that the amplitude B, which is absent to first order in $A_{yy}(\theta)$, is poorly described by the present model.

The results of this calculation and the comparison with the data seem to indicate that the wavefunctions of Ref. 9 with D_2 = -0.16 or -0.24 fm^2 are quite reasonable. Since to a "good" approximation $\rho = \alpha^2 D_2$, where ρ is the asymptotic D-to-S state ratio and $\alpha = 1.072$ fm^{-1}, these results suggest $\rho \sim -0.2$. We will come back to this point later. These calculations also confirm our previous interpretations which indicated that while the vector analyzing powers arise from interference of the dominant E2 radiation with small M2 (E1) admixtures, these additional multipoles have only a minor effect on the tensor analyzing powers.

Experimental efforts to study the energy dependence of the vector and tensor analyzing powers of the ^{2}H(d,γ)^{4}He reaction are underway. Preliminary results for A_{yy}(E) and A_y(E) at θ_{lab} = 130° are shown in Figs. 8 and 9. As seen here, A_{yy} is relatively constant with energy to within experimental error with a typical value of about 0.07 for E of 3-to-12 MeV. However, the vector analyzing power $A_y(\theta)$ changes

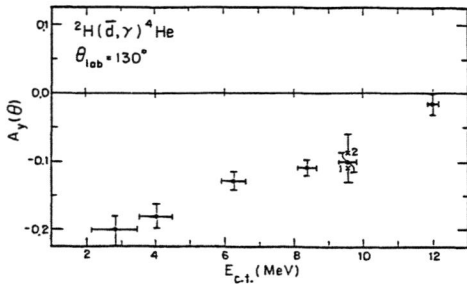

Fig. 8. A_y data, TUNL (·), WISC (x): 1 (127°), 2 (136°). (See Ref. 23).

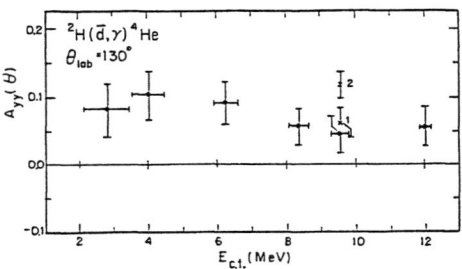

Fig. 9. Same as Fig. 8, but A_{yy}.

almost linearly from -0.2 at 3 MeV to near zero at 12 MeV. This energy dependence was measured to test the calculation of Tostevin[9] which attributes A_y to the interference of the dominant E2 radiation with an M2/E1 "background." Preliminary results of his calculation of $A_y(E)$ indicate qualitative agreement with the data of Fig.8.

While the dominant $^0S_0$ component of the α-particle can be readily populated via E2 transitions only from the $^0D_2$ continuum, the $^2D_0$ component can be populated via E2 transitions from the $^2S_2$, $^2D_2$ and $^2G_2$ continuum states. The important point here is that S-wave capture is allowed in the case of the $^2H(d,\gamma)^4He$ reaction leading to the D-state of ^4He. We can therefore expect the effect of the D-state to increase in this reaction as the incident energy is lowered to the

point where the centrifugal barrier begins to seriously suppress the D-wave capture leading to the S-state.

The angular distribution of the outgoing γ-rays will be sensitive to the presence of this S-wave capture in that such a term will give rise to an isotropic component. This component will add to the usual $\sin^2 2\theta$ term arising from D-wave capture to the S-state. For point geometry, the $^2H(d,\gamma)^4He$ γ-ray angular distribution would have a zero 90° yield if the reaction were pure E2 and the ^4He ground state had no D-state admixture. The addition of D-state should increase the 90° yield. Therefore, the ratio R = $\sigma(90°)/\sigma(135°)$

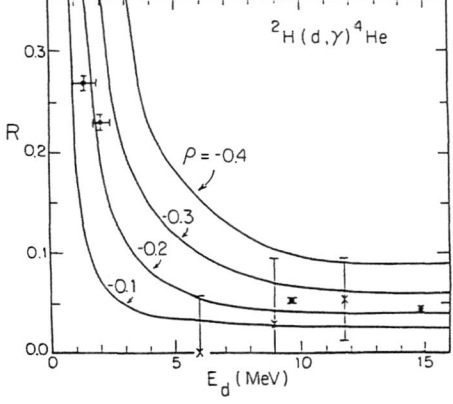

Fig. 10. R=$\sigma(90°)$ cm / $\sigma(135°)$ cm vs E_d(lab). Curves are calculations (see text).

should be a good measure of the D-state contribution to the capture reaction. R would be expected to increase rapidly with decreasing energy at very low energies where the S-wave capture starts to dominate. Experimental values of R are shown in Fig. 10.[11]

The curves shown in Fig. 10 are the results of calculations which are similar to those previously described. Once again the scattering wavefunctions were generated with separable potentials constrained to give the energy dependence of the phase shifts of the wavefunctions obtained using the RGM.[10] At these low energies, however, Coulomb repulsion, previously neglected, must be taken into account. Each E2 amplitude was given an energy dependent Coulomb penetration factor which was calculated as the square of the ratio of the E2 amplitude obtained with a Coulomb distorted wave to that obtained with a plane wave of the same energy. The bound state wavefunctions were generated in Woods-Saxon potentials, as before. Only E2 radiation was considered. The results are shown for various values of ρ, the asymptotic D-to-S state ratio ($\rho = N_2/N_0$). Figure 10 suggests that $\rho = -0.20 \pm 0.05$, in good agreement with the calculated value obtained in Ref. 9 and shown (see above) to be consistent with the 9.6 MeV polarized and unpolarized data.

The value of ρ obtained from these low-energy data is expected to be quite reliable. For one thing, the behavior of the wavefunctions inside the nucleus do not affect the results at these energies; the asymptotic part of the wavefunction determines the result. Furthermore, nuclear distortions should be small at these energies so that, for example, tensor force mixing in the continuum can be safely neglected. The value of ρ obtained here corresponds to a D-state probability of between 5 and 13% for the ^{4}He ground state in the two deuteron model employed in the calculation.

An important implication of the presence of S-wave capture to the D-state of ^{4}He in the ^{2}H(d,γ)^{4}He reaction concerns the absolute cross sections. Figure 11 shows the available cross section data for this reaction for deuteron lab energies ranging from 50 keV to 50 MeV.[11] The calculated curves are the predictions of the same model used above to compute R. It can be seen that an order of magnitude enhancement occurs at E_d (lab) = 0.5 MeV as a result of the presence of the D-state. Additional measurements at very low energies have been performed recently by the Cal Tech group.[12] In this work cross section and angular distribution data were measured for E_d (lab) of 100 keV-to-1.0 MeV. The results of their angular distribution measurements are shown in Fig. 12, which

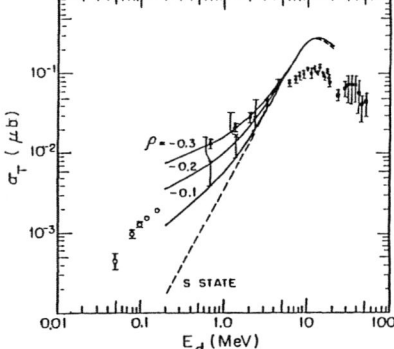

Fig. 11. Angle integrated cross-section vs E_d(lab). See Ref. 11 for original references to data. Curves are described in the text.

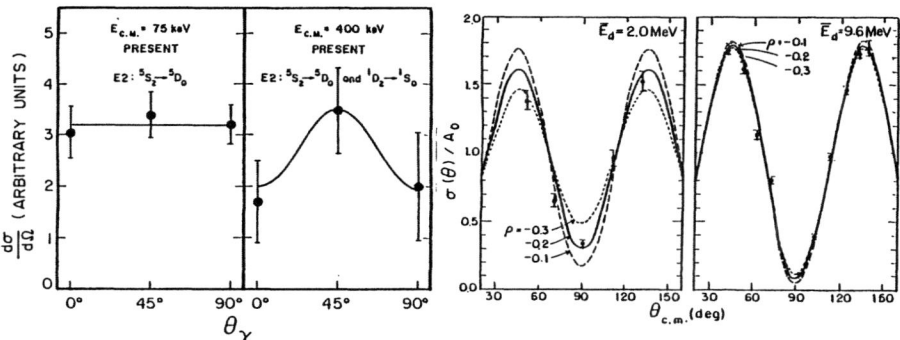

Fig. 12. Data on left are from Ref. 12. Data on right ($\overline{E}_d$ is center-of-target lab energy) are from Ref. 11. Curves are direct capture calculations with D-state included. (Calculations on right same as Figs. 10 and 11).

also includes data from a previous work. These results show that at $E_d(\text{lab})=0.8$ MeV the ratio of $\sigma(90°)$-to-$\sigma(45°)$ is about 0.6, while at E_d (lab) = 150 keV this ratio is 1.0. This indicates that isotropy is indeed achieved at $E_d(\text{lab})=150$ keV, as expected for pure S-wave E2 capture to the D-state component of ^{4}He. These authors used their low energy cross section measurements, which agreed well with those of Ref. 13, to evaluate the astrophysical factor $S(E)$: $S(E) = \sigma(E) \, E \, \exp(2\pi\eta)$, where $2\pi\eta = 31.40/E_{c.m.}^{1/2}$ ($E_{c.m.}$ in keV). An E2 direct capture (two-point deuteron) calculation which used Wood-Saxon potentials with $r_0=1.4$ fm and $a=0.7$ fm adjusted to bind the S and D states independently was able to fit these data very well. The factor $S(E)$ obtained with this model is almost three orders of magnitude larger when calculated with the D-state of ^{4}He included compared to that for a pure S-state ^{4}He ground state. The resulting rate due to this new evaluation is ~32 times larger than recommended previously.[14] The quantitative consequences of this result for various astrophysical problems will, of course, require detailed calculations.

It is worth noting that the direct capture calculations performed at these low energies were used to estimate a D-state admixture within the framework of the model. The result obtained, $1.4 \pm 0.8\%$, is considerably lower than previous estimates extracted at higher energies. However, the model used here is considerably less sophisticated than those discussed above. Furthermore, finite size effects, deuteron polarizability, and deuteron D-state effects could be sizeable and need to be accounted for in a good calculation.

A definite prediction can be made for the tensor analyzing power in the case of very low energy ^{2}H(d,γ)^{4}He capture. If we assume that the ^{2}H(d,γ)^{4}He reaction is dominated by S-wave capture to the D-state of ^{4}He at, say, $E_d(\text{lab})=150$ keV where the isotropic angular distribution suggests that this is the case, then there is only one E2 transition matrix element:

($^2S_2$ (E2)→$^2D_0$) ≡ Se$^{i\phi}$s. The tensor analyzing powers, assuming only E2 radiation, are now determined since $|S|^2$ appears in the numerator and denominator of the expressions for these observables: $T_{20}(\theta)$ = -0.35 $P_2(\cos\theta)$, $T_{22}(\theta)$ = -0.144 $P_2^2(\cos\theta)$, and $A_{yy}(\theta)$ = 0.62 - 0.38 $\cos^2\theta$. It will be interesting to test the assumptions of pure S-wave E2 capture to the D-state by measuring the tensor analyzing powers at these low energies. Such experiments are presently underway at TUNL.

Additional data on the $^2H(d,\gamma)^4He$ reaction at higher energies are becoming available. Data obtained at TUNL at E_{ct}(lab)=14.85 MeV for $\sigma(\theta)$ and $T_{20}(\theta)$ are shown in Fig. 13. We[15] have also recently measured $A_{yy}(\theta)$ and $A_y(\theta)$ at E_d = 20 MeV at the LBL 88-inch Cyclotron Laboratory. The preliminary data from this experiment indicate that if the $A_{yy}(\theta)$ data are fitted by a constant, the value of A_{yy} = 0.14 ± 0.05 is obtained. Finally, measurements of $\sigma(\theta)$ and $A_{yy}(\theta)$ have been performed at IUCF[16] for E_d(lab) ≅ 90 MeV. The preliminary results of this work appear to indicate that the assumption of dominant E2 radiation is valid even at this rather high energy. The hope here, of course, is that the reaction mechanism will remain relatively simple as the energy increases. The measurements at the higher energies, when performed with adequate precision, should be sensitive to the small r behavior of the wavefunctions and should therefore be able to test the theoretical predictions more thoroughly, especially effects such as those due to the 3-body force.

Another experimental means of measuring the D-state of ^4He consists of measuring the tensor analyzing powers in (d,α) transfer reactions. Measurements of the $^{89}Y(\vec{d},\alpha_0)^{87}Sr$ reaction have been made at 9, 12 and 16 MeV[17,18]. Exact finite-range distorted-wave Born approximation analyses including the effects of L and J mixing were used to extract a D_2 parameter. The d-d relative motion in the α-particle was described by Woods-Saxon potentials consistent with their separation energies and with the α-particle rms radius. A D_2 value of -0.3 ± 0.1 fm^2 was extracted. This result is in reasonable agreement with

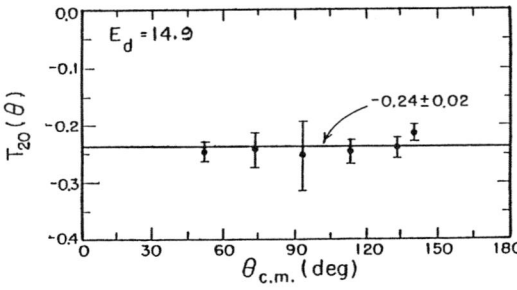

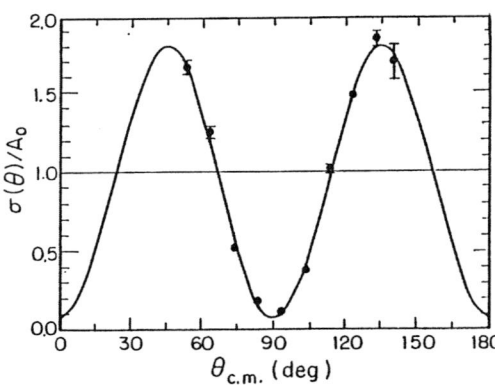

Fig. 13. TUNL data at E_d=14.9MeV. T_{20} fit to a constant, $\sigma(\theta)$ fit to constant + $\sin^2 2\theta$.

the value calculated in Ref. 9 and shown to be consistent with the 9.6 MeV capture data, as well as the value of ρ (~D_2) found from the behavior of the low energy capture angular distributions.

Measurements of a similar nature have also been applied to ^{3}He to study its D-state by means of the ^{1}H($\vec{d}$,γ)^{3}He reaction. In this case the tensor analyzing power is zero unless there exists a finite S=3/2 transition matrix element. The reaction in this case is dominated by E1 radiation for deuteron energies below 50 MeV. If the E1 transition is ΔS=0, then an S=3/2 transition matrix element will arise primarily from the existence of S=3/2 configurations in the ground state. Since L+S=J and J=1/2, this implies L=2.

The tensor analyzing power data $T_{20}(\theta)$ were measured at E_d(lab)=19.8 MeV by means of a magnetic spectrograph which detected the recoiling ^{3}He nuclei[19]. Since the ^{3}He recoils are confined to a 5.2° cone about the beam axis, they can all be accepted by the spectrograph at once. The ^{3}He energy determines the γ-ray angle, so that the entire angular distribution of $T_{20}(\theta_\gamma)$ can be observed at once. These data are shown in Fig. 14. The curves are the result of two-body direct capture calculations which employed Faddeev-generated wavefunctions to describe the ground state of ^{3}He. Two-body wavefunctions were projected out of these three-body wavefunctions. These Faddeev wavefunctions were obtained by inputting the two-body properties. When the deuteron D-state probability was taken to be 4-7%, the ^{3}He D-state was calculated to be between 5 and 9%. The solid curve in Fig. 14 corresponds to this range of ^{3}He D-state. If the deuteron D-state is changed to 2% or 9%, the dashed and dot-dashed curves of Fig. 14 are obtained, respectively. Although it is not a very strong effect, a better result is obtained when the accepted deuteron D-state probability is input. Unfortunately these comparisons are complicated by the fact that the different Faddeev wavefunctions have, in addition to different D-state probabilities, different binding energies ranging (for ^{3}H) from 9.46 to 7.13 MeV. This, of course, has a strong effect on the asymptotic wavefunction which plays the dominant role in determining the observables at these energies. Although the calculated $T_{20}(\theta)$ [with no free parameters] gives a reasonably nice description of the observed $T_{20}(\theta)$ data, it also appears to be too small in magnitude

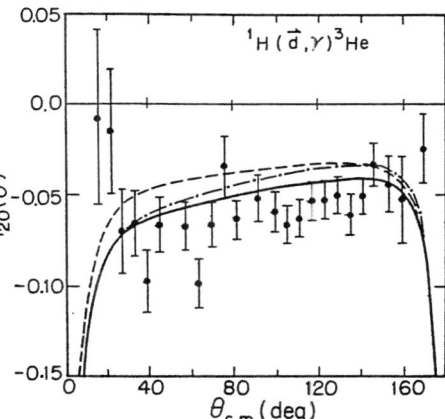

Fig. 14. $T_{20}(\theta)$ data from Ref. 19. Calculations are described in text; Solid: 9.1%, Dashed 2.2%, Dot-Dashed 11.5% D-state in ^{3}He.

by about 20%. Whether this reflects a problem with the three-body wavefunction or a problem with the approximate two-body calculation done here must await a full three-body model calculation which treats the bound and continuum states in a consistent manner.

If asymptotic forms of the bound state wavefunctions are used, it is possible to obtain an estimate of the asymptotic D-to-S state ratio (η) for ^{3}He by fitting the calculated $T_{20}(\theta)$ to the data. In this case there is a direct proportionality between η and T_{20}. A good fit is obtained for a D-to-S state ratio of $\eta = 0.032 \pm 0.014$, where the error reflects the experimental uncertainties[19].

An additional measurement and calculation of this reaction have been reported recently.[20] In this case the tensor analyzing power $A_{yy}(\theta)$ was measured for the ^{1}H(d,γ)^{3}He reaction at $\theta_{lab}=90°$ and projectile energies of 29.2 and 45.3 MeV. Careful measurements yielded the values of 0.0282 ± 0.0016 and 0.0113 ± 0.0014 for $A_{yy}(90°)$ at these two energies, respectively. The results were compared to an exact Faddeev calculation which treated both the bound and continuum states using the Reid-Soft-Core (RSC) interaction. The result indicated that in the region of 2-5 fm, the RSC calculated D-state wavefunction is too large by ~20%.

The calculations mentioned above have been examined to answer several heretofore outstanding questions. First, the calculations showed[20] that distortions in the entrance channel are important. A plane-wave approximation cannot lead to a quantitative understanding of the data. The authors[20] were also able to estimate the contributions of multipolarities with L>2, and showed that their contributions are "entirely negligible" in the angular range of 40°-to-150°. In addition it was shown that the effects of the deuteron D-state are minor. A_{yy} changes by only 4.7% if the deuteron D-state is omitted from their calculations. Finally, it was found that 89% of A_{yy} is due to transitions into D-state components with $\ell=\lambda=1$, where ℓ denotes the relative angular momentum between the "interacting" particles, and λ denotes the angular momentum of the "free" particle relative to the center-of-mass of ^{3}He. Since this part of the D-state is not contained in a projected two body wavefunction, it should be that a two-body calculation misses 89% of A_{yy} (or T_{20}). This, however, does not appear to be the case, as seen in Fig. 14. This apparent contradiction must await a new full Faddeev calculation for its resolution.

One final new experimental result which I would like to mention is the measurement of the ^{4}He($\vec{\mathrm{d}}$,^{3}He)^{3}H reaction performed with tensor polarized deuterons at $E_d=35$ and 45 MeV[21]. This reaction is dominated by proton transfer at forward angles of the outgoing ^{3}He, and by neutron transfer at backward angles (or forward angles of the outgoing ^{3}H). Hence one can extract η_{3H} and η_{3He} from the same data by analytic continuation to the pole region. The results of the present experiment were $\eta_{3H}=0.045 \pm 0.005$ and $\eta_{3He}=0.048 \pm 0.005$. These results overlap the result obtained from the ^{1}H($\vec{\mathrm{d}}$,γ)^{3}He experiment at 19.8 MeV ($\eta=0.032 \pm 0.014$). The model calculations of Ref. 22 suggest a value of η of between

0.025 and 0.029.

Tensor polarized deuteron capture experiments have proven to be a valuable source of information for the D-state components of ^{4}He and ^{3}He. Present analyses of the data indicate that the asymptotic D-to-S state ratio for ^{4}He is about $\rho=-0.2 \pm 0.05$, in good agreement with recent theoretical estimates of this quantity and with the value of D_2 obtained from a study of (d,α) transfer reactions. A simple two-point-deuteron model suggests that the two-deuteron configuration in the ground state of ^{4}He contains between 5 and 13% D-state. The existence of this D-state allows for S-wave capture which dominates the reaction at very low energies, giving an order of magnitude enhancement to the calculated cross section at $E_d=0.5$ MeV. New measurements of these very low energy cross sections have resulted in a major revision of the astrophysical factor for this reaction. The results for ^{3}He indicate that Faddeev wavefunctions which have 5-to-9% D-state probability for ^{3}He give a reasonable description of the data, although not exact. The extracted D-to-S state asymptotic ratio of $\eta=-0.032 \pm 0.014$ is in reasonable agreement with theory and with the results obtained from the ^{4}He$(\vec{d},^3$He$)^3$H reaction ($\eta=0.048 \pm 0.005$). A critical need for full Faddeev calculations is obvious. The sensitivity of these observables to the presence of three-body-forces remains to be exploited.

This work was partially supported by the U.S. Department of Energy, Office of High Energy and Nuclear Physics, under Contract No. DE-ACO5-76ER01067.

References

1. R. Zurmuhle, W. Stephens and H. Straub, Phys. Rev. 132, 751 (1963).

2. W.E. Meyerhof, W. Feldman, S. Gilbert and W. O'Connell, Nucl. Phys. A131, 489 (1969).

3. D.M. Skopik and W.R. Dodge, Phys. Rev. C6, 43 (1972).

4. M. Poutissou and W. Del Bianco, Nucl. Phys. A199, 517 (1973).

5. H.R. Weller, P. Colby, N.R. Roberson and D.R. Tilley, Phys. Rev. Letts. 53, 1325 (1984)

6. S. Mellema, T.R. Wang and W. Haeberli, Phys. Letts. 166B, 282 (1986); and private communication.

7. F.D.Santos, A. Arriaga, A.M. Eiró and J.A. Tostevin, Phys. Rev. C31, 707 (1985).

8. J.A. Tostevin, to be published in Phys. Rev. C (private communication).

9. R. Schiavilla, V.R. Pandharipande and R.B. Wiringa, Nucl. Phys. A449, 219 (1986).

10. F.S. Chwieroth, Y.C.Tang and D.R. Thompson, Nucl. Phys. A189, 1 (1977).

11. H.R. Weller, P. Colby, J. Langenbrunner, Z.D. Huang, D.R. Tilley, F.D. Santos, A. Arriaga and A.M. Eiró, Phys. Rev. C34, 32 (1986). (note: Fig. 12 comes from this

paper, but plotting errors were corrected for the present paper).

12. C.A. Barnes, K.H. Chang, T.R. Donoghue, C. Rolfs and J. Kammeraad, (private communication), submitted to Phys. Rev. Letts.

13. J.F. Wilkerson III and F.E. Cecil, Phys. Rev. C31, 2036 (1985).

14. W.A. Fowler, G.R. Caughlan and B.A. Zimmerman, Ann. Rev. Astr. Ap. 5, 525 (1967).

15. H.R. Weller, R.M. Whitton, S. Kuhn, E. Hayward and W. Dodge, private communication.

16. K. Pitts, IUCF (private communication).

17. B.C. Karp, E.J. Ludwig, W.J. Thompson and F.D. Santos, Phys. Rev. Letts. 53, 1619 (1986).

18. B.C. Karp, E.J. Ludwig, J.E. Bowsher, B.L. Burks, T.B. Clegg, F.D. Santos and A.M. Eiró, Nucl. Phys. A457, 15 (1986).

19. M.C. Vetterli, J.A. Kuehner, A.J. Trudel, C.L. Woods, R. Dymarz, A.A. Pilt and H.R. Weller, Phys. Rev. Letts. 54, 1129 (1985).

20. J. Jourdan et al., Nucl. Phys. A453, 220 (1986).

21. B. Vuaridel et al., Invited talk at the International Symposium on the Three-Body Force in the Three-Nucleon System, Washington, D.C., 1986 (and W. Grüebler, private communication).

22. B.F. Gibson and D.R. Lehman, Phys. Rev. C29, 1017 (1984).

23. Drs. L.C. Biedenharn (Duke U.) and M. Danos (NBS) have suggested that the behavior of A_y as a function of energy is correlated with the behavior of the ratio of the ^{4}He(γ,p)-to-^{4}He(γ,n) cross sections. This ratio has been previously viewed as evidence for charge symmetry breaking (CSB). However, Biedenharn and Danos have pointed out that charge polarization of the deuterons can be the mechanism for this apparent CSB. Their model also leads to a dipole moment and therefore to the existence of E1 radiation in the d+d capture channel. This E1 radiation should disappear as the (γ,p)-to-(γ,n) ratio approaches 1 (i.e., near $E_x=30$ MeV or $E_d=12$ MeV). If A_y arises from E1-E2 interference, then its energy dependence is consistent with this model.

THE TWO NUCLEON GROUND STATE

M. Rosa-Clot

Departement of Physics, UNiversity of Pisa I-56100 Italy

INFN Sezione di Pisa, I-56100 Italy

ABSTRACT : The properties of the deuteron are reviewed with particular enphasis to the D-state observables. The dominance of the tensor force and of the one-pion exchange potential (OPEP), together with constraint on the asymtptotic behaviour, allow to determine the S- and D- wave functions from 1 fm to infinity .

The modern theory of N-N interaction begins, in the fourties, with the analysis of the ground state properties of the two nucleon system: the deuteron. This statement sound a little provocative, but is what actually happened when the measurement of the deuteron quadrupole moment [1] provided a decisive evidence for a tensor component of the nuclear forces [2,3].

It is interesting that even in this first pioneering works, the attention was focoused on the D-state properties of the deuteron and the binding energy in itself was just only a parameter to take into account to get consistent calculations.

The situation looks not too different now.

The deuteron basic parameters are given in Table I [4,5]. They are the binding energy ε and magnetic moment μ which depend on the short range behaviour of the system, the radius r_d, the effective range ρ and S- normalization A_S which essentially fix the size of the deuteron and the quadrupole moment and the asymptotic D/S ratio η which characterizes the D-state properties . Apart from ε and μ these quantities are considered "external observable" since they depend on the long range charactersitics of the deuteron system.

TABLE I: DEUTERON PROPERTIES (Data are taken from ref. [4,5])

ε = 2.224575 (9) MeV	μ = 0.857406 (1) n.m.	
r_d= 1.963 (4) fm.	$\rho(-\varepsilon-\varepsilon)$ =1.737 (12) fm	A_S=0.8802 (20) fm$^{-1/2}$
η = 0.0271 (4)	Q = 0.2859 (3) fm^2	

We analyze these quantities in the general framework of the N-N potential model.
The normalized deuteron wave function can be written, using the S and D-state radial wave functions u(r) and w(r) as:

$$\Phi_{JM}(r,\theta) = 1/(\sqrt{4\pi}) \cdot [u(r) + S_{12}/\sqrt{8} \cdot w(r)]/r \cdot \chi_{JM} \qquad (1)$$

where χ_{JM} is the normalized spin wave function (J=1, M=±1,0).

The S and D-wave functions satisfies the coupled equations (hereafter h=c=1):

$$u''(r) = [\alpha^2 + U_{00}(r)] \cdot u(r) + U_{02}(r)\, w(r) \qquad (2a)$$

$$w''(r) = [\alpha^2 + 6/r^2 + U_{22}(r)] \cdot w(r) + U_{20}(r)\, u(r) \qquad (2b)$$

where $\alpha^2 = |M\epsilon|$, M = 938.9 MeV and the terms $U_{ik}=MV_{ik}$ are linear combinations of the central, tensor, spin-orbit and quadratic spin-orbit potentials V_C, V_T, V_{LS} and V_{LL}: $V_{00}=V_C$ $V_{02}=V_{20}=\sqrt{8} \cdot V_T$; $V_{22} = V_C -2V_T -3V_{LS} -3V_{LL}$.

Outside the range of nuclear interaction, the radial wave functions are completely determined by the deuteron binding energy ϵ and the asymptotic normalization constants A_S and $A_D=\eta\, A_S$:

$$u(r) = A_S\, \tilde{u}(r) \;\text{---}\!\!> A_S e^{-\alpha r} \qquad (3a)$$

$$w(r) = A_S \tilde{w}(r) \;\text{---}\!\!> \eta\, A_S e^{-\alpha r}[1+3/\alpha r+3/(\alpha r)^2]. \qquad (3b)$$

<u>Binding Energy</u>: Its value can be expressed through the integrals:

$$\epsilon = <T> + <V> = \int [u'(r)^2 + w'(r)^2 + 6/r^2 \cdot w(r)^2]\, dr +$$

$$+ \int [u(r)^2 \cdot V_{00}(r) + w(r)^2 \cdot V_{22}(r) + 2 \cdot u(r) \cdot w(r) \cdot V_{02}(r)]dr. \qquad (4)$$

The value of ϵ is a small number resulting from the cancellation of kinetic energy term (about 20 MeV) and potential attraction. Its exact value results from a fine, rather arbitrary, tuning of the short range part of the potential.

To show this in more details we plot in fig.1 the first term of the second integral in Eq. 4 (the central contribution) using OPEP, Paris [6] and a V14 potential [7]).

This contribution is very different in the three cases. What is surprising is that the two modern potentials are rather different also at intermediate range (between 1 and 2 fm.) where in principle we should expect an analogous qualitative behaviour. The only information we get is that the OPEP central is 2-3 times too weak in this range. The repulsive behaviour of the potential near the origin is rather arbitrary and it can be modified to get the right binding energy.

The main contribution of the potential to ϵ comes from the S-D interference term (-18 MeV for Paris and - 20 MeV for V14 potential), well reproduced by the OPEP too.

Finally the diagonal D-state contribution is rather weak in all the three cases. The OPEP overstimates the repulsive contribution in the intermediate range, however in solving the differential equation (2b) details of V_{22} are not relevant due to the dominance of the centrifugal barrier.

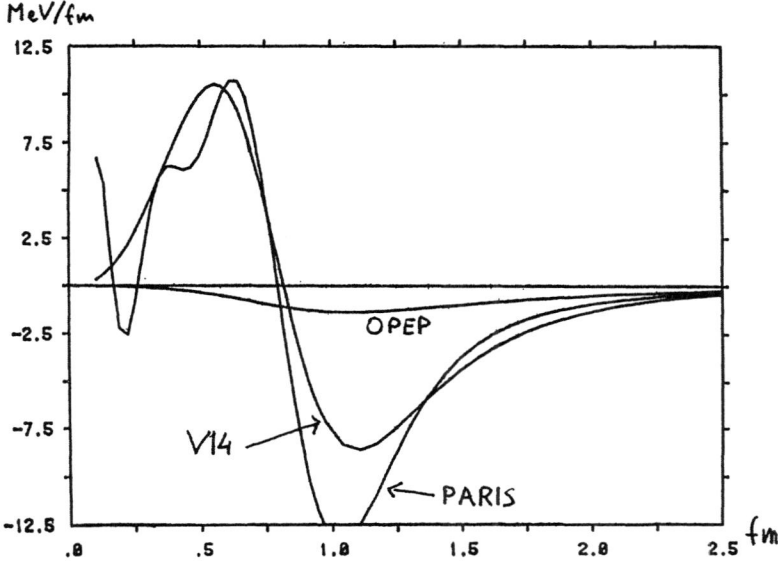

FIG 1. Plot of the potential density function $u^2 V_{00}$, using OPEP, Paris and V14 potentials respectively.

Magnetic moment: The main part of the magnetic moment μ is given by the addition of the proton and neutron magnetic moments, $(\mu_p + \mu_n) = 0.879696$ n.m. The remaining small difference $\delta\mu$ (only -2.6%) comes from many other effects [4]: D-state component, LS and LL potentials contribution due to the minimal coupling, isobar component in the deuteron wave function and exchange effects.

All this effects are small, with different sign and their evaluation is rather uncertain, so that the original hope to get information on the D-state probability P_D from the value of $\delta\mu$ was deceived by the complexity of the problem.

Deuteron size: The external quantities $\rho(-\varepsilon-\varepsilon)$ and A_S are related by [8]:

$$(1+\eta^2) A_S^2 = 2\alpha/[1-\alpha \, \rho(-\varepsilon, -\varepsilon)] \tag{5}$$

where

$$A_S = 2\int \left[\exp(-2\alpha r)-(\tilde{u}^2(r)-\tilde{w}^2(r))/(1+\eta^2)\right] dr \tag{6}$$

More recently the connection between r_d and A_S has been explored. Since the radius is defined as $1/4\int r^2(u^2(r)+w^2(r))dr$ the overwelming bulk of the contribution comes from the asymptotic S-wave amplitude which alone gives $r_d= A_S/4\,\alpha^{-3/2}=1.98$ fm. i.e. the experimental value to about 1% [9].

A careful analysis of the model dependent contribution and of the exchange effects allows to conclude that the measurement of r_d is the best way to get A_S from experiments. Even if there are discrepancies between different experimental results and method of analyisis we can conclude that A_S is experimentally know to a precision of 0.1 % [9,5].

D-state observables: The situation for the D-state observables is clear now at a level of 1%. Ericson and Rosa-Clot [10,11,12] have developed a method to reliably estimate the value of Q and η. They have also evidentiate that these quantities are dominated by the tensor OPEP:

$$V_\pi^T=f^2[1+3/(\mu r)+3/(\mu r)^2]e^{-\mu r}/r \tag{7}$$

where $f^2 =0.078$ and μ is the pion mass.

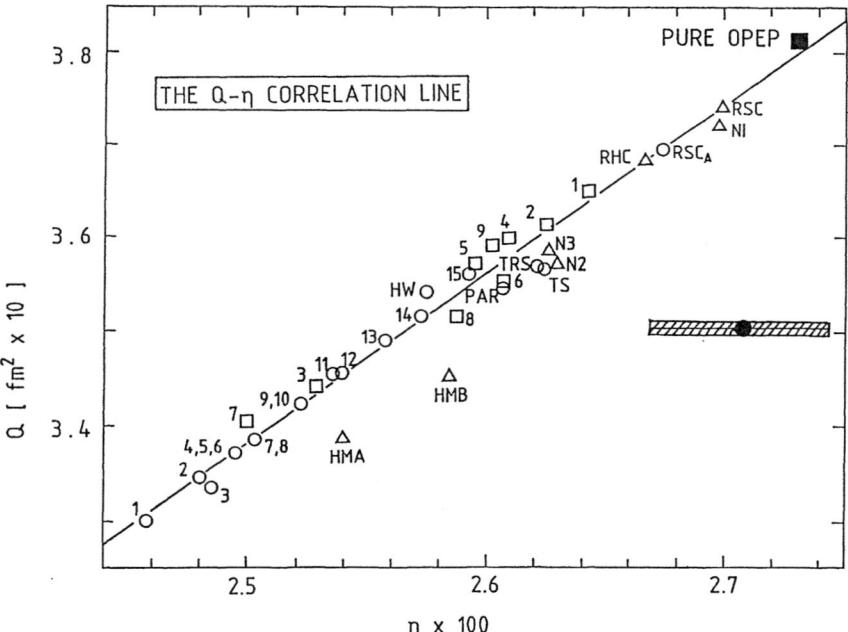

Fig. 2 Variation of Q/A_S^2 versus η for different potentials. Both Q and η are rescaled to the value $f^2=0.078$. The bar refers to the experimental values.

The main results of this analysis is given in the two equations for η and Q:

$$\eta = \int \eta(r)dr = M\sqrt{8} \int r J_2(r)\, V_T(r)\, \tilde{u}(r)\, dr \tag{8}$$

$$Q = 1/\sqrt{50}\, (A_S^2/\alpha^3) \int F(r)\, \eta(r)\, dr \tag{9}$$

where $J_2(r)$ is the regular solution of the omogeneous equation 1b and F(r) is a smooth function regular at the origin.

Furthermore Eq.s 7 and 8 imply a almost linear relation between η and Q which has to hold for any potential model. This relation is shown in fig. 2.

The wave functions: Let us assume now that the total binding energy is taken form experiments and not calculated as eigenvalue of the Schrœdinger equation. The coupled equations (2a,b) can be integrated inwards from large r for different values of the D/S ratio η. To do this we use explicitly the OPEP.

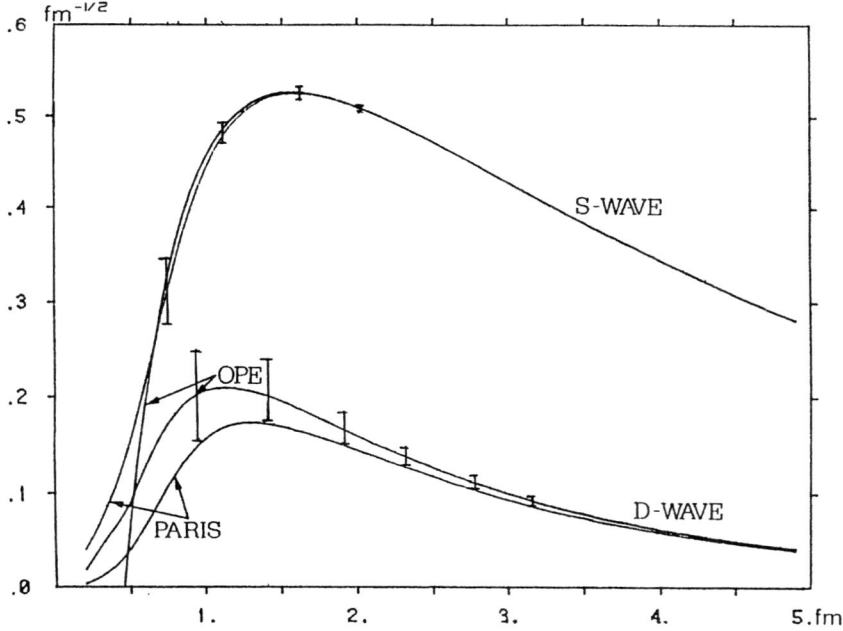

Fig 3. Plot of the S and D- wave functions obtained with OPEP non regularized. The value of $\eta = 0.0274$ has been chosen to get the D wave regular. The Paris wave functions are superposed. The error bar are the limit obtained exployting mathematical inequalities [16,17]

Fixing the S-wave asymptotic amplitude and varying the parameter η, we try to construct $\tilde{u}(r)$ and $\tilde{w}(r)$. The complete solution Φ_{JM} , being the energy eigenvalue fixed, is irregular. However the S and D-wave functions, $\tilde{u}(r)$ and $\tilde{w}(r)$, are regular respectively for $\eta=0.0274$ and $\eta=0.0271$ and that these two theoretical values are near to the physical one.

Fixing the value of A_S to the physical one we can see that the solution found above is approximately that obtained with a realistic potential from the eigenvalue equation.

In fig. 3 we plot the S and D wave for the value $\eta=0.0274$ which does the D-wave regular at the origin using the right value of A_S [13].

This result is rather stable and can be improved by regularizing the singular behaviour of the tensor potential. Actually, the introduction of a 'short range' form factor a' la Glendenning and Kramer [14] allows to get a regular solution, generally decreasing the value of η ($\eta=0.02635$ and a cut-off $\Lambda=900$ MeV [13,15]).

We can now impose the unitarity condition $\int [u(r)^2+w(r)^2]\ dr= P_S + P_D = 1$ to get the value of A_S which, in this case, turns out to be about 2.8% too small. This is a small discrepancy, but has to be qualitatively understood, in view of the level of precision used in this analysis.

The physical way to handle this patology is to introduce a more reliable central potential. If the central OPEP is mutiplied by a factor 2.5 in the region between 1 and 2 fm and smoothly matched to the right asymptotic behaviour we get, with $\Lambda=900$ MeV, $\eta=0.0262$. Asymptotic ratio is now correct and the agreement with the most sophisticated wave functions is substantially improved. In particular the value of the D-state pourcentage is 6.0%.

The D state probability P_D: The value P_D found above is in good agreement with the results of more sophistycated potentials. Since in the literature a lot of work has been done around the problem of the value of P_D it is worth to analyse the validity of this result.

We first note that the S-wave u(r) is remarkably stable and is very little affected by changes in the central part of the potential. So any variation in A_S can be seen as an effect of a modification either of the S-wave or of the D-wave function. Exploiting now the stability of u(r) and neglecting the term η^2 which is numerically irrelevant we can write [12]:

$$P_D= \int w^2(r)\ dr = \int [\ A_S^2 \exp(-2\alpha r) - u^2 (r)]\ dr - A_S^2\ \rho(-\epsilon,-\epsilon)/2 \qquad (9)$$

This relation allows to understand the stability of P_D in many recent calculations and in our oversimplified approach: once the effective range is fixed the value of P_D is strongly contrained by the stability of the S-wave solution. In this way we get the value $P_D = 6\% \pm 1\%$.

Schwartz inequalities: the same problem of building up the wave function was discussed by Klarsfeld, Martorell and Sprung [5,16] from a different point of view. They explored the consequences of OPEP when the experimental value of Q and the deuteron radius $<r^2>$ are imposed as constraints. Then, establishing Schwartz inequalities, it's possible to get the value of η within narrow limits, independent from direct experiments: $0.0262 < \eta < 0.0264$. This approach also gives a good description of the S-wave function outside 1 fm. and determines the D-wave function as well, although with larger uncertainties.

The main merit of this approach is to be model independent and to give rigorous bound on the wave function and on the parameter η, starting from very weak assumtions on the short range behviour of the potential.

The short range behaviour of the wave function: The problem of the deuteron wave function in the short range region (r<0.8 fm) is still open. A lot of work have been done using modern approaches which involve quark model and cromodynamics. Even if we are rather sceptic respect to these issues, the possibility to evidentiate anomalous effect in the short range region still exists.

Recently it was suggested that the wave function can have a node at about 0.5 fm. [17]. To explore this possibility, even in the framework of the potential model, we need some more experimental inputs.

The data obtained from the electron scattering at low momentum transfer ($q < 4$ fm^{-1}) can be used to improve the analysis and to explore the details of the deuteron wave function up to 0.5 fm. from origin. Work is in progress [18].

The present status of the analysis can be summarized looking at the plot of the isodensity curves for the deuteron probability function given by $|\Phi_{11}(r,\theta)|^2$ (fig. 4).

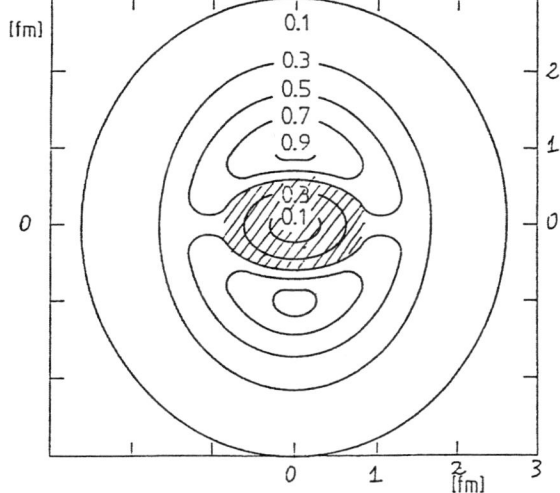

Fig.4 Isodensity for the deuteron probability function

The dashed area represent the region where uncertainties in the square of the wave function exceed the 5%. The plot evidentiates the present level of knowledge and in particular how small is the physical space allowed for search of anaomalous effects in the deuteron wave function.

REFERENCES

1. Kellog,J.M., Rabi, I.I., Ramsey, N.F., Zacharias,J.R. Phys.Rev. 56 728 (1938)
2. Schwinger, J. Phys.Rev 57 728 (1938)
3. Bethe, H.A., Phys.Rev. 57 390 (1940)
4. Ericson,T.E.O. Nucl.Phys. A416: 281c-296c (1984)
5. Klarsfeld, S., Martorell, J., Sprung, D.W.L. Nucl.Phys.A352: 113-24 (1981)
6. Lacombe, M.,Loiseau, B., Richard, J.M., Vinh Mau, R., R.Cote, J. Phys.Rev.C21: 861-73 (1983)
7. Wiringa, R.B., Smith, R.A., Ainsworth, T.L. Phys.Rev.C29: 1207 (1983)
8. Noyes, N.P., Ann. Rev. Nucl. Sci. 22 465 (1972
9. Ericson, T.E.O., Mahalanabis, J. Z.Phys. A 322 237 (1985)
10. Ericson, T.E.O., Rosa-Clot, M. Phys.Lett.110B: 193-98 (1982)
11. Ericson, T.E.O., Rosa-Clot, M. Nucl.Phys.A405: 497-533 (1984)
12. Ericson, T.E.O., Rosa-Clot, M. Ann.Rev.of Nucl.Sc.36: 233 (1985)
13. Righi, S., Rosa-Clot, M. Preprint IFUP-16-86
14. Glendenning, N.K., Kramer, G. Phys.Rev. 126: 2159-68 (1962)
15. Friar, J.L. Phys.Rev.C22: 796-812 (1980)
16. Klarsfeld, S., Martorell, J., Sprung, D.W.L. Jour.Phys.G10: 165 (1984)
17. Klarsfeld, S., Martorell, J., Sprung, D.W.L. Preprint 1986
18. Work is in progress in collaboration with Cambi,A., Mosconi, B., Ricci, P.

ON THE THEORY OF EXCHANGE CURRENTS

E. Truhlík and J. Adam, Jr.

Institute of Nuclear Physics, CS 250 68 Řež, Czechoslovakia

Abstract

The form of the continuity equation for the electromagnetic
meson exchange currents is discussed. The problem of including
the hadron form factors is considered.

1. Introduction

The stumbling block of the theory of the meson exchange
currents (MECs) has been the inclusion of the effects due to
the finite size of the nucleons - the electromagnetic (elm.)
and strong formfactors. Since there is no definite description
of these effects based on the fundamental theory of hadrons
(QCD), the formfactors are usually introduced in a phenomenological way. The important constraint to be obeyed by any
particular model of MECs follows from the continuity
equation:

$$\vec{k} \cdot \vec{j} = \left[H, \rho \right] \quad , \tag{1}$$

where H is the Hamiltonian of the nuclear system. Only if
eq. (1) is satisfied, the nuclear elm. transition amplitudes
are gauge invariant.

Although sophisticated technique has been developed [1,2] ,

which allow to construct the NN-potential and the MECs consistently in the same framework, it is rather difficult to
satisfy eq. (1) in the actual calculations, involving the use
of the realistic NN-potentials [3,4] in the description of
the nuclear states. It is not easy to construct the MECs
corresponding to the inner part of the potential, that is represented by multimeson exchange or treated in a phenomenological way. Moreover, the realistic potentials do not include
all contributions of the relativistic origin, necessary to
ensure approximate Lorentz invariance [5,6] . Recent studies
of the MECs have, therefore, concentrated mainly on the contributions due to those meson exchanges, that have been
firmly established in the theory of the NN-potential (π, ρ, ω).
Advanced techniques [1,2] have been used for the construction
of the exchange charge density (ρ (2)), that is of the relativistic origin. The spatial part of the exchange current
($\vec{j}(2)$) is usually treated in the static non-relativistic
approximation. In what follows we will deal with the isovector spatial exchange current $\vec{j}^3(2)$ (the isospin index will
be suppressed.)

The effects due to this current are most distinctly recognized in the description of the backward deuteron electrodisintegration [7] and the trinucleon isovector magnetic form
factor [8,9,10] . The data available cover the region of high
momenta transfer, where the calculations of the MECs contributions are rather sensitive to the elm. and strong formfactors used.

As for the elm. formfactors: the standard MECs [10] have been
multiplied either by the Dirac formfactor $F_1^V(k^2)$ or by the
Sachs one $(G_E(k^2))$. The predictions of the calculations with
$F_1^V(k^2)$ are in much better agreement with experimental data.
Since the difference between $F_1^V(k^2)$ and $G_E(k^2)$ is of the
relativistic origin, a number of authors [11-13] advocated
the use of $F_1^V(k^2)$ with the non-relativistic current $\vec{j}(2)$.

The modification of the meson-nucleon vertices BNN(B= π, ρ ...)
in the MEC operators has also been widely discussed [8,10,12,
14-17] . Following the theory of the NN-potential, the finite
size of the nucleons has been currently simulated by the mono-

pole formfactors:

$$\Gamma_B(q^2) = (\Lambda^2 - m_B^2)/(\Lambda^2 + q^2) \ , \qquad (2)$$

in each BNN vertex of the nucleon Born and seagull diagrams. Then, in order to satisfy eq. (1), the mesic current (as a whole) is to be multiplied by a rather complicated function of momenta [10,16]. For the Paris potential, which involves multiple meson exchange, Riska [18] and Arenhövel with colla- borators [19] generated π- and ρ-meson - like currents from the (non-relativistic) spin-spin and tensor parts of the po- tential. The Paris potential [3] includes, however, also A_1-meson exchange. The method, given in refs. [18,19], does not allow to identify unambigously the corresponding part of the potential and to construct the MECs of the A_1-meson range. Moreover, such a construction of the interaction dependent currents from the potential is basically a non-relativistic technique [20].

In this paper we discuss our choice of the elm. and strong formfactors in $\vec{j}(2)$ for the chiral model of the MECs [13].

2. The electromagnetic formfactors

In the consistent theory [1,2] one gets the elm. formfactor accompanying the MEC operator by taking into account every diagram (up to the desired order) in the perturbation theory considered (see the treatment of $\rho(2)$ in ref. [21]). In this approach, some velocity dependend contributions to the MECs (namely those due to the retardation in meson exchange) are fixed only if the corresponding terms in the NN-potential are treated in the same framework. This is usually not case for the semiphenomenological realistic NN-potentials. For this reason only the static current $\vec{j}(2)$ is commonly considered. In the determination of the static relativistic corrections to $\vec{j}(2)$, the continuity equation (1) plays an important heuristic role.

Keeping only the contributions up to the first order in the (pair) interaction and taking into account the leading relativistic corrections, eq. (1) splits into (for each operator the superscripts indicate the order in $1/m$):

$$\vec{k}.\vec{j}^{(1)}(1)=\left[T^{(1)}, \rho^{(0)}(1)\right] , \tag{4a}$$

$$\vec{k}.\vec{j}^{(1)}(2)=\left[V^{(1)}, \rho^{(0)}(1)\right] , \tag{4b}$$

$$\vec{k}.\vec{j}^{(3)}(1)=\left[T^{(1)}, \rho^{(2)}(1)\right]+\left[T^{(3)}, \rho^{(0)}(1)\right] , \tag{4c}$$

$$\vec{k}.\vec{j}^{(3)}(2)=\left[T^{(1)}, \rho^{(2)}(2)\right]+\left[V^{(1)}, \rho^{(2)}(1)\right]+\left[V^{(3)}, \rho^{(0)}(1)\right] \tag{4d}$$

The eqs. (4a,b) concern the standard non-relativistic operators [10], the S-matrix method leads to the Dirac formfactor $F_1^V(k^2)$ adjoined to each contribution to $\vec{j}^{(1)}(2)$ as well as $\rho^{(0)}(1)$ and the convection part of $\vec{j}^{(1)}(1)$ [10]. It is, however, the Sachs formfactor $G_E(k)$, which is connected with the charge distribution in nucleon. The one-body charge density, derived from the relativistic operator

$$\hat{\jmath}_\lambda = \frac{i}{2} \tau^3 (F_1^V \gamma_\lambda - \frac{1}{2m} F_2^V \sigma_{\lambda\mu} k_\mu) \tag{5}$$

by the Foldy-Wouthuysen transformation, reads up to the order $1/m^2$ as follows:

$$\rho^{(0)}(1) = F_1^V(k^2) \delta(\vec{k}+\vec{p}-\vec{p}') \frac{\tau^3}{2} = F_1^V(k^2) \rho_{point}(1) , \tag{6a}$$

$$\rho^{(2)}(1) = \vec{k}.\vec{s}(1), \quad \vec{s}(1) = -\frac{F_1^V+2F_2^V}{8m^2}\left[\vec{k}+i\vec{\sigma}\times(\vec{p}'+\vec{p})\right]\rho_{point}(1) . \tag{6b}$$

Dropping the velocity dependent spin-orbit part of $\rho^{(2)}(1)$ one gets the static charge density:

$$\rho_{st}(1) =\left[F_1^V - \frac{F_1^V+2F_2^V}{8m^2} \vec{k}^2\right] \rho_{point}(1) = G_E(k^2)(1-\frac{\vec{k}^2}{8m^2}) \tag{7}$$

$$\cdot \rho_{point}(1) = \tilde{G}_E(k^2) \rho_{point}(1) \cong G_E(k^2) \rho_{point}(1) .$$

The question is how to modify the non-relativistic currents $\vec{j}^{(1)}(1)$ and $\vec{j}^{(1)}(2)$, so that the commutators in eqs. (4a,b) would involve $\rho_{st}(1)$ instead of $\rho^{(0)}(1)$. Several authors [9,10,15,22,23] have done this just by replacing in the convection part of $\vec{j}^{(1)}(1)$ and in each contribution to $\vec{j}^{(1)}(2)$ $F_1^V(k^2)$ by $G_E(k^2)$. Let us point out that in this approximation the relativistic effects connected with the motion of the nucleon as a whole ($V^{(3)}$, $T^{(3)}$) are neglected as well as the two-body exchange charge density $\rho(2)$ and the non-static part of $\rho(1)$. Only the effects of the short range oscillations of the nucleon (Darwin-Foldy part of $\rho^{(2)}(1)$) are kept (often only part $\sim F_2^V$) to modify the interaction with the elm. field.

Delorme [12] has demonstrated another modification of the theory. To illustrate it let us remind the redefinition of the elm. current considered by Liou and Sobel [24] and Ohta [25] in their study of the low energy theorems. The relativistic one-body current can be presented as follows:

$$\vec{j}^{(3)}(1) = k_o \vec{s}(1) + i\vec{k} \times \vec{\mu}(1) \ , \ k_o = \frac{\vec{p}'^2}{2m} - \frac{\vec{p}^2}{2m} \ . \quad (8)$$

In the nuclear calculation it is natural to replace $k_o \rightarrow k_o = E_f - E_i$, where $E_f(E_i)$ is the energy of the final (initial) nuclear state. This is equivalent to the redefinition of the one-body current (containing now effectively an interaction dependent part):

$$\tilde{j}^{(3)}(1) = \left[H, \vec{s}(1)\right] + i\vec{k} \times \vec{\mu}(1) = \vec{j}^{(3)}(1) + \left[V, \vec{s}(1)\right] . \quad (9)$$

To keep the total current unchanged one should define:

$$\tilde{j}^{(3)}(2) = \vec{j}^{(3)}(2) - \left[V, \vec{s}(1)\right] \ . \quad (10)$$

The relativistic part of the continuity equation now reads:

$$\vec{k} \cdot \tilde{j}^{(3)}(1) = \left[H, \rho^{(2)}(1)\right] + \left[T^{(3)}, \rho^{(0)}(1)\right] \ , \quad (11a)$$

$$\vec{k} \cdot \tilde{j}^{(3)}(2) = \left[V^{(3)}, \rho^{(0)}(1)\right] + \left[T^{(1)}, \rho^{(2)}(2)\right] \ . \quad (11b)$$

Passing now to the static limit described above, it is easy to see that the relativistic correction to the r.h.s. of eq. (1) $\left[H, \vec{k} \cdot \vec{s}_{st}(1)\right]$ is compensated by the divergence of the modified one-body current $\tilde{j}^{(3)}(1)$. In fact, Delorme suggested that the whole exchange current $j(2)$ proportional to F_2 could be simulated by the redefinition of the one-body current. Here we present some arguments justifying this idea, limiting ourselves to the static limit.

Let us first specify the current $\vec{j}_{st}(2)$ in both approaches. Replacing $F_1^V \rightarrow G_E$ leads to:

$$\vec{j}_{st}^{G}(2) = G_E(k^2)\vec{j}_{point}^{(1)}(2) = \vec{j}^{(1)}(2) - \frac{\vec{k}^2}{4m^2} F_2^V(k^2)\vec{j}_{point}^{(1)}(2) \quad (12)$$

In the second approach:

$$\vec{j}_{D,st}(2) = \vec{j}^{(1)}(2) + \left[V, \vec{s}_{st}^{t}(1)\right] = \vec{j}^{(1)}(2) - \frac{F_2^V(k^2)}{4m^2}\vec{k} \quad (13)$$

$$\cdot (\vec{k} \cdot \vec{j}_{point}^{(1)}(2)) \quad , \quad \vec{s}_{st}^{t}(1) = - \frac{F_2^V(k^2)}{4m^2}\vec{k} \, \rho_{point}(1) \quad \cdot$$

For the moment we consider only the part of $\rho^{(2)}(1)$ proportional to $F_2^V(k^2)$. In our construction of the exchange currents in the S-matrix method we have shown [13], that the interaction dependent part of the current satisfies the continuity equation (see also [26]):

$$\vec{k} \cdot \vec{j}_F(2) = \left[V, \rho^1(1)\right] + k_o \, \rho_F(2) \quad , \quad (14)$$

with $\rho^1(1)$— the part of $\rho(1)$ proportional to $F_1^V(k^2)$. This equation is neither of the form (4) nor of the form (11). The reason is that the identification of the positive-frequency Born (PFB) term with the iteration of the one-body current is valid only in the non-relativistic approximation. The leading contributions of the diagrams considered add up just to the non-relativistic current $\vec{j}^{(1)}(2)$ (except for the PV πNN and A_1- meson negative Born diagrams, which are of the order $1/m^3$, and the additional ρ-meson transverse contribution). The higher order corrections to these contributions are velocity dependent and should be dropped in the static limit. The only source of the additional static contribution is the part of PFB term, which cannot be included into the iteration of the impuls approximation (IA).

According to ref. [10], in the static approximation only the singular part of PFB term is to be identified with the iteration of IA. The singularity due to the nucleon propagator can be compensated either by the energy dependence of the γNN-vertex (the part $\sim F_2^V \sigma_{\mu 4} k_0$) or by the energy dependence of the BNN vertex (couplings with the derivative). All terms are to be treated in the lowest order in $1/m$.

It is easy to show, that the first non-singular contribution is just:

$$\delta \vec{j}_F = \left[V, \vec{s}_{st}^t(1)\right] = -\vec{k} \frac{F_2^V(k^2)}{4m^2} (\vec{k} \cdot \vec{j}_{point}^{(1)}(2)) \ . \tag{15}$$

Adding $\delta \vec{j}_F$ to the non-relativistic current we arrive at:

$$\vec{j}_{F,st}(2) = \vec{j}^{(1)}(2) + \left[V, \vec{s}_{st}^t(1)\right] = \vec{j}_{D,st}(2) \ . \tag{16}$$

For the PV πNN coupling we put the full Feynman current $\vec{j}_F$ into the form: $\vec{j}_F(PV) = \vec{j}_F(PS) + \Delta \vec{j}_F(PV)$. The correction $\Delta \vec{j}_F(PV)$ is then cancelled (in the order considered) by the second nonsingular contribution of the PFB diagram with the PV πNN coupling. The full static current $\vec{j}_{F,st}^t(2)$ is, therefore, the same for both πNN couplings, in agreement with the relation $V_{st}(PS) = V_{st}(PV)$.

For the ρ-exchange currents, the situation is complicated by the presence of a number of velocity dependent contributions of the order of the standard ρ-exchange current [10], [13]. The importance of these terms should be estimated before considering higher order corrections.

We would like to point out that the discussion above cannot replace the consistent treatment of the relativistic corrections. It is, however, clear that it is not necessary to multiply $\vec{j}_{point}^{(1)}(2)$ by the $G_E(k^2)$ in order to satisfy the continuity equation involving the realistic nucleon charge density $\rho_{st}(1)$ eq. (7). The longitudinal ($\sim \vec{k}$) correction $\delta \vec{j}_F$ eq. (15) seems to be more relevant. The form of $\delta \vec{j}_F$ implies, that in the calculation of the magnetic multipole transitions $\vec{j}^{(1)}(2) = F_1^V(k^2)\vec{j}_{point}^{(1)}(2)$ should be used.

3. The strong formfactors

In this section we discuss the introduction of the strong
formfactors into the MECs of the hard pion method [13] so
that the eq. (1) still holds. Our exchange currents differ
from the standard ones [10] by the presence of the currents
of the A_1-meson range and by the anomalous transverse mesic
current of the ρ-meson range, which is twice as large as the
standard one.

The introduction of the monopole nucleon formfactors into the
standard MECs has been discussed at length by Riska [10]. Let
us remind, that in the Born and seagull diagrams each BNN
vertex is multiplied by the monopole formfactor (2), whereas
the whole mesic current is multiplied by the function [10,16]:

$$F_B(q_1^2, q_2^2) = \Delta_F^{m_B}(q_1^2) \Delta_F^{m_B}(q_2^2) + \left[-1 + (\Lambda_B^2 - m_B^2) \frac{d}{d\Lambda_B^2} \right]. \tag{17}$$

$$\cdot \Delta_F^{\Lambda_B}(q_1^2) \Delta_F^{\Lambda_B}(q_2^2) \,, \qquad \Delta_F^X(q^2) = 1/(x^2 + q^2) \,.$$

The same can be done without violating the gauge invariance
already on the level of the relativistic Feynman diagrams.
This prescription works both for the one π- or ρ-exchange.
This is no longer true for the MECs of the A_1- meson range.
Introducing the formfactors in the way described above one
discovers, that in order to maintain gauge invariance, the
additional current ΔJ_λ should be added, with divergence:

$$k_\lambda \Delta J_\lambda = i \left(\frac{g_\rho \, g_A}{m_A} \right)^2 F_1^V(k^2) (\vec{\tau}_1 \times \vec{\tau}_2)^3 \bar{u}(p_1') \gamma_\alpha \gamma_5 u(p_1) \tag{18}$$

$$\cdot \bar{u}(p_2') \gamma_\beta \gamma_5 u(p_2) \cdot (q_{1\alpha} q_{1\beta} - q_{2\beta} q_{2\alpha}) \Gamma_A(q_1^2) \Gamma_A(q_2^2)$$

$$\cdot \left[\Delta_F^{\Lambda_A}(q_1^2) + \Delta_F^{\Lambda_A}(q_2^2) \right] \,.$$

From this equation the current ΔJ_λ can be determined up to
a transverse part. The possible choice is:

$$\Delta J_\lambda = i\left(\frac{g_\rho \, g_A}{m_A}\right)^2 F_1^V(k^2)(\vec{\tau}_1 \times \vec{\tau}_2)^3 \, \bar{u}(p_1')\gamma_\alpha \gamma_5 u(p_1) \qquad (19)$$

$$\cdot \, \bar{u}(p_2')\gamma_\beta \gamma_5 u(p_2)(q_{1\alpha}\delta_{\lambda\beta} - q_{2\beta}\delta_{\lambda\alpha}) \, \Gamma_A(q_1^2) \, \Gamma_A(q_2^2)$$

$$\cdot \left[\Delta \stackrel{\Lambda_A}{F}(q_1^2) + \Delta \stackrel{\Lambda_A}{F}(q_2^2)\right] \quad .$$

One gets easily another current, obeying (18), by replacing
$(q_{1\alpha}\delta_{\lambda\beta} - q_{2\beta}\delta_{\lambda\alpha}) \rightarrow (q_{1\beta}\delta_{\alpha\lambda} - q_{2\alpha}\delta_{\beta\lambda})$.

In order to fix the transverse part a model, similar to the one used by Mathiot [17], has been developed. We simulate the short-range effects by the introduction of fictitious heavy ($m_B* = \Lambda_B$) particles π^*, ρ^*, A^* (for details see [27]). In this model, the standard results for the π- and ρ-meson MECs are reproduced. For the A_1-meson we have obtained:

$$\Delta J_\lambda^{model} = \Delta J_\lambda + \Delta J_\lambda' \quad , \qquad (20)$$

where ΔJ_λ is given in eq. (19) and $\Delta J_\lambda'$ reads:

$$\Delta J_\lambda' = i\left(\frac{g_\rho \, g_A}{m_A}\right)^2 F_1^V(k^2)(\vec{\tau}_1 \times \vec{\tau}_2)^3 \cdot \bar{u}(p_1')\gamma_\alpha \gamma_5 u(p_1) \qquad (21)$$

$$\cdot \, \bar{u}(p_2')\gamma_\beta \gamma_5 u(p_2) \cdot \left\{\Gamma_A^2(q_2^2)\Delta \stackrel{\Lambda_A}{F}(q_1^2) \left[q_{1\lambda} k_\alpha - (k \cdot q_1)\delta_{\lambda\alpha}\right]\right.$$

$$\cdot \, q_{2\beta} - (1 \rightleftharpoons 2) \Big\} \quad .$$

References:

1. Friar, J.L., Ann. Phys. (N.Y.) 104, 380 (1977).
2. Gari, M., Hyuga, H., Z. Phys. A277, 291 (1976).
3. Lacombe, M., Loiseau, B., Richard, J.M., Vinh Mau, R., Coté, J., Pirès, P., de Tourreil, R., Phys. Rev. C21, 861 (1980).
4. Machleidt, R., Holinde, K., Elster, C., to be published.
5. Machleidt, R., The meson theory of nuclear forces and nuclear matter, preprint TRIUMF, Sep. 1985, TRI-PP-85-68.
6. Foldy, L.L., Phys. Rev. 122, 275 (1961).

7. Auffret, S., Cavedon, J.-M., Clemens, J.-C., Frois, B., Goutte, D., Huet, M., Juster, F.P., Leconte, P., Martino, J., Mizuno, Y., Phan, X.H., Platchkov, S., Phys. Rev. Lett. 55, 1362 (1985).

8. Gerard, A., Samour, C. (eds.): Proceedings of the XIth Europhysics Divisional Conference on Nuclear Physics with Electromagnetic Probe. Paris, July 1 - 5, 1985. Nucl. Phys. A446, Nos. 1,2 (1985).

9. Sauer, P.U., Progr. in Part. and Nucl. Phys. 16, 35 (1985).

10. Riska, D.O., Progr. in Part. and Nucl. Phys. 11, 199 (1984).

11. Hadjimichael, E., Phys. Lett. B172, 156 (1986).

12. Delorme, J., Nucl. Phys. A446, 65c (1985).

13. Adam, J.,Jr., Truhlík, E., Czech. J. Phys. B34, 1157 (1984).

14. Hadjimichael, E., Goulard, B., Bornais, R., Phys. Rev. C27, 831 (1983).

15. Maize, M.A., Kim, Y.E., Nucl. Phys. A420, 365 (1984).

16. Thakur, J., Lock, J.A., Phys. Lett. 67B, 29 (1977).

17. Mathiot, J.F., Nucl. Phys. A412 , 201 (1984).

18. Riska, D.O., Phys. Scripta 31, 471 (1985).

19. Buchmann, A., Leidemann, W., Arenhövel, H., Nucl. Phys. A443, 726 (1985).

20. Blunden, P.G., A consistent treatment of nuclear currents for scalar and vector fields. Preprint, Center for Theoretical Physics, MIT, No. 86-644.

21. Friar, J.L., Phys. Rev. C22, 796 (1980).

22. Friar, J.L., Fallieros, S., Phys. Rev. C13, 2571 (1976).

23. Fabian, W., Arenhövel,H., Nucl. Phys. A341, 253 (1979).

24. Liou, M.K., Sobel, M.I., Phys. Rev. C7, 2044 (1973).

25. Ohta, K., Phys. Rev. C19, 965 (1979).

26. Bentz, W., Nucl. Phys. A446, 678 (1985).

27. Adam, J., Jr., Truhlík, E., in preparation.

PHOTO AND ELECTRODISINTEGRATION OF THE FEW-BODY SYSTEMS

J.M. Laget

Service de Physique Nucléaire-Haute Energie, CEN-Saclay,
F-91191 Gif-sur-Yvette Cedex, France

Abstract

The present status of the analysis of very inelastic photo and electronuclear reactions is reviewed, whith a special emphasis on the ^{3}He(e,e'2p)
and ^{3}He(γ,pp)n reactions, which are the best tools to study the two body
correlations and the three-body exchange currents.

Introduction

The study of the photo and electrodisintegration of the few-body systems
is a very active field of research, which I cannot review completely today.
I will rather organize my talk around one of the major issue of modern
nuclear physics and try to answer the following question :

How can we study and determine the short range behaviour of nuclear
systems ?

However two problems immediately arise, and we must answer the two other
questions first :

What are the relevant (one, two or three-body) mechanisms ?

How can we disentangle the study of the reaction mechanism and the study of the wave function itself ?

Since I extensively dealt with those problems in my Banff lectures [1] last summer, I do not reproduce here the corresponding discussion but I summarize the main topics of my talk.

I. The one-nucleon wave function

Provided that interaction effects are fully taken into account [2], the study of the (e,e'p) reaction [3,4], for small values of the virtual photon four momentum, has led to strong constraints on the D and ^{3}He wave functions, up to momenta of the order of 600 MeV/c.

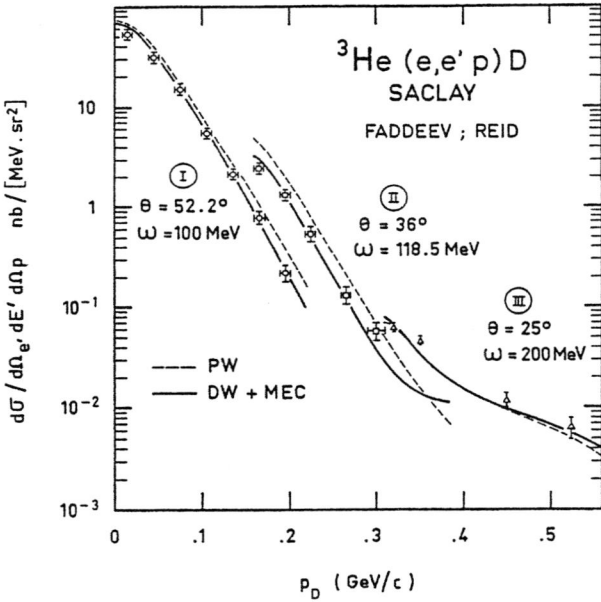

Fig. 1. The cross section of the ^{3}He(e,e'p) reaction recently measured at Saclay [3,4], in three different kinematics, is plotted against the momentum of the undetected deuton. The electron scattering angle θ and the energy ω of the virtual photon are given in each case.

As an example Fig. 1 shows the analysis of the ^{3}He (e,e'p)d reaction in the kinematics achieved at Saclay [3,4]. Similar studies have been performed at Amsterdam [5,6]. The effects of final state interactions are important and must be considered in the analysis.

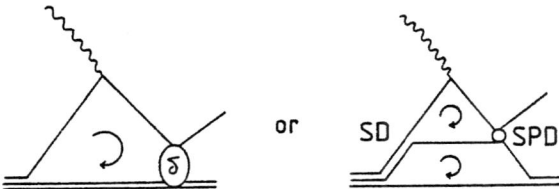

Fig. 2. The two ways of computing the final state interactions in the $^3\text{He}(e,e'p)d$ reaction.

They can be treated according to the two prescriptions depicted in Fig. 2. The first consists to parametrize the pd rescattering matrix element in terms of the corresponding phase-shifts : all the results already published [1, 2, 5, 6] have been computed in that way. The second prescription is used in Fig. 1, and consists in computing microscopically the nucleon rescattering diagram. The two-loop integral is computed numerically, and the nucleon-nucleon scattering amplitude is expanded in terms of the numerical values of the half off-shell S, P and D scattering matrix elements which correspond to the same potential (Reid Soft-Core) which generates the bound state wave function. So far only the transitions from the S-waves of the two active nucleons to the SPD scattering states, and from the D-wave to the S-scattering states have been considered. The transitions between the D-wave and the P and D scattering states, have not yet been taken into account. While they might change a little bit the cross section value in the third kinematics, which probes the D-wave content of ^3He, they are not expected to change significantly the cross sections of the first two kinematics.

It is worthwhile to note that both prescriptions lead to the same results when the relative kinetic energy of outgoing pd pair is high enough ($T_{pd} \gtrsim 50$ MeV).

Note also that a subtle cancelation, between the effects of final state interactions and meson exchanges, makes the result of the full calculation very close to the plane-wave result in the third kinematics. This plane wave cross-section contains the contribution of the np exchange graph (where the deuteron is directly ejected, the proton being spectator). While its contribution is negligible in the first two kinematics, it ranges from 10 % at $p_D = 420$ MeV/c to 50 % at $p_d = 600$ MeV/c. It is dominant in the

Amsterdam kinematics [5, 6].

II. The two nucleon relative wave function

Today the cleanest signature of two-nucleon correlations is the spectrum of the protons emitted in the continuum of the reactions ^{3}He(e,e'p)x [4] or ^{3}He(Y,p)x [7], which have been recently measured at Saclay. The top of the peak which appears in Fig. 3 corresponds to the electrodesintegration of a nucleon pair at rest and its width is due to its Fermi motion inside ^{3}He.

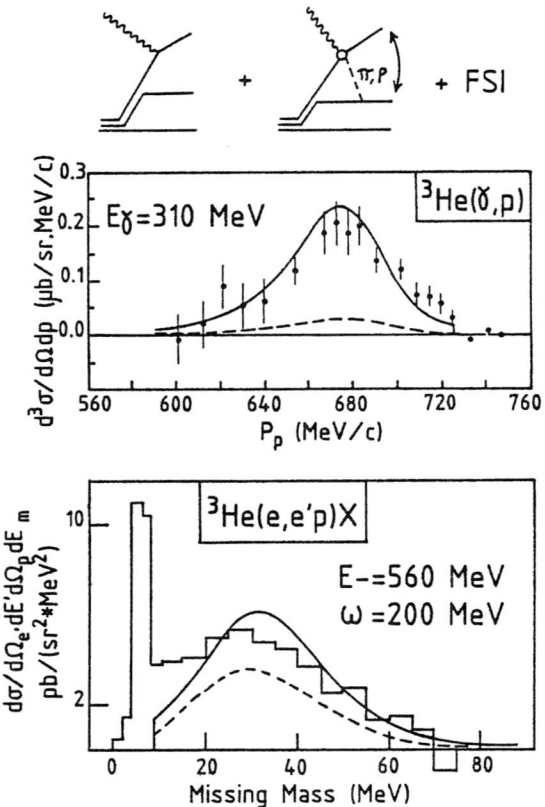

Fig. 3. In the upper part, the spectrum of the protons emitted at θ_p = 23° in the reaction ^{3}He(Y,p) is plotted against the proton momentum. In the lower part, the spectrum emitted in the ^{3}He(e,e'p) at θ_p = 60° is plotted against the missing mass of the undetected system. The full curves are the result of the complete calculation [1,2]. Meson exchange currents and Δ-formation mechanisms are not included in the dashed curves.

While the (Y,p) spectra are dominated by the exchange current contribution and the Δ-formation mechanism, the (e,e'p) spectra are more directly sensitive to the relative wave function of the two active nucleons.

Note that these experiments probes the high momentum components of this relative wave function in ^{3}He up to 600 MeV/c. However the separation between the transverse and the longitudinal cross sections is still lacking. It will allow to get rid of meson exchange and Δ-formation mechanisms

III. The pd capture reactions

Therefore no freedom is left to play with the high momentum components of the three-body wave function. They can can be used to analyse other channels as, for instance, the pd → ^{3}HeY and the pd→tπ$^+$ reactions [1, 8] in the Δ-energy range. While the two-nucleon mechanisms represent the major part of the cross-section, they alone cannot reproduce all the data. The discrepancies, which remain between the theory and experiment, are the hint that three-body mechanisms might also be considered in these channels.

IV. Three-body forces

The most promising way to study these three-body mechanisms is the study of the kinematically complete ^{3}He(Y,pp)n experiment ; its cross section is compared to the ^{3}He(Y, pn) reaction cross-section in Fig. 4. One of the

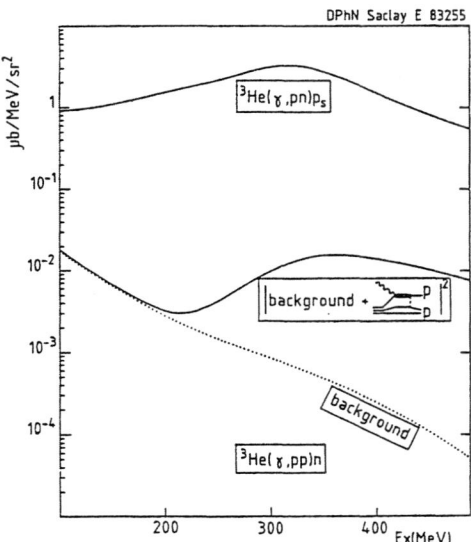

Fig. 4. The photodisintegration cross section of the pn (upper part) and pp pair (lower part) at rest in ^{3}He (see text).

276

detected proton is assumed to be emitted at θ_p = 90°, with respect to the incoming photon, in the center of mass frame of the active pair. In the pp channel the background is due to all the graphs due to final state interaction and corresponding to pion reabsorption in a pn active pair. It does not include the pion reabsorption graph in the pp active pair, which dominates, the (γ,pp) cross section. See ref. [2] for details of the model. Selection rules [9] strongly suppress the two-body mechanisms in the (γ, pp) channel, and makes it very sensitive to the three-body currents [1,9]. As an example Fig. 5 shows their relative importance, with respect to the two-body current, in a kinematics where they are enhanced : the effect of the triangular singularity [10], which corresponds to the onshell propagation of one of the pions is maximum here.

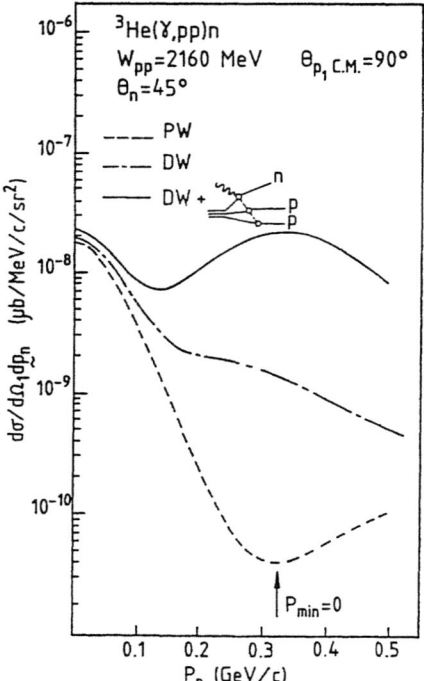

Fig. 5. The photodisintegration cross section of a pp pair in ³He is plotted against the values of the neutron momentum. The invariant mass of the pair is assumed to be constant at W = 2160 MeV, and each proton is assumed to be emitted at 90°, with respect to the incoming photon, in their c.m. frame. The meaning of the curves is given on the figure 4. The arrow shows the place where the triangular singularity is maximum.

V. Two-nucleon correlations

Finally, the (e,e'2p) reaction is the most promising way to study two-nucleon correlations. The transverse cross section is strongly suppressed, for the same reasons as the ^{3}He(γ,2p) reaction. Charged meson exchange currents and Δ-formation mechanisms do not contribute at all to the longitudinal cross-section : this is the best place to study in details, the two-body correlations, provided that final state interactions are carefully taken into account. Fig. 6 clearly illustrates this very important point, which I fully discussed in ref. [11].

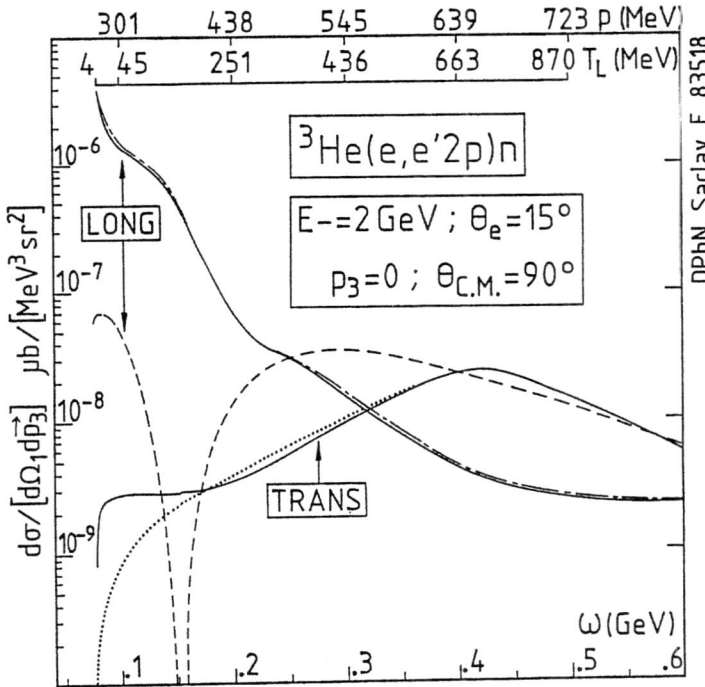

Fig. 6. The excitation functions of the transverse and longitudinal reduced cross section of the electro-disintegration of a pp pair at rest in ^{3}He, when each proton is emitted at $\theta_{c.m.}$ = 90° with respect to the incoming virtual photon. The common value of the proton momentum P, as well as the proton relative kinetic energy T_L, is plotted on abcissa. Dashed lines : plane wave without meson exchange contribution. Dotted lines : plane wave with meson exchange amplitude. Dot-dashed lines : pp rescattering included. Solid lines : all final state interactions included.

VI. Conclusion

To summarize, the present generation of electron accelerators have allowed us to strongly constrain the one-body momentum distributions and to start the study of the two-body correlations in nuclei. The complete determination of the kinematics of very inelastic electronuclear reaction, where few particles are emitted in the continuum, is the only way to disentangle the various reaction mechanisms, and pin down the various parts of the nuclear wave function. Two typical examples are the $^3He(\gamma,2p)$ and $^3He(e,e'2p)$ reactions, which allow us to look at the three nucleon force and the two nucleon correlations respectively. Their study, at high momentum transfer, will be the heart of the research program at the new generation of multi-GeV electron accelerators, and will allow us to improve our knowledge of the short-range behaviour of nuclear systems.

References

[1] J.M. Laget, New Vistas in Electronuclear Physics, E. Tomuziak, H. Kaplan and E. Dressler, Eds., Plenum Press, New-York, p. 361 (1986)

[2] J.M. Laget, Phys. Lett., 151B, 325 (1985).

[3] E. Jans, P. Barreau, M. Bernheim, J.M. Finn, J. Morgenstern, J. Mougey, D. Tarnowski, S. Turck-Chieze, S. Frullani, F. Garibaldi, G.P. Capitani, E. de Sanctis, M. K. Brussel and I. Sick, Phys. Rev. Lett. 49, 974, 1982.

[4] J. Morgenstern, Second Workshop on perspective in nuclear physics at intermediate energy, S. Boffi et al., Eds World Scientific, Singapore, p. 355 (1986).
J. Morgenstern, this issue.

[5] P. de Witt Huberts, this issue.

[6] P.H.M. Keizer, P.C. Dunn, J.W.A. den Herder, E. Jans, A. Kaarsgaarn, L. Lapikas, E.N.M. Quint, P.K.A. de Witt Huberts, H. Postma, J.M. Laget, Phys. Lett. 157B, 255 (1985).
P.H.M. Keizer, Thesis, Amsterdam (unpublished).

[7] N. d'Hose, G. Audit, A. Bloch, N. de Botton, L. Ghedira, L. Jammes, J.M. Laget, J. Martin, E. Mazzucato, C. Schull, G. Tamas, E. Vincent, P. Argan and P. Stoler, 8ème Session d'Etude Biennale de Physique Nucléaire, Aussois, J. Meyer Ed., LYCEN 8502 (Lyon University) p. S10.1, 4-8 feb. 1985.

[8] J.M. Laget, Second Workshop on perspective in nuclear physics at intermediate energy, S. Boffi et al., Eds World Scientific, Singapore,

p. 247 (1986).

[9] J.M. Laget, Nucl. Phys., A446, 489c, (1985).

[10] J.M. Laget, Phys. Rep. 69, 1 (1981).

[11] J.M. Laget, Preprint DPhN Saclay n°2367, July 1986.

Y-SCALING*

C. Ciofi degli Atti°, E. Pace°+ and G. Salmè°

°Istituto Nazionale di Fisica Nucleare, Sezione Sanità
Istituto Superiore di Sanità, Physics Laboratory
Viale Regina Elena, 299, I-00161 Rome, Italy

+Dipartimento di Fisica, Università di Roma "La Sapienza"
Piazzale Aldo Moro 2, I-00185 Rome, Italy

Abstract

A theory of y-scaling which properly takes into account nucleon momentum and separation energy is illustrated. The fsi in the continuum states of the residual (A-1) system are correctly taken into account and a model independent criterion for extablishing the relevance of the fsi between the struck nucleon and the residual nucleus is given. The differences between longitudinal momentum distributions and scaling functions due to binding effects are presented and the possibilities offered by y-scaling for the investigation of nucleon dynamics (high momentum components and nucleon swelling in the medium) are discussed.

1. Introduction

Following the well known work by West [1], an appreciable number of papers on y-scaling has appeared in the last few years. We have developed an approach where, unlike previous treatments, nucleon separation energy and momenta are correctly taken into account and the effect of final state interaction (fsi) is properly investigated [2]. Following Ref. 2, we discuss in this paper,: i) the definition and the physical meaning of the scaling function and the scaling variable; ii) the effects of nucleon binding and fsi in y scaling; iii) the possibilities offered by y scaling for the investigation of nucleon dynamics. In order to introduce our approach, we consider it useful to review some of the conclusions reached in Ref. 1.

2. y-scaling within West's approach

The concept of y-scaling in deep inelastic electron scattering by nuclei has been introduced by West [1] with the aim of establishing a relation between y-scaling and the form of

* Presented by C. Ciofi degli Atti

the NN interaction. In the light of recent work in this field [2-4], some of the conclusions of Ref. 1 have to be reconsidered. We start from very general equations for quasi elastic (q.e.) electron scattering by nuclei and then will discuss the assumptions made by West to obtain y-scaling.

The inelastic cross section in Born Approximation reads

$$\frac{d^2\sigma}{d\,\Omega_2\,d\,\epsilon_2} = \sigma_{Mott} \left[(\frac{q_\mu^2}{q^2})^2\,W_L(q,\omega) + \frac{1}{2}\,(\frac{q_\mu^2}{q^2} + 2tg^2\frac{\theta}{2}\,)W_T(q,\omega) \right] \quad (1)$$

q and ω being the trimomentum and energy transfer and $q_\mu^2 = q^2 - \omega^2$. The structure functions $W_{L(T)}$ are

$$W_{L(T)}(q,\omega) = \sum_f |<f|\hat{J}_{L(T)}(\vec{q})|0>|^2\delta(\omega + E_o - E_f) \quad (2)$$

with E_o and E_f being nuclear eigenvalues and $J_{L(T)}$ the current operators. By closure over final states and using the time evolution formalism, eqn(2) reduces, at high momentum transfer, to the following form (for ease of presentation only W_L will be considered; the same conclusions apply to W_T if the convective current is disregarded)

$$W_L(q,\omega) = \frac{1}{2\pi}\sum_i \int_{-\infty}^{\infty} dt\,e^{-i\omega t}<0|e^{i(\hat{H} + \frac{(\vec{P}_i+\vec{q})^2 - P_i^2}{2M} - E_o)t}|0> \quad (3)$$

where $\vec{p}_i$ is the momentum operator of the i-th nucleon and $\hat{H}$ the nuclear Hamiltonian. Although eqn (3) has the advantage of being free from the final nuclear states |f>, its *practical* evaluation critically depends upon the form of $\hat{H}$ (specifically, upon the commutation rules between $\hat{H}$ and $\vec{q}.\vec{p}/M$). In Ref. 1 the following relation

$$<0|e^{i(\hat{H}+ \frac{(\vec{P}_i + \vec{q})^2 - P_i^2}{2M})t}|0>=<0|e^{i(E_o + \frac{(\vec{P}_i + \vec{q})^2 - P_i^2}{2M})t}|0> \quad (4)$$

has been used to reduce eqn (3) to the following form

$$W_L(q,\omega) = \sum_i<0|\delta(\omega- \frac{(\vec{P}_i + \vec{q})^2 - P_i^2}{2M})|0> \quad (5)$$

which, in momentum representation, becomes

$$W_L(q,\omega) = \int d\vec{k}\,n(k)\delta(\omega- \frac{(\vec{k}+\vec{q})^2}{2M} + \frac{k^2}{2M}) \quad (6)$$

where n(k) is the nucleon momentum distribution. Eqn (6) is the starting point for y scaling. In fact, by introducing the longitudinal $(k_{\parallel})$ and perpendicular $(k_\perp)$ components of $\vec{k}$ along $\vec{q}$ $(k_{\parallel}= k$

$\cos\alpha$, $k_\perp = k \sin\alpha$, α being the angle between $\vec{q}$ and $\vec{k}$) one gets [1]

$$W_L(q,\omega) = \int d\vec{k}\, n(k)\, \delta(\omega - \frac{q^2}{2M} - \frac{\vec{q}\,\vec{k}}{M}) \tag{7a}$$

$$= 2\pi \int dk_\| \, \delta(\omega - \frac{q^2}{2M} - \frac{qk_\|}{M}) \int_0^\infty d\, k_\perp^2 \, n(k_\|, k_\perp) \tag{7b}$$

$$= |\frac{d\omega}{dk_\|}|^{-1} 2\pi \int_0^\infty dk_\perp^2 \, n(k_\|, k_\perp) \tag{7c}$$

$$= |\frac{d\omega}{dk_\|}|^{-1} 2\pi \int_{|\dot{y}_0|}^\infty kdk \, n(k) \tag{7d}$$

where

$$y_0 = k_\| = \frac{M}{q} (\omega - \frac{q^2}{2M}) \tag{8}$$

and $d\omega/dk_\| = d\omega/dy_0 = q/M$ is a phase space factor arising from the dependence of the energy

δ function upon $k_\|$. From eqn (7) one obtains that the quantity

$$f(y_0) = \frac{d\omega}{dy_0} W_L(q,\omega) = 2\pi \int_{|y_0|}^\infty kdk \, n(k) \tag{9}$$

is a function of one variable only y_0, i.e. it remains constant independently of the values of q

and ω, provided these correspond to the same value of y_0; concisely, one says that q

$W_L(q,\omega)/M$ "scales" in the variable y_0, the "scaling function" $f(y_0)$ being the nucleon

longitudinal momentum distribution . Such a scaling property of the longitudinal structure

function ("y_0-scaling") has been obtained by West in Ref. 1, but a word of caution is necessary

concerning the relationship between y_0 scaling and the form of the nuclear Hamiltonian. Eqn.

(5) follows directly from eqn. (4) and therefore its validity relies on the validity of eqn. (4).

Relation (4) is certainly true if V = const or V = 0, but it can be shown that, unlike what stated

in Ref. 1, it does not hold in the most general case of a *local* potential energy-operator in H.

Thus West's attempt to relate y_0 -scaling to the form of the nuclear Hamiltonian, so as to

ascribe deviations from scaling to, e.g., the nonlocality of the NN interaction, has to be

reconsidered. When the potential energy operator (even in the form of a single particle

harmonic oscillator well) is taken into account in the Hamiltonian appearing in eqn (3), then an

exact evaluation of such an equation is rather problematic, the more so when nuclear wave

functions resulting from realistic two-body interactions are adopted to describe the nuclear

ground state $|0\rangle$. In such a case the evaluation of the cross section does require a direct

evaluation of the transition matrix element in eqn. (2), with explicit assumptions about the final state wave function |f>. It is clear therefore that any formulation of y-scaling depends, to some extent, upon the theoretical framework within which the final state |f> is treated. In our approach we start from the Plane Wave Impulse Approximation (PWIA) and, subsequently, the modifications induced by the fsi will be investigated.

In summary, West's assumptions leading to y_0-scaling can be listed as follows: i) one photon exchange approximation; ii) only nucleon degrees of freedom are considered; iii) after interaction with the photon, nucleon undergoes a transition from the "free" state $k_i^2 / 2M$ to the "free" state $(\vec{k}_i + \vec{q})^2 / 2M$; iv) the convective current is disregarded; v) non relativistic kinematics is used. In our theory of y scaling approximations i) and ii) are still kept, but the other ones are released, namely: realistic Spectral Functions are used to describe nucleon dynamics, the convective current is taken into account, relativistic kinematics is used. As a result scaling will occur in a variable which differs from $k_\parallel = (\vec{q} \cdot \vec{k})/q$ and only in some special cases can the scaling function be considered a longitudinal momentum distribution.

3. y-scaling within the PWIA

3.1 The PWIA cross section. Spectral Functions and momentum distributions

It can be shown that at high momentum transfer, the inclusive quasi elastic (q.e.) cross section in PWIA can be written in the following form (see e.g. Ref. 5)

$$\sigma_2 \equiv \frac{d^2\sigma}{d\varepsilon_2 d\Omega_2} = \{Z\sigma_{ep} + N\sigma_{en}\} \cdot \left| \frac{\partial\omega}{k\partial\cos\alpha} \right|^{-1} 2\pi \int_{E_{min}}^{E_{max}(q,\omega)} dE \int_{k_{min}(q,\omega,E)}^{k_{max}(q,\omega,E)} P(k,E) k dk \quad (10)$$

where $\vec{k} = \vec{k}_N - \vec{q}$ is the momentum of the bound nucleon ($\vec{k}_N$ being the momentum of the struck nucleon in the lab system) and $E = |E_A| - |E_{A-1}| + E^*_{A-1}$ its removal energy (E_A and E_{A-1} being the ground-state binding energies of the A and (A-1) nucleon systems and E^*_{A-1} the excitation energy of the latter); σ_{eN} is the free electron-nucleon cross section (without the recoil factor); $d\omega/kd\cos\alpha$ is the phase space factor which arises from the integration over the direction of the unobserved nucleon (α is the angle between q and k); E_{max}, k_{min} and k_{max} are the limits of integration which are fixed by the energy conservation; $P_N(k,E)$ is the nucleon Spectral Function which, for the sake of presentation will be considered to be the same for protons (p) and neutrons (n) ($P_p = P_n = P$)

$$P(k,E) = \sum_f | (2\pi)^{-3/2} \int e^{i\vec{k}\cdot\vec{z}} G_{f0}(\vec{z}) d\vec{z} |^2 \delta(E - (E^f_{A-1} - E_A)) \quad (11a)$$

where the overlap integral is

$$G_{f0}(\vec{z}) = \int d\vec{x}\ldots d\vec{y}\ \psi_{A-1}^{f*}(\vec{x}\ldots\vec{y})\ \psi_{A}^{0}(\vec{x}\ldots\vec{y}\ \vec{z}) \tag{11b}$$

In eqn (11a) the sum over f stands for summation over the discrete energy states of the A-1 system and integration over its continuum states. Moreover, the wave functions of the initial target nucleus (Ψ_A) and the final (A-1) system (Ψ_{A-1}^f) are considered to be exact eigenvalues of Hamiltonians containing the same two-body interaction; in this sense, our Spectral Function automatically and exactly includes all "final state interactions" in the continuum states of the A-1 system; the only plane wave in our approach is that describing the relative motion of the knocked-out nucleon and the (A-1) system. The Spectral Function represents the joint probability to find in the nucleus A a nucleon with momentum k and energy E or, equivalently, the probability that the system (A-1) is left with excitation energy E^*_{A-1} after a nucleon with momentum k has been removed. Therefore, we can write

$$P(k,E) = P_{gr}(k) + P_{ex}(k,E)\ (1 - \delta_{E,E_{min}}) \tag{12}$$

where $P_{gr}(k) = P(k, E_{min})$ ("two-body channel" spectral function), yields the probability that the final (A-1) system is left in its ground state (corresponding to $E^*_{A-1}= 0$ and $E = E_{min}$), whereas $P_{ex}(k,E)$ ("break-up channel" spectral function) yields the probability that the final (A-1) system is left in all possible excited states (with $E^*_{A-1} \neq 0$, $E = E_{min} + E^*_{A-1}$). The following relation [6] between the Spectral Function and the momentum distributions n(k) will be useful in what follows

$$n(k) = \int e^{i\vec{k}\cdot(\vec{z}-\vec{z}')}\rho(\vec{z},\vec{z}')d\vec{z}d\vec{z}' = \int_{E_{min}}^{\infty} P(k,E)\,dE = n_{gr}(k) + n_{ex}(k) \tag{13}$$

where $\rho(z, z')$ is the one body density matrix and

$$n_{gr}(k) = (2\pi)^{-3}|\int d\vec{z}\ e^{i\vec{k}\cdot\vec{z}}G_{00}(\vec{z})|^2$$

$$n_{ex}(k) = (2\pi)^{-3}\sum_{f\neq 0}|\int d\vec{z}\ e^{i\vec{k}\cdot\vec{z}}G_{f0}(\vec{z})|^2 \tag{14}$$

3.2 Definition and physical meaning of the scaling function and the scaling variable

Eqn (10) trivially tells us that if the PWIA is an appropriate description of the scattering, then the following quantity

$$F_1 = \frac{\sigma_2}{\{Z\sigma_{ep} + N\sigma_{en}\}}\left|\frac{\partial\omega}{k\partial\cos\alpha}\right| \tag{15}$$

represents the nuclear structure function, to be denoted $F(q,\omega)$, i.e.

$$F_1 = F(q,\omega) = 2\pi \int_{E_{min}}^{E_{max}(q,\omega)} dE \int_{k_{min}(q,\omega,E)}^{k_{max}(q,\omega,E)} P(k,E)\, k\, dk \qquad (16)$$

Let us introduce a "scaling variable" y, which, for the time being, will be only required to be a

function of q and ω, so that any q and ω dependence can be expressed as a q and y dependence.

Thus the nuclear structure function can be rewritten as

$$F(q,y) = 2\pi \int_{E_{min}}^{E_{max}(q,y)} dE \int_{k_{min}(q,y,E)}^{k_{max}(q,y,E)} P(k,E)\, k\, dk \qquad (17)$$

In what follows the quantity F_1 (eqn (15)) will be called "scaling function"; it is a function of q

and y, which "scales" in y if it becomes q-independent. Since within our framework (PWIA)

the scaling function F_1 coincides with the nuclear structure function $F(q,y)$ (eqn (17)), the

scaling properties of F_1 will be investigated by analysing the scaling properties of $F(q,y)$. Such

an analysis should answer the following questions: i) "which is the explicit expression of the

scaling variable y?" ; ii) "under which kinematical conditions does $F(q,y)$ become independent

of q?" ; iii) "what is the physical meaning of the q-independent (or asymptotic) scaling

function?". The scaling properties of F (q,y) come clearly out from the following observation:

since P(k,E) is a rapidly decreasing function of k and E, by looking at the explicit expressions

of E_{max}, k_{max} and k_{min} one easily realizes that $E_{max} \approx k_{max} \approx \infty$ already for moderate values

of q, whereas there is still a sizeable dependence of k_{min} upon q. Therefore the scaling

properties of eqn (17) are mainly governed by the q dependence of k_{min}. The latter is

determined from the energy conservation

$$\omega + M_A = \{ M^2 + (\vec{k} + \vec{q})^2 \}^{1/2} + \{ (M_{A-1} + E^*_{A-1})^2 + k^2 \}^{1/2}$$
$$= \{ M^2 + q^2 + k_{\parallel}^2 + k_{\perp}^2 + 2q\, k_{\parallel} \}^{1/2} + \{ (M_{A-1} + E^*_{A-1})^2 + k_{\parallel}^2 + k_{\perp}^2 \}^{1/2} \qquad (18)$$

in correspondence of $\cos\alpha = -1$ (corresponding to $y < 0$), i.e. from

$$\omega + M_A = \{ M^2 + (q - k_{min})^2 \}^{1/2} + \{ (M_{A-1} + E^*_{A-1})^2 + k_{min}^2 \}^{1/2} \qquad (19)$$

From the above equation we see that k_{min} depends upon q, ω and E^*_{A-1}, $k_{min} = k_{min}$ (q,ω,

E^*_{A-1}), so that it cannot be adopted as a scaling variable. Let us suppose that in the process

under consideration $E^*_{A-1} = 0$ (a physical case will be discussed later on) ; then k_{min} will

depend only upon q and ω and therefore can represent a scaling variable; one has therefore

$$k_{min} = k_{min} \ (q, \ \omega, E^*_{A-1} = 0) = |y| \tag{20}$$

with the equation defining y being just eqn (19) with $E^*_{A-1} = 0$, i.e.

$$\omega + M_A = \{ M^2 + (q + y)^2 \}^{1/2} + \{ M^2_{A-1} + y^2 \}^{1/2} \tag{21}$$

The condition $E^*_{A-1} = 0$ leads therefore to an important consequence, namely, because of eqn (20), the structure function (17) scales in y provided q is large enough that $k_{max} \approx \infty$. Moreover, since E^*_{A-1} has been assumed to be equal to zero, only P_{gr} (k) survives in the Spectral Function (12) so that the structure function reduces to

$$f_{gr}(y) = 2\pi \int_{|y|}^{\infty} n_{gr}(k) k dk \tag{22}$$

The scaling variable y and the quantity (22) have a very clear physical meaning: by comparing eqns (21) and (19) one easily realizes that $|y|$ is the lowest longitudinal momentum of a nucleon bound with minimal removal energy $(E^*_{A-1} = 0, E = E_{min})$; correspondingly f_{gr} is the longitudinal momentum distribution of these nucleons. The condition $E^*_{A-1} = 0$, however, is only satisfied in electrodisintegration of deuteron.

3.3 y-scaling in deuteron

An ideal system to study y-scaling effects is the deuteron [7], for which $E^*_{A-1} = 0$ (E = E_{min} = 2.225 MeV), P(k, E) = n(k) δ $(E-E_{min})$, so that $k_{min} = |y|$ for any value of q. The structure function (17) then considerably simplifies

$$F(q,y) = 2\pi \int_{|y|}^{k_{max}(q,y)} n(k) k dk \tag{23}$$

Thus scaling in deuteron is only governed, within the PWIA, by the q-dependence of k_{max}. For large values of q such that $k_{max} \approx \infty$, eqn (23) simply becomes the longitudinal momentum distribution

$$f(y) = 2\pi \int_{|y|}^{\infty} n(k) k dk \tag{24}$$

from which the momentum distribution n(k) can be obtained

$$n(k) = \frac{1}{2\pi} \frac{1}{y} \frac{df}{dy} \qquad k = |y| \tag{25}$$

3.4 y-scaling in complex systems and the role played by nucleon binding

For a complex system $E^*_{A-1} \neq 0$, so that $k_{min} = k_{min} (q, \omega, E^*_{A-1})$ and, as already pointed out, cannot be assumed as a scaling variable. The latter, however, can still be chosen in the form given by eqn (21) and deviations from scaling in this variable will be due (within the PWIA) to the effect of binding. Indeed for a complex nucleus the structure function (17) can be rewritten, using eqn (12), in the following form

$$F(q,y) = 2\pi \int\limits_{|y|}^{k_{max}(q,y,E_{min})} n_{gr}(k)\,k\,dk + 2\pi \int\limits_{E_{min}}^{E_{max}(q,y)} dE \int\limits_{k_{min}(q,y,E)}^{k_{max}(q,y,E)} P(k,E)\{1-\delta_{E,E_{min}}\}k\,dk \quad (26)$$

where the first term in the rhs represents the contribution from the transition to the ground state of the final (A-1) system ($E = E_{min}$) and the second term is the contribution from the knock out of nucleons with $E > E_{min}$. For large values of q such that $k_{max} \approx E_{max} \approx \infty$ one has

$$F(q,y) = 2\pi \int\limits_{|y|}^{\infty} n_{gr}(k)\,k\,dk + 2\pi \int\limits_{E_{min}}^{\infty} dE \int\limits_{k_{min}(q,y,E)}^{\infty} P(k,E)\{1-\delta_{E,E_{min}}\}k\,dk \quad (27)$$

The first term trivially scales as in deuteron case, whereas the second term yields the scaling violation due to binding effects. In the asymptotic limit (q → ∞) eqn (27) becomes [5]

$$F(y) = 2\pi \int\limits_{E_{min}}^{\infty} dE \int\limits_{|y-(E-E_{min})|}^{\infty} P(k,E)\,k\,dk \quad (28)$$

Thus, in the asymptotic limit, the structure functions of both deuteron and complex systems scale in y. However, the former (eqn (24)) is a longitudinal momentum distribution, whereas the latter (eqn (28)), because of the presence of nucleon binding E, is not. If the dependence of k_{min} upon E is disregarded in eq. (28), then the asymptotic scaling function (28) becomes the longitudinal momentum distribution. In facts (cf. eqn. (13)) one has

$$2\pi \int\limits_{|y|}^{\infty} k\,dk \int\limits_{E_{min}}^{\infty} P(k,E)\,dE = 2\pi \int\limits_{|y|}^{\infty} n(k)\,k\,dk = f(y) \quad (29)$$

The quantitative difference between the longitudinal momentum distribution f(y) (eqn. 29) and the asymptotic scaling function F(y) (eqn. 28) has been investigated in Ref. 8 for the case of the three-body system.

3.5 *On the difference between the momentum distribution and the scaling function in the three-body system*

In Fig. 1 the (proton) longitudinal momentum distribution f(y) (eqn (29)) and the asymptotic scaling function (eqn (28)) for ^{3}He calculated with a Spectral Function corresponding to the Reid Soft Core interaction are shown. It appears that only for $|y| < 400$ MeV/c can the asymptotic scaling function be approximated with the longitudinal momentum

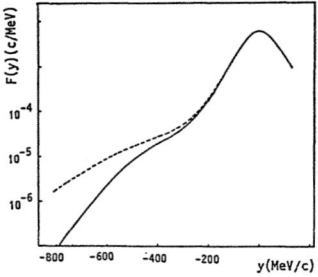

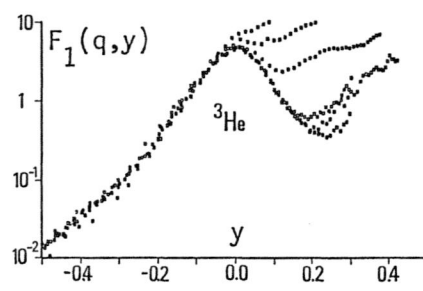

FIGURE 1

The (proton) longitudinal momentum distribution f(y) in ^{3}He (eqn (29), dashed line), compared with the asymptotic scaling function F(y) (eqn (28), full line) calculated with the variational Spectral Function corresponding to the RSC interaction [13] (after Ref. 8).

FIGURE 2

The scaling function F_1 (q,y) eqn (15) of ^{3}He obtained from the experimental data of Ref.9.

distribution, whereas at higher values of y the two quantities sharply differ because of binding effects. These, therefore, play a very important role in y scaling at high values of y; in particular taking into account nucleon binding *decreases* the asymptotic scaling function at high values of

y (|y|> 400 MeV/c). It is therefore difficult to extract the high momentum components from y scaling particularly for complex nuclei (A > 3), whose Spectral Function is not known.

3.6 *The approach to scaling and the effect of fsi in y-scaling. A model independent criterion for establishing the limits of validity of the PWIA*

We reiterate that in our approach all "final state interactions" in the continuum states of the A-1 system is properly taken into account. What is left out, according to the PWIA is the fsi between the knocked out nucleon and the A-1 system, and only the effect of this fsi will be considered in what follows. In Ref. 2 the following model-independent criterion to establish the limits of validity of the PWIA by means of y scaling, has been given. Since, for fixed values of y, E_{max} and k_{max} increase with q, whereas k_{min} decreases, then, independently of the form of the spectral function P(k, E), the structure function F(q,y) (eqn (17)) *increases with q until it reaches its asymptotic value given by eqn (22) for deuteron and eqn (28) for a complex system. Therefore, if the PWIA holds, the experimental scaling function should approach its asymptotic value by increasing with q; a different approach to scaling would therefore represent a proof of the breaking down of the PWIA and a clear signature of the presence of fsi.*

4. Experimental scaling functions of ^{2}H, ^{3}He and ^{4}He

4.1 Two-body and three-body systems

In Fig. 2 the scaling function F_1 (q,y) of ^{3}He (eqn (15)) obtained from the experimental data of Ref. 9, is shown. The well known scaling behaviour for y < 0 can be observed, which means that the low energy transfer side of the q.e. peak can indeed be interpreted mainly as due

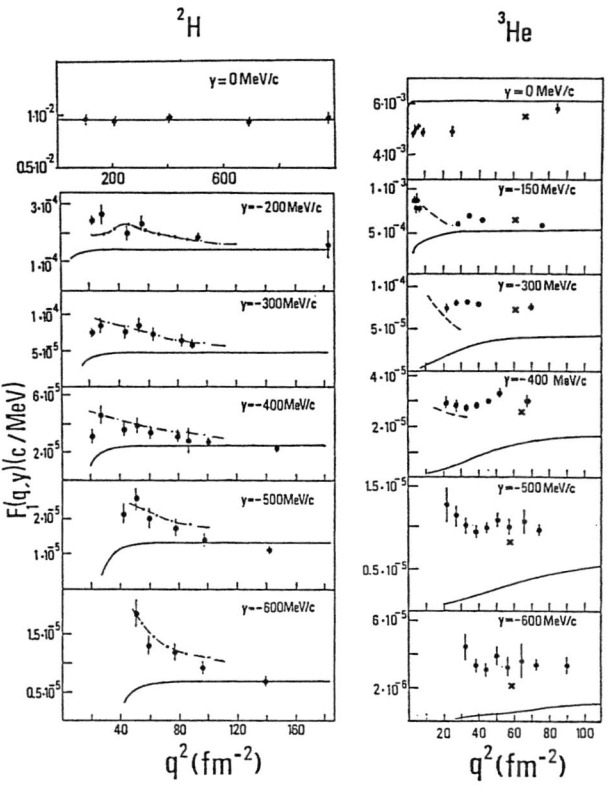

FIGURE 3

The scaling function F_1 (eqn (15)) of ^{2}H and ^{3}He vs. momentum transfer for various values of y obtained (Ref.2) from the experimental data of Refs. 16 and 9 respectively. The full lines represent the PWIA results (eqn(17)) and the dashed lines include fsi, evaluated according to Refs 11 and 12. The crosses include effects from three-body forces (Ref. 26). The momentum distribution of ^{2}H and the spectral function of ^{3}He correspond to the RSC interaction (Ref.13).

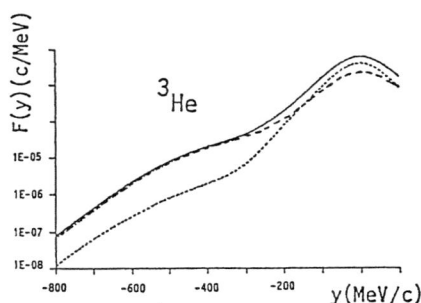

FIGURE 4

The asymptotic scaling function of ^{3}He calculated with the variational spectral function. The full line is eqn (28); the dotted and dashed lines are the two-body (^{3}He p+d) and three-body (^{3}He p + (np)) channel contributions, respectively (after Ref. 2)).

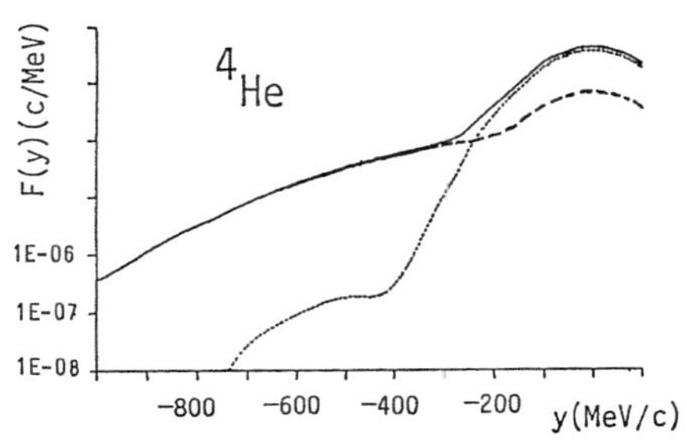

FIGURE 5

The longitudinal momentum distribution for ^{4}He (eqn. 30) calculated with the momentum distributions of Ref. 14. The dotted line is the contribution from the two-body channel and the dashed line the contribution from all possible break up channels of the three-body final state (after Ref. 15).

to the coupling of the photon with nucleon degrees of freedom only [10]. However for a quantitative check of the extent to which y scaling holds, such a log plot does not suffice: the q dependence of the experimental data for fixed values of y should be plotted for establishing the limits of validity of the PWIA reaction mechanism. These plots are shown in Fig. 3. It can be seen that, apart from y ≈ 0, the behaviour of F_1 is far from that predicted by the PWIA, i.e. an

increase of F_1 (q, y) with q. The observed *decrease* of the experimental data with q is a clear *model independent* proof of the breaking down of the PWIA. An explicit introduction of fsi using the exact continuum wave function for deuteron [11], and using the results of Ref. 12 to approximatively describe nucleon deuteron rescattering in the three-body system, show indeed that they are very large. A careful treatment of fsi is therefore a prerequisite for extracting reliable information on nucleon dynamics from y scaling, particularly in the region of high values of y. Concerning the results presented in Fig. 3 two comments are in order. The theoretical F_1 (q,y) (eqn. 17) increases with q, both for ^{2}H and ^{3}He as predicted by the PWIA. However a rather different rate of increase for the two systems can be noticed. This has a clear explanation: the q dependence of F_1 for deuteron is only governed by k_{max} (eqn (23)) which rapidly increases with q so that F_1 sharply reaches its asymptotic value (eqn. 24). The much slower increase of F_1 for ^{3}He is due to the q dependence of k_{min}, i.e. to the binding effect. Concerning the effect of fsi, it should be noticed that, as expected, they are very important at small q (where the energy of the struck nucleon relative to the A-1 system is very low) and seem to die out with increasing momentum transfer. The effects of fsi are much larger in deuteron and less important in ^{3}He because in the kinematics of Ref. 9 the CM energy of the system "nucleon-spectator pair" is very large. Finally it is worth pointing out that in ^{3}He the region of high y is almost entirely determined by the three-body final states as shown in Fig. 4.

4.2 The four-body system

The full Spectral Function for ^{4}He is not yet available. Recently the nucleon momentum distribution (eqn 13) for ^{4}He has been calculated with realistic interactions[14]; the quantity n_{gr} (eqn (14)) has also been evaluated and n_{ex} was obtained as the difference between n(k) and n_{gr}(k) (cf eqn. (13)). The longitudinal momentum distribution

$$f(y) = 2\pi \int_{|y|}^{\infty} n_{gr}(k) k dk + 2\pi \int_{|y|}^{\infty} n_{ex}(k) k dk = f_{gr}(k) + f_{ex}(y) \qquad (30)$$

is shown in Fig. 5. It can be seen that many particle effects (leading to f_{ex}) are even more important than in ^{3}He and sharply increase with y. We know that the longitudinal momentum distribution does not take into account binding effects and, according to the results presented in Fig. 1, at high values of $|y|$ it is much higher than the asymptotic scaling function (eqn. (28)), where binding effects are properly taken into account. As shown in Ref. 15 the quantity (30) strongly overestimate the old SLAC data [27]. Recently, new SLAC experimental data have

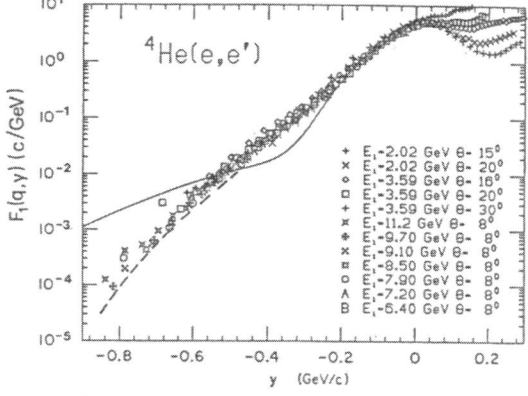

Figure plot axes: vertical $F_1(q,y)$ (c/GeV) from 10^{-5} to 10^1; horizontal y (GeV/c) from -0.8 to 0.2. Label: $^4He(e,e')$

Legend:
+ E_1-2.02 GeV θ- 15°
× E_1-2.02 GeV θ- 20°
◇ E_1-3.59 GeV θ- 16°
□ E_1-3.59 GeV θ- 20°
+ E_1-3.59 GeV θ- 30°
× E_1-11.2 GeV θ- 8°
✦ E_1-9.70 GeV θ- 8°
× E_1-9.10 GeV θ- 8°
⊠ E_1-8.50 GeV θ- 8°
○ E_1-7.90 GeV θ- 8°
A E_1-7.20 GeV θ- 8°
B E_1-6.40 GeV θ- 8°

FIGURE 6

The scaling function F_1(eqn. 15) of ^{4}He obtained [17] from the experimental data of Refs. 27 and 28. The full line correspond to equation 29 and the dashed line to equation 28 (see text) (adapted from Ref. [17]).

been obtained [17, 28] and the scaling function has been properly plotted (See Fig. 6). The finding of Ref.15 is confirmed by the new data: the longitudinal momentum distribution (30) overestimates the experimental scaling function at high y. The origin of such discrepancy is clear: it is due to the lack of binding effects in f(y). In facts the correct quantity to be compared with experimental data should be eqn (17), which requires the unknown Spectral Function of ^{4}He. If we consider this to be the same as in ^{3}He and use eqn (28), the disagreement with the experimental data at high y is strongly reduced. This simple example shows the importance of binding effects in ^{4}He and points to the difficulty of obtaining information on high momentum components in complex nuclei if y scaling is simply interpreted in terms of longitudinal momentum distributions.

5. An approximate version of y scaling where nucleon binding and perpendicular momentum components are disregarded

Our theory of y scaling does differ from the one used by several authors [10,18,19] to analyse the experimental data. Our approach is only based upon the PWIA; the nucleon momentum and its separation energy are properly taken into account. From eqn (19) it follows that $k_{||} = k_{||}$ (q, ω, $k_{\perp}$, E^*_{A-1}) which means, in particular, that all nucleons with all values of $k_{||}$ $k_{\perp}$ and E allowed by the energy conservation (18) contribute to the cross section (10) (and to the scaling function (15)). In Refs. 10 and 18, an approximate version of y-scaling has been proposed based (besides the PWIA) upon the additional assumptions that $k_{\perp}$ and E can be disregarded in the energy conservation (18); in this case $k_{||}$ will depend only upon q and ω , $k_{||}$ = $k_{||}$ (q, ω) and can therefore be assumed as a scaling variable. Moreover, since $k_{||}$ is the only relevant nucleon quantity appearing in the scattering process, the following relation can be written

$$\sigma_2(q,\omega) \, d\omega \; = \; \{Z\sigma_{ep} + N\sigma_{en}\} \; f(y) \; dy \tag{31}$$

f(y) being the probability distribution to find a nucleon with momentum $k_{||}$ = y . From eqn

(31) the definition of the scaling function will be [10, 18]

$$F_2(q,y) = \frac{\sigma_2}{\{Z\sigma_{ep} + N\sigma_{en}\}} \frac{d\omega}{dy} \qquad (32)$$

The above equation is an approximation of eqn (15), the approximation of disregarding $k_\perp$ and

E resulting mainly in the presence of the factor $d\omega/dy$ instead of the factor $d\omega/k \, d\cos\alpha$. Several formal and numerical arguments have been given in Ref. 2 to illustrate the limits of validity of eqn (32). The conclusion reached there is that, apart from $y \approx 0$, F_2 sensibly differ from F_1 (see Fig. 7). Therefore, the extraction of information about nucleon dynamics by using F_2 might be very unreliable. An illuminating example is provided by the momentum distribution of deuteron which will be discussed in the next Section.

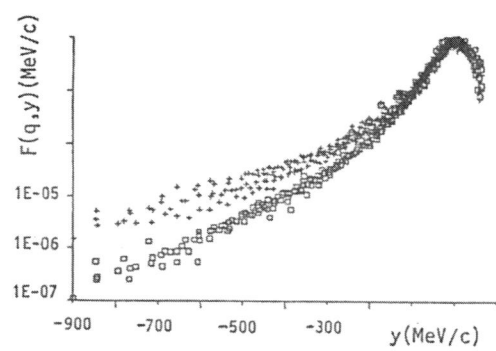

FIGURE 7

The scaling function of ^{2}H obtained from the experimental data of Ref. 16. Crosses: F_1 (q,y), eqn (15). Squares: F_2 (q,y), eqn (30). The same difference between F_1 and F_2 has been found for ^{3}He and heavier nuclei (after Ref. 2).

6. y-scaling and nucleon dynamics

6.1   The high momentum part of the nucleon momentum distribution in deuteron and consistency between exclusive and inclusive scattering

Although at low momentum transfer fsi is relevant, nevertheless it appears that both for ^{2}H and ^{3}He nuclei, F_1^{exp} seems to reach a scaling behaviour at large momentum transfer where the effects of fsi is strongly reduced (cf. Fig. 3). Therefore, the points at large momentum transfer could be considered, in principle, to represent the asymptotic values of F_1^{exp} (q, y). Using these points the q-independent scaling functions for ^{2}H and ^{3}He can be obtained. These are shown in Fig. 8 (the points for ^{2}H have also been corrected for fsi). It can be seen that in the case of deuteron experimental points and theoretical calculations agree fairly well up to y = -600 MeV/c, whereas for ^{3}He F_1^{th} is sensibly lower than F_1^{exp} for y < -200 MeV/c. Using eq. (25), the nucleon momentum distribution in deuteron can be obtained, the same cannot be done for ^{3}He since the asymptotic scaling function (eqn (28)) is not directly related to a momentum distribution). The results are presented in Fig. 9. It is worth pointing out that; i) in the region of overlap the momentum distribution obtained by this way agree with the exclusive d(e,e' p) n data; ii) at high momenta there is a qualitative agreement with predictions of conventional nucleon-nucleon interactions. The extraction of n(k) from inclusive experimental data using the concept of y scaling has been first proposed by Bosted et al [16].

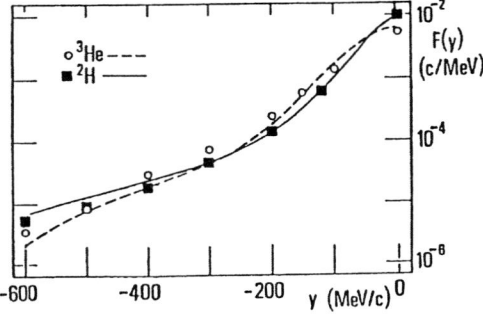

FIGURE 8

Experimental scaling function for ^{2}H and ^{3}He constructed by considering only the points at high values of q^2 shown in Fig. 3. The data points for deuteron have been corrected for fsi [11], whereas for ^{3}He no correction has been applied. The theoretical curve for ^{2}H corresponds to the longitudinal momentum distribution (eqn. 24) and the theoretical curve for ^{3}He to the asymptotic scaling function (eqn. 28) calculated with RSC interaction (after Ref. 20).

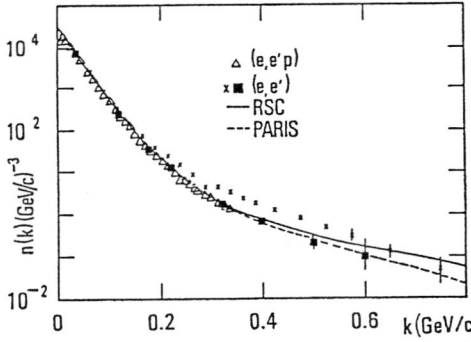

FIGURE 9

Proton momentum distribution in ^{2}H. Triangles: results of exclusive (e,e' p) reaction (Ref. 21) corrected for fsi and MEC (Ref. 22). Black squares: results obtained from the scaling function F_1 (eqn (15)) extracted from the inclusive (e,e') data of Ref. 16 corrected for fsi. The crosses where obtained in Ref. 16 using the approximate scaling function F_2 (eqn (32)) and disregarding the fsi. The curves correspond to the deuteron non relativistic momentum distribution obtained from RSC and Paris interactions (after Ref. 20).

There, however, fsi has been disregarded and, more important, the definition F_2 for the scaling function has been used (eqn. 32). The results of the analysis of Ref. 16 are given by the crosses in Fig. 9 and show a strong disagreement with exclusive data and with Paris interactions. Such a disagreement has been sometimes interpreted as an evidence of 6q admixtures in deuteron, but it is now clear that its origin has to be sought in the use of an approximate definition of the scaling function as well as to the neglect of fsi. For such a reason the attempts at obtaining the momentum distribution of A>2 nuclei using the scaling function F_2 are not reliable.

6.2 Investigation of "nucleon swelling" in the nuclear medium by y-scaling

Several investigations have recently appeared aimed at the study of possible change of "nucleon radii" in nuclei by means of y scaling [18, 19] . The idea is sketched below. If one assumes that nucleons bound in nuclei differ from the free ones, then within the PWIA eqn (10) reduces in the scaling regime to

$$\sigma_2 = [Z\sigma_{ep}^{bound} + N\sigma_{en}^{bound}] \frac{\partial\omega}{k\partial\cos\alpha}^{-1} F(y) \qquad (33)$$

Therefore if the PWIA is a correct description of the scattering, then the observation of a scaling behaviour of

$$F_1^{exp} = \frac{\sigma_2^{exp}}{[Z \sigma_{ep}^{free} + N \sigma_{ep}^{free}]} \left| \frac{\partial \omega}{k \, \partial \cos \alpha} \right| \tag{34}$$

is an indication that $\sigma^{bound} = \sigma^{free}$. Therefore some information on nucleon properties in the medium could in principle be obtained by looking at the variations of the denominator of eqn (36) which are still compatible with scaling. Such a reasoning has been adopted in Ref. 18 where a limit of 6% for the increase in the "radius" of the nucleon in ^{3}He has been obtained. In this respect we would like to point out that these investigations were based upon eqn (32) with fsi and binding effects disregarded, so that, apart from the point $y \approx 0$, (where $F_1 = F_2$) they are not conclusive, particularly in the light of the small effect found which might be of the same order of the theoretical uncertainties in the interpretation of y scaling. Moreover it is improper to talk about "change of nucleon radii". In fact these analyses were performed by modifying the "radius" parameter of the dipole form factor, strongly changing by this way the nucleon form factor in the whole q^2 range. In fact, it is the strong modification of the nucleon form factor in the region of $q^2 > 25$ fm^{-2} which deteriorates the scaling behaviour in ^{3}He. On the other hand side, the nucleon "radius" can in principle be strongly modified by mechanisms which change the low momentum part of G_E^N without affecting its high momentum part. In Ref. 23 a model of the nucleon form factor has been used where, guided by recent investigations on a chiral invariant bag model for the nucleon [24], both the low (given by the pion cloud), and the high (given by the bag radius), momentum parts of G_E^N can independently be modified. The result is that, even a, e.g., 20% increase of the radius due to the polarization of the pion cloud is compatible with y scaling. It should also be pointed out that the quasi elastic peak (y=0) at high momentum transfer can be affected by MEC; therefore, it is only with a longitudinal-transverse separation [25] that some reliable information on the modification of nucleon form factors in the nuclear medium could be obtained. Experiments of this kind are being planned at SLAC [17].

7. Conclusions : usefulness of y scaling plots of inclusive experimental data

As a conclusion, the following arguments should illustrate the usefulness of y-scaling plots of inclusive experimental data:

1) many q.e. peaks corresponding to rather different kinematical conditions can be unified in a plot like the one shown in Fig. 2 (as first proposed in Ref. 10). Such a plot allows one to grossly select all the experimental data which mainly result from the coupling of photons to nucleon degrees of freedom only (y<0), from those resulting from coupling with non nucleonic degrees of freedom (y > 0). Such a possibility is obviously not offered by the usual analysis of separate quasi elastic peaks;

2) by a q-plot of the data for fixed values of y (Fig. 3), all cross sections corresponding to different q and ω but to the same value of $k_{min} = |y|$ are unified and the q dependence of the scaling function is obtained. Such a q dependence, which cannot be inferred by analysing

separate q.e. peaks can be very useful. In facts it yields model independent experimental evidences about the breaking down of the PWIA cross section (particularly about its factorized form), and, more important yields the q *dependence of such a breaking down,* which represents a unique and very stringent test of various approaches which go beyond the PWIA. By longitudinal -trasverse separation such a plot would be even more significant [25].

3) if the data eventually scale we have experimental evidences about the correctness of the PWIA reaction mechanism and could be confident that the quantities we are plotting are indeed related to a nuclear structure function. Then important properties of nucleon dynamics could in principle be investigated. For example it has been shown [20] that the nucleon momentum distribution of deuteron can be studied by using the concept of y scaling up to very high momenta (eqns (24) and (25)). The fact that in the region of overlap the results presented in Fig. 9 coincide with exclusive data makes such an analysis rather reliable.

References

1. G. B. West, Phys. Rep. 18 (1975) 263.
2. C. Ciofi degli Atti, E. Pace and G. Salmé, "Theory of y-scaling with proper treatment of nucleon momentum and separation energy", Preprint INFN 86/6, July 1986 submitted for publication.
   C. Ciofi degli Atti, E. Pace and G. Salmé, in "The Three-Body Force in the Three-Nucleon System", B.L. Berman and B.F. Gibson eds., Lecture Notes in Physics, Vol. 260 (Springer - Verlag, 1986) p. 349.
   C. Ciofi degli Atti, E. Pace and G. Salmé, III International Symposium"Mesons and Light Nuclei", Cechoslovakia 1985, Czech. J. Phys. B36 (1986) 960.
3. A. Yu. Korchin and A. V. Shebeko, Z. Phys. A299 (1981) 131.
4. S.A. Gurvitz, S. Wallace and J. Tjon Phys. Rev. C34 (1986) 648.
5. E. Pace and G. Salmé, Phys. Lett. 110B (1982) 411.
6. C. Ciofi degli Atti, E. Pace and G. Salmé, Phys. Lett. 141B (1984) 11.
7. C. Ciofi degli Atti, Lett. Nuovo Cimento 41 (1983) 330.
8. C. Ciofi degli Atti, E. Pace and G. Salmé, Phys. Lett. 127B (1983) 303.
9. D. Day et al, Phys. Rev. Lett. 43(1979) 1143.
10. I.Sick, D. Day and J. S. Mc Carthy, Phys. Rev. Lett. 45 (1980) 871.
11. W. Leidemann and H. Arenhoevel, private communication.
12. J.M. Laget, Phys. Lett. 151B (1985) 325 and private communication.
13. C. Ciofi degli Atti, E. Pace and G. Salmé, Phys. Rev. C21 (1980) 805.
14. Y. Akaishi, Nucl. Phys. A416(1984) 4096 and private communications.
    Y. Akaishi, these Proceedings.
15. C. Ciofi degli Atti, T. Katayama, C. Salmé, O. Benhar and S. Liuti in "2nd Workshop on Perspectives in Nuclear Physics at Intermediate Energies" S. Boffi, C. Ciofi degli Atti and M. Giannini Eds., World Scientific 1985 p. 309, and to be published.
16. P. Bosted et al, Phys. Rev. Lett. 49 (1982) 1380.
17. Z.E. Meziani, these Proceedings.
18. I. Sick, Phys. Lett. 157B (1985) 13.
19. R.D. McKeown, Phys. Rev. Lett. 56 (1986) 1452
20. C. Ciofi degli Atti, in "Few Body Problems"1984, World Scientific, L.D. Faddeev and T.I. Kopaileshvili eds., 1986, p. 463.
21. M. Bernheim et al, Nucl. Phys. A365 (1981) 349.
22. H. Arenhoevel, Nucl. Phys. A384 (1982) 287.
23. C. Ciofi degli Atti et al., to be published.
24. E. Oset, R. Tegen and Weise, Nucl. Phys. 1426 (1984) 456.
25. C. Ciofi degli Atti and G. Salmè, Proc. of the 6th Seminar "Electromagnetic Interactions of Nuclei at Low and Medium Energies", Moscow Dec. 11-12, 1984, p. 224.
26. T. Sasakawa, these Proceedings.
27. S. Rock et al, Phys. Rev. C26 1592 (1982)
28. D. Day et al, NE3 SLAC Experiment. Preliminary results. Quoted in Ref. 17

CONTINUUM CALCULATIONS [*]

B. Goulard and T. Pochet

Laboratoire de Physique Nucléaire, Université de Montréal, P.Q., Canada

G. Cory-Goulard and D. Hennequin

Collège Militaire de Saint-Jean, P.Q., Canada

We present an approach to the calculation of transition amplitudes from bound state to the continuum for three nucleons nuclei. The formalism for transitions induced by electron scattering is carried out and some numerical results about the tritium muon capture are shown.

During the last few years, there have emerged two levels of description of the nuclei, the first is the traditional description in terms of nucleons and mesons, while the second is based on QCD, and uses quarks and gluons. On the theoretical side, QCD and related concepts such as bags and skyrmions, get closer and closer to the phenomenology of nuclear physics. On the experimental side, high duty cycle electron accelerators and high energy proton accelerators with high intensity will soon become available Experiments with these machines will provide informations on the detailed structure of nuclei, and this should provide guidance for a deeper conceptualization which will overcome the shortcomings of the nucleon-meson theories. In order to identify any shortcomings of nucleon-meson theories, one must clearly distinguish between the intrinsic limitations of this traditional approach and the inevitable approximations required in a practical calculation; such approximations should therefore be self-consistent. For processes involving three nucleons undergoing a transition from a bound to a continuum state, the complexity of the three-body dynamics makes exact calculations extremely difficult. As a

---

[*]Paper submitted in absentia

consequence, a number of experiments involving a continuum of three nucleons (one deuteron - one neutron) are still in need of a quantitative interpretation. This is even the case at low energies where the traditional nuclear description is supposed to be valid at least in principle and still more so at higher energies where new nuclear degrees of freedom are expected to show up.

We present a review of our approach to the calculation of transition amplitudes from bound state to the continuum for trinucleonic nuclei. In this approach transition amplitudes are consistently connected to the originating Hamiltonian using the Faddeev method [1]. As already mentionned, these calculations will be of use at two levels : i) to interpret experiments at low energies which are not interpretable yet, even in the framework of nucleons only, because of the intricacies of the three-body dynamics, ii) to speculate about effects involving new degrees of freedom in high momentum transfer processes.

In a first section, the formalism as applied to electron scattering will be presented. Then a second section will be devoted to the case of muon capture by the triton.

1. Electron Scattering

Calculations of nuclear transition amplitudes involve the expression $<\varphi_f|H_{em}|\varphi_{bd}>$ where $|\varphi_f>$, H_{em} and $|\varphi_{bd}>$ are respectively the final scattering state (3 nucleons, 1 deuteron - 1 neutron), the nuclear electromagnetic operator and the initial bound trinucleonic state [2].

Set $j_\mu(\vec{x})$ be the 4-vector nuclear electromagnetic current, the aim is to calculate the cross-section $d\sigma$ corresponding to the following diagram :

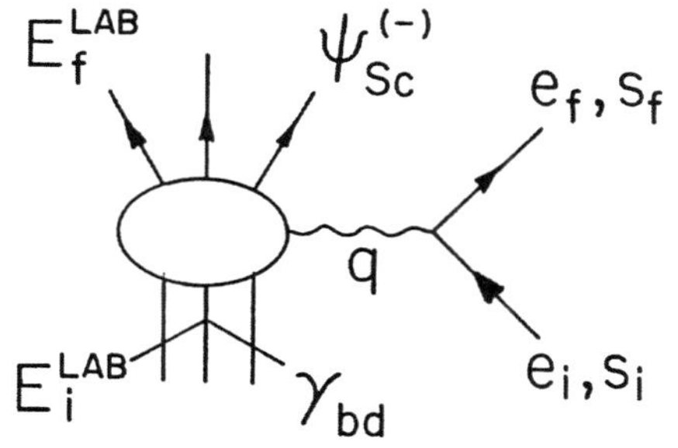

Fig. 1 - Diagram of electron scattering
inducing a nuclear transition.

In obvious notations, $d\sigma$ is written as :

$$d\sigma = 2\pi\delta(E_f^{lab} - E_i^{lab} + e_f - e_i) \frac{e^4}{q^4} \frac{m^2}{e_i e_f} \frac{1}{2} \cdot$$

$$\cdot \sum_{s_i s_f} |\bar{u}_{P_f s_f} \gamma_\mu u_{P_i s_i} < \varphi_f | \int j(\vec{x}) e^{i\vec{q}\cdot\vec{x}} d\vec{x} |\varphi_{bd}>|^2 \frac{d\vec{P}_f}{(2\pi)^3} \frac{d\vec{P}_i}{(2\pi)^3} \tag{1}$$

The traditional way is to find $|\psi^{(-)}_{scat}>$ which is corresponding asymptotically to an incoming scattered wave. One is then confronted with an homogeneous differential equation :

$$(E_q - H_N) |\psi^{(1)}_{scat}> = 0 \tag{2}$$

coupled with a difficult asymptotic behaviour. This concept has been the underlying base of several pionneering works on photo - and electro - disintegration of Helium three and Triton with the use of separable forces for the continuum [3]. Only a few investigations on cases involving neutron - deuteron states at fixed energy are based on a complete Faddeev calculation up to now [4,5]. In the approach advocated in the following, the nuclear bound state acted upon by the perturbative operator is considered as a source $|S^\mu_q>$ which yields a purely outgoing function $|X^\mu_q>$. One is then confronted with an inhomogeneous differential equation :

$$(E_q - H_N)|X^\mu_q> = |S^\mu_q> = \int j(\vec{x}) e^{i\vec{q}\cdot\vec{x}} d\vec{x} |\varphi_{bd}> \tag{3}$$

together with an asymptotic behaviour more tractable than for eq. (2). The coordinates $(\vec{x}, \vec{y})$ of the nucleons in the system of the center of mass are related to their coordinates $(\vec{R}_1, \vec{R}_2, \vec{R}_3)$ in the laboratory by the relation :

$$\vec{x} = \vec{R}_2 - \vec{R}_3$$
$$\vec{y} = \frac{2}{\sqrt{3}} \left(\frac{\vec{R}_2 + \vec{R}_3}{2} - \vec{R}_1 \right) \tag{4}$$

and the associated momenta $(\vec{p}, \vec{s})$ are related to the momenta in the laboratory according to the relations :

$$\vec{p} = \frac{1}{2} (\vec{P}^{(2)} - \vec{P}^{(3)})$$
$$\vec{s} = \frac{1}{2\sqrt{3}} (\vec{P}^{(2)} + \vec{P}^{(3)}) - \frac{1}{\sqrt{3}} \vec{P}^{(1)} \tag{5}$$

The spherical coordinates $\vec{x} \equiv (x, \theta_x, \varphi_x)$, $\vec{y} \equiv (y, \theta_y, \varphi_y)$ and the generalized coordinates $\vec{X} = \{\vec{x}, \vec{y}\}$, $\vec{k} = \{\vec{p}, \vec{s}\}$ will be used often in the following lines.

Indeed, equations such as eq. (3) have been investigated in the coordinate representation [6,7]. The source $< \vec{X} \mid S_q>$ is spatially localized and yields a well defined asymptotic behaviour for $< \vec{X} \mid X_q>$ which is now going to be briefly described.

For the sake of clarity, the spin and isospin variables will be omitted, unless stated otherwise. For example, the expression $\mid \varphi^- \vec{p} \vec{k}>$ introduced two lines ahead stands for $\mid \varphi^- \vec{p} \, m_2 \, m_3 \, \tau_{z2} \, \tau_{z3}, \, \vec{k} \, m_1 \, \tau_{z1}>$.

a) Three nucleons case.

$$|\varphi_f> = \frac{A}{\sqrt{6}} |\varphi^-_{\vec{p}\vec{k}}> \tag{6}$$

where A is the antisymmetry operator $(1+P^++P^-)(1-P_{23})$, and $\mid \varphi^- \vec{p} \vec{k}>$ is the non antisymmetric solution of the Schrödinger equation for the potential $V_1+V_2+V_3$. Also the initial and final total energies are written :

$$E^{lab}_f = \frac{\hbar^2 \vec{p}^2}{m_N} + \frac{3}{4} \frac{\hbar^2 \vec{k}^2}{m_N} + \frac{1}{6} \frac{\hbar^2 \vec{q}^2}{m_N} \tag{7a}$$

$$E^{lab}_i = -|E_{3He}| \tag{7b}$$

where E_{3He} represents the binding energy of the Helium three. The asymptotic behaviour of the nuclear Green function yields :

$$<\vec{X}|X^\mu_q> = <\vec{X}|G^+_N(E_q)|S^\mu_q> \quad \begin{array}{c} |\vec{X}| \to \infty \\ \frac{|\vec{x}|}{|\vec{y}|} = \text{Cst} \end{array}$$

$$\sim a^\mu_q(\hat{x}, \hat{y}, x/y) \frac{e^{ikx}}{x^{5/2}} \tag{8}$$

with

$$a^\mu_q(\hat{x}, \hat{y}, \theta) = \frac{m_N}{\hbar^2} \frac{e^{i\pi/4}}{2(2\pi)^{5/2}} K^{3/2} < \psi^-_{p\hat{x}, q\hat{y}} |S^\mu_q> \tag{9}$$

b) One deuteron - one nucleon case.

$$|\varphi_f> = \frac{1+P^++P^-}{\sqrt{3}} |\varphi^-_{\vec{k}}> \tag{10}$$

where $\mid \varphi^- \vec{k}>$ is the non antisymmetric solution of the Schrödinger equation for the potential $V_1+V_2+V_3$. Also

$$E^{lab}_f = \frac{3}{4} \frac{\hbar^2 \vec{k}^2}{m_N} - |E_d| + \frac{1}{6} \frac{\hbar^2 \vec{q}^2}{m_N} \tag{11a}$$

$$E^{lab}_i = -|E_{3He}| \tag{11b}$$

where E_d represents the binding energy of the deuteron. The asymptotic behaviour of the nuclear Green function yields :

$$<\vec{x}\,|\,\chi_q^\mu> \; = \; <\vec{x}\,|\,G_N(E_q)\,|\,S_q^\mu> \; \underset{\substack{|\vec{x}|\,\to\,\infty\ \ |\vec{y}|\,\to\,\infty \\ |\vec{x}|\,=\,\text{cst}}}{\sim} \; a_q^\mu \, \frac{e^{\,iqy}}{y} <\vec{x}\,|\,\varphi_{deu}> \tag{12}$$

with

$$a_q^\mu(\hat{y}) \; = \; - \, \frac{m_N}{4\pi\hbar^2} <\varphi_{k\hat{y}}^-\,|\,S_q^\mu> \tag{13}$$

So that the cross-sections take the following explicit forms :

$$d\sigma_{3N} \; = \; 2\pi\delta(E_f^{lab} - E_i^{lab} + e_f - e_i) \, \frac{e^4}{q^4} \, \frac{m_N^2}{e_i e_f} \, 24 \, \frac{\hbar^4}{m_N^2} (2\pi)^5 k^{-3} \cdot$$

$$\cdot \frac{1}{2} \sum_{\pm s_i s_f} \left| \bar{u}(p_f,s_f)\gamma_\mu u(p_i,s_i)\,\mathcal{A}_q^\mu(\hat{p},k,\theta,m_2,m_3,\tau_{z2},\tau_{z3},m_1,\tau_{z1}) \right|^2 \cdot \tag{14}$$

$$\cdot \; \frac{d^3\vec{p}_f\, d^3\vec{p}_i\, d^3\vec{k}}{(2\pi)^9}$$

$$d\sigma_{1N-1D} \; = \; 2\pi\delta(E_f^{lab} - E_i^{lab} + e_f - e_i) \, \frac{e^4}{q^4} \, \frac{m_N^2}{e_i e_f} \, 3(4\pi)^2 \frac{\hbar^4}{m_N^2} \cdot$$

$$\cdot \frac{1}{2} \sum_{\pm s_i s_f} \left| \bar{u}(p_f,s_f)\gamma_\mu u(p_i,s_i)\, a_q^\mu(k) \right|^2 \frac{d^3\vec{p}_f\, d^3\vec{k}}{(2\pi)^6} \tag{15}$$

The summation over s_i, s_f and μ yields :

$$\frac{m_N^2}{e_i e_f} \sum_{\pm s_i s_f} \left| \bar{u}(p_f,s_f)\gamma_\mu u(p_i,s_i) A^\mu \right|^2 =$$

$$\left| A^0 \right|^2 (1 + \cos\theta) + \left| \vec{A} \right|^2 (1 - \cos\theta) + \frac{e^2}{e_i e_f} \left| \vec{p}_i \cdot \vec{A} \right|^2 \tag{16}$$

$$-\frac{2(e_i + e_f)}{e_i e_f} \, \text{Re}\{A^0 \vec{p}_i \cdot \vec{A}\}$$

with $\vec{p}_i \cdot \vec{p}_f = e_i e_f \cos\theta$ and $A_\mu = \mathcal{A}_\mu$ for $d\sigma_3$, a_μ for $d\sigma_{1N-1D}$.

Angular decomposition

At the level of eqs (14), (15), (16), it is appropriate to carry out angular decompositions of the various expression. Final formal expressions are now available for computer programming. In view of the cumbersome nature of the intermediate formulae, only an outline will be given below. A detailed report of the calculations will be published elsewhere.

The starting point is the angular expansion of the ^{3}He bound state :

$$< \vec{x}\,\vec{y}\,|\,m_{3_{He}}1/2> \;=\; \sum_{\lambda j L \sigma J t} \frac{\varphi_{\lambda j L \sigma J t}(x,y)}{xy} \; Z_{\lambda j L \sigma J t}^{1/2\;m_{3_{He}}}(\hat{x},\hat{y})\;\eta^{\frac{1}{2}\frac{1}{2}}_{\frac{1}{2}\,t} \qquad (17)$$

with $<m_{3_{He}}\tfrac{1}{2}|\,m_{3_{He}}\tfrac{1}{2}> = 1$, $\{\lambda j L\sigma J t\} \equiv \alpha$ = set of quantum numbers of each component. With the introduction of the operators :

$$C^{LM}(q^2) \;=\; \sum_{i=1}^{3} OP^{LM}(i, Cb)$$

and $E^{LM}(q^2)$ and $M^{LM}(q^2)$ corresponding to the Coulomb, electric and magnetic operators of angular momentum L(M) respectively, one gets an angular decomposition of $<\vec{x}\,\vec{y}|S^0_q{}^{(1)}m_{3_{He}}>$ in terms of Coulomb operators, and also of $<\vec{x}\,\vec{y}\,|S_q{}^{(1)}m_{3_{He}}>$ in terms of electric and magnetic operators. This procedure is similar to the decomposition of eq. (17), the superscript (1) refers to the Faddeev amplitude. These new expressions incorporate (J MT) which corresponds to the final continuum state quantum numbers and are written $E_{\lambda j L\sigma J t}(q^2)^{L\,JT\,(1)}$, C and M.

Finally, in the same way as one defines $|S\,\mu_q{}^{(1)}m\,_{3_{He}}\tfrac{1}{2}>$,

$$|S^{\mu}_{q\;m_{3_{He}}}> \;=\; (1+P^+ + P^-)\,|S^{\mu\,(1)}_{q\;m_{3_{He}}}> \qquad (18)$$

one can defines $|X\mu_q{}^{(1)}m\,_{3_{He}}\tfrac{1}{2}>$ which satisfies the equations :

$$(E_q - H_0 - V_1(1+P^+ + P^-))\,|X^{\mu\,(1)}_{q\;m_{3_{He}}}> \;=\; |S^{\mu\,(1)}_{q\;m_{3_{He}}}> \qquad (19)$$

The components $<\vec{x}\,\vec{y}\,|\,X\mu_q{}^{(1)}m_{3_{He}}>$ are expanded similarly to the expressions $<\vec{x}\,\vec{y}\,|S\mu_q{}^{(1)}\,m_{3_{He}}>$ with the substitution of the functions C, E and $M_{\lambda j L\sigma J t}(q^2)^{L\,JT\,(1)}(x,y)$ by the functions GC, GE and $GM_{\lambda j L\sigma J t}(q^2)^{L\,JT\,(1)}(x,y)$. The later are obtained by resolution of the Faddeev equations with the corresponding sources.

Finally, a very general analytical expression is available for the cross-sections of electrodisintegration into a three nucleons channel and/or one deuteron - one nucleon channel

Resolution of equations

The most time consuming part of the calculations of the cross-sections is determining the solution of eqs (19) which satisfies the asymptotic conditions described by eqs (8), (12). The equation which determines the complex function $X(\vec{x},\vec{y})^{(1)}$ has the following structure[*] :

$$\left\{ E - [H_0 - L(\vec{x},\vec{y})] - V_1(1+[H\text{-Integral}]) \right\} X^{(1)}(\vec{x},\vec{y}) = S^{(1)}(\vec{x},\vec{y}) \qquad (20)$$

The continuous function $X(\vec{x},\vec{y})$ can be given in a dense discrete set of points. So, searching $X(\vec{x},\vec{y})$ amounts to searching for a finite number of values of $X(\vec{x},\vec{y})$ over a grid. Eq. (20) is transformed into :

$$(L(\vec{x},\vec{y})-E) X_i(\vec{x},\vec{y}) + \sum_{i=1}^{M} V_1^{ij}(x) X_j(\vec{x},\vec{y})$$

$$+ \sum_{j=1}^{M} \left\{ V_1(x) \int_{-1}^{+1} du \, h(x(u), y(u)) X_j(x(u), y(u)) \right\} = -S_i(\vec{x},\vec{y}) \qquad (21)$$

One searches the $X_i(\vec{x},\vec{y})$, i = 1,..., M and E over a specific area and satisfying the same limit conditions. The problem is written in terms of ρ and θ .

The Laplacian $L(x,y) = L(\rho,\theta)$ is written with the help of splines in order to approximate partial derivatives while for each value of X_j, the integral expression is expressed with splines :

$$\int_{-1}^{+1} h(\rho(u), \theta(u)) X_j(\rho(u), \theta(u)) \, du \approx$$

$$\sum_{i_\theta=1}^{N_\theta} \sum_{i_\rho=1}^{N_\rho} X_j(\theta_{i_\theta}, \rho_{i_\rho}) \int_{-1}^{+1} h(\rho(u), \theta(u)) S_{i_\theta}(\theta(u)) S_{i_\rho}(\rho(u)) \, du \qquad (22)$$

so that the calculation involves many integrals to be computed. Those integrals are computed following the Gauss-Legendre method.

The eqs (20) have now turned to a set of algebraic equations $AX = B$ where A and B are known, X is the unknown vector with components $X_j(\theta_{i\theta}, \rho_{i\rho})$; B corresponds to the source term $s(\vec{x},\vec{y})$ and A corresponds to the left-hand side of eq. (21) once it is put in a matricial form. For a grid with $N_\theta = N_\rho = 20$, the A matrix has MxMx160 000 dimensions. Explicitation of that matrix yields a matrix D constituted by a sequence of blocks centered along the diagonal of A and a matrix R constituted by residual terms outside these blocks : $A = D + R$. Iterative methods are available to treat

---

[*] In the present section, the superscripts μ and 1 will be omitted.

these matricial equations. The Padé method has been used to accelerate the convergence of the solution [8].

2. Muon Capture by a triton

Results on muon capture by a Triton are given in the following not only because it represents a simple illustration of our method (only one channel, three electrically neutral particles, . . .) but also because of the intrinsic interest of muon capture by very light nuclei. For the sake of example , let us mention the interest of measuring the Triton recoil energy in the reaction $\mu^- + {}^3He \to {}^3H + \nu_\mu$ to obtain useful informations on the muon neutrino [9]. Several aspects of Triton capture make it different from hydrogen capture and should make information from the former process complementary to the Hydrogen information. Once the muon is captured by the Triton, the three neutrons cannot come out in a s-state because of the Pauli principle, this somewhat "first forbiden" transition will yield transition matrix elements different from those corresponding to Hydrogen. Closure estimates of the reduction factors have been studied in the past and compared to Hydrogen and Helium [10]. Because of this suppression of the muon capture process, the presence of mesonic exchange currents might be much more significant then for neighbour nuclei (Hydrogen, Deuteron, Helium, . . .).

Our approach, when applied to the calculation of muon capture rate yields expressions containing the integral representation of the amplitude :

$$(2\pi)^3 \frac{d\Lambda}{d^3 v}\mu \, ({}^3H \to 3n) = \frac{2\hbar}{m_N} K |\varphi_\mu|^2 \int |a_{\vec{\nu}}(\hat{x}, \hat{y}, x/y)|^2 \cdot \\ \cdot \cos^2\theta_x \sin^2\theta_x d^2x d^2y d\theta \tag{23}$$

Finally, knowing that the detectors are located far away from the origin, $\hat{p} = \hat{x}$, $\hat{s} = \hat{y}$, $p/s = x/y$, $p = K\cos\theta_\chi$, $s = K\sin\theta_\chi$, so that :

$$\Lambda_\mu({}^3H \to 3n) = \frac{2\hbar}{m_N} \int \frac{d^3v}{(2\pi)^3} d^3p \, d^3s \, \frac{1}{K^3} |\varphi_\mu|^2 |a_{\vec{\nu}}(\hat{p}, \hat{s}, p/s)|^2 \tag{24}$$

The figures below correspond to the use of a Sprung-de Tourreil super soft core nucleon-nucleon interaction (SSC). The weak Hamiltonian is given by :

$$\mathcal{H}_{WI} = \frac{3}{2} \tau^+ (1 - \vec{\sigma} \cdot \hat{\nu}) \tau_N^- \Big[G_V 1 \cdot 1_N + G_A \vec{\sigma} \cdot \vec{\sigma}_N - G_P \vec{\sigma} \cdot \hat{\nu} \, \vec{\sigma}_N \cdot \hat{\nu}_N \\ - g_V \vec{\sigma} \cdot \hat{\nu} \vec{\sigma}_N \cdot \vec{P}_N - g_A \vec{\sigma}_N \cdot \hat{\nu} \, \vec{\sigma}_N \cdot \vec{P}_N \Big] \delta(\vec{r} - \vec{r}_N) \tag{25}$$

The expression between accolades in a sum of five terms that will be referred to as O_1, O_2, O_3, O_4 and O_5 ; O_1, O_2, O_3 are non relativistic, O_4 and O_5 are relativistic to the order $1/m_N$. The constants in the right-hand of eq. (25) are standard combinations of elementary weak interaction constants. The initial $\{\mu^- -^3H\}$ state is either in a singlet state denoted by the subscript 0 or a triple state denoted by the subscript 1 so that the total muon capture $\Lambda = 1/4\,\Lambda^0 + 3/4\,\Lambda^1$.

Fig. 2 displays the spectacular effect of a (SSC) nucleon-nucleon interaction on the neutrino spectrum and the capture rate which increases from 50 sec$^-$ within the Born approximation to 69 sec$^-$ with final state interaction. The transition takes place between an initial singlet state and a $J^\pi = 3/2^-$ final state, the five operators are taken into account.

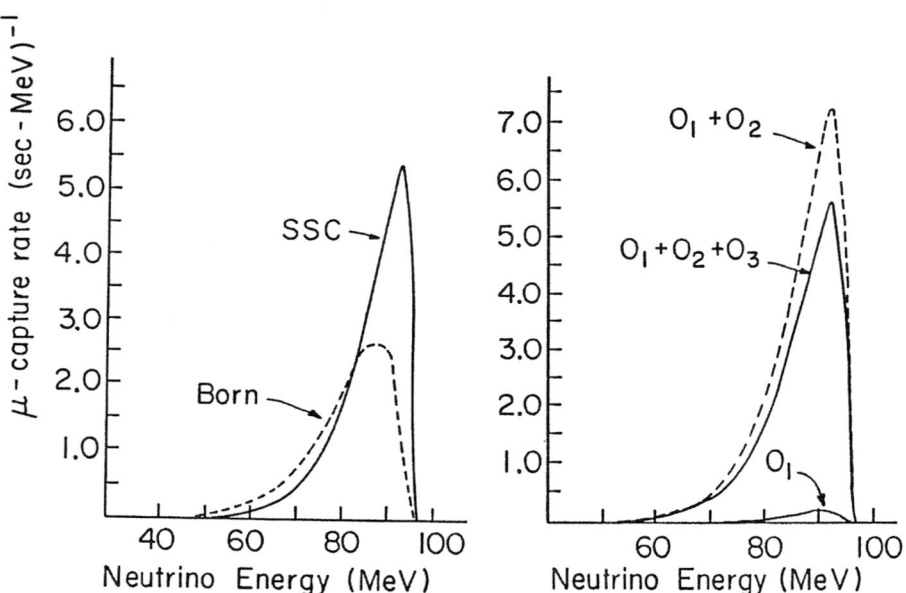

Fig. 2 Neutrino energy spectrum corresponding to a singlet initial state and a $J^\pi = 3/2^-$ final state.

Fig. 3 Contribution of the weak operators shown in eq. (25).

The relative importances of these operators are shown in fig. 3. The continuous line corresponds to the action of the operator O_1 only while the dashed line corresponds to the sum of operator O_1 and O_2. This difference between the continuous and dashed lines shows that the axial term $G_A \vec{\sigma} \cdot \vec{\sigma}_N$ is making the dominant contribution to the muon capture. The shaded area which corresponds to three curves very close to one another; the first one corresponds to the sum of operators O_1, O_2 and O_3, and the two last ones include the relativistic corrections O_4 and O_5. The quasi-identity between those three curves illustrates the small contribution of the so-called relativistic terms.

Some application among others

a) Muon capture by Helium three. This reaction, a natural extension of the muon capture by Triton, presents two outgoing scattering channels [11] :

$$\mu^- + {}^3He \longrightarrow p + n + n + \nu_\mu$$
$$\longrightarrow d + n + \nu_\mu$$

A treatment, based on Faddeev equations with realistic two-nucleon interaction of the final states at the level of nucleons only, is currently being carried out at Montréal. This treatment can be followed by a study of the meson exchange currents.

b) Photo-disintegration of Helium three and Triton. Here again a group of data involving low energy incoming photons (< 35 Mev) yields an agreement between theory and experiment which is subject to improvement [12]. Hoping that the introduction of final state interactions brings a better situation, another group of data at higher photon energies (> 150 Mev) might then be investigated in a formalism involving new degrees of freedom.

c) Electron quasi-elastic scattering [13]. Disagreement between experiment and theory becomes quite significant when one gets away from the elastic peak. The most obvious reason seems to come from the ignorance of the final state interaction between the expulsed nucleon and the two other nucleons.

Summary and conclusion

We have described our calculations of the transition amplitudes from bound states to the continuum for three nucleon nuclei, and the corresponding electron scattering formalism has been completed. Previous numerical results for muon capture have also been reviewed. The solution of the basic equations involves a great deal of sophisticated applied mathematics and computer programming. Presently, we are systematically developing numerical tools and restructuring an old program, before concentrating on a specific physical process.

Acknowledgements

This works was supported in part by the Natural Sciences and Engineering Research Council of Canada. J. Torre (Institut des Sciences Nucléaires de Grenoble) had participated to the preliminary stages of this work.

References

[1] Torre, J. and Goulard, B. : Phys. Rev. Lett. 43, 1222 (1979).

[2] Goulard, B. et al., contributed paper at the International Symposium on the "Three-Nucleon Force in the three-Nucleon System", The George Washington University, Washington DC, 24-26 April 1986.

[3] Lehman, D.R., Prats, F. and Gibson, B.F. : Phys. Rev. C19, 310 (1979); Lehman, D.R. talk presented at the symposium quoted in ref. [2].

[4] Torre, J. and Goulard, B. : Phys. Rev. C28, 529 (1983).

[5] Jourdan, J. et al : Phys. Lett. 162B, 269 (1985).

[6] Newton, R.G. : Ann. Phys. (N.Y.) 74, 324 (1972).

[7] Merkuriev, S.P., Gignoux, C. and Laverne, A. : Ann. Phys. (N.Y.) 99, 30 (1976).

[8] Hennequin, D. : communication at the Eight Autumn School on Few-Body Problem, Lisbon, Portugal, October 13-18 (1986); G. Cory-Goulard at al. : communication at the same school.

[9] Deutsch, J. et al. : "Neutrino 82", Balatonfured, Hungary, June 16-18 (1981).

[10] Torre, J., Gignoux, C. and Goulard, B. : Phys. Rev. Lett. 40, 511 (1978).

[11] Wang, I.T. : Phys. Rev. B139, 1544 (1965).

[12] Faul, D.D., Berman, B.L., Meyer, P. and Olson, D.L. : Phys. Rev. C24, 849 (1981).

[13] Meier-Hajduk, H., Hajduk, C., Sauer, P.U. and Theis, W. : Nucl. Phys. A395, 332 (1983).

CONTINUUM STATE CALCULATIONS FOR THE REACTION ^{3}He(e,e'p)d

E. van Meijgaard and J.A. Tjon

Institute for Theoretical Physics
Princetonplein 5
P.O. Box 80.006
3508 TA   Utrecht, The Netherlands

Final state interactions are accounted for in the coincidence cross sections of electrodisintegration of ^{3}He by solving the Faddeev equations for the scattering states. Local s-wave spin dependent potentials of the Yukawa type are used as two-nucleon input.

In this contribution we report on a calculation of the electrodis-integration process of the trinucleon system. In particular we are inter-ested to go beyond the standard PWIA analysis and to study the effect of final state interactions (FSI) between the three nucleons in the case of both two-body and three-body breakup processes. Neglecting MEC effects, the electrodisintegration amplitude is given by the sum of the PWIA amplitude, the other Born type diagrams and the connected three-body diagrams in the final state. The latter class of diagrams describes the FSI between all the three particles. In calculating this we assume that the dynamics of the three-momentum system can essentially be treated nonrelativistically. Rela-tivistic kinematics are only used to relate the four momentum of the photon to the three-momentum and energy of the trinucleon system.

For the case of the two-body breakup the contributions to the electro-disintegration process of ^{3}He are shown in Fig. 1. The diagrams correspond to the various terms occurring in the resolvent series of the three-particle S-matrix. In the PWIA (diagram 1a) the coincidence cross section factorizes into the half off-shell electron-nucleon cross section $(d\sigma/d\Omega)_{ep}$ and the nucleon momentum distribution ρ_2. However for the other diagrams the factorization is no longer valid. In calculating the current matrix elements we in general need an off-shell extrapolation of the one

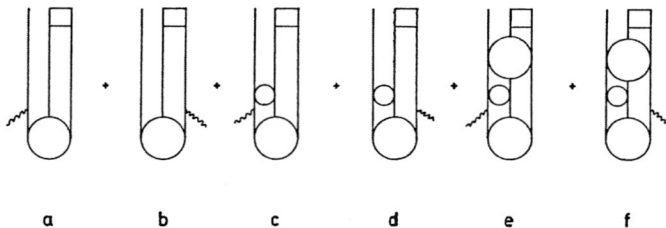

a   b   c   d   e   f

Fig. 1. Diagrammatic representation of the various contributions to the electrodisintegration process. (1a) is the PWIA approximation, while (1b) is the deuteron knockout process. The connected diagrams, describing the FSI, consist of the multiple scattering series, the first order being given by (1c) and (1d) and all the higher order diagrams by (1e) and (1f), which can be determined by solving the Faddeev equations.

nucleon current operator, which depends explicitly on the nucleon momenta of the trinucleon system in the intermediate state. Here we adopt the choice made by Dieperink et al. [1], which has implicitly built in current conservation [2]. It is constructed from the one nucleon current operator

$$J_\mu = i\left[\gamma_\mu(F_1+\kappa F_2) + i(p+p')_\mu \, \kappa F_2/2m_N\right] \tag{1}$$

For the electromagnetic form factors F_n we use the parametrization of Höhler et al. [3]

As two-nucleon input we have used spin dependent s-wave local potentials of the Yukawa type

$$V(r) = -\lambda_A \frac{e^{-\mu_A r}}{r} + \lambda_R \frac{e^{-\mu_R r}}{r} \tag{2}$$

which give a good reproduction of the s-wave NN-phase shifts up to 250 MeV lab energy for both spin channels [4]. Using these potentials the properties of the trinucleon bound state and the neutron-deuteron scattering are qualitatively well described [5]. The one nucleon momentum distribution function for the two-body breakup of ^{3}He, given by

$$\rho_2(\vec{p}) = \frac{3}{2} \sum_{s_t} \sum_{s_n, s_d} |_1\langle\vec{p}' \, s_n; \, \phi_d s_d |\phi_t s_t\rangle|^2 \tag{3}$$

is shown in Fig. 2 for the case of these s-wave potentials. At low momenta the agreement with the experimental data [6] is good, but at larger momenta the calculated results are too high. This is due to the fact that our wave

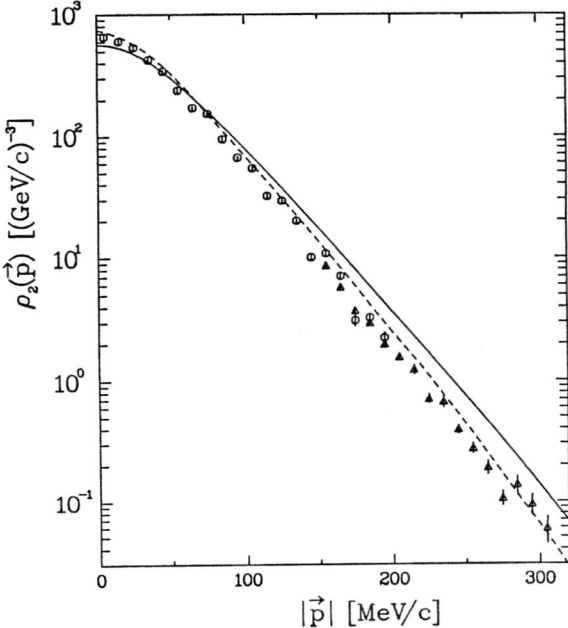

Fig. 2. Nucleon momentum distribution of ^{3}He for the two-body breakup. The solid curve represents our calculation and the dashed curve shows the result of Laget. The experimental date are from Jans [6].

functions contain too little correlation at high momenta. The Reid soft-core results of Laget [7], which are also shown in Fig. 2, indeed show a steeper falloff.

To calculate the connected three-body diagrams we first solve the Faddeev equations at the energies dictated by the (e,e') process. To determine the electromagnetic current matrix elements we need the half off-shell solutions for the continuum states. These are obtained by applying the Padé-approximant method to the multiple scattering series of the Faddeev equations [5]. The actual calculations are done using the unitary pole approximation for the two-nucleon potential.

As an example the results of a calculation for the Amsterdam kinematic set up [8] are shown in Fig. 3. In this kinematic region direct deuteron knock out (diagram 1b) is the dominating process, whereas a pure PWIA calculation underestimates the cross section by three orders of magnitude. However the coherent sum of the connected diagrams (1c-1f) and the Born-

310

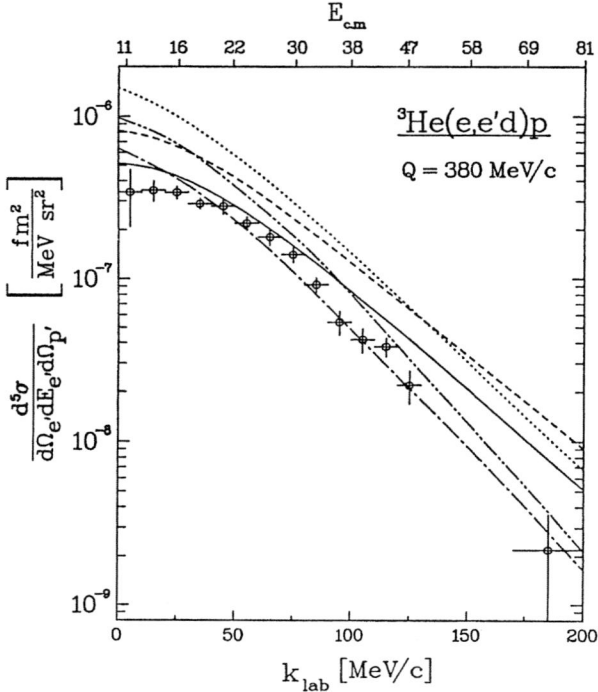

Fig. 3. Coincidence cross section at fixed momentum transfer. The dashed and solid curves correspond to a Born and Born + full FSI calculation. The dotted curve represents a calculation involving only the Born type and lowest order connected diagrams (Fig. 1a-1d). The other curves are results from Laget [7], representing a PWIA (—··—) and full (—·—) calculation. The full calculation of Laget includes effects of MEC. Experimental data points are from Amsterdam [8].

type diagrams (1a-1b) gives rise to a considerable reduction of the cross section. This demonstrates that FSI are important in this kinematic configuration. One possible approximation to the connected three-body amplitude is to keep only the lowest order connected diagrams, given in Fig. 1c and 1d. The result is also shown in Fig. 3. It is clear that this approximation is not correct in this kinematic situation. Therefore it is necessary to determine the complete multiple scattering series solution.

This contribution was financially supported by the Consiglio Nazionale delle Ricerche (Italia) on the proposal of the Netherlands Organization for the Advancement of Pure Research (Z.W.O.).

References

1. A.E.L. Dieperink, T. de Forest Jr., I. Sick and R.A. Brandenburg, Phys. Lett. 63B, 261 (1976).

2. T. de Forest Jr., Nucl. Phys. A392, 232 (1983).

3. G. Höhler et al., Nucl. Phys. B114, 505 (1976).

4. R.A. Malfliet and J.A. Tjon, Nucl. Phys. A127, 161 (1969).

5. W.M. Kloet and J.A. Tjon, Ann. Phys. 79, 407 (1973).

6. E. Jans et al., Phys. Rev. Lett. 49, 974 (1982).

7. J.M. Laget, Phys. Lett. 151B, 325 (1985) and private communication.

8. P.H.M. Keizer et al., Phys. Lett. 157B, 255 (1985).

ELECTRON SCATTERING AND FEW-NUCLEON SYSTEMS

Bernard Frois

Service de Physique Nucléaire - Haute Energie

CEN Saclay, 91191 Gif-sur-Yvette Cedex, France

I. Introduction

In the past few years, considerable progress has been achieved in theoretical calculations of few-nucleon systems. This meeting has shown in particular that calculations with a two-body force are now reliable and numerically accurate. The amount of data that are both quantitative and interpretable is scarce. Traditionnally, one has considered binding energies,
charge radii, magnetic dipole and charge quadrupole moments as a primary
source of information on bound-states. These observables are no longer
sufficient to test the state of the art in theoretical calculations. The
major problem is that it is now clear that nuclear physics cannot be described in terms of nucleons only. There are many more degrees of freedom
due to the presence of mesons and excitations of the nucleons. These
effects modify charge and current distributions, and introduce many-body
forces. It is impossible with a few numbers to disentangle these effects.
Electron scattering has proved to be the best experimental approach to
solve this problem.

Electrons penetrate nuclei with an interaction sufficiently weak so that
there is no significant perturbation of the nucleus. There is no need to
unfold complex multiple scattering processes. Electrons are point particles
and their interaction with the nucleus is well understood. Since they can
transfer independently momentum and energy, their spatial resolution can be
easily adjusted to the scale of processes that have to be studied. This

scale is simply related to the maximum momentum transfer. A rule of thumb, based on a standard bandwidth argument of Fourier transforms, is that 500 MeV electron single arm scattering data probe the structure of nuclei with a spatial resolution of the order of 0.8 fm. This is ideal for probing nucleon distributions in configuration space since the details of their shell structure is limited by the finite size of nucleons, which is just of this order of magnitude. Higher energies are needed for coincidence experiments in order to map out momentum distributions.

II. Electromagnetic Structure of the Deuteron

The elastic electron scattering from deuterium is a combination of three form factors, monopole and quadrupole charge form factors and a magnetic form factor. The magnetic form factor $B(Q^2)$ can be separated by performing a series of experiments at forward and backward angles ; but it is not possible to separate the monopole and quadrupole form factors without polarization experiments. The magnetic form factor has been measured recently [1] up to 30 fm^{-2}. The experimental data show a smoth fall-off compatible with the theoretical prediction of Gari and Hyuga [2] which includes a large effect of the $\rho\pi\gamma$ exchange current. This calculation is not fully relativistic and does not take into account $\Delta\Delta$ components in the deuteron wave function, therefore the good agreement with the experimental data might be fortuitous. At higher momentum transfers, the parton model predicts also a smooth fall-off. Thus, it would be natural to imagine that the magnetic form factor of the deuteron provides a natural link between a description in terms of nucleons and mesons, and a description in terms of quarks and gluons. A recent experiment at SLAC [3] has used the new 4 GeV injector to measure the magnetic form factor of the deuteron between 25 and 65 fm^{-2}. This experiment is discussed in details by Arnold in this session. These data are very exciting since they show clearly the existence of a diffraction minimum in the region of 50 fm^{-2}. This diffractive structure shows that even at such very high momentum transfers, one has not reached the asymptotic regime. The interpretation of these data will require a consistent relativistic treatment of two-nucleon dynamics. At present, the magnetic form factor of the deuteron is not well understood theoretically. Different calculations are able to reproduce the experimental results, assuming either $\Delta\Delta$ components, relativistic effects or quark interchange mechanisms; The main difficulty is related to the absence of low energy theorems for isoscalar processes. Isovector processes are constrained by chiral symmetry, and can be described in a model independent way at low energy, thus one has a certain confidence in their theoretical description. For isoscalar processes, the situation seemed much more difficult. Until recently there was little hope to understand these processes because theoretical predictions were highly model dependent. The situation has changed recently when Nyman and Riska [4] showed that isoscalar currents can be derived from the chiral anomaly term of the effective Lagrangian and therefore can be predicted in a completely model independent way. Nyman and Riska have shown that an excellent agreement is obtained with experimental data up to 30 fm^{-2}. Since their calculation has not yet been performed beyond this limit, it is not yet known if this new approach is able to describe the region of the diffraction minimum around and beyond 50 fm^{-2}. We are eagerly waiting their calculation ...

Relativistic effects have also been investigated in high accuracy measurements of the structure function $A(Q^2)$ of the deuteron, reported in this session by Platchkov. The new precise measurements of $A(Q^2)$ and $B(Q^2)$ should help to find the best theoretical approach for a relativistic description of the deuteron. A very promising avenue seems to be the use of relativistic front-form two-nucleon dynamics. Chung, Coester, Keister and Polyzou [5], have calculated the relevant matrix elements of single nucleon currents to obtain the deuteron form factors, using Reid-Soft-Core, Paris and ANLV14 wave functions. They find that the absence of two-body current operators produces only a small violation of the invariance which is cured by the ad hoc addition of a small two-body current which does not affect the magnetic form factor and makes very small contributions to the monopole and quadrupole charge form factors. With this approach, one is able to describe correctly the entire range of experimental measurements of the magnetic form factor. The front-form two-nucleon dynamics approach is different from the covariant wave-function models used by Arnold, Carbon and Gross [6] or by Zuilhof and Tjon [7]. However, it is disturbing that they predict relativistic effects of very different magnitudes and of opposite signs.

III. ^{3}H and ^{3}He Form Factors

The experimental situation is reviewed at this meeting by Platchkov and Turchinetz. The new measurements [14] of the tritium form factors up to ~ 25 fm^{-2} have shown, that one observes the same discrepancies between experiment and theory when only nucleons are taken into account. Sauer has shown at this meeting that the one-body contribution, coming from three non-relativistic nucleons, is now reliably calculated. Thus, the charge form factors of ^{3}H and ^{3}He, show clearly the breakdown of the conventional theory, which assumes that nucleonic degrees of freedom are sufficient to describe nuclear properties.

Considerable discussions have been held in the Washington 1986 meeting on the effect of three-body forces. The two-pion component of the three-body force is reasonably well-understood. The main difficulty is found at the level of the short-range components. Carlson, Pandharipande and Wiringa [8] have considered a purely phenomenological approach in order to investigate the effects of a three-body force, in light nuclei, but also in nuclear matter. The Hannover group, in a more ambitious approach, has treated explicitly isobar degrees of freedom, but this does not describe the complete three-body force. A general discussion can be found in the paper of Friar, Gibson, Payne and Chen [9]. Three-body forces tend to increase the binding energy of ^{3}H and ^{3}He, by about 1 MeV, which corresponds to 2 % of the potential energy. This is sufficient to reconcile experiment and theory at the level of binding energies and charge radii. However, the effects of three-body forces and isobar componens in the ground state wave function are too small to explain the discrepancies found in the charge and magnetic form factors of ^{3}H and ^{3}He. One has to take into account explicitly non nucleonic degrees of freedom. The most complete calculations have been performed by the Hannover group [10] ; both the charge and the magnetic form factors are well reproduced by taking into account meson-exchange currents. These theoretical predictions are reliable for the magnetic form factors, because the pion exchange current appears on the same footing as nucleonic currents and is constrained by chiral symmetry. The situation is different for the charge form factors. Corrections to the impulse approximation are of relativistic orders. Since the wave functions are not corrected for relativistic effects, it is clear that a perturbative expansion limited to an arbitrarily selected class of diagrams, is not reliable. It is exactly the ideal situation to test the prediction based on the chiral anomaly of the effective Lagrangian, proposed by Nyman and Riska [4].

IV. Nucleon Momentum Distributions

The (e,e'p) reaction is a very good probe of nucleon momentum distributions. For quasielastic kinematics, final state interactions and meson exchange calculations are usually small. Furthermore Laget has shown at this meeting that such corrections are reasonably well understood. The main difficulty is that at present the incident electron energies and the duty cycle of existing accelerators give access to only a limited range of missing momentum. New results from NIKHEF are reported by de Witt Huberts and from Saclay by Morgenstern. These new data extend now to missing momentum up to 600 MeV/c. This is the beginning of the region where correlations are expected to play an important role in nucleon momentum distributions. The central question is whether or not nucleon-nucleon correlations are strongly dependent on the atomic mass number. At present, "exact" calculations have been done only for the three-nucleon bound states. These calculations serve as a calibration for other methods, which are more appropriate for the study of heavier systems. Schiavilla, Pandharipande and Wiringa [11] in particular have developed a general Monte Carlo method to study momentum distributions of nucleons and nucleon clusters in nuclei. Traini and Orlandini have also studied momentum distributions with a phenomenological model which includes short-range and tensor correlations. Both groups predict a similar behavior for the high momentum components of the momentum distribution of ^{4}He. Schiavilla et al.[11] have calculated A = 2,3,4 and infinite nuclear matter, while Traini and Orlandini [12] have calculated ^{4}He, ^{16}O and ^{40}Ca. Momentum distributions are predicted by the two theoretical approaches to be proportional to each other for $k > 2.5$ fm^{-1}. However, Schiavilla et al.[11] predict an increase of the momentum distribution with A, while Traini and Orlandini predict a decrease. The available data for ^{3}He are in reasonable agreement with theoretical predictions up to 600 MeV/c. There is no evidence for any serious discrepancy between experiment and theory up to this relatively high momentum. However, this is the only nucleus for which, data have reached such a high nucleon momentum, this is one of the obvious justifications for a high duty cycle accelerator of E > 1 GeV. Some data at lower missing momentum have been measured at NIKHEF for ^{4}He ; preliminary results are in good agreement with the results of Schiavilla et al. Higher momentum data are being measured at Saclay.

At present the only source of informations on nucleon momentum distributions, at very high momentum, is quasi-free single arm electron scattering. Sick et al.[13] have shown that the data can be described by a single variable y when $q \to \infty$ and $\nu \to \infty$. Ciofi Degli Atti has analyzed at this meeting the effect of final states interaction. In this session, Meziani is going to discuss the interest of a longitudinal and transverse separation of the quasielastic response function which is planned at SLAC.

V. Conclusion

This session on recent electron scattering data shows the great wealth of experimental results that are now available. Several laboratories have completed very important experiments which have considerably increased our understanding of few-nucleon systems. These results have shown the importance of non-nucleonic degrees of freedom. Up to momentum transfers $Q^2 = 1$ $(GeV/c)^2$ the theoretical situation seems to be on firm grounds, and there is a good agreement with experimental data. The information on nucleon momentum distributions is more limited, but the experimental data are now reaching the region where the short-range behavior of the two-body wave function plays an important role.

References

[1] S. Auffret et al., Phys. Rev. Lett. 54, 649 (1985).

[2] M. Gari and H. Hyuga, Nucl. Phys. A264, 409 (1976).

[3] P. Bosted, 2nd Conference on the Intersections Between Particle and Nuclear Physics 1976, AIP Proceedings 150 (1986).

[4] E.M. Nyman and D.O. Riska, University of Helsinki, preprint 86-25.

[5] F. Coester et al., private communication.

[6] R.G. Arnold, C.E. Carlson and F. Gross, Phys. Rev. C21, 1426 (1980).

[7] J. Zuilhof and J.A. Tjon, Phys. Rev. C24, 736 (1981).

[8] J. Carlson, V.J. Pandharipande and R.B. Wiringa, Nucl. Phys. A401, 59 (1983).

[9] J. Friar et al., Phys. Rev. C34, 1463 51986).

[10] C. Hajduk et al., Nucl. Phys. A405, 581 (1983) ;
W. Strueve et al., Nucl. Phys. A405, 620 (1983).

[11] R. Schiavilla, V.J. Pandharipande and R.B. Wiringa, Nucl. Phys. A449, 219 (1986).

[12] M. Traini and G. Orlandini, Z. Phys. A321, 479 (1985).

[13] I. Sick et al., Phys. Rev. Lett. 45, 871 (1980).

[14] F.P. Juster et al., Phys. Rev. Lett. 55, 2261 (1985) and references therein.

RECENT STUDIES OF TWO AND THREE-NUCLEON ELASTIC FORM FACTORS

S. Platchkov
Service de Physique Nucléaire-Haute Energie, CEN-Saclay,
F-91191 Gif-sur-Yvette Cedex, France

Abstract

Recent experiments at Saclay involving two and three-nucleon elastic form factors are reviewed. New precise data for the deuteron $A(q^2)$ structure function are presented. These data are very sensitive to the nucleon-nucleon potential and to the neutron electric form factor. For helium-3 and tritium form factors separation of isoscalar and isovector components is performed. Both charge components disagree with theoretical predictions.

I. Introduction

The two and three-nucleon form factors play a very important role in the understanding of the fundamental properties of the nucleon-nucleon interaction. Since they can be calculated exactly, they are used as a quantitative testing ground for present models and new concepts and ideas. From an experimental point of view it is important to have precise and reliable data in the kinematical region that is available with the present electron accelerators. Such a program was initiated few years ago at the Saclay electron accelerator (ALS). In this contribution we report on two different topics related to this program. In the first one we present new precise measurements on the deuteron $A(q^2)$ structure function. In the second one the separation of the isoscalar and isovector components of the three-nucleon elastic form factors is discussed.

II. The deuteron $A(q^2)$ structure function

The deuteron $A(q^2)$ structure function is considered to be one of the most important observables in a few-nucleon system for several reasons. First it provides a standard check for modern realistic potentials, whose parameters are fitted to nucleon-nucleon scattering data. Second, within the expense of some assumptions, it can be used to extract the neutron electric form factor, a quantity which is of upmost importance for quark models of the nucleon. Finally, good knowledge of $A(q^2)$ at medium momentum transfers will allow more reliable separation of the charge and quadrupole form factor of the deuteron, once the planned MIT-Bates T_{20} experiment (see contribution from W. Turchinetz) is completed.

The $A(q^2)$ structure function is experimentally known up to $q^2 = 100$ fm^{-2}. In the low and medium momentum transfer region (0 - 25 fm^{-2}) measurements from several laboratories exist [1]. The deviation of each of these measurements from a purely phenomenological fit to them is shown in Figure 1. Also plotted are impulse approximation predictions from few realistic nucleon-nucleon potentials [2]. All the predictions are calculated using the neutron charge form factor parametrization of Galster et al. [1]. Apart from the low momentum transfer region ($q^2 < 4$ fm^{-2}), the data are not precise enough to distinguish between different model predictions. In particular, the measurements of Galster et al. [1] disagree by about 15 % with the measurements of Buchanan et al. [1], while in the same q^2 region the

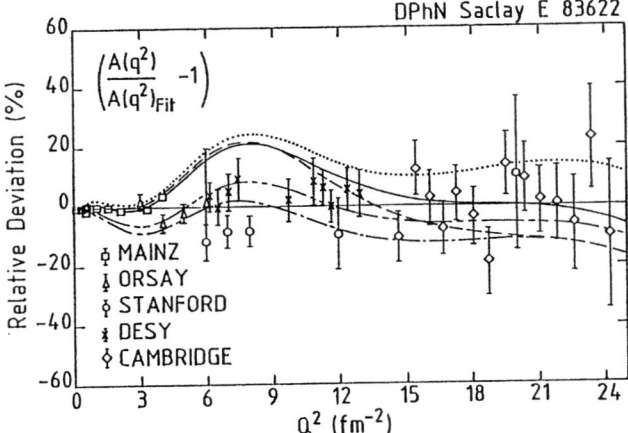

Fig. 1. Available experimental data for $A(q^2)$. The reference line is a phenomenological fit to the world data up to $q^2 = 35$ fm^{-2}. The quantity plotted is a percentage deviation versus this fit. The data are compared with theoretical predictions computed in impulse approximation : Paris (solid), RSC (dotted), HM1 (dash-dotted,), FL5 (dashed), and V14 (dashe-double dotted).

different calculations lie within 20 %. For these reasons we have taken much more precise data between $q^2 = 1$ and $q^2 = 18$ fm^{-2}.

The experiment was performed at the Saclay electron linac (ALS) using the standard equipment of the HE1 experimental hall. Data were taken at 200, 300, 500 and 650 MeV. We have used the same target and detectors than those used for the measurement of the $B(q^2)$ structure function [3]. Particular attention has been paid to the accuracy of these new data. Most of the data measured have statistical errors smaller than 1 %. Detector efficiency, solid angle, incident electron current and target density as its dependence on the incident intensity were carefully measured. In the present stage of the analysis, the total experimental uncertainty, which adds quadratically to the statistical error, is 2 %. Our measurements are absolute ; however, for an independent check of our procedure, we have also measured hydrogen cross sections. They were found to agree within 1 % with the proton form factor parametrization of Simon et al. [4].

The magnetic contribution to the cross sections was determined from a fit to the previous $B(q^2)$ measurements [3]. Its highest value is 25 % at $q^2 = 18$ fm^{-2}. The resulting $A(q^2)$ data are plottred in Fig. 2 as deviations versus the fit to the older measurements. The overall uncertainties are significantly improved. In the momentum transfer region near $q^2 = 4$ fm^{-2} our new data disagree by up to 10 % with the reference fit. If compared with the impulse approximation predictions [2] of Fig. 1, our data follow closely the shape of the Bonn HM1 and Argonne V14 potentials, and seem to disagree with Paris and RSC. However, the $A(q^2)$ structure functions is also very sensitive to the neutron charge form factor G_{en}. Fig. 2 also shows the model predictions for the Paris potential obtained using four different possibilities for G_{en} : the parametrizations of Galster et al. and Höhler et al. [1, 13], the extreme value $G_{en} = 0$ and the recent prediction of Gari and Krumpelmann [5], based on the vector dominance model and constrained by QCD at high q^2. At $q^2 = 15$ fm^{-2} the difference between the two extreme predictions is 60 %, much larger than the 20 % effect due the use of dirrerent NN potentials.

All these predictions have been calculated in impulse approximation. Neither meson exchange currents (MEC) nor relativistic effects have been considered. Below $q^2 = 20$ fm^{-2} the meson exchange currents [6] slightly increase $A(q^2)$. Although small, their effect depends on the assumptions about the meson-nucleon coupling. In the same region, the relativistic effects, which consistently include part of the MEC contribution, have been

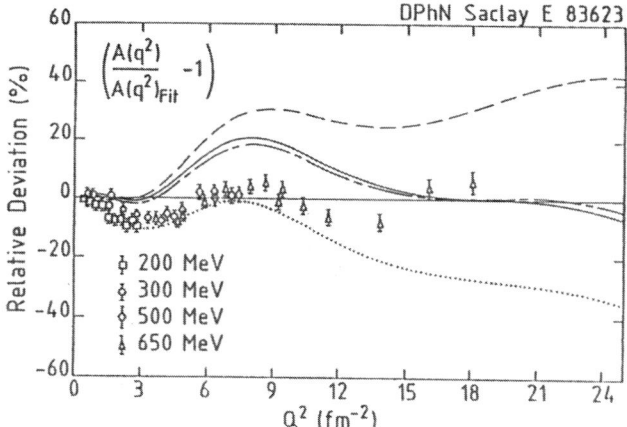

Fig. 2. Our new measurements plotted versus the reference line. The model predictions are computed in impulse approximation using the same NN potential (Paris), but different neutron electric form factors [1, 5, 13]: Galster et al. (solid), Gari and Krumpelmann (dashed), Höhler et al. (dash-dotted) or G_{en} = 0 (dotted).

found by two different groups of authors [7] to decrease $A(q^2)$ by about 10 %. We therefore expect the net effect of both MEC and relativistic corrections to be relatively small. Yet consistent calculation of these two effects remains to be performed in order to allow more reliable conclusions on G_{en} form factor and/or NN potential models.

III. Isospin structure of the three-nucleon system

Knowledge of both ^{3}He and ^{3}H form factors allows an independent study of both isoscalar (IS) and isovector (IV) components of the electromagnetic current operator. Separation of these components gives better insight on the presently available theoretical calculations. It also provides a direct means of comparison with the deuteron form factors which have a well defined isospin nature. Such a separation became possible only recently, with the measurements of the ^{3}H charge and magnetic form factors[8].

In order to determine their isoscalar and isovector components, knowledge of the ^{3}He and ^{3}H form factors at the same values of q^2 is required. Obviously this experimental condition is not always satisfied. An easy way to perform the separation is to use a functional representation of the data. We fitted all the three-nucleon form factors using a Sum of Gaussian expansion [9]. The analysis of the world data sets and the fitting procedure are described in Ref. [10].

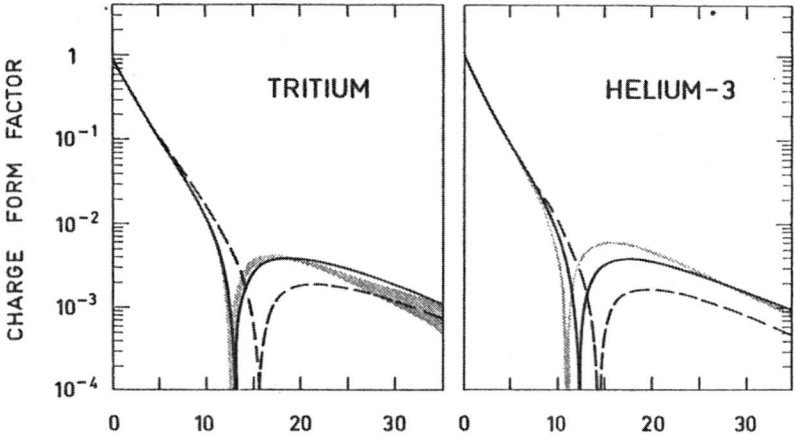

Fig. 3. Charge form factors of ^{3}H and ^{3}He. The shadowed band is a fit to the experimental points. The predictions of the Hannover group Ref. [11] are evaluated with (solid line) or without (dashed line) meson exchange currents.

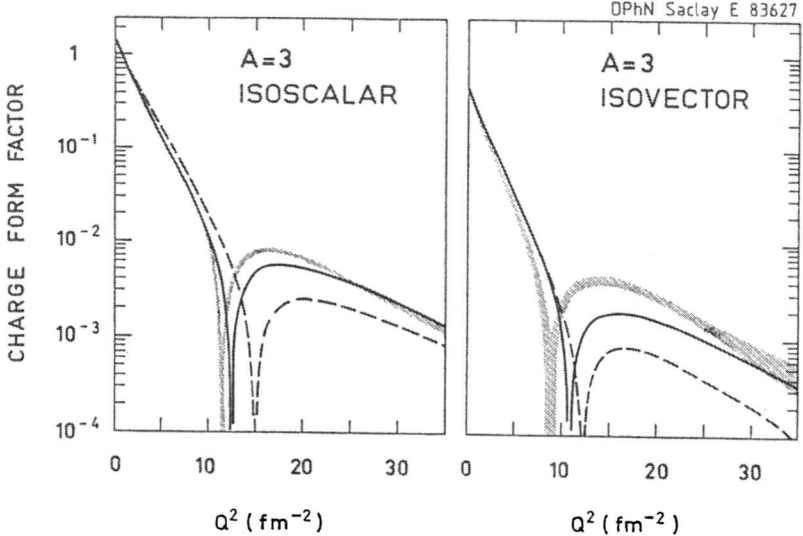

Fig. 4. Isoscalar-isovector separation for the A = 3 charge form factors. (same notation as for Fig. 3).

Figure 3 shows the ^{3}He and ^{3}H charge form factors. The experimental results (the shadowed band includes both statistical and systematical errors) are compared with the calculation of the Hannover Group [11]. In this calculation the Δ-isobar is consistently included in the nucleonic wave function, thus accounting for the main part of the three-body force. The exchange current contribution, computed with pseudovector πNN coupling, brings the results for ³H into almost perfect agreement with the data, while for ³He significant difference remains. Both isospin form factors (Fig. 4) also disagree with the experiment, thus showing that neither the IS nor the IV components are well reproduced. In the region of the secondary maximum of the IV form factor the experimental value is almost twice the theoretical one.

This disagreement between theory and experiment is most likely due to an incorrect treatment of the meson-exchange currents and relativistic effects. Other effects such as three-body forces significantly improve the binding energy, but have only a minor effect on the form factors. We also believe that this disagreement can hardly be ascribed to a poor knowledge of the one-body current. Quite surprisingly, the reason for this latter remark comes from the IS/IV separation of the magnetic form factors (Fig. 5). The IS part comes out to be in perfect agreement with experiment up to $q^2 = 20$ fm^{-2}. The theoretical MEC contributions for ³He and ³H magnetic form factors are isovector only ; it is therefore natural that they do not contribute in the IS part. The remaining difference between the two theoretical curves shows that the effect of the Δ-isobar is also negli-

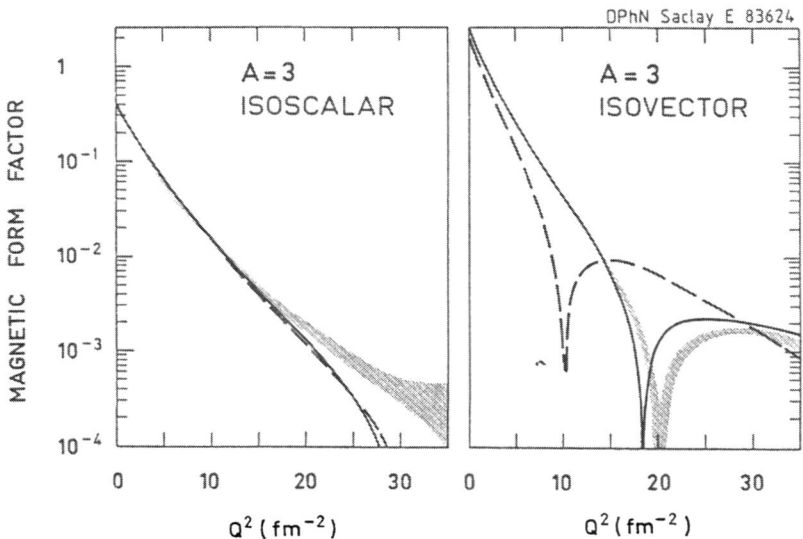

Fig. 5. Isoscalar-isovector separation for A = 3 magnetic form factors (same notation as for Fig. 3).

gible. The missing strength beyond 20 fm^{-2} could come from the absence of isoscalar MEC such as $\rho\gamma\pi$ current, whose contribution to the deuteron magnetic form factor at 25 fm^{-2} is 50 % of the impulse approximation result.

The good agreement for the IS magnetic form factor below 20 fm^{-2} is achieved using the one-body current alone. The theoretical one-body current is computed with the same wave function, which is used to compute the one-body contribution to the charge form factors. Thus, the good agreemment for the IS magnetic part indicates that the one-body contribution to the charge form factor is also reasonable.

Figure 5 also shows the IV magnetic form factor. As expected, this form factor cannot be explained by the one-body current alone. The MEC dramatically improve the theoretical prediction. This result is to be compared with the backward electrodisintegration of the deuteron at threshold [12], another pure IV transition where the MEC also dominate the cross section. In both cases good agreemment with the data is obtained using the value Λ = 1.2 GeV for the meson-nucleon hadronic vertices. Using the value Λ = 0.7 GeV would strongly suppress the MEC contribution and would completely destroy the agreement between theory and experiment. This observation is at variance with the present three-nucleon bound-state calculations, in which the use of three-body force requires Λ to be 0.7 GeV in order to fit the binding energy of the triton. The same calculations, if performed with Λ = 1.2 GeV, would overbind the triton by more than few MeV.

IV. Conclusion

New precise data for the deuteron $A(q^2)$ structure function and for tritium charge and magnetic form factors have been presented. The precision of the measurements on $A(q^2)$ is now of an order of magnitude better than the spread of the various theoretical predictions. Thus, they serve as a stringent test for nucleon-nucleon models or for a more reliable determination of the uncertainties on the neutron electric form factor. The tritium data have been used to separate the isoscalar and isovector components for both charge and magnetic three-nucleon form factors. Comparison with theory shows that neither IS nor IV charge pieces are well reproduced. Separation of the magnetic form factors indicates that the one-body current and the IV meson exchange contribution are reasonably well understood.

References

[1] G.G. Simon, Ch. Schmitt and V.H. Walther, Nucl. Phys. A364, 285 (1981) ;

D. Benaksas, D. Drickey and D. Frèrejacque, Phys. Rev. 148, 1327 (1966) ;

S. Galster, H. Klein, J. Moritz, K.H. Schmidt, D. Wegener and J. Bleckwenn, Nucl. Phys. B32, 221 (1971) ;

C.D. Buchanan et M.R. Yearian, Phys. Rev. Lett. 15, 305 (1965) ;

J.E. Elias, J.I. Friedman, G.C. Hartmann, H.W. Kendall, P.N. Kire, M.R. Sogard, L.P. Van Speybroeck and J.K. de Pagter, Phys. Rev. 177, 2075 (1969) ;

R.G. Arnold, B.T. Chertok, E.B. Dally, A. Grigorian, C.L. Jordan, W.P. Schütz, R. Zdarko, F. Martin and B.A. Mecking, Phys. Rev. Lett. 35, 776 (1975) ;

[2] K. Holinde et R. Machleidt, Nucl. Phys. A247, 495 (1975) ;

M. Lacombe, B. Loiseau, R. Vinh Mau, J. Côté, P. Pirès and R. de Tourreil, Phys. Lett. 101B, 139 (1981) ;

R.B. Wiringa, R.A. Smith and T.L. Ainsworth, Phys. Rev. C29, 1207 (1984).

E. Lomon and H. Feshbach, Ann. Phys. (N.Y.) 48, 94 (1968).

[3] S. Auffret, J.M. Cavedon, J.C. Clemens, B. Frois, D. Goutte, M. Huet, Ph. Leconte, J. Martino, Y. Mizuno, X.-H. Phan, S. Platchkov and I. Sick, Phys. Rev. Lett. 54, 649 (1985).

[4] G.G. Simon, Ch. Schmitt, F. Borodowski and D.H. Walther, Nucl. Phys. A333, 381 (1980).

[5] M. Gari and W. Krumpelmann, Phys. Lett. 141B, 295 (1984).

[6] M. Gari et Hyuga, Nucl. Phys. A264, 409 (1976).

[7] R.G. Arnold, C.E. Carlson et F. Gross, Phys. Rev. C21, 1426 (1980)

[8] F.-P. Juster, S. Auffret, J.-M. Cavedon, J.-C. Clemens, B. Frois, D. Goutte, M. Huet, P. Leconte, J. Martino, Y. Mizuno, X.-H. Phan, S. Platchkov, S. Williamson and I. Sick, Phys. Rev. Lett. 55, 2261 (1985).

[9] I. Sick, Nucl. Phys. A218, 509 (1974).

[10] J. Martino, Proc. Int. Conf. on three-body force in three-nucleon system, Washington (1986) Lecture Notes in Physics, 260, 129 (1986).

[11] P. Sauer, in Progr. in Part. and Nucl. Phys., 16 (1985) and references therein

[12] S. Auffret, J.-M. Cavedon, J.-C. Clemens, B. Frois, D. Goutte, M. Huet, F.P. Juster, P. Leconte, J. Martino, Y. Mizuno, X.H. Phan, S. Platchkov and I. Sick, Phys. Rev. Lett. 55, 1362 (1985).

[13] G. Höhler, E. Pietarinen, I. Sabba-Stefanescu, F. Borkowski, G.G. Simon, V.H. Walther and R.D. Wendling et al., Nucl. Phys. B114, 505 (1976).

FEW BODY STUDIES AT BATES: RECENT PAST AND NEAR FUTURE

W. Turchinetz
Bates Linear Accelerator Center
LNS and Physics Department
Massachusetts Institute of Technology
Middleton, Massachusetts 01949

Abstract

A brief review is given of elements of the Bates program of electromagnetic probing of few nucleon systems. A few topics on electron scattering are discussed in some detail at the expense of the rest of the program. The selection is based on the relative amount of laboratory resources used in the recent past and planned for the near future. An attempt is made to provide an adequate list of references for those who wish to find out more about work which is mentioned only in passing.

1.Introduction

In this report I have combined my remarks given verbally in two separate sessions of this workshop. For the most part, I will discuss work in progress, which includes preliminary analysis of data recently taken, preparations to perform experiments which have been accepted by the laboratory and improvements to the accelerator and experimental facilities which made the experiments possible. As a consequence, any results shown must be considered as preliminary and should not be quoted further without checking with the authors who are listed in the references.

After a brief outline of accelerator and facility developments in Section 2, I will discuss the Bates program; on elastic form factors in

Section 3; on deep inelastic electron scattering in Section 4, and photo-reactions, including pion production and Compton scattering, in Section 5. Already from this Table of Contents it is apparent that experiments on few body systems at Bates have moved from the category of occasional experiments to a prominence that is now comparable with and threatens to overwhelm the high resolution spectroscopy which is usually associated with Bates.

2. Accelerator and Facility Development

As is well known, since there is a history of 50 years in studying few body systems with electromagnetic probes, significant further progress requires access to some combination of greater precision, higher momentum transfer, higher excitation energy or new observables such as the various polarization observables and coincidences among products of many body final states. In the past five years or so, there has been in progress at Bates a systematic program of development of the facilities to improve this access on a broad front, including: 1) increased beam energy, and hence, momentum transfer, 2) spectrometer systems, 3) polarized electrons, 4) cryogenic target systems, and 5) recoil polarimeters. In this section, I give a status report on items 1, 2, and 3 and postpone discussion of 4 and 5 to later where specific experiments are discussed.

2.1 Accelerator

The accelerator comprises 160 M of S Band waveguide powered by 12 klystrons rated at 4 MW peak and 100 KW average. The maximum energy, for one pass, is about 450 MeV, and a recently completed recirculator[1] has resulted in 790 MeV. A project, now in progress, will increase the klystron's power output to about 5.4 MW and result in about 1 GeV maximum energy by mid-1987. In one-pass operation, 80 µA or more of average current is available and has occasionally been used, but more typical is 40 to 50 µA. In principle, the same average current should be available at recirculated energies, but to date, only 30 µA has been used. Approximately 80% of the accelerated beam lies within 0.3% of momentum spread. The maximum duty factor used to date has been 1.7%, with 0.75% being more typical. High duty factor operation is expensive in power and components, and is not encouraged; but as the demand for coincidence experiments increases, it becomes more and more difficult to resist. As a result, the highest priority for further accelerator improvement is to achieve 100%

duty factor as soon as possible. To this end, a proposal[2] was submitted to the U.S. Department of Energy in 1984 to construct at Bates a pulse stretcher ring and related facilities to achieve extracted beams of about 90% duty factor and capability to do experiments with thin internal targets intercepting the circulating beam of 80 mA. This proposal has been favorably reviewed by various committees of the government and it is our hope that it will appear as a budget item in the not too remote future.

Finally, the polarized electron source which has been under construction by a collaboration involving scientists from Yale, Syracuse, CUNY, Harvard and MIT delivered in June of this year its first high intensity electron beam and was accelerated to 250 MeV with approximately the expected polarization. Plans are being developed to make polarized electrons available on any beam line and any beam energy within the range of the accelerator.

2.2 Experimental Facilities

Figure 1 shows schematically the layout of the beam switchyard and experimental halls, and in Table 1 are listed the properties of the spectrometer systems which are in various stages of development at Bates.

In the North Hall is located the high resolution energy loss spectrometer system (ELSSY) which is used mainly for high resolution electron scattering from complex nuclei.[3] The resolution available today is routinely about 10^{-4}, with special care has been as good as 4×10^{-5}, and further improvements may be possible and are being pursued. In addition to the high resolution, this facility has a number of auxiliary features which make it possible to perform absolute measurements with high precision (<1%). These include: a) a low emittance (<10^{-3} mc-cm) incident beam, b) high precision, redundant monitoring of beam intensity, c) extensive internal baffling in the dipoles, d) a long distance (2 M) between the target and solid angle defining slits, and e) capability to measure the absolute beam energy with a precision of $<10^{-3}$. These auxiliary features are particularly important for few body studies as will be seen below. The beam transport system to this spectrometer may be operated as an achromat, but is usually tuned to provide a beam dispersed in momentum (7 cm/%) in the vertical direction, with a space focus for a monochromatic component. An auxiliary chicane makes possible detection of 180° scattering.

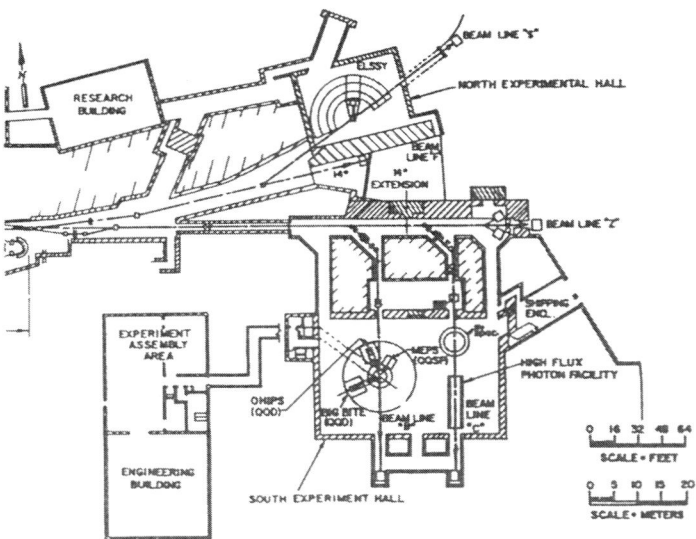

FIGURE 1. Schematic diagram of the Bates Experimental Area

TABLE 1

Design Specifications for Bates Spectrometers

	ELSSY	MEPS[a]	OHIPS[a,b]	BIGBITE[a]	π°
Solid Angle (msr)	5.4	35	20	2-15	~ 0.5
Momentum Acceptance (%)	10	20	16	50-10	15
Maximum Momentum (MeV/c)	900	400	1300	800	~ 350
Resolution ($\Delta p/p$)	4×10^{-5}	5×10^{-4}	5×10^{-4}	10^{-3}-10^{-2}	10^{-1}-10^{-2}
Angular Resolution (mr)	1	< 5	< 15	> 2	-
Angular Range	20°-180°	0°;30°-150°	30°-150°	20°-160°	0°-180°
Flight Path (m)	11	5	9	4	-
Radius of Curvature (m)	2.2	0.75	2.54	2	-
Configuration	DD	QQSP	QQD	QQD	-

[a]Each of these spectrometers features a variable first drift space; a larger angular range is then possible with reduced solid angle and increased flight path (see also Table III.B.1).

[b]Construction of specially designed quadrupoles for OHIPS is assumed.

In the South Hall, which was completed in 1979, there is a complex of
five operating spectrometers on three beam lines and further development
in progress. The design objectives for these spectrometers emphasized
solid angle and momentum range at the cost of resolution, although in some
cases reasonable resolution ($<10^{-3}$) may be recovered by software corrections.

On the B line, which is a 90° achromat, there is located a target
region at the common pivot of three spectrometers: OHIPS, MEPS and
BIGBITE. These spectrometers may be used singly, in pairs or in triple
coincidence. To date, only the first two possibilities have been required. All three spectrometers are mounted on air pads and it is possible to move BIGBITE, without dis-assembly, between beam lines B and C.
Beam line C is also a 90° achromat and has been used, to date, to produce
a collimated photon beam to a π° opening angle spectrometer and a NaI
gamma ray spectrometer. A low intensity γ-tagger ($\sim 10^4$ sec^{-1}) is
available for calibration and test purposes. This beam line has recently
been further developed to produce a high intensity photon beam ($\sim 5 \times 10^{11}$
photons/sec/50 MeV). This beam is needed for experiments in high resolution photo-neutron spectroscopy and for measurement of the neutron polarization in D(γ,n).

3. Elastic Form Factors. A = 3, 2, and 1

The ordering in A is as indicated since for A = 3 we have recently
concluded data taking for both ^{3}H and ^{3}He and for A = 2 an approved
experiment is in preparation while for A = 1 only feasibility studies have
been approved.

3.1 A = 3

Electron scattering data for the isodoublet ^{3}H and ^{3}He is of great
importance in testing our most basic models of nuclear structure and the
nuclear current. Both elastic form factors for ^{3}He have been available
for several years now and their impact on theory has been prodigious.
With the recent publication of the Saclay measurements on ^{3}H of both Fm
and Fc, the data base for the doublet has been beautifully extended.[4]
Already there has been much discussion of these data which continues at
this workshop. More or less in parallel with the Saclay experiment,
measurements of these form factors were completed last April at Bates and

the data are under analysis. The Bates data check and complement the Saclay data in a number of important ways.

In each of these experiments scattered electrons were detected using the high resolution electron scattering spectrometers: ELSSY at Bates and the 900 at Saclay. The major difference in the experimental plans lies in the target design. At Saclay the tritium was at 20°K and in the liquid phase, while at Bates it was at 45°K in the gaseous phase at a density of 25 mg-cm^{-3}. The volume of the Saclay target was about 4 cm^3 and contained about 10 kCi of ^{3}H, while the Bates target had a volume of about 390 cm^3 and at the operating pressure of 15 bar contained about 125 kCi of ^{3}H. A schematic diagram of the Bates target is shown in Figure 2. It is similar to, but more complicated than, the one at Saclay.

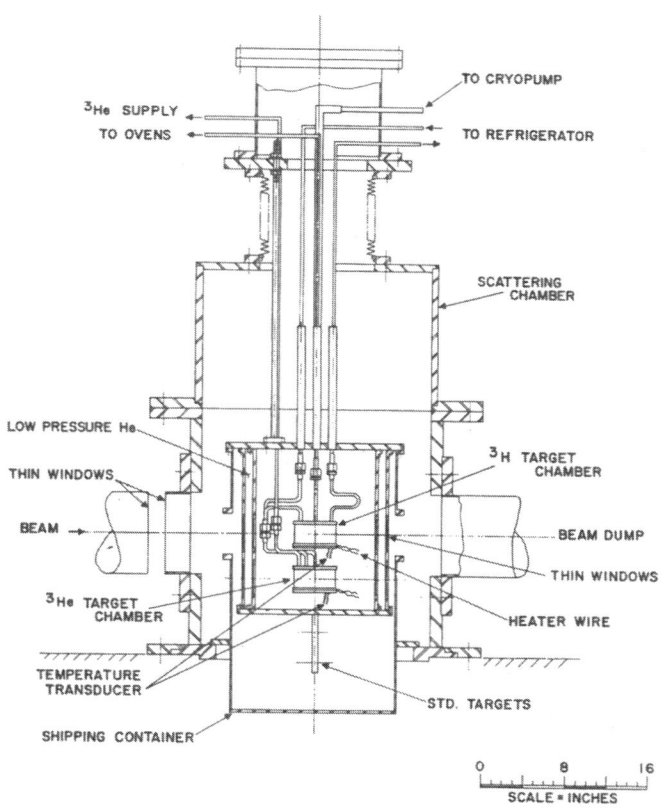

Figure 2. Schematic diagram of Bates Tritium Target System.

The Saclay target was optimized for measurement of the elastic form factors to the highest momentum transfers, while the Bates target was designed to meet a number of objectives in addition to the elastic form factors. Foremost among these was the measurement of the inelastic cross section including the quasi-elastic and delta excitation as will be discussed in Section 4.1 below. In addition, the Bates target was designed to permit a precise comparison of the ^{3}H scattering with ^{3}He in the same experiment by moving remotely identical targets of ^{3}H, ^{3}He, empty and solid targets into the beam. Direct comparisons with ^{1}H were also made. Our aim was to measure absolute cross sections to within an uncertainty approaching 2%. At this stage of the analysis, no reason has yet appeared to abandon this objective. The ultimate precision will probably be determined by the uncertainty in the target density. Within the uncertainties of the Saclay data, there is good agreement between the two data sets for both form factors. In addition, it was possible at Bates to extend the range in momentum transfer of F_c, down to q = 0.3 fm^{-1} because of the target design and out to q = 5.6 fm^{-1} because a 790 MeV beam was available at Bates while 695 MeV was the maximum available at Saclay. The new Bates data in the extended q range are consistent with the published fit to the Saclay data.

3.2 A = 2 Elastic Form Factors

As is well known, three form factors must be measured to achieve a complete experimental determination of the deuteron current. The status of the Rosenbluth separation has been reviewed elsewhere at this meeting where there is reported evidence for a minimum in Fm^2 near q^2 = 2 GeV/c^2. Separation of the monopole charge form factor F_c from the quadrupole F_Q requires the measurement of a polarization observable. Of the various possibilities[5], at the present time t_{20} is the most promising and a number of laboratories[6] are devoting much effort to this measurement.

A convenient expression for t_{20} is

$$t_{20} = - 2\alpha \frac{(2X+X^2)}{1+2X^2}$$

where $X = 2/3\eta \ G_Q/G_C$, $\eta = q^2/4M_d^2$ and $\alpha = A(q^2)/(A(q^2)+B(q^2)\tan(\theta/2))$.

The quantity α is known accurately from existing experimental data. The quantity X is independent of the nucleon charge form factors, in particular the poorly known neutron electric form factor. Thus, the quantity $t_{20/\alpha}$ will be very sensitive to the model of the deuteron current and relatively insensitive to the nucleon form factors.

In Fig. 3 we show the presently available data[7,8] compared with a number of predictions. In the region of q where there are data, the range of theoretical variation is smaller than the experimental uncertainties; although certain classes of separable potentials are excluded by the early Bates data[7]. At q > 4 Fm^{-1} apparently anything is possible. The curve labelled IA refers to an impulse approximation (nucleons only) calculation[9] using the Paris potential. Other realistic potentials yield similar results. IA and MEC refers to a calculation[9] in which meson exchange currents have been included following the prescription of Gari and

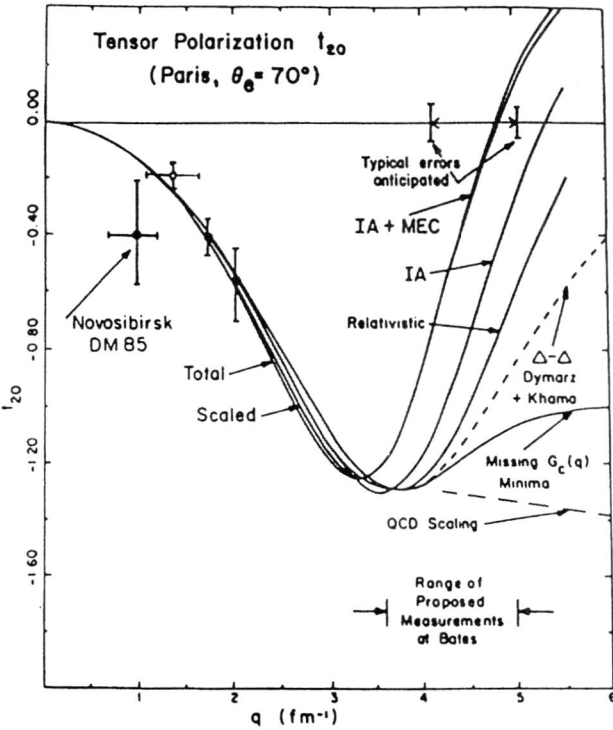

Figure 3. T_{20} as a Function of q. The experimental data are from Ref.[7] and Ref.[8]. For a discussion of the theoretical curves, see the text.

Hyuga[10]. Relativisitc refers to a calculation[9] in which some relativistic corrections have been made to the IA following the prescription of
Friar[11] but with no MEC. Δ-Δ comes from Dymarz and Khanna[12]. The
missing G_c (q) minimum is filled in according to the recipe of Kobushkin
and Shelest[13] which considers the possibility of a 6q component in the
deuteron; while QCD scaling is the "smoking gun" of Carlson and Gross[14].

Also shown in this figure is the region in q^2, ($14fm^{-2} < q^2 < 25fm^{-2}$)
in which measurements will be made at Bates in an experiment[15] which is
under preparation and, if all goes well, should run in 1987. This experiment, like the earlier one at Bates, will measure t_{20} of the recoiling
deuterons in coincidence with the scattered electrons. It requires:
1) 50μA at 950 MeV, 2) a 7 cm long liquid deuterium target which can tolerate this beam current, 3) a deuteron transport channel which separates
deuterons from protons of the same momentum and 4) a deuteron polarimeter
which has a high efficiency and analyzing power for polarized deuterons in
the energy range 100 MeV < E_d < 170 MeV. The first three of these require
only the expenditure of enough time and money to achieve and are in
progress. The fourth, the polarimeter, in addition to time and money,
requires nature's help in providing a suitable nuclear reaction as an
analyzer.

An extensive survey of the tensor-analyzing power of exclusive and
inclusive reactions of deuterons with several targets has been carried out
at LNS-Saclay[16]. Similar measurements have been carried out at IUCF at
E_d = 80 MeV[17]. The result of these investigations has led to the selection of (d,p) elastic scattering from 100 < E_d < 250 MeV.

The essential features of the reaction as seen in Figure 4 are:
- All tensor-analyzing powers are appreciable and show
  large variation with center of mass scattering angle,
- The shape and magnitude are very similar at E_d = 80
  and 191 MeV, indicative of little energy dependence.

Major advantages of a polarimeter based on this reaction are:
  (i) The existence of angular bins where T_{20} = o will allow for
      continuous monitoring of the unpolarized efficiency, ε_o.
 (ii) The weak energy dependence of the analyzing power will
      reduce errors due to uncertainty in the momentum distri-
      bution of the deuteron flux.
(iii) From the φ dependence of the yield, it will be possible to
      extract t_{21} and t_{22}.

(iv) As G_M is already well known in the momentum range
$14 < q^2 < 26$ fm^{-2}, we may use the t_{22} measurement to confirm that the polarimeter is operating in a reliable fashion.

A polarimeter based on dp scattering between 100 and 250 MeV has been
built at the University of Alberta. Above $T_d = 100$ MeV, the scattered
particle's energies are large enough to escape a thick target, allowing
trajectory determination; below 250 MeV, the same energies are small
enough to allow a total energy measurement. Thus, the elastic event is
identified by the energy-angle correlation of the scattered particles and
by the angular correlation as well. The AHEAD polarimeter (Alberta High
Efficiency Analyzer for Deuterons) has been calibrated at Saturne[18] and
will be used at Bates for ed scattering. An efficiency of 2×10^{-3} around
150 MeV is expected.

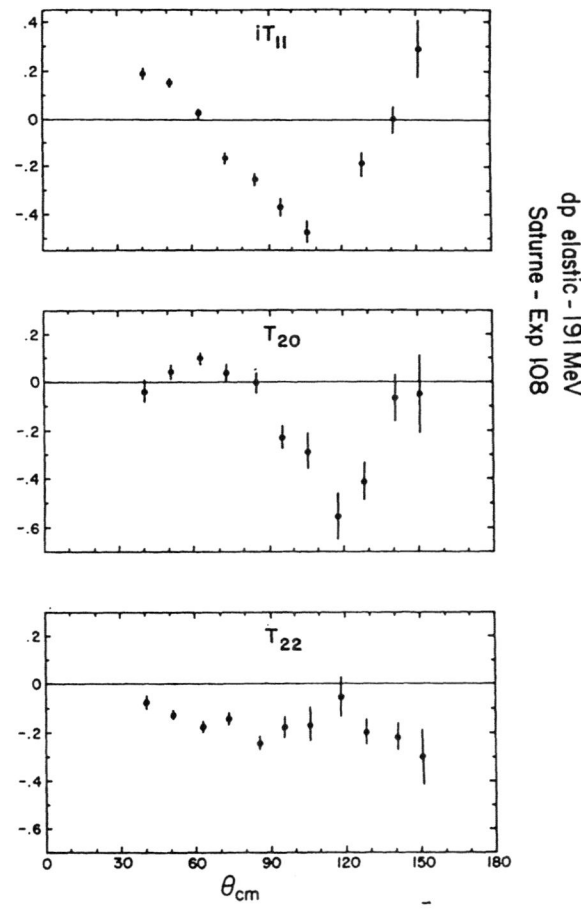

Figure 4. Tensor analyzing powers for d,p elastic scattering

3.3 A = 1 Neutron Charge Form Factor

Of the four nucleon form factors, the least well determined experimentally is F_C^N, the neutron charge form factor. To date it has been extracted from analysis of elastic or quasi-elastic scattering from deuterium where it appears quadratically with the other, much larger, form factors. With the recent development of a polarized electron beam at Bates, a proposal has been submitted[19] to measure the spin transfer from the polarized electrons to the neutrons in quasi-elastic from deuterium. This initiative proposes the use of the Kent State U. polarimeter in which the second scattering site is a large area (~ 1 M^2) plastic scintillator array. The difficulties of using such devices in the environment of a high intensity electron beam are notorious, hence prior to undertaking the experiment, a feasibility study has been recommended by the PAC and is planned. Some preliminary results from an earlier feasibility study for another D(e,e'n) experiment[20] (see Section 4.2 below) suggest that the spin transfer experiment may be possible at Bates even with the present low duty cycle.

4. Deep-Inelastic Electron Scattering

4.1 Inclusive Scattering

In the range of nuclear excitation energies available at Bates, deep-inelastic scattering, DIS, means mainly quasi-elastic scattering from the bound nucleons and quasi-free excitation of the delta. There have been three separate inclusive experiments at Bates on DIS from few nucleon systems in recent years. One experiment[21] was a comparatively low statistics survey of DIS from ^{2}H, ^{3}He and ^{4}He. A sample of the quality of the data is shown in Figure 5 in which the measured differential cross section for the three targets at 597 MeV and 60° is compared with recent calculations.[22,23]

Already from this simple comparison one has exposed many features of the present status of this subject. For ^{2}H, the calculations of Arenhovel[22] give a good account of this cross section. Isobar components (IC) and MEC give significant effects on the high and low ω side of the peak. For A = 3 and 4, the data are compared with calculations of the Rome group[23]. In this calculation there are no final state interactions, MEC or IC's. The deficiency of strength in the calculated cross section in both the low ω and dip regions is already clearly evident in ^{3}He and is much larger in ^{4}He. In fact, in ^{4}He the dip disease is already about as large as it gets in ^{12}C or medium weight nuclei[24].

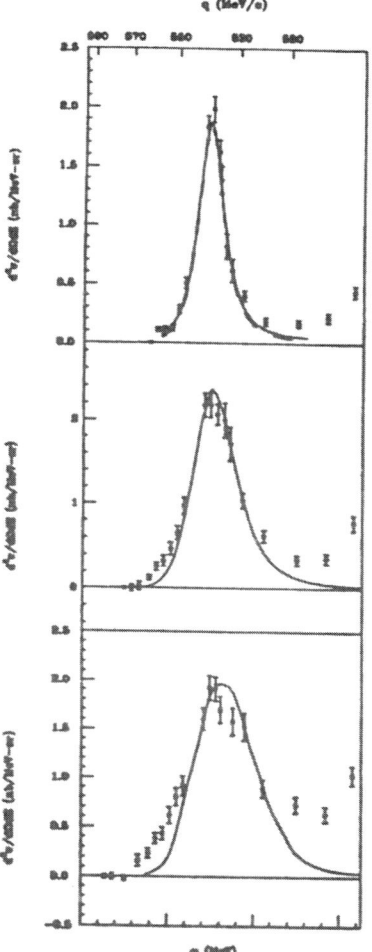

Figure 5. Differential cross
sections for quasi-elastic electron
scattering of 596.8 MeV electrons
at 60° for (a) ^{2}H, (b) ^{3}He and
(c) ^{4}He. The experimental points
are compared to the calculations of
Arenhovel[22] for ^{2}H and of the Rome
group[23] for ^{3,4}He. In part (a) for
^{2}H the dashed curve is the IA result
and the solid curve is the full calculation including MEC and IC.

These early data have also been analyzed to yield a separation into
longitudinal and transverse response functions and are being prepared for
publication. For ^{3}He they agree with the recent data from Saclay[25]
and with new Bates data which are presently being analyzed.

For the case of ^{2}H, a UMass group has made high statistics measurements at 180° at beam energies of 220, 270 and 320 MeV. Figure 6 shows
their data compared with a calculation of Leideman and Arenhovel[27].
One notes that at the peak of the cross section the theory overestimates
the cross section by about 10%. At 270 MeV and 320 MeV, the complete
calculation fits the data better than the "nucleons only" calculation

338

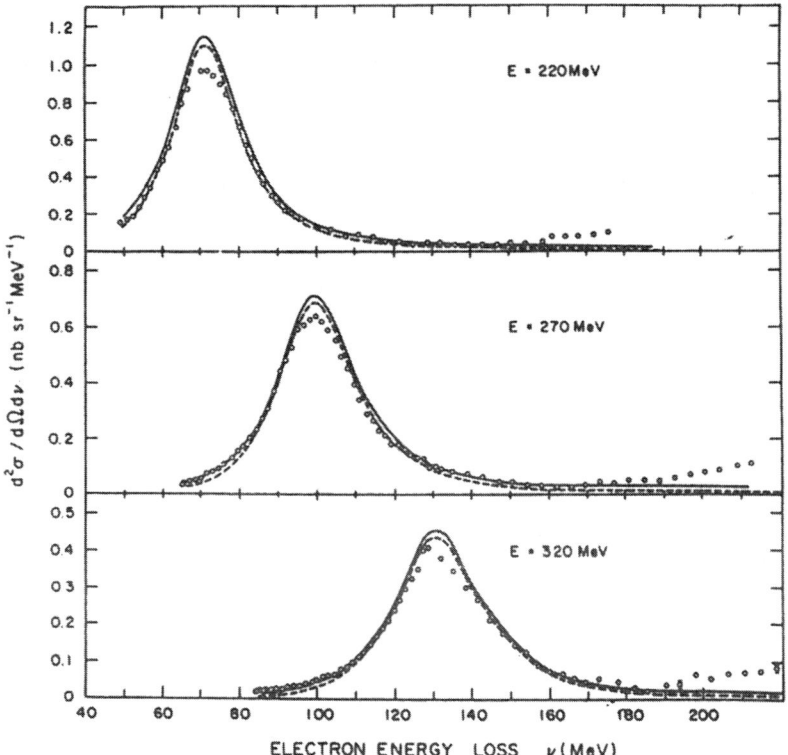

Figure 6. Quasi elastic scattering cross sections from ^{2}H at 180°.

while at 220 MeV it is vice-versa. Figure 7 shows the data in a slightly different way in which the relative importance of the nucleons-only, MEC and Isobar Configurations are indicated. The agreement between the data and the calculation is very good.

During the period at Bates that the elastic data on ^{3}H and ^{3}He were being taken, high precision data were also taken on the DIS. Figure 8 shows the region in q-ω space covered by the experiment.

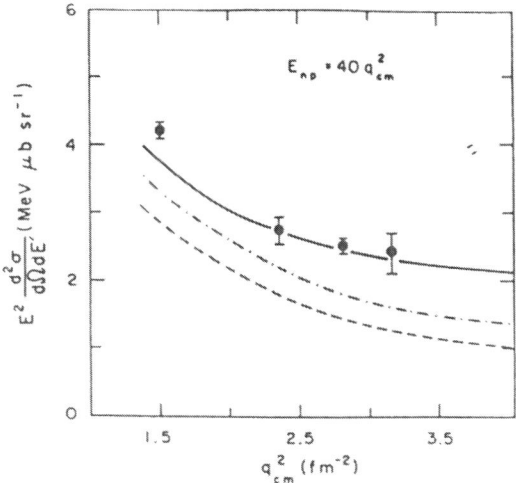

Figure 7. Quasielastic cross section multiplied by E^2 for the kinematic condition $E_{np} = 40 \, q^2_{cm}$. The cross section is differential in the scattered electron energy E' and the solid angle Ω for a scattering. The curves represent the theoretical preditions of Leidemann and Arenhovel[27]. The dashed curve is the nucleons-only calculation, the dashed-dot curve includes MEC contributions, and the solid curve is the total including IC.

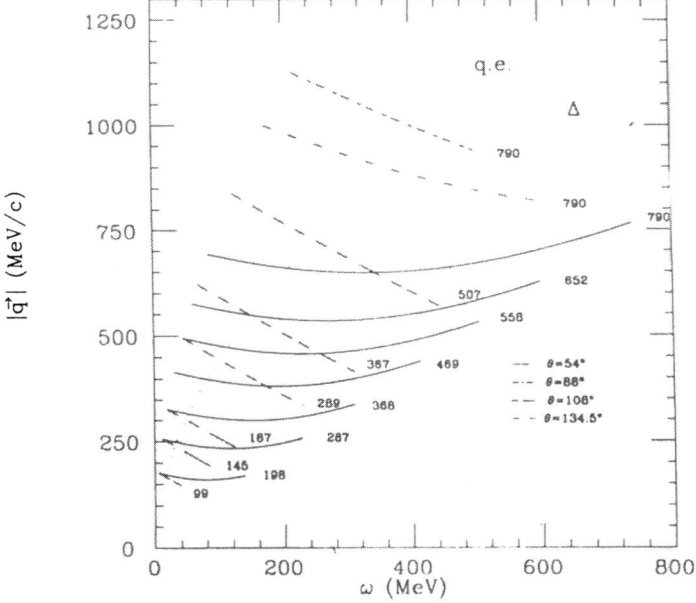

Figure 8. Kinematics for Bates ^{3}H and ^{3}He quasi-elastic scattering experiment.

340

Figures 9 and 10 show quasi-elastic scattering spectra from ^{3}H and ^{3}He at 370 MeV at 54° and 134.5°. The data have not yet been radiatively unfolded. The values of $|q|$ at the quasi-elastic peak are 320 and 550 MeV/c respectively. Also shown on these figures for the purposes of orientation only are the predictions of a simple theoretical model. The inputs to the model are as follows: the ground state momentum distribution of Cioffi degli Atti et al.[23]; plane wave final state and the impulse approximation formalism (including relativistic kinematics) of Moniz[28]. The curves shown correspond to the single photon exchange cross section folded with the radiators represented by the target and enclosures ("simple prediction"). Finally, some systematic shortcomings of this theoretical framework should be noted. In order to model the response for a given incident energy and scattering angle at a given ω (i.e. to fold in the effects of radiation), the cross sections for all values of $|q|$ and ω smaller than those at the desired point must be known. The framework described above is used to calculate these cross sections as well. It tends to overestimate the longitudinal response at low $|q|$, resulting in excess strength, both at the quasi-elastic peak and in the "dip" region (higher ω). Note that the responses at 54° are about 75% longitudinal.

A=3 CROSS SECTIONS AT E_i=367.7 MeV

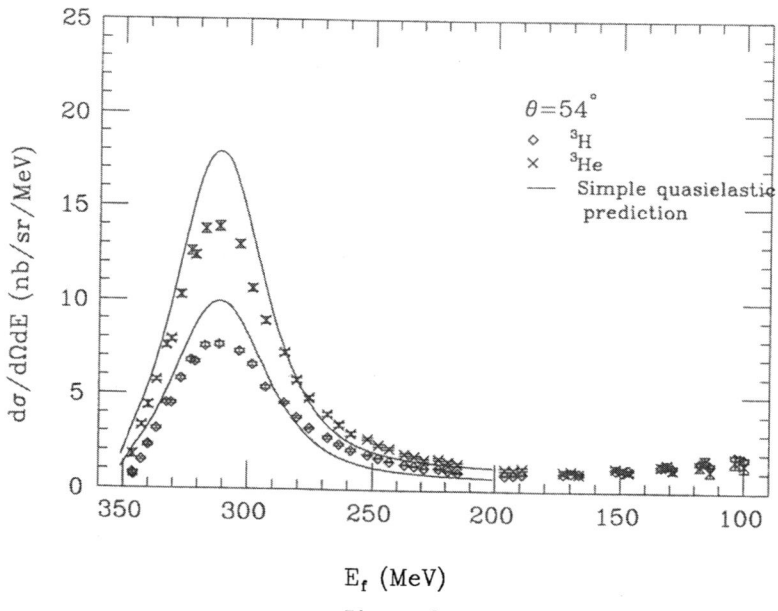

Figure 9

A=3 CROSS SECTIONS AT E_i=366.8 MeV

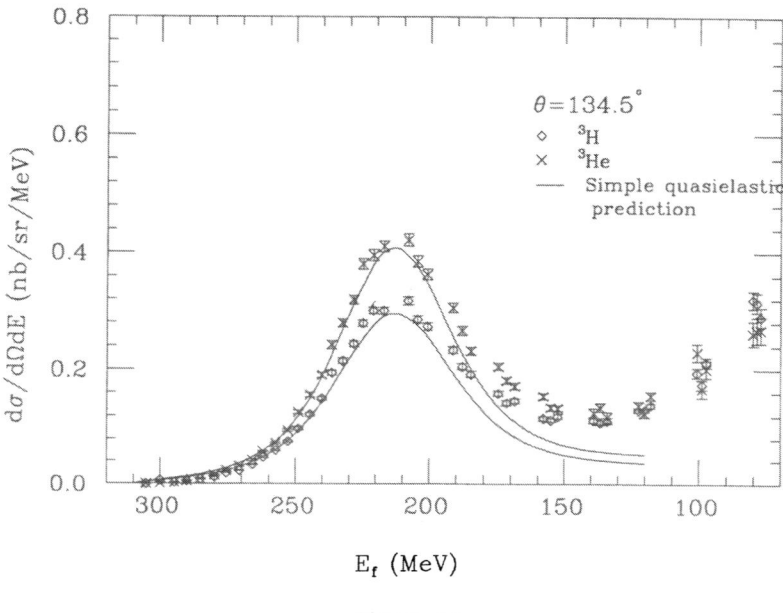

Figure 10

4.2 Exclusive Deep Inelastic Scattering

In contrast to the extensive measurements of inclusive DIS discussed above, so far there has been very little work done at Bates on exclusive reactions. In part, this is because high enough beam energy and suitable targets have not been available until only recently and, in part, because for many of the experiments which have been proposed, considerable development of the most appropriate detector systems is still required.

The only approved experiment in this category is a measurement[29] of the two-body break-up in ^{4}He(e,e'p). Transverse-longitudinal separations will be made in kinematic regions which supplement recent measurements at NIKHEF and Saclay. A preliminary run to check out the apparatus was made in June this year and the experiment should run sometime in early 1987.

Another experiment for which some feasibility studies have been made is a proposal[20] to measure the angular distribution of neutrons out of the scattering plane in the d(e,e'n) reaction in order to extract the interference response functions R_{LT} and R_{TT} and further to elucidate this elementary reaction. To develop an appropriate detector for the neutrons, tests have been made of the response of plastic scintillator detectors and different shielding configuration to a 240 MeV electron beam incident targets of varying thickness. To help separate the neutrons from the γ flash, the electron beam was chopped in a train of 20 ns pulses separated by 200 ns. Figures 11 and 12 show the effect of different shielding thicknesses in the line of sight flight path and suggest that practical shielding can be designed to make such measurements possible. These data also suggest that the large area Kent State polarimeter might be suitable for the spin transfer measurement discussed in Section 3.3 above.

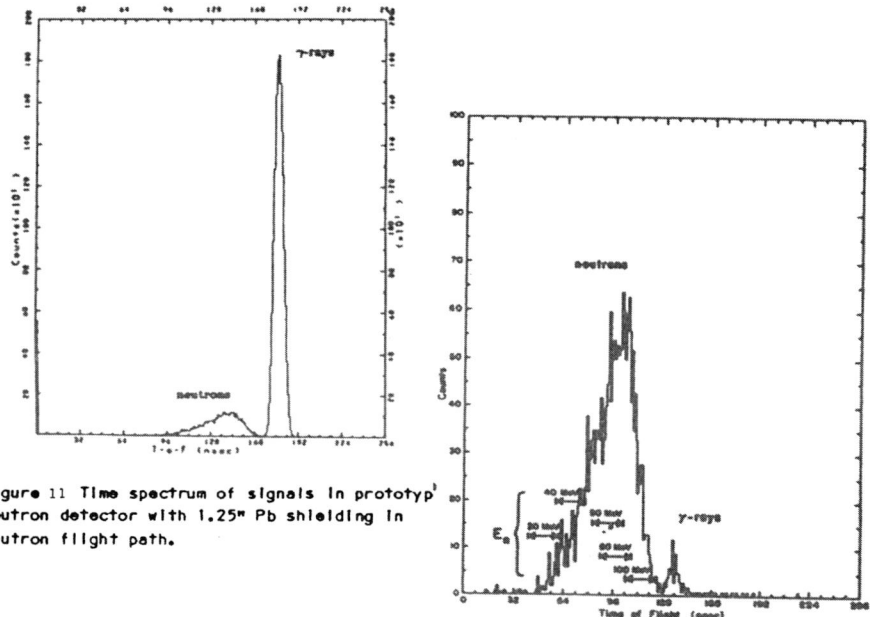

Figure 11 Time spectrum of signals in prototype neutron detector with 1.25" Pb shielding in neutron flight path.

Figure 12. Time spectrum of signals in prototype neutron detector with 3.25" Pb shielding in neutron flight path.

The likely shape of the future program in this area is indicated by the number of new proposals which were reviewed at the last PAC meeting in June of this year. There were three directed to the E2/M1 ratio in the NΔ transition; two[30][31] by p(e,e'pγ) and one[33] by p(e,e'p)$\pi°$. There was one proposal[33] to measure ^{3}He(e,e'2p) aimed at short range correlations and another[34] to get at the Roper resonance through both inclusive p(ee') and exclusive (e,e'π^+) reactions. Although none was approved for various reasons, I am sure that many, if not most, will be re-submitted.

5. Photoreactions

The Bates program on photo-reactions on few body systems has been reviewed recently[35][36] and I will not include that material in this paper. Since that time, the work on high energy $^4He(\gamma,p)$ and $^4H(\gamma,n)$ has been published[37] as has the work on $^4He(\gamma,\gamma)$ at 190 MeV[39], and $^4He(\gamma,\pi°)$ at E_γ = 290 MeV[38].

This latter experiment is part of a systematic experimental check of the delta-hole model for which Compton scattering data at the same energy as the $\gamma\pi°$ data will provide a stringent test. Such an experiment, for which a large NaI crystal detector has been constructed, is scheduled to run this autumn. The detector, with dimensions approximate 19" x 22" and an expected resolution of about 2% at 300 MeV, has been delivered to Bates and is being prepared for the experiment by a Boston University group[40].

Two other new experiments, $^3He(\gamma,\pi^+)$[41] and η photoproduction[42] on the proton and more complex nuclei, are also in preparation.

Acknowledgement

I would like to thank my colleagues for their permission to discuss their work before it has been published. Karen Dow, Doug Beck, Steve Dytman, Aron Bernstein and Michel Garcon were particularly helpful but bear no responsibility for any misconception. Of these colleagues at Bates whose work is so briefly mentioned, or overlooked completely, I beg indulgence for this somewhat biased summary. Audrey Iarocci produced this manuscript with evident skill and good humor for which she has my thanks.

This work is supported by DOE Contract #DE-AC02-76ER03069.

References

1. J. B. Flanz, S. B. Kowalski and C. P. Sargent, IEEE Trans. on Nucl. Sci. NS28, 2847 (1981).
2. Bates Proposal for a CW Upgrade (1984).
3. W. Bertozzi et al., Nucl. Instr. & Meth. 162, 211 (1979).
4. F. P. Juster et al., Phys. Rev. Lett. 55, 2261 (1985).
5. R. G. Arnold, C. E. Carlson and F. Gross, Phys. Rev. C23, 363 (1981).
6. M. Garcon, Contribution to the CEBAF Summer Workshop 1986.

7. M. E. Schulze et al., Phys. Rev. Lett. 52, 597 (1984).

8. Y. F. Dmitriev et al., Phys. Lett. 157B, 143 (1985) and preprint INP - Novosibirsk 86-75.

9. M. E. Schulze, Ph.D Thesis, MIT, unpublished.

10. M. Gari and H. Hyuga, Nucl. Phys. A264, 409 (1976); A274, 333 (1976); A278, 372 (1977).

11. J. L. Friar, Ann. Phys. (NY) 81, 332 (1973) and Phys. Rev. C12, 685 (1975).

12. R. Dymarz and F. C. Khanna, Phys. Rev. Lett. 56, 1448 (1986).

13. A. P. Kobushkin and V. P. Shelest, Sov. J. Part. & Nucl. 14, 483 (1983).

14. C. E. Carlson and F. Gross, Phys. Rev. Lett. 53, 127 (1984).

15. Bates Proposal 84-17, J. M. Cameron, M. E. Schulze and W. Turchinetz, co-Spokesmen.

16. M. Garcon et al., preprint DPhN/Saclay 2258 (1986), to be published in Nucl. Phys.

17. E. J. Stephenson et al., IUCF Scientific and Technical Report, 58 (1983).

18. M. Garcon and J. M. Cameron, private communication.

19. Bates Proposal 85-05, R. Madey and S. Kowalski, co-Spokesmen.

20. Bates Proposal 81-06, W. Bertozzi, J. M. Finn and M. A. Kovash, co-Spokesmen.

21. Bates Proposals 82-06, A. Bernstein and B. Quinn, co-Spokesmen and 80-17, R. Whitney, Spokesman.

22. H. Arenhovel, Nukleonika 24, 273 (1979); W. Fabian and H. Arenhovel, Nucl. Phys. A314, 253 (1979).

23. C. Ciofi degli Atti, E. Pace and G. Salme, Phys. Lett. 141B and private communication.

24. Z. E. Mezziani, Nucl. Phys. A446, 113C (1985).

25. C. Marchand et al., Phys. Lett. 153B, 29 (1985).

26. Bates Proposal 76-05, G. Peterson, Spokesman; B. Parker et al., UMass preprint, Sept. 1986, submitted to Phys. Rev. C.

27. W. Leidemann and H. Arenhovel, Nucl. Phys. A393, 385 (1983).

28. E. J. Moniz, Phys. Rev. 184, 1154 (1969).

29. Bates Proposal 84-15, M. Epstein and S. Margaziotis, co-Spokesmen.

30. Bates Proposal 86-04, M. Kovash, Spokesman.

31. Bates Proposal 86-13, C. Papanicolas, Spokesman.

32. Bates Proposal 86-12, G. Chang, Spokesman.

33. Bates Proposal 86-10, J. Lightbody, Spokesman.

34. Bates Proposal 86-05, J. M. Finn, Spokesman.

35. J. L. Matthews, Can. J. Phys. 62, 1100 (1984).

36. W. Turchinetz, Nucl. Phys. A446, 23C (1985).

37. R. A. Schumacher et al., Phys. Rev. C33, 50 (1986).

38. D. R. Tieger et al., Phys. Rev. Lett. 53, 755 (1984).

39. E. J. Austin et al., Phys. Rev. Lett. 57, 972 (1986).

40. J. Miller, private communication.

41. Bates Proposal 83-11, A. Bernstein and S. Dytman, co-Spokesmen.

42. Bates Proposal 86-07, S. Dytman and J. C. Peng, co-Spokesmen.

PROTON AND DEUTERON FORM FACTORS
AT LARGE MOMENTUM TRANSFER*

R. G. Arnold

The American University, Washington, D. C. 20016

and

Stanford Linear Accelerator Center, Stanford CA 94305

Abstract

Two recent experiments at SLAC are described, one on electron-proton elastic scattering out to $Q^2 = 31.3$ $(GeV/c)^2$, and another on 180° elastic scattering from the deuteron out to 2.7 $(GeV/c)^2$. The proton data yield new values for the magnetic form factor G_M^p which are compared to predictions from perturbative QCD. Preliminary results from the deuteron measurement indicate the presence of a diffraction feature in the magnetic form factor $B(Q^2)$ at high Q^2. Recent speculation on the neutron form factors at high Q^2 and the implications for experiment are briefly described.

Introduction

The common thread linking the measurements described here is the desire to probe the nucleons and the deuteron out to the largest accessible momentum transfer in electron scattering to examine their structure at short distances or high internal momentum. A major goal of such measurements is to discern what are the relevant degrees of freedom operable at a given momentum transfer. We know that nucleons are complex extended objects that are bound states of more elementary constituents (the quarks), and that the force which binds nucleons into nuclei is a remnant of the force that binds the quarks into nucleons. For the past decades nuclear physics has progressed by building a phenominological picture of nuclei in terms of interacting mesons and nucleons without explicitly considering the quark constituents. A major task today, and a primary motivation for the experiments described here, is to find out in which kinematic regime and to what level of accuracy nucleon structure and eventually nuclei can be understood directly in terms of interactions among quarks. This is an enormous task with many complex aspects, and though many have worked some years on it, answers come slowly. The purpose of this talk is to describe the results of two recent experiments that bring new evidence to the discussion. This is necessarily a progress report, first because only preliminary results are available for one experiment (deuteron), and also because we are far from achieving a satisfactory answer to some of the most basic questions.

*Work supported by the U.S. Department of Energy contract DE-AC03-76F00515 (SLAC) and the National Science Foundation Grant PHY85-10549 (American University).

The Proton

Elastic electron scattering has been a source of information about nucleon structure since the first experiments more that three decades ago. The earliest measurements revealed that the proton and neutron are not point particles. For the first two decades the data were primarily interpreted in models where the virtual photon is pictured to interact with the nucleon via exchange of intermediate vector mesons[1] (vector meson dominance models – VMD). There are various versions of these models containing different numbers, masses, decay widths and coupling constants for the mesons, and they generally give a fair representation of the data (most extensively known for the proton). The VMD models provide a convenient parameterization of the electromagnetic interaction which summarize what in the quark language must in fact be due to a very complicated nonperturbative long range interaction of the virtual photon with the quarks and gluons in the nucleons.

Following the discoveries in deep inelastic lepton scattering, where it was revealed that at Q^2 as low as 1 $(GeV/c)^2$ nucleons behave as though made of a finite number of quasi-free, charged, spin-1/2, pointlike constituents, the question arises, when do we have to describe elastic scattering in terms of the interactions of these constituents? The first attempts to explain elastic form factors from quark constituents were the dimensional scaling laws.[2] They predict the shape of the electromagnetic form factors, and other exclusive reactions, versus momentum transfer to follow a power of Q^2 determined by the number of constituents in the interaction. The power law dependence follows from the scale invariant interaction between quarks at short distance (quarks are pointlike modulo logarithmic corrections from gluon radiation). The nucleon is pictured to be a complex quark state containing three valence quarks and some number of gluons and extra quark-antiquark pairs. According to the dimensional scaling laws the probability for elastic scattering at large Q^2 will eventually depend only on the component containing just the valence quarks; the other components will fall away faster with increasing Q^2.

The underlying mechanism of the dimensional scaling laws in terms of hard quark rescattering has subsequently been computed directly from perturbative QCD (PQCD). Brodsky and Lepage[3] showed that at large momentum transfer, where large is determined by the onset of power law behavior in the data, the nucleon form factor G_M is given by the convolution of three amplitudes

$$G_M = \int dx dy \phi(x, Q^2) T_H(x, y, Q^2) \phi(y, Q^2). \tag{1}$$

The factor $\phi(y, Q^2)$ is the distribution amplitude in the infinite momentum frame for finding the nucleon in the state with just the three valence quarks having longitudinal momentum fractions $y = y_1, y_2, y_3$. The hard scattering amplitude $T_H(x, y, Q^2)$ gives the chance for scattering three quarks from the initial state y to the final state x, and $\phi(x, Q^2)$ is the amplitude for leaving the three quarks nearly parallel and bound in the final state proton after absorbing momentum transfer Q^2. The distribution amplitudes $\phi(x, Q^2)$ are obtained by integrating the full wave function $\psi(x, k_\perp)$ over transverse momentum $k_\perp$. The PQCD result[3] for G_M is

$$G_M^p = \frac{\alpha_s^2(Q^2)}{Q^4} \sum_{n,m} d_{nm} \left(\ln \frac{Q^2}{\Lambda^2} \right)^{-\gamma_m - \gamma_n}. \tag{2}$$

The factors $\alpha_s^2(Q^2)$ and $1/Q^4$ come from the quark-gluon coupling to three quarks in T_H, and the factors d_{nm} describe the initial and final quark distribution amplitudes.

The basic philosophy of the perturbative QCD approach is that at some value of Q^2 we believe that the form factor comes to depend primarily on the hard scattering of the valence quarks and depends only weakly on the other wave function components or on the details of the distribution amplitude for the three quarks. In this picture the old problem of how to deal with the complexities from the nonperturbative structure are all swept into the factors $\phi(x, Q^2)$ and $\phi(y, Q^2)$. They cannot be calculated directly from PQCD, because that would be equivalent to solving for the quark bound state including the nonperturbative components. There are constraints on the forms that the distribution amplitudes can take from the laws of QCD (at $Q^2 \to \infty$, $\phi_{as} = 120x_1x_2x_3$ and the quarks uniformly share the momentum, and $\phi(x, Q^2)$ at two different finite Q^2 are related because they evolve with Q^2 according to fixed rules), but in general their form is not known. The hope in the early PQCD calculations was that the first order terms would dominate and that uncertainties from lack of knowledge of $\phi(x, Q^2)$ would be sufficiently small that possible deviations from pure $1/Q^4$ behavior in the data might be blamed on the Q^2 dependence of $\alpha(Q^2)$ due to QCD effects (gluon radiation). In that case we might hope to determine, or constrain, the value of Λ_{QCD} from the data.

This was the situation at the time the present experiment was begun, and a primary motivation was to obtain high quality data that could be compared to detailed PQCD calculations. Meanwhile the theoretical picture has continued to evolve, and at the present time the situation is unclear. To begin with, some people have argued[4,5] that the perturbative QCD approach will not be applicable until very high Q^2, certainly above 10 to 30 $(GeV/c)^2$ accessible in experiments. The arguments come in various forms, but they essentially all boil down to the same point: that even up to very high Q^2 the elastic scattering probability receives large contributions from components of the proton quark wave function in which there are significant soft components. Another way to say this is that it may be very unlikely for the incoming virtual photon to find the proton in a state with just three nearly colinear (in the IMF) valence quarks. From the critics point of view we are being fooled by the shape of the data into believing that the PQCD regime has been reached; they argue that the observed approximate $1/Q^4$ behavior of G_M^p is built up out of scattering on a complex system involving many soft contributions with valence quarks having significant transverse momentum. In that case until (if ever) there is a more comprehensive description of the nonperturbative structure, we will have to contend with more phenomenological descriptions of this regime, such as parameterizations in terms of vector mesons. One counter argument in favor of PQCD is that there is a lot of compelling experimental evidence[6] that the hard scattering regime has been reached if we look at the data for other exclusive reactions, for example $\gamma p \to \pi^+ n$.

There are two major and related problems with the PQCD approach. The first is to understand how to obtain the distribution amplitudes $\phi(x, Q^2)$ to use in the calculations. The second is to calculate higher order $\mathcal{O}(m^2/Q^2)$ terms. In the early calculations it was assumed that the dependence of G_M on $\phi(x, Q^2)$ was small. There is expected to be a very slow $\log Q^2$ variation from the QCD evolution, but that will be much smaller that the Q^2 variation from T_H. Subsequently it was discovered that the normalization of the 3-quark contribution to G_M depends significantly on the shape of $\phi(x, Q^2)$ versus $x_1x_2x_3$. It turns out that for amplitudes symmetric about $x_1x_2x_3 = 1/3$ (valence quarks have equal fraction of the nucleon longitudinal momentum) the first order calculations give identically zero for G_M^p due to cancellation among the terms in T_H from the allowed diagrams. For the 3-quark component to contribute significantly, and thus for PQCD to be applicable, the longitudinal momentum must be distributed asymmetrically among the quarks.

Meanwhile others[7-10] have been working to solve for $\phi(x, Q^2)$ from more-or-less first principles of QCD. They use constraints from what are called QCD sum rules along with normalization and momentum conservation conditions to put limits on the first few moments (integrals of $\phi(x, Q^2)$ wrt $x_1x_2x_3$) of $\phi(x, Q^2)$. The QCD sum rules have theoretical uncertainty, since they depend upon assumptions and models for the incalculable nonperturbative interactions of the quarks, and therefore they only roughly constrain the shapes of $\phi(x, Q^2)$ with some error. In this procedure the $\phi(x, Q^2)$ are expanded in a series of Appell polynomials orthogonal on the $x_1x_2x_3$ triangle, and models are proposed for the first few (up to 6) of these polynomials consistent with the sum rules. The present results [7-10] all give $\phi(x, Q^2)$ with asymmetric momentum distributions. The model of Chernyak and Zhitnitsky,[7] for example, has 65% of the longitudinal momentum of the proton in the IMF carried by one up quark with its spin aligned with the spin of the proton. Using asymmetric shapes for $\phi(x, Q^2)$, the values for G_M^p are finite and in the range of the data.

The second problem for PQCD is to evaluate the higher order terms. This is important in the case of the nucleon form factors because a quantitative comparison with the data requires knowledge of both the leading order Dirac form factor $F_1 \sim 1/Q^4$ and the higher order Pauli form factor $F_2 \sim 1/Q^6$. Helicity is conserved for massless quarks, and since the Pauli term involves a helicity flip of the interacting quark, it contains an extra factor $\mathcal{O}(m^2/Q^2)$. The perturbative calculations so far only evaluate G_M to leading order, which effectively only includes the F_1 part.

Let us now look at the results from the new experiment. I will not comment on the details of the experiment as they are described in Ref. 11, except for the following points. This experiment was optimized for measurement of ep elastic scattering at large Q^2 with low background and small systematic errors. The cross section was measured by detecting only the scattered electrons in a single arm spectrometer. The major factor which governed the experimental design and the final errors was the low counting rate at high Q^2 (as low as three counts per day). The major potential sources of background were from electrons scattering from the end walls of the target and from the high flux of soft stray particles that could confuse the identification of the infrequent elastically scattered electrons. The end cap background was eliminated by a shield near the target, and a new powerful detector package with sufficient redundancy to reject junk particles was built for the spectrometer. This measurement, taken with a 60 cm long target and some months of running time at full SLAC beam power, has extended ep elastic scattering to the present practical maximum in Q^2.

The results for G_M^p, extracted from the cross sections assuming $G_E^p = G_M^p/\mu_p$, are shown in Figure 1, along with the previous data and two theoretical curves from PQCD. The new data show $Q^4G_M^p$ reaching a nearly constant value near Q^2 around 5 $(GeV/c)^2$. The hint in the previous data, albeit with large errors, of an increase over $1/Q^4$ at high Q^2 is ruled out. The curve by Brodsky and Lepage[3] is calculated with a simple nonrelativistic quark wave function (all quarks at $x_i = 1/3$) and is arbitrarily normalized to the data at $Q^2 = 10 \ (GeV/c)^2$. The slow decrease with Q^2 arises from the $\alpha(Q^2)$ variation using $\Lambda = 100$ MeV. The curve by Chernyak and Zhitnitsky[7] is an absolute prediction based on their asymmetric model for $\phi(x, Q^2)$ with a factor of 2 theoretical uncertainty in normalization.

From the point of view of perturbative QCD we could conclude that the experimental verification of nearly $1/Q^4$ behavior for G_M^p justifies the use of perturbative methods, and that now the main problem is to use the data to guide our understanding of the properties of the 3-quark component of the wave function, and to look for evidence of higher order effects. It is also conceivable that the present data has not much to do with perturbative QCD. The large sensitivity to $\phi(x, Q^2)$ and the surprising result that for the three quark component the momentum is not symmetrically distributed could be an indication that the nonperturbative effects still dominate at this Q^2.

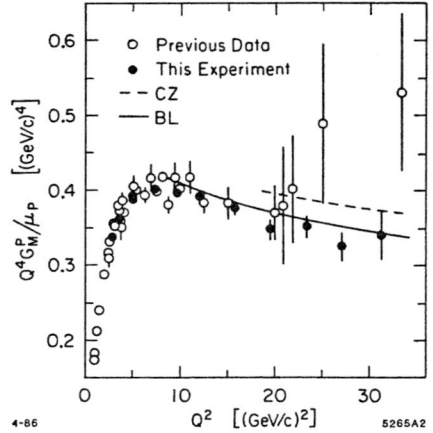

Fig. 1. New results for the proton form factor G_M^P from Ref. 10. The perturbative QCD curves are: BL (Ref. 3), CZ (Ref. 7).

It is also possible that the present measurements are in a transition region between the nonperturbative and the hard scattering regimes, and that aspects of both are necessary to understand the data. A recent work by Gari and Krümpelmann[12] adopted this point of view. They have constructed a phenomenological merger of VMD and the asymptotic constraints given by the dimensional scaling laws. The usual VMD expressions for nucleon form factors in terms of photon-meson and meson-nucleon couplings and meson propagators are augmented with a momentum dependent form factor containing a new scale parameter Λ_2 The meson-nucleon and the Pauli and Dirac e.m. form factors are written as

$$F_1 = \frac{\Lambda_1^2}{\Lambda_1^2 + \hat{Q}^2} \frac{\Lambda_2^2}{\Lambda_2^2 + \hat{Q}^2} \qquad (3.a)$$

$$F_2 = F_1 \frac{\Lambda_2^2}{\Lambda_2^2 + \hat{Q}^2}. \qquad (3.b)$$

The value of $\Lambda_1 = 0.8$ GeV is determined from meson-nucleon data; the effective momentum transfer $\hat{Q}^2$ contains the logarithmic variation $\log(Q^2/\Lambda_{QCD}^2)$. The parameter Λ_2, to be determined from fits to the form factor data, adjusts the transition from VMD to QCD. For $Q^2 \ll \Lambda_2^2$ the momentum dependence follows the monopole form factors F_1 and F_2 plus the vector meson propagators. For $Q^2 \gg \Lambda_2^2$ the meson propagators die away and $F_1 \sim 1/Q^4$ while $F_2 \sim 1/Q^6$.

Gari and Krümpelmann fit their model to all the available form factor data (present experiment excluded). Their value of $\Lambda_2 = 5.15 \ (GeV/c)^2$ suggests that the transition takes place around $5 \ (GeV/c)^2$, and their value of $\Lambda_{QCD} = 0.29$ GeV is consistent with values extracted from scaling violations in deep inelastic lepton scattering. Two examples from their fit shown in Figure 2 illustrate several points. In their model, like other VMD models, G_M^p gets a significant contribution from F_{2p} in the range of the present data. This means that we must be cautions in comparing first order PQCD calculations to this data. The slow decrease in $Q^4 G_M^p$ above 5 $(GeV/c)^2$ may be mostly due to the fall off of the F_{2p} contribution and may not all be blamed on the logarithmic variation of $\alpha(Q^2)$ in the F_{1p} term.

The second point is that F_{1n} is very small (near zero) in the Gari-Krümpelmann model. This arises naturally in VMD models when $m_\rho = m_\omega$, and seems to be required to get a consistent fit to the limited data[13] on the neutron cross section at high Q^2.

There are several consequences from $F_{1n} \simeq 0$:

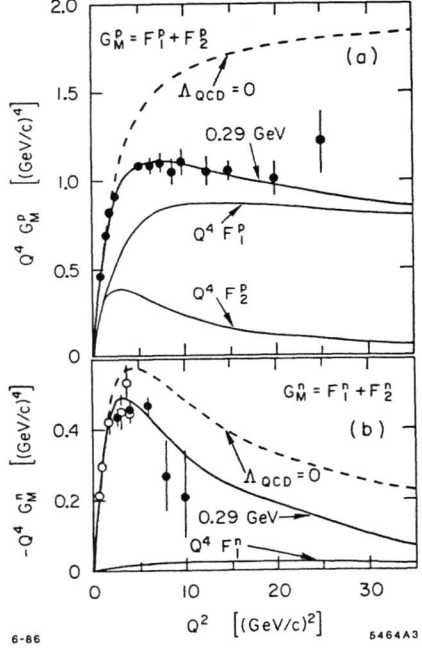

6-86   5464A3

Fig. 2. Results from the VMD + QCD model of Ref. 12. a) the proton G_M^P, b) the neutron G_M^n.

a) The neutron $G_M^n = F_{1n} + F_{2n}$ would be dominated by the higher order F_{2n} term, in which case it does not make sense to compare data for G_M^n to first order PQCD calculations. Data on G_M^n at high Q^2 might provide a useful testing ground for the higher order terms. Within the context of PQCD it would be necessary to find asymmetric wave function models for $\phi(x, Q^2)$ which simultaneously give a large $G_M^p \simeq F_{1p}$ and a small F_{1n}. Some attempts in this direction have been made.[9]

b) The neutron $G_E^n = F_{1n} - Q^2 F_{2n}/4m^2$ may be comparable in size to G_M^n around $Q^2 = 4$ $(GeV/c)^2$. It could be that the folklore that G_E^n is small, or even zero, is true at low Q^2, but at higher Q^2 the situation is completely reversed. This prediction can be tested in a standard Rosenbluth measurement of quasi-elastic ed scattering.[14]

c) Knowledge of the neutron form factors is essential for interpretation of other electromagnetic data. One key example is the deuteron forward angle elastic form factor $A(Q^2)$, which in the impulse approximation is mostly proportional to the isoscalar charge form factor $A \sim (G_E^p + G_E^n)^2 \phi_{body}^2$. The smaller G_E^n beats against the larger G_E^p and small changes around zero give big effects in $A(Q^2)$. It is possible that a long standing inability of the impulse calculations[15] to give large enough values for $A(Q^2)$ could be traced to using the wrong neutron form factor.

The Deuteron

Measurements of the deuteron form factors at large Q^2 provide important tests of reaction mechanisms and the nature of the nucleon constituents at short distance. The deuteron magnetic form factor $B(Q^2)$ is expected to be especially sensitive to ingredients in the description. The traditional starting point is the impulse approximation.[15-17]

There the virtual photon couples just to the nucleons moving in the wave function. Impulse models with various assumptions for deuteron wave functions and bound nucleon form factors all give sharp diffractive shapes for $B(Q^2)$. The location of the minimum and height of the second maximum are sensitive to the shape of the short range part of the deuteron S and D state wave functions. Other models augment the impulse approximation with scattering from meson exchange currents[17] and nucleon isobar currents[18] that can shift or even obliterate the diffraction feature. There are estimates of deuteron form factors using the assumptions of hard quark scattering[19] that is the basis of perturbative QCD; they predict form factors falling smoothly like powers of Q^2 with no diffraction features. The hybrid quark models[20] incorporate scattering from 6-quark components of the wave function supposed to exist at the core when nucleons are within a certain overlap radius from each other. They give widely varying predictions for the deuteron form factors. Recently calculations have been made using skyrmions[21] as the operable degrees of freedom. The meson exchange currents are effectively included directly in such models, and they fill in the diffraction minimum in $B(Q^2)$.

It is clear that data on $B(Q^2)$ at large Q^2 will be useful for discriminating between the widely varying predictions, and hopefully will give some indication as to which objects the virtual photon couples to.

A new experiment has recently been completed in the NPAS program at SLAC in which electron-deuteron elastic scattering was measured by detecting the electrons scattered near 180° in coincidence with the recoiling deuterons. The $B(Q^2)$ was expected to be as much as two orders of magnitude smaller than the forward angle function $A(Q^2)$, therefore a large backward angle is required. The measurement was done in a double arm 180° spectrometer system specially constructed on the floor in SLAC End Station A. Data was taken in two installments, one in June–July 1985 and again in June 1986. Measurements were made at eight values of momentum transfer from 1.2 to 2.7 $(GeV/c)^2$. The cross sections at high Q^2 were very low ($\sim 10^{-42}$ cm^2/sr), leading to counting rates of events per week using a 20 cm long liquid deuterium target and a spectrometer solid angle of 18 msr.

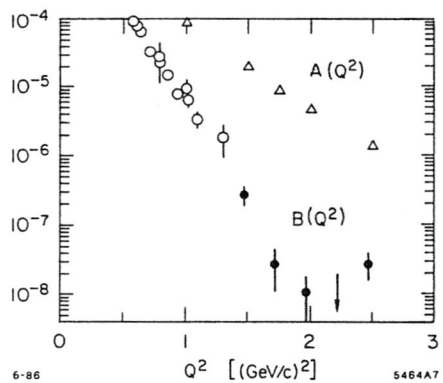

6-86 Q^2 [$(GeV/c)^2$]   5464A7

Fig. 3. Preliminary results for $B(Q^2)$ (Ref. 22) and previous data (Ref. 23).

The analysis of data is still in progress. The preliminary results from the 1985 data are shown in Figure 3. The new points fall very quickly with Q^2 and then indicate a deviation from a smooth decrease at Q^2 around 2 to 2.5 $(GeV/c)^2$. The 1986 data with points at 2.25, 2.5 and 2.7 $(GeV/c)^2$ confirm and strengthen the definition of an apparent diffraction feature in that region.

This experiment has nearly doubled the measured range in Q^2 for $B(Q^2)$. Like the proton experiment describe above, it has extended the measurement out in momentum transfer to the practical maximum set by low counting rates. The apparent existence of a diffraction feature around 2 to 2.5 $(GeV/c)^2$ will sharply discriminate among the competing models. When the final results are available they will provide a quantitative test of the short range deuteron structure.

REFERENCES

1. G. Höhler et al., Nucl. Phys. B114, 505(1976); S. Blatnik, N. Zovko, Acta Phys. Austriaca 39, 62(1974), F. Iachello, A. Jackson, A. Lande, Phys. Lett. 43B, 191(1973).

2. S. Brodsky, G. Farrar, Phys. Rev. D11, 1309(1975).

3. S. Brodsky, G. Lepage, Phys. Rev. D22, 2157(1980).

4. N. Isgur, C. Llewellyn Smith, Phys. Rev. Lett. 52, 1080(1984).

5. V. A. Nesterenko, A. V. Radyushkin, Sov. J. Nucl. Phys. 39, 811(1984)

6. See Ref. 2 and Refs. therein.

7. V. L. Chernyak, I. R. Zhitnitsky, Nucl. Phys. B246, 52(1984), V. L. Chernyak, A. R. Zhitnitsky, Phys. Rep. 112, 173(1984).

8. C. Carlson, Proceedings of the Nato Advanced Study Inst., Banff, Canada, Aug 22 - Sept 4 1985.

9. M. Gari, N. G. Stefanis, Ruhr Universität Bochum Print RUB TPII-85-16; RUB TPII-85-23 (1986).

10. I. D. King, C. T. Schrajda, University of Southampton Print SHEP 85/86-15 (1986)

11. R. Arnold et al., SLAC PUB 3810 (1986); Phys. Rev. Lett. 57, 174(1986).

12. M. Gari, W. Krümpelmann, Z. Phys. A322, 689(1986); Ruhr Universtät Bochum Print RUB TPII-86-5 (1986).

13. S. Rock et al., Phys. Rev. Lett. 49, 1139(1982); R. J. Budnitz et al., Phys. Rev. 173, 1357(1968); W. Albrecht et al., Phys. Rev. 26B, 642(1968).

14. R. Arnold et al., "A Proposal to Separate the Charge and Magnetic Form Factors of the Neutron at Momentum Transfer $Q^2 = 2$ to 5 Q^2 " , SLAC-NPAS Proposal NE11 (1986).

15. Examples of impulse approximation results for deuteron form factors including relativistic effects are given in R. Arnold, C. Carlson, F. Gross, Phys. Rev. C21, 1426(1980). The effect of assuming $F_{1n} = 0$ was noted.

16. R. S. Bhalerao, S. A. Gurvitz, Phys. Rev. Lett. 47, 1815(1981).

17. M. Gari, H. Hyuga, Nucl. Phys. A264, 409(1976); Nucl. Phys. A278, 372(1977).

18. M. Gari, H. Hyuga, B. Sommer, Phys. Rev. C14, 2196, (1976).

19. S. Brodsky, B. Chertok, Phys. Rev. D14, 3003(1976); S. Brodsky, C. Ji, G. P. Lepage, Phys. Rev. Lett. 51, 83(1983).

20. A. P. Kobushkin, V. P. Shelest, Sov. J. Part. Nucl. Phys. 14, 483(1983);

21. D. Riska, Proc. 2nd Conference on the Intersections Between Particle and Nuclear Physics, Lake Louise, Canada, May 26-31, 1986.

22. P. Bosted, Proc. 2nd Conference on the Intersections Between Particle and Nuclear Physics, Lake Louise, Canada, May 26-31, 1986.

23. S. Auffret, et al., Phys. Rev. Lett. 54, 649(1985); R. Cramer, et al, Z. Phys. C29, 513(1985).

ELECTRODISINTEGRATION OF FEW BODY SYSTEMS AT SLAC AND THE Y SCALING APPROACH*

Z. E. Meziani

Department of Physics,

and

Stanford Linear Accelerator Center

Stanford University, Stanford, CA 94035

It is proposed that extraction of the scaling function $F(y)$ from the transverse and longitudinal response functions in inclusive quasi-elastic electron scattering from ^3He and ^4He is a powerful method to either study the validity regime of the impulse approximation by allowing the access to the high nucleon momentum components in these nuclei, or the electromagnetic properties of bound nucleons.

INTRODUCTION

The prediction of possible scaling behaviour of the response function in quasi-elastic electron scattering from nuclei with respect to the longitudinal component of the nucleon momentum aroused great interest in the nuclear physics community. The idea, proposed by West [1] as a result of an analogy with atomic physics phenomena, provides a suitable approach in attempting to extract the momentum distribution of nucleons in nuclei, exploiting the scaling phenomenon as a signature of one-nucleon knock-out in the scattering process.

The reaction mechanism of electron scattering in the quasi-elastic region is usually described in the impulse approximation as a one-step process where the virtual photon knocks-out a single nucleon. Under this approximation, West formulated the expression

---

*Work supported by the Department of Energy, contract DE–AC03–76SF00515.

of the scaling function $F(y)$ for the case of a non relativistic Fermi gas system. Since then, a great effort has been devoted not only to experimental testing of the original idea [2–4], but also to improve the definition of the scaling variable and to understand the relation between the scaling function $F(y)$ and the nucleon momentum distribution in realistic cases [5–8]. On one hand the theoretical progress achieved in clarifying the ambiguities of the early analysis of ^{3}He data in the quasi-elastic region has been significant, showing that higher momentum transfer data [in the range of $Q^2 = 5$ (GeV/c)^2] are needed to reach the perfect scaling regime. On the other hand the scaling behavior of the response in quasi-elastic electron scattering was proposed by Sick [9] also as a sensitive way to test the nucleon electromagnetic properties in nuclei since the European Muon Collaboration (EMC) effect [10,11] posed several questions about the size of nucleons in the nuclear medium.

In light of the new inclusive electron scattering data [12] on ^{4}He, which measured momentum transfers up to 2.5 (GeV/c)^2, we would like to propose a method of experimental analysis which we think would be a better test of the validity regime of the impulse approximation and also a powerful method to study the nucleon properties in the nuclear medium.

SCALING FUNCTION AND VARIABLE

The inclusive electron scattering cross section in the one-photon exchange approximation is a function of two independents variables, the four momentum transfer Q^2, and the energy transfer ω:

$$\frac{d\sigma}{d\Omega d\omega} = \sigma_M \left\{ \left(\frac{Q}{|\vec{q}|}\right)^4 R_L(Q^2, \omega) + \left[-\frac{1}{2}\left(\frac{Q}{|\vec{q}|}\right)^2 + tg^2\frac{\theta}{2}\right] R_T(Q^2, \omega) \right\}, \qquad (01)$$

$$Q^2 = \omega^2 - \vec{q}^{\,2}. \qquad (02)$$

where $\vec{q}$ is the three momentum transfer carried by the virtual photon, σ_M is the Mott cross section, and R_L and R_T are the longitudinal (charge) and the transverse (convection and magnetization currents) response functions respectively.

A further step may be taken in the description of the electron-nucleus cross section in the quasi-elastic region, if, the Plane Wave Impulse Approximation is assumed. Under this approximation the relation between the spectral function $S(k,\epsilon)$ and the measured inclusive cross section is given by [5–8]:

$$\frac{d\sigma}{d\Omega d\omega} = \left\{ Z \overline{\frac{d\sigma}{d\Omega_p}} + N \overline{\frac{d\sigma}{d\Omega_n}} \right\} \left|\overline{\frac{\partial \omega}{k\partial cos\alpha}}\right|^{-1} \frac{1}{(2\pi)^2} \int_{\epsilon_-}^{\epsilon_+} d\epsilon \int_{k_{min}(q,\omega,\epsilon)}^{k_{max}(q,\omega,\epsilon)} S(k,\epsilon)k dk \qquad (03)$$

where $|\overline{d\sigma/d\Omega_{p(n)}}|$ is the electron proton (neutron) cross section evaluated at $k_{min}(q, \omega, \epsilon_{min})$, $\cos\alpha = \vec{q} \cdot \vec{k}/|q \cdot k|$ defines the angle between the struck nucleon momentum $\vec{k}$ and the incoming virtual photon momentum $\vec{q}$. k_{max} and k_{min} are defined by pure kinematical conditions. $S(k, \epsilon)$ is a probability of finding a nucleon in the nucleus with the momentum k and binding energy ϵ

Experimentally, we are interested in extracting the so called scaling function $F(y)$ expressed as the following ratio [5–8]:

$$F(y) = \frac{d\sigma}{d\Omega d\omega} \left/ \left\{ Z\,\overline{\frac{d\sigma}{d\Omega_p}} + N\,\overline{\frac{d\sigma}{d\Omega_n}} \right\} \left| \overline{\frac{\partial\omega}{k\partial\cos\alpha}} \right|^{-1} \frac{1}{(2\pi)^2} \right.$$

$$= \int_{\epsilon_-}^{\epsilon_+} d\epsilon \int_{k_{min}(q,\omega,\epsilon)}^{k_{max}(q,\omega,\epsilon)} S(k, \epsilon) k dk$$

(04)

where y is the momentum solution of the total energy conservation equation evaluated at $\epsilon = \epsilon_{min}$ and $\cos\alpha = -1$. In other words y is the minimal momentum of the struck nucleon verifying the energy conservation of the process as follow;

$$\omega + M_A = (M^2 + q^2 + y^2 + 2yq)^{1/2} + (M_{A-1}^2 + k^2)^{1/2}$$

(05)

where M_A and M_{A-1} are respectively the total mass of the initial and the recoil nucleus, k and q are the magnitudes of the nucleon and the virtual photon momenta, respectively.

It is important to notice that the phase space factor defined as $d\omega/dy$ in the early scaling analysis of ^{3}He is incorrect unless one uses West's definition of the scaling variable. The correct phase space factor needed, independently of the definition of the scaling variable used, is $\left| \partial\omega/k\partial\cos\alpha \right|_{k=k_{min}}$ [8]. This factor arises naturally from the angular integration performed using the full energy conservation δ function.

The ^{3}He data in the quasi-elastic region have been reanalyzed in Refs. [6,7] showing almost the same scaling behavior of the data with a different shape of the scaling function compared to the early analysis due to the different phase space factor used. We want in this case to concentrate on the ^{4}He data obtained SLAC's new Nuclear Physics Facility. Inelastic cross sections have been measured at two energies and three angles covering a range of momentum transfer from 0.5 (GeV)^2 to 2.5 (GeV)^2. We have analyzed these new preliminary data combined with the previous data from Ref. [13] allowing to cover a range of longitudinal momentum y from 0.0 GeV/c to -0.8 (GeV/c). Fig. 1 shows the experimental results compared with a theoretical scaling function obtained using the following relation:

$$F(y) = \int_{y}^{\infty} n(k) k dk$$

(06)

where $n(k)$ is a momentum distribution given in Ref. [14] generated by solving the Shrœdinger equation using the ATMS method.

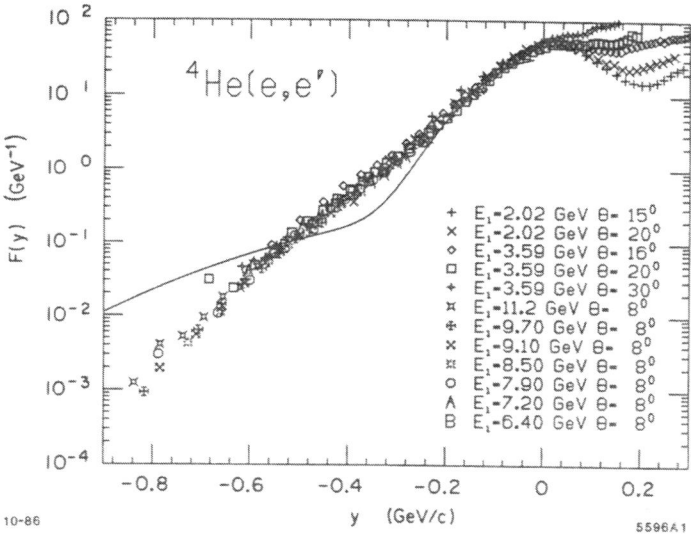

Fig. 1. Scaling function of ^{4}He obtained using (04) with data of Refs. [12,13] compared to theoretical scaling function obtained using (06).

We observe mainly the same features shown in the analysis of ^{3}He data from Ref.[7], namely that the theoretical calculation using no interaction in the final state overestimates the data $|y| > 0.25$ GeV/c, and we also notice a pronounced change of slope at that y value. On the contrary experimental data behave more like a straight line. From a detailed calculation by Laget [15] on ^{3}He we expect that final state interactions of the residual nucleus and the ejected nucleon to have both significant contribution. It is thus, very important not to draw quik conclusions about our understanding of the high nucleon momentum components. The reaction mechanism in a nucleus such as ^{4}He is complicated. One has to wait for a more complete four body calculations, in which the continuum solutions of the Bethe-Salpeter equation are provided to understand the region beyond $|y| > = 0.25$ GeV/c. However it is clear that in the region where the data and the theoretical curve give a unique answer, the dominant process is one-nucleon knockout ($|y| < 0.25$ GeV/c). In this region the PWIA works quite well and the momentum distribution can be extracted safely from the data. It is then a matter of preference performing exclusive $(e, e'p)$ or inclusive (e, e') experiments to access the momentum distribution in nuclei. Exclusive experiments are a powerful tool in these studies. However reaching the high component of the momentum distribution requires electron beam machines with high duty cycle factors.

TRANSVERSE AND LONGITUDINAL SCALING FUNCTIONS

One further step can be advanced in inclusive experiments, in studying either the momentum distribution or the electromagnetic properties of the nucleon in the nucleus, by expressing the equation (03) in such a way that the electric and magnetic contributions of the electron nucleon cross section are explicitly separated. If the electric and the magnetic parts of the resulting separated equation are compared with equation (1.1) we can obtain expressions of the scaling function $F(y)$ in terms of the transverse and longitudinal response functions:

$$R_T(y) = \left| \frac{\partial \omega}{k \partial \cos \alpha} \right|^{-1} \frac{-q_n^2}{2 E_k E_{k\prime}} \widetilde{G}_M^2 \, F_T(y) \tag{07}$$

$$
\begin{aligned}
R_L(Q,\omega) = \left| \frac{\partial \omega}{k \partial \cos \alpha} \right|^{-1} \Big\{ &\widetilde{G}_E^2 \frac{(E_k + E_{k\prime})}{4 E_k E_{k\prime}(1+\tau)} \, F_L(y) \\
&- \frac{1}{2 E_k E_{k\prime}} \left(q^2 - \frac{(E_k + E_{k\prime})^2}{2(1+\tau)} \right) \widetilde{G}_M^2 \, F_L(y) \Big\}
\end{aligned}
\tag{08}
$$

$$1 + \tau = \left(1 + \frac{Q^2}{4M^2} \right)$$

where (E_k, k) and $(E_{k\prime}, k')$ are respectively the energy-momentum of the struck and outgoing nucleons. $\widetilde{G}_E^2$ and $\widetilde{G}_M^2$ are the effective electric and magnetic form factors of the nucleus;

$$
\begin{aligned}
\widetilde{G}_E^2 &= Z G_E^{p2} + N G_E^{n2} \\
\widetilde{G}_M^2 &= Z G_M^{p2} + N G_M^{n2}
\end{aligned}
\tag{09}
$$

and

$$q_n^2 = (E_K - E_{K\prime})^2 - (K - K')^2 .$$

We want to emphasize that besides the PWIA no further approximations are needed to obtain the relations (07),(08). This consequently imposes the following relation:

$$F_L(y) = F_T(y) = F(y)$$

This relation can be checked experimentally if one has data of the transverse and longitudinal response functions obtained by the Rosenbluth technique. These separated response functions are not available in the region of high momentum transfer. However, if one restricts the range of momentum transfers from .2 $(\mathrm{GeV/c})^2$ to .5 $(\mathrm{GeV})^2$ the results of the existing data analyzed following (07) (08) show an interesting behavior. Fig. 2 shows the extracted longitudinal $F_L(y)$ and transverse $F_T(y)$ scaling functions from the data of ^3He according to Ref. [17] It is suprising to see that these two functions are different but tend to converge to the same value at q=0.5 (GeV/c). These results show that the impulse approximation is not valid for this nucleus at transfers lower than about 0.5 (GeV/c). As an example, a heavier nucleus [18] (^{12}C) has been analyzed

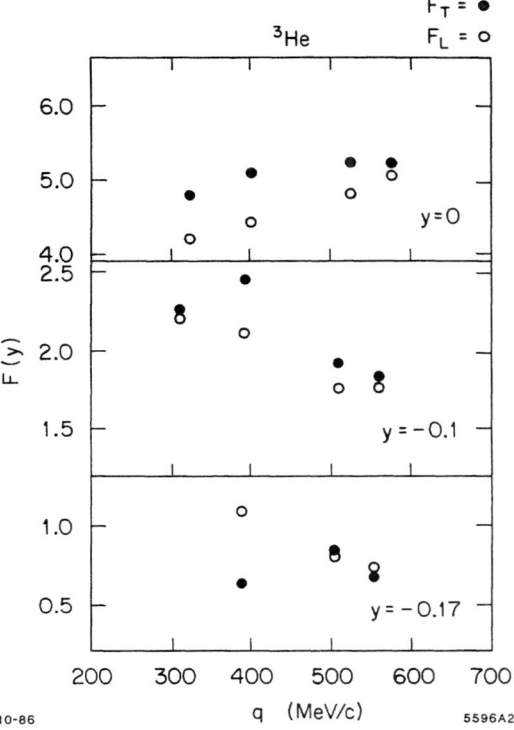

Fig. 2. Separated transverse $F_T(q,y)$ (triangles) and longitudinal $F_L(q,y)$ (circles) extracted from ^3He data of Ref. [16] using formulae (7,8). (After ref. 17)

the same way and the result are shown in Fig. 3. The situation in this case is more critical, since the scaling regime seems to be reached around 0.5 (GeV/c) in momentum transfer; however, no convergence of the two scaling functions is observed.

At this stage it is important to notice that if one assumes that the free nucleon form factors that we have used in the analysis are correct, then this result is an obvious breakdown of the impulse approximation. However the separate scaling behavior of each function is disturbing and can lead to the following question: Could a modification of the nucleon electromagnetic form factors lead to a convergence of these functions and maintain their scaling behavior? The answer to this question is yes. As suggested by Mulders for ^{12}C in Ref. [19] if one modifies the electric and magnetic form factors as follow:

$$G_E^\star = \left(1. + \frac{Q^2}{0.54(GeV/c)^2}\right)^{-2}$$

$$G_M^\star = \mu_{p,(n)}^\star \left(1. + \frac{Q^2}{0.69(GeV/c)^2}\right)^{-2}$$

(010)

$$\mu_{p,(n)}^\star = 1.12\ \mu_{p,(n)}$$

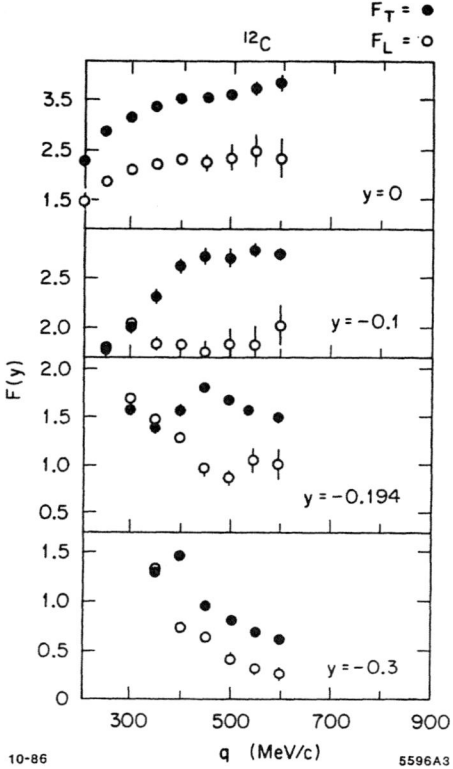

Fig. 3. Same as Fig. 2, but for ^{12}C data from Ref. [18]. (After ref. 17)

the overlap of the two scaling functions can be obtained. One cannot make the same statement about ^{3}He since the effect seems to be density dependent, it must be small in this nucleus. However it is important to know that this behavior is pronounced in ^{4}He compared to ^{3}He since the former is strongly bound. These issues can be studied as soon as separated response functions data become availaible for ^{4}He at high momentum transfers.

We do not recommend to study this problem as suggested in Ref. [9] using the scaling behavior of the total response function without performing the separation, the main reason being that at high momentum transfer the total response function is dominated by its transverse part. By examining the suggested modification of the nucleon form factors for ^{12}C (equation 010) one can clearly see that no change in the momentum dependence of the magnetic form factor is needed to explain the observed difference between the transverse and longitudinal scaling functions. In this case the total response function at high momentum transfer will always display scaling behavior. We illustrate this statement in Fig. 4 by showing the scaling function obtained using modified electromagnetic form factors following relation (010). It is evident that no change is

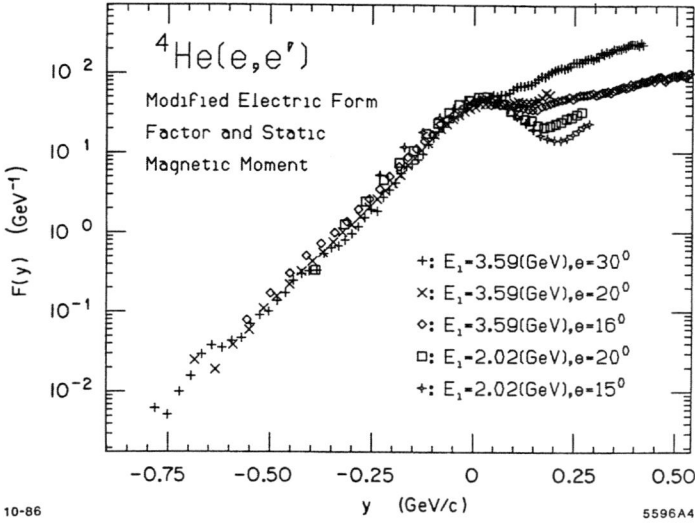

Fig. 4. Scaling function of ^{4}He obtained using the suggested modified nucleon electromagnetic form factors. The data are preliminary data of Ref. [12]. This scaling function is to be compared to the one of Fig. 1.

observed in the quality of the scaling if one compare this result to Fig. 1 where the free electromagnetic form factors have been used. It is thus very important to understand that the best way to study the modification of the electromagnetic form factors is by performing first the separation of the electric and magnetic components in the total cross section then analysing the data using relations (07),(08).

A new experiment (NE-9) is planned at the Nuclear Physics Injector at SLAC to perform the separations of the two response functions R_L and R_T for ^{3}He,^{4}He and ^{56}Fe around $Q^2 = 1$ (GeV/c)^2. Our aim for this experiment is to have a comparative study between few body (calculable) and many body systems in order to examine in a powerfull way the different issues discussed.

REFERENCES

[1] G. B. West, Phys. Rep. $\underline{18C}$, 269 (1975).

[2] P. D. Zimmerman, C. F. Williamson and Y. Kawazowe, Phy. Rev. $\underline{C19}$, 279 (1979).

[3] I. Sick, D. Day and J. S. McCarthy, Phys. Rev. Lett. $\underline{45}$, 871 (1980).

[4] P. Bosted et al., Phys. Rev. Lett. $\underline{49}$, 1380 (1982).

[5] E. Pace and G. Salmè, Phys. Lett. $\underline{110B}$, 411 (1982).

[6] C. Ciofi Degli Atti, E. Pace and G. Salmè, Phys. Lett. $\underline{127B}$, 303 (1983)

[7] S. A. Gurvitz, S. Wallace and J. Tjon, Phys. Rev. $\underline{C}$, (1986).

[8] C. Ciofi degli Atti, in "Few-body problems in Physics", L. D. Faddeev and T. I. Kopaleishvili Eds, World Publishing, Singapore 1984, pag 463.

C. Ciofi degli Atti, E. Pace and G. Salmè, Czech. J. Phys B36, 960(1986).

C. Ciofi degli Atti, E. Pace and G. Salmè, These Proceedings.

[9] I. Sick, Phys. Lett. 157B, 13 (1985).

[10] J. J. Aubert et al., Phys. Lett. 123B, 275 (1983).

[11] R. G. Arnold et al., Phys. Rev. Lett. 52, 727 (1984).

[12] D. Day et al., NE3 Experiment Preliminary results

[13] S. Rock, R. Arnold, B. Chertok, Z. Szalata, D. Day, J. S. McCarthy, F. Martin, B. A. Mecking, I. Sick, and G. Tamas, Phys. Rev. C26, 1592 (1982).

[14] Y. Akaishi, Contribution to this conference.

[15] J. M. Laget, Phys. Lett. 151B, 325(1985).

[16] J. C. Marchand et al., Phys. Lett. 153B,(1985).

[17] C. Ciofi degli Atti and G. Salmè, Proc. of the 6th Seminar "Electromagnetic Interactions of Nuclei at Low and Medium Energies", Moscow Dec. 10–12 1984, p. 224.

[18] P. Barreau et al., Nucl. Phys. A402, 515(1983).

[19] P. J. Mulders, Phys. Rev. Lett. 54, 2560 (1985).

HIGH MOMENTUM COMPONENTS AND MANY BODY EFFECTS IN ³He(ee'p) AND ³He(ee') EXPERIMENTS

J. Morgenstern

Service de Physique Nucléaire-Haute Energie, CEN-Saclay,
F-91191 Gif-sur-Yvette Cedex, France

Abstract

The proton momentum distribution has been masured at Saclay, up to 600 MeV/c in a new coincidence ³He(ee'p) experiment [16] following a previous measurement up to 300 MeV [1]. Evidence of two-body correlations observed in the ³He(ee'p)pn channel is shown. These results are compared to previous inclusive ³He(ee') performed at SLAC [2] and Saclay [3].

I. Introduction

Three-body systems can be described now with realistic wave functions in their ground state. They are the simplest cases where we can study the short range behaviour of the nuclear forces as short range correlations, three-body forces etc. It is of particular interest to measure the form factors of three-body nuclei at high momentum transfer [4, 5]. On the other hand, the determination of the momentum distribution of the nucleons, especially for values > 1.5 fm^{-1} where we expect short range effects is of particular importance. For this purpose ee'p experiments produce a direct information on this distribution. We can also use results from inclusive ee' measurements which produce an information on the proton distribution in the nuclei integrated on the momentum and the binding energy of the different proton states.

II. Cross Section

In the framework of Born approximation to describe electron scattering, the ee'p cross section can be written [6] :

$$\frac{d^6\sigma}{dE'd\Omega_e, dT_p, d\Omega_{p'}} = \Gamma\sigma_\gamma \qquad (1)$$

with :

$$\sigma_\gamma = \sigma_T + \varepsilon\sigma_L + \varepsilon\cos 2\alpha\sigma_{TT} + [\varepsilon(\varepsilon + 1)]^{\frac{1}{2}}\cos\alpha\sigma_{TL} \qquad (2)$$

where the four quantities σ_T, σ_L, σ_{TT}, σ_{TL} are functions of q, ω, p', γ the momentum and energy transfer of the electron, the momentum of the emitted proton and the angle of this momentum with the electron transfer momentum. A similar expression can describe the electron scattering on a moving off-shell proton inside the nucleus [7].

If we assume Plane Wave Impulse Approximation (PWIA), we can write :

$$\frac{\sigma_T}{\sigma_T^p} = \frac{\sigma_L}{\sigma_L^p} = \frac{\sigma_{TL}}{\sigma_{TL}^p} = \frac{\sigma_{TT}}{\sigma_{TT}^p} = KS(E_{m,p}) \qquad (3)$$

or

$$d^6\sigma = Kd^6\sigma^p S(E_{m,p}) \qquad (4)$$

where K is a kinematical factor and $S(E_{m,p})$ the spectral function which depends on the momentum p and separation energy E_m of the proton in its initial state. In that case we have a simple way to determine the spectral function by measuring the e,e'p cross section and using an appropriate expression for the scattering of the electron on an off-shell moving proton ; we use the expression proposed by de Forest [7] called $K\sigma_{ep}^{cc1}$.

But PWIA is not exact and we must do corrections for final state interactions and exchange currents [8].

To obtain the expression for the inclusive cross-section we integrate eq. (2) on the proton variables p', γ and we obtain :

$$\frac{d^3\sigma}{dE'd\Omega_e,} = \Gamma[S_T(q,\omega) + \varepsilon S_L(q,\omega)].$$

III. Inclusive ³He(ee') experiments

Fig.1 shows three spectra taken at Saclay for an incident electron energy equal to 667 MeV at 60°, 90° and 145°. The theoretical curves are calculated by Laget [8], using for ³He a wave function obtained with a five channel Faddeev calculation [9]; final state interaction, exchange currents and real pion production are taken into account.

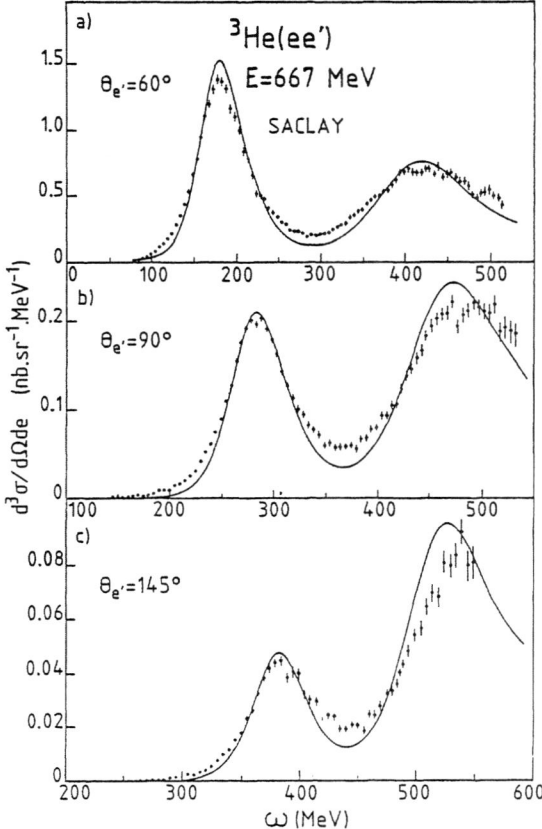

Fig. 1. Inelastic electron scattering spectra measured at Saclay, —— Laget calculation [8] (see text).

We can see that the agreement is very good for the quasi-elastic peak ; but the theoretical curve is below the experiment for small ω values and in the dip region between the quasi-elastic peak and the Δ-resonance (we don't discuss the real pion production here). The same kind of agreement is obtained by others groups [10, 11] for the quasi-elastic peak. The discrepancy in the dip region can be due to an inaccurate evaluation of meson exchange processes or rescattering effects in real pion production.

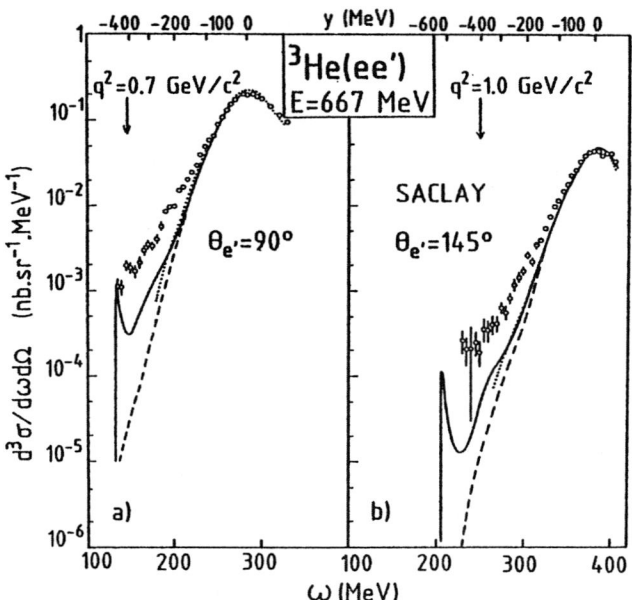

Fig. 2. The same as Figs. 1b) and 1c) at small ω values on logaritmic scale. --- Laget PWIA ; •••• Ciofi et al. PWIA and Meier-Hadjuk et al. PWIA (the two calculations are very close) ; —— Laget full calculation.

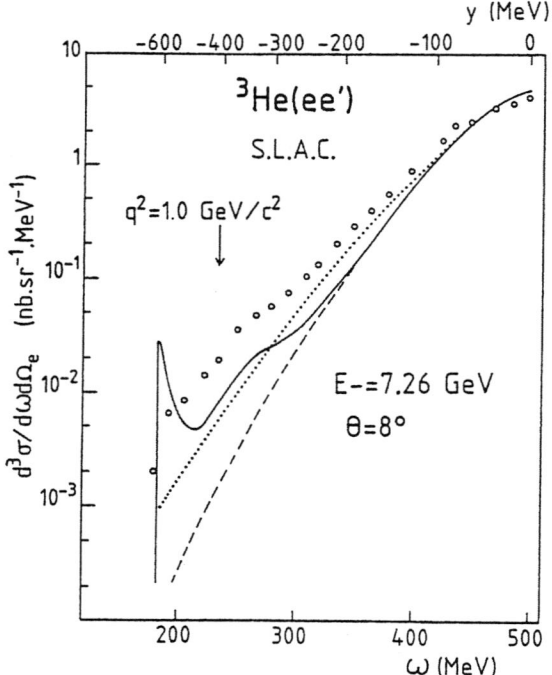

Fig. 3. The same as Fig. 2 for the SLAC experiment

Table I

E : incident electron energy, θ_e: electron scattering angle, P: initial proton momentum, E_m : missing energy, q^2 : square of the 4-momentum transfer.

Kinematics	E (MeV)	$\theta_{e'}$ (deg)	P (MeV/c)	2-body break-up		3-body break-up	
				E_m (MeV)	q^2 (fm^{-2})	E_m (MeV)	$\langle q^2 \rangle$ (fm^{-2})
I [1]	528	52.2	0 - 200	5.5	4.5	0 - 60	4.2
II [1]	509	36.0	160 310	5.5	1.9	0 - 60	1.8
III [16]	560	25.0	320 600	5.5	1.0	0 - 90	1.0

Using PWIA and $K\sigma_{ep}^{cc1}$ of T. de Forest [7], we obtain the spectral function. For kinematics I and II we have used de Forest formula instead of Priou one as it was done in ref. [1] to obtain the spectral function. In Fig. 4 we show this spectral function $S(e_{m,p})$ for the two-body break-up

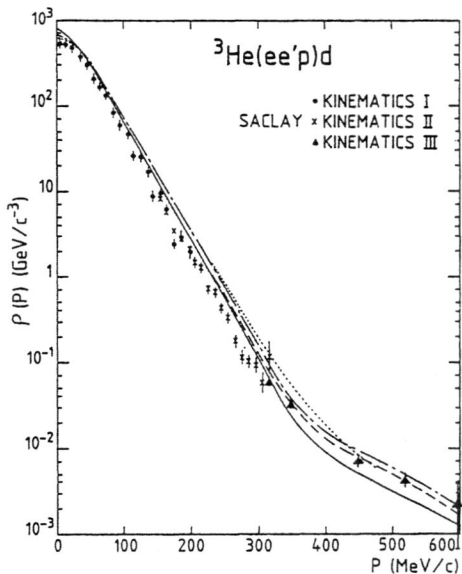

Fig. 4, ^{3}He(ee'p)d spectral function measured at Saclay. PWIA calculations : —— Laget [8] ; ---- Meier-Hadjuk et al. [10] ; — · — Ciofi et al. [11] ; ···· Schiavilla et al. [13].

For the left-hand side of the peak the contribution from exchange current and real pion production is very small and cannot explain the discrepancy. For a given value of ω, only the proton momenta greater than a certain minimum value $|y|$ contribute to the cross section ; $|y|$ is given by :

$$\omega + \Gamma_A = (M^2 + q^2 + y^2 - q|y|)^{\frac{1}{2}} + (M^2_{A-1} + y^2)^{\frac{1}{2}} \qquad (11)$$

this value is near zero on the top of the quasi-elastic peak and increases when ω decreases. Fig. 2 shows the left hand-side of the two last spectra of Fig. 1. We have plotted the y value corresponding to each ω value. For comparison we have shown the theoretical curves obtained in PWIA with a Reid Soft Core potential by Laget [8] using a wave function calculated by the Faddeev method [9], by Meier Hadjuk et al. with a more recent Faddeev calculation [10] and by Ciofi et al. [11] with a variational calculation ; these two last calculations agree between each other and lie a factor ~ 2 above the first one for $|y| > 300$ MeV/c, but they underestimate the experiment by a factor ~ 3. This Saclay experiment confirms qualitatively the results obtained at SLAC [2] some years ago at higher electron energy, smaller scattering angle but similar momentum transfer as shown on Fig. 3.

The effect of final state interaction has been evaluated by Laget [8] and increases the cross section by a factor ~ 3 for $|y| > 300$ MeV. With this kind of correction the wave function of ref. [9] produces a cross-section a factor ~ 3 below the experiment while the wave functions from refs. [10] and [11] come closer to the experiment.

The effect of final state interaction is very important for high $|y|$ values and the sensitivity of the inclusive experiment to the momentum distribution above 300 MeV/c is questionable.

IV. Coincidence ^{3}He(ee'p) experiment

To get a better understanding of the problem of momentum distribution in ^{3}He, especially the high values, we have measured in coincidence the scattered electron and the ejected proton. In PWIA, we can identify the modulus of recoil momentum P_R and of the initial proton momentum P in the nucleus, and extract the spectral function according to eq.(4). We have determined the momentum distribution between 0 and 600 MeV. For that purpose we have measured three different kinematics as shown in Table I.

channel ³He(ee'p)d which reduces to the momentum distribution ρ(p) (E_CM = 5.5 MeV). For comparison we have plotted the spectral function computed by Brandenburg et al. [9] (Faddeev five channels), Meier et al. [10] (Faddeev eighteen channels), Ciofi et al. [11] (variational method), Schiawilla et al. [13] (variational method including three-body forces). We can see that the different models agree better between each other at small momentum. The Schiawilla et al. calculation which reproduces the ³He binding energy come closer to the experiment near p = 0. J.M. Laget has calculated [8, 14] the effects of final state interaction and meson exchange ; Fig. 5 shows different corrections for the third kinematics ; at 500 MeV/c the overall correction is equal to 40 %. Among these corrections the contribution of the deuton pole is ~ 25 %. This last correction which is negligible at small momentum, increases with the emitted proton angle and depends crucially on the incident energy ; at 500 MeV/c with an incident energy of 390 MeV this contribution is seven times bigger than the PWIA estimation [8]. With these corrections and the Brandenburg et al. wave function [9], Laget reproduces the coincidence data up to 600 MeV/c ; but with the same model he underestimates the inclusive cross section at

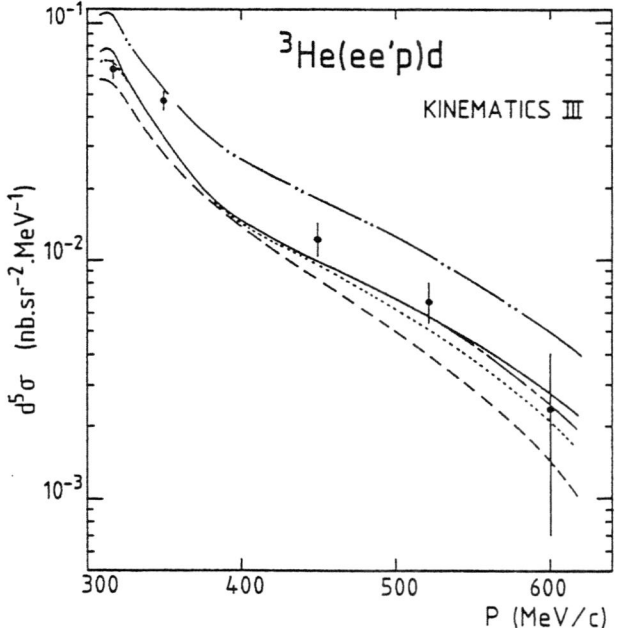

Fig. 5. Measured cross-section : Saclay kinematics III Laget calculations (Brandenburg et al. [9] wave function) ---- PWIA ; ••••• PWIA + deuton pole ; — • — PWIA + deuton pole + FSI ; —— PWIA + deuton pole + FSI + MEC ; — •• — Laget full calculation with Ciofi et al. wave function [11]

$|y| > 300$ MeV by a factor 3. Using the Ciofi et al. wave function he overestimates the coincidence cross-section by a factor two at high momentum, but in this latter case the agreement with the inclusive data is better. Perhaps the calculation of the final state interaction is not accurate enough especially for the inclusive (small ω) experiment. On the other hand the Saclay coincidence experiment was done at $q^2 \sim 1$ fm^{-2} for kinematics III while the inclusive ones shown in III were done at $q^2 \sim 25$ fm^{-2}. At this high momentum transfer relativistic effects can occur in the nuclear medium and the above interpretations cannot be applied simply to these inclusive experiments.

V. Many-body effects in the continuum

We look now for the three-body channel ^{3}He(ee'p)pn. Fig. 6 shows the missing energy spectra obtained at three different proton angles. The

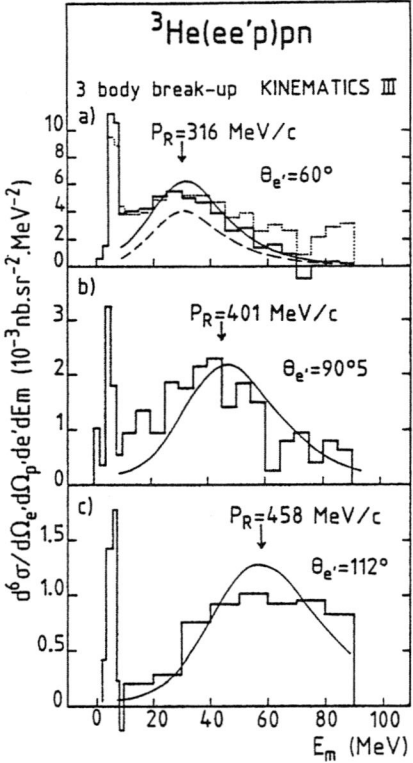

Fig. 6. Missing energy spectra for the three-body break-up : Saclay kinematics III •••• without radiative corrections ; ——— with radiative corrections. Laget calculations with Brandenburg et al. wave function : ---- PWIA ; ——— full calculation.

narrow peak at 5.5 MeV corresponds to the two-body break-up, the recoiling being a bound deuteron. The large bump can be interpreted as the quasi elastic electron scattering on a correlated proton neutron pair, the second proton being spectator. In that case the recoil momentum P_R is taken by the neutron. The recoiling system formed by this neutron and the spectator proton has a mass $M_R^* = \left[(E_n + M_p)^2 - p_R^2 \right]^{1/2}$ with $E_n = (M_n + P_R)^{1/2}$, if we assume that the proton-neutron pair is at rest relative to the spectator proton, in the initial state. The arrows correspond to this calculation. In fact the proton-neutron pair is not at rest and this movement produces the width of the bump. The theoretical curves were computed by J.M. Laget [8, 14]. The dashed curve correspond to PWIA, with the Brandenburg et al. [9] wave function. The solid curve was obtained adding the meson exchange effects and the final state interaction. He obtains a good agreement with the experiment. If we compare the integral of the missing energy spectrum with the calculation of Ciofi for the three-body channel, we find a good agreement in PWIA as for the two-body break-up ; we have not the complete calculation but if the effect of final state interactions and meson exchange currents is similar to the Laget calculation, this will lead to an overestimate of the cross section by a factor ~ 2, the same factor as for the two-body break-up.

VI. Conclusion

In the two-body break-up channel ^{3}He(ee'p)d, we get the proton momentum distribution $\pi(p)$ in ^{3}He until 600 MeV/c. The agreement with current calculations remains within a factor two within six decades for $\rho(p)$. Corrections to PWIA are $\leq$ factor two [15] ; so we can get the momentum distribution with an accuracy ~ 20-30 %. To improve the situation, we need a complete treatment of the three-nucleon system in the continuum including exchange currents.

Above 300 MeV/c the value of $\rho(p)$ obtained from coincidence experiment is smaller by a factor ~ 3 than $\rho(p)$ extracted from inclusve measurements at small ω and high q ; for the two experiments the same model is used for the analysis. In the inclusive experiment the final state interactions are more important than in the coincidence experiment and one has to take relativistic effects into account.

References

[1] E. Jans, P. Barreau, M. Bernheim, J.M. Finn, J. Morgenstern, J. Mougey, D. Tarnowski, S. Turck-Chieze, S. Frullani, F. Garibaldi, G.P. Capitani, E. de Sanctis, M. K. Brussel and I. Sick, Phys. Rev. Lett. $\underline{49}$, 974, (1982).

[2] I.Sick, D. Day and J.S. McCarthy, Phys. Rev. Lett. $\underline{45}$, 970 (1980)

[3] C. Marchand, P. Barreau, M. Bernheim, P. Bradu, G. Fournier, Z.E. Meziani, J. Miller, J. Morgenstern, J. Picard, B. Saghai, S. Turck-Chieze, P. Vernin, M.K. Brussel, Phys. Lett. $\underline{153B}$, 29 (1985).

[4] J.M. Cavedon, B. Frois, D. Goutte, M. Huet, Ph. Leconte, J. Martino, X.-H. Phan, S.K. Platchkov, S.E. Williamson, W. Boeglin, I. Sick, P. de Witt-Huberts, L.S. Cardman and C.N. Papanicolas ; Phys. Rev. Lett., $\underline{49}$, 986 (1982).

[5] F.-P. Juster, S. Auffret, J.-M. Cavedon, J.-C. Clemens, B. Frois, D. Goutte, M. Huet, Ph. Leconte, J. Martino, Y. Mizuno, X.-H. Phan, S. Platchkov, S. Williamson and I. Sick, Phys. Rev. Lett. $\underline{55}$, 2261 (1985).

[6] C.W. Akerlof, W.W. Ash, B. Berkelman and M. Tigner, Phys. Rev. Lett. $\underline{14}$, 1036 (1965).

[7] T. de Forest, Nucl. Phys. $\underline{A392}$, 232 (1985).

[8] J.M. Laget, Phys. Lett. $\underline{151B}$, 325 (1985).

[9] R.A. Brandenburg, Y.E. Kim and A. Tubis, Phys. Rev. $\underline{C12}$, 1368, (1975).

[10] H. Meier-Hajduk, Ch. Hajduk, P.U. Sauer and W. Theis et al. Nucl. Phys. $\underline{A395}$, 332 (1985).

[11] C. Ciofi degli Atti, E. Pace and G. Salmé, Phys. Lett. $\underline{141B}$, 14 (1984).

[12] E. Pace and G. Salmé, Phys. Lett. $\underline{110B}$, 411 (1982).

[13] R. Schiavilla, V.R. Pandharipande and R.B. Wiringa, Nucl. Phys. $\underline{A449}$, 219 (1986).

[14] J.M. Laget, New Vistas in Electronuclear Physics, E. Tomuziak et al., Eds. Plenum Press, New-York, 361 (1986).

[15] J.M. Laget, communication to this Conference.

[16] C. Marchand, M. Bernheim, P. Dunn, A. Gérard, A. Magnon, Z.E. Meziani, J. Morgenstern, J. Mougey, J. Picard, D. Reffay, S. Turck-Chieze, P. Vernin, M.K. Brussel, S. Frullani, F. Garibaldi, G.P. Capitani, E. de Sanctis, to be published.

ELECTRODISINTEGRATION OF ³He STUDIED WITH THE PROTON KNOCKOUT REACTION (e,e'p)

P.K.A. de Witt Huberts

NIKHEF-K, P.O. Box 41882, 1009 DB Amsterdam

Abstract

Recent data of quasi-elastic proton knockout from ³He in various different kinematical conditions are discussed. The question of high-momentum components is addressed up to recoil momentum $p_m \approx 500$ MeV/c in the two body breakup channel. The role of the final state interaction is investigated by measuring the spectral function at $p_m \approx 100$ MeV/c for a range of relative (p - d) kinetic energy in the recoil c.m.-system. We discuss some aspects of the reaction mechanism: virtual photon (γ_v)-deuteron coupling and γ_v-proton coupling. The absolute calibration of the (e,e'p) reaction may be gauged in the elementary ³He system for which realistic wavefunctions, calculated with the Faddeev technique, are available.

1. Introduction

In the trinucleon system the simplest static properties, the binding energy E_b and the rms charge radius, have not been reproduced by three-body theory in a fully satisfactory fashion [1] so far. The observed discrepancies, though rather small, presumably reflect some not well-understood aspects of the trinucleon wavefunction. A highly selective probe of the wavefunction is given by the quasi-free knockout process either inclusive $(e,e')_{qf}$ or the coincidence proton knockout-reaction (e,e'p). In such reactions the momentum probability distribution of nucleons is being probed. A matter of prime importance concerns the amount of high-momentum components in the wavefunction. From the inclusive data on ³He$(e,e')_{qf}$ measured at high electron energies (e = 2.8 - 14.7 GeV) at SLAC [2] and analyzed in a y-scaling approach it has been concluded [3] that an appreciable excess over the Faddeev prediction of nucleon-momentum components at $p_m > 250$ MeV/c exists. The relatively large effects should be easily detectable in the exclusive reaction in the two-body breakup (2bbu) channel $^3He(e,e'p)^2H$ and/or the three-body breakup (3bbu) channel $^3He(e,e'p)pn$. However, the recent Saclay data [4] do not support the presence of an excess of high momentum components up to $p_m \approx 300$ MeV/c. Part of the excess cross section observed in $(e,e')_{qf}$ may be due, at least at relatively small values of momentum transfer ($q \leq 500$ MeV/c), to the coupling of the virtual photon to a (p,n) pair. Clearly, in the exclusive reaction, a reliable treatment of final state interactions (FSI) is of great importance in order to extract from the data unambiguous information on the momentum probability distributions or, equivalently, the spectral functions.

We discuss here a series of recent measurements [5],[6] carried out at NIKHEF-K, of the $^3He(e,e'p)^2H$ and $^3He(e,e'd)p$ reactions that were designed to address the questions raised above. In the 2bbu channel at recoil momentum values $p_m < 300$ MeV/c a special effort has been invested to obtain precise data with an absolute calibration of better than ten percent. This is discussed in section 2. In addition the role of the FSI at $p_m \approx 100$ MeV/c in the 2bbu channel is investigated in a broad range of relative (p-d) kinetic energies (T_{pd}). In section 3 the $^3He(e,e'd)p$ reaction out to high recoil momentum ($p_m = 500$ MeV/c) is presented and the results discussed with consideration of effects beyond the plane-wave impulse approximation. A summary and an outlook to future experimental and theoretical developments are given in section 4.

2. Spectral function for 2 body-breakup ($0 \leq p_m \leq 140$ MeV/c)

Coincidence cross sections have been measured with the dual spectrometer setup at NIKHEF-K [7] at primary electron energies e = 367 and 390 MeV. A specially developed flat target cell [8] containing liquid ^3He at a temperature below 2.2° K and a pressure smaller than 400 mbar was used. With beam loads up to 0.6 W target thickness decreased by less than 10 % from zero-load thickness. The effective thickness was calibrated to ≈ 4 % by comparing measured elastic scattering cross sections with literature values. Repeated calibrations, interleaved with the coincidence measurements, showed excellent stability of typically one percent. Taking account of solid angle calibration (2 %), coincidence efficiency stability (≈ 1 %) and dead time corrections an overall systematic error of 7 % is found. Coincidence data were measured in parallel kinematics with 3-momentum transfer q = 433 MeV/c. Two ranges of recoil momentum (p_m) were covered: i) p_m = 10 - 50 MeV/c at relative (p-d) kinetic energy $T_{pd} = 69.5$ MeV and ii) p_m = 75 - 125 MeV/c at $T_{pd} = 68.5$ MeV. The flat target-cell geometry and the use of a dispersion-matched beam on target resulted in a missing mass resolution of $\Delta E_m = 435$ keV. A typical E_m-spectrum, corrected for radiative effects, is shown in fig. 1 that illustrates the clean separation of 2-body and 3-body breakup channels. The spectral function $S_2(p_m)$ for the 2bbu channel was obtained by dividing the coincidence cross sections by the off-shell electron proton cross section σ^*_{ep} [9] and the phase space factor. Our results for $S_2(p_m)$ are shown in fig. 2 together with the Saclay results [4]. Note

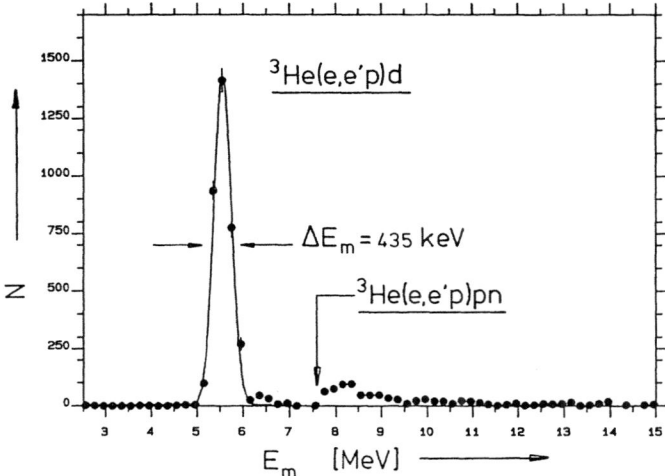

Fig. 1   Missing mass spectrum of the ^{3}He(e,e'p) reaction corrected for radiative effects.

that this is a sensible composite picture of spectral functions since nearly the same kinematics were employed, notably $T_{pd} = 64.7$ MeV in the Saclay data. Good mutual consistency of the data sets is observed.

In the figure two theoretical spectral functions are shown, calculated with the Paris nucleon-nucleon potential [10] and with the Tjon-potential [11]. In the latter approach the correct binding energy E_b of the trinucleon system is achieved and consequently an improved description of the (e,e'p) data is obtained at low p_m ($\leq$ 50 MeV/c), where $S_2(p_m)$ is acutely sensitive to E_b. One notes that both calculations overshoot the data appreciably at p_m-values larger than 50 MeV/c. Most likely the prime mechanism responsible for the major part of the discrepancy is the final state interaction (FSI). Indeed recent developments in the treatment of secondary reaction processes beyond the plane wave impulse approximation indicate non-negligible effects of FSI in the present kinematical domain. Such processes (rescattering in the final state and meson exchange effects) have been calculated by J.M. Laget [12] in a diagrammatic approach. The (p-d) rescattering process, dominant in the (0 - 300) MeV/c range of p_m, leads to a reduction of the plane wave cross section σ_{pw} and thus of the spectral function. Casting the calculations in a sumrule representation $\Sigma(p_m) = 4\pi \int_0^{P_m} S_2(p) p^2 dp$ and treating the data likewise the results shown in fig. 3 are obtained.

Two observations ensue: i) The FSI reduces the integrated spectral function by $\approx$ 20 %, ii) the data fall (15 $\pm$ 10) % below the distorted spectral function calculation. Given the freedom of $\approx$ 5% in the various brands of Faddeev calculations for $S_2(p_m)$, in the prescription for the electron proton off-shell cross section σ_{ep} ($\approx$ 4 %) [9] and upon consideration of the approximations involved in the diagrammatic expansion of the three-body reaction

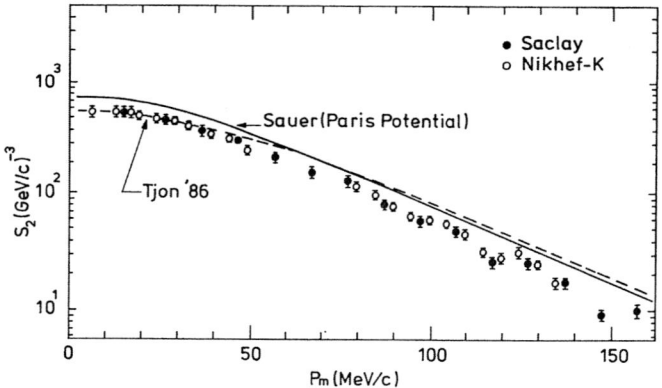

Fig. 2　Spectral function data for 2-body breakup and two Faddeev calculations are shown.

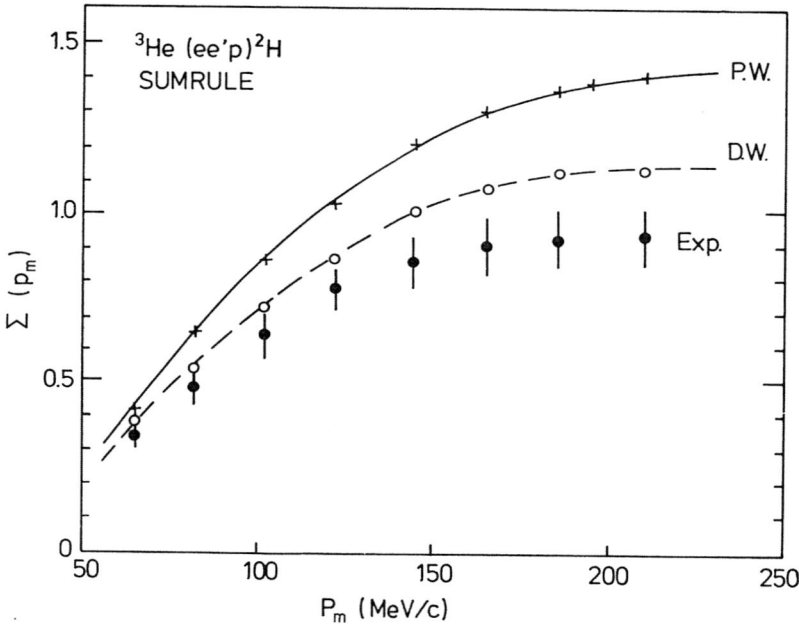

Fig. 3　Sumrule for 2 bbu as a function of p_m: experimental and theoretical (plane wave (PW) and distorted wave (DW).

mechanism, the conclusion is justified that the calculated (e,e'p) cross section is in good agreement with the data.

In order to further investigate the importance of effects beyond the plane wave impulse approximation, three additional measurements were done in which the p_m range covered was kept constant and the relative (p-d) kinetic energy T_{pd} ranged between 23.4 MeV and 103.5 MeV. By fixing the 3-momentum transfer at q = 434 MeV/c and by varying both the energy transfer ω and the angle between initial proton momentum (p) and q three spectral functions were obtained at T_{pd} = 23.4, 68.5 and 103.5 MeV. In heavier nuclides it has been observed experimentally that the energy dependence of the absorption factor R ($R \equiv \sigma_{DW}(e,e'p)/\sigma_{PW}(e,e'p)$) due to the final state interaction follows that of the p-nucleus reactive cross section σ_R. The σ_R decreases by at least a factor two in heavier systems in going from T_{pA} = 30 MeV to T_{pA} = 100 MeV and by approximately 40 % in the (p-d) system. Conversely one might hypothesize that, if the T_{pd} dependence of the spectral function is observed to be small, the final state interaction is small. The data are shown in fig. 4.

Each of the three spectral functions in the range p_m = 75 - 120 MeV was fitted with $S(p_m, T_{pd}) = S(p^0{}_m, T_{pd}) \exp (- \alpha(p_m - p^0{}_m))$ where $p^0{}_m$ = 100 MeV/c. Within the error the same value for α was found $\alpha = - 32 \pm 2 (\text{GeV/c})^{-1}$. The fitted values for $S(p^0{}_m, T_{pd})$ are 52(3), 51(3) and 50(2) $(\text{GeV/c})^{-3}$ for T_{pd} = 23.4, 68.5 and 103.5 MeV, respectively. Within the error of 5 % no energy dependence is observed. This observation is in agreement with the Laget calculation [12] of FSI effects as is shown in fig. 5. A very interesting development has occurred recently in that continuum Faddeev calculations for the electrodisintegration of ^{3}He have become available [13]. As input to the 3-body equations a local S-wave NN potential of the Yukawa type has been used [11]. The preliminary result is shown in fig. 5 as the dash-dotted curve. Significant deviations occur relative to the result of the calculation in which a diagrammatic approach is used. The T_{pd} dependence, however, is again smaller than $\approx 6 \%$, consistent with the empirical observation.

3. Two-body breakup at large recoil momentum (p_m = 225 - 500 MeV)

With the relatively small duty factor of $\approx$ one percent a measurement of the spectral function $S_2(p_m)$ beyond p_m = 300 MeV/c becomes quite difficult due to the level of accidental coincidences. However, by employing a novel recoil detection technique in which the deuteron is detected this impediment can be overcome and the 2bbu spectral function can be mapped out up to p_m = 500 MeV/c. The accidental rate is appreciably reduced (see fig. 6) in the ^{3}He(e,e'd)p channel because the proton singles rate in the hadron spectrometer is suppressed by particle-identification filters derived from the multiple-scintillator stack of the trigger detector. The detection efficiency of the (e,e'd) channel was calibrated by measuring elastic e-^{2}H scattering in coincidence with the recoiling deuteron. The recoil-momentum range p_m = 225 - 500 MeV/c was covered by the (e,e'p) channel from (225 - 320) MeV/c and by the (e,e'd) channel from (315 - 500) MeV/c. The typical kinematics in the two reaction channels

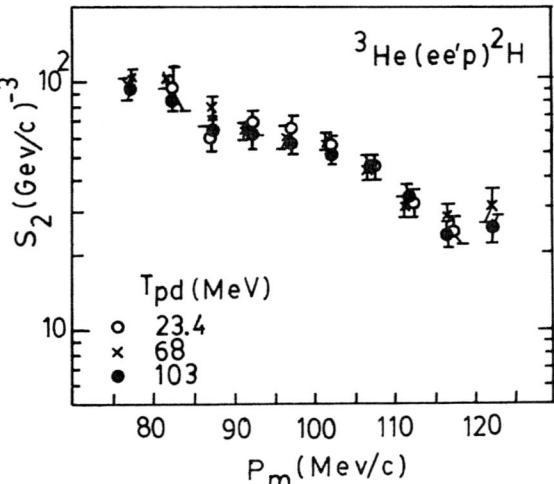

Fig. 4   Dependence of the spectral function on relative (p - d) kinetic energy T_{pd}.

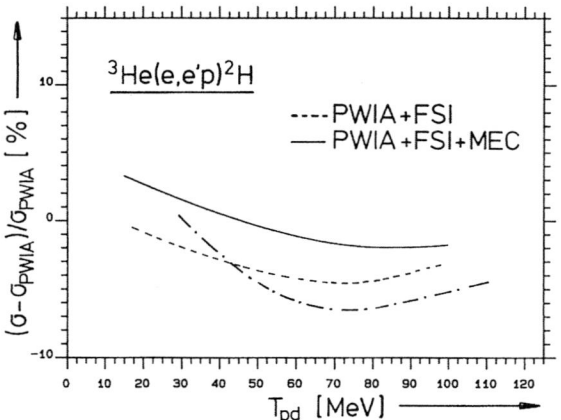

Fig. 5   Calculated effect of the final state interaction on the PWIA spectral function.

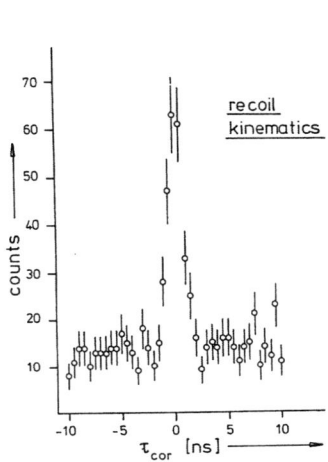

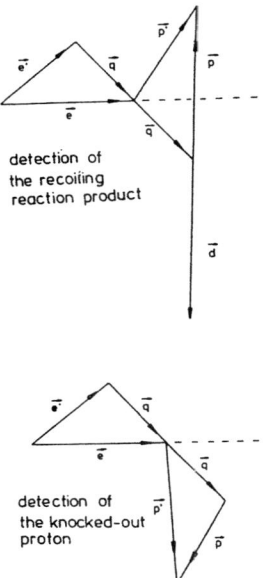

detection of
the recoiling
reaction product

detection of
the knocked-out
proton

Fig. 6   Coincidence timing spectrum in
           the deuteron channel.

Fig. 7   Kinematic diagrams illustrating the
           complementary reactions (e,e'p)
           and (e,e'd).

are illustrated in fig. 7. In order to enhance the cross section a relatively small value of q = 250 MeV/c was selected. The relative T_{pd} was kept near constant at T_{pd} = 92.9 MeV in the entire p_m-range covered.

The spectral function of the ^{3}He(e,e'p)^{2}H reaction obtained from three overlapping kinematics is shown in fig. 8. The agreement with the Saclay data [4] is fair in view of the 8(11) % systematic error of the NIKHEF (Saclay) data. In comparison with the Faddeev-calculation a pronounced quenching of the experimental $S_2(p_m)$ is observed. The composite data-set ((e,e'p) and (e,e'd)) is shown in fig. 9. The notable feature is a change of slope in the falloff of the cross section beyond p_m = 350 MeV/c, suggestive of enhanced high momentum components, if the plane wave impulse approximation (PWIA) would be strictly valid. However, one should be more careful in the interpretation. The two data sets are replotted in figs. 10 and 11 to allow a clear comparison with theory. The PWIA calculation, with photon (γ_v)-proton coupling only, overshoots the data up to p_m = 350 MeV/c but undershoots appreciably at values of p_m > 350 MeV/c (see fig. 10). In a diagrammatic approach Laget [12] has calculated the effects of (γ_v - pn) coupling (i.e. including proper

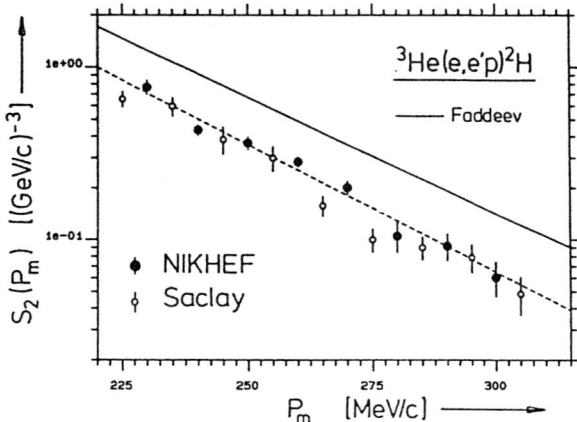

Fig. 8   Spectral function data in the range p_m = 200 - 300 MeV/c.

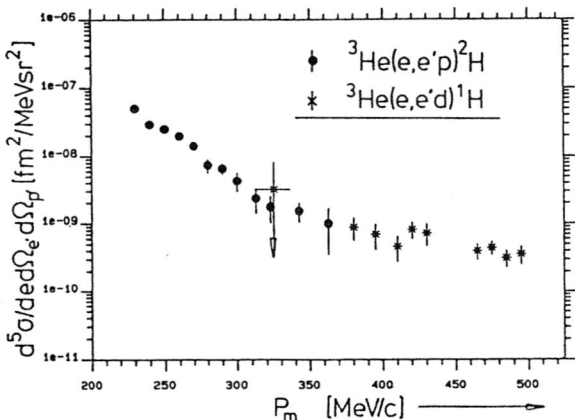

Fig. 9   Cross sections in the high recoil momentum region.

antisymmetrization in the initial and final states), rescattering in the final state (FSI) and meson exchange current (MEC) coupling of the virtual photon. Below p_m < 300 MeV/c the FSI reduces the calculated cross sections appreciably (20 - 50 %). For p_m > 350 MeV/c an inverse effect, due to interfering amplitudes, is observed in the sense that the rescattering process enhances the cross section relative to the PWIA result. In this kinematical domain the role of MEC is not very important in the calculation. The full calculation overshoots the data by a factor two, however. In view of the approximations involved in this approach it would be extremely interesting to avail of continuum Faddeev calculations and to also study the effect of the various brands of spectral functions available in the present kinematic domain.

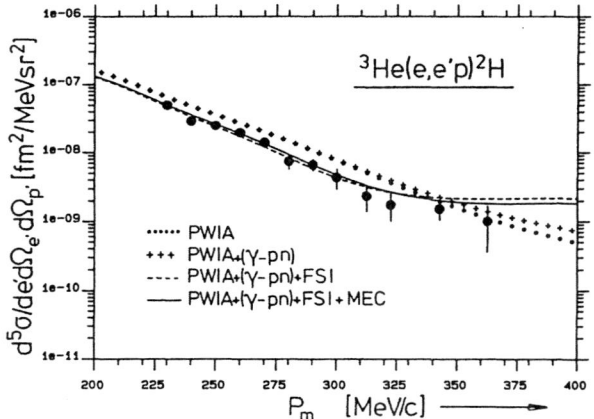

Fig. 10 Differential cross sections at high p_m of the ^{3}He(e,e'p)^{2}H reaction and theoretical distorted wave predictions .

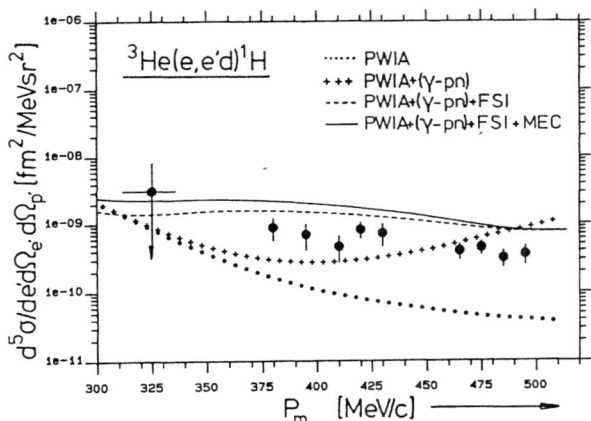

Fig. 11 Differential cross sections at very high p_m of the ^{3}He(e,e'd)p reaction.

4. Summary and outlook

The precise (e,e'p) coincidence data available nowadays provide a selective testground for the calculation of proton spectral functions in the trinucleon systems. In the two-body breakup channel of ^{3}He, that contains 75 % of the total knockout strength, fair agreement with calculations is achieved provided that the Final State Interaction (FSI) is taken properly into account. A very important theoretical development is the ability to perform 3-body continuum Faddeev calculations implying a fundamental treatment of FSI. In the range of recoil momenta larger than 350 MeV/c the cross sections are substantially larger than the PWIA prediction. As suggested by a diagrammatic treatment of the reaction mechanism, the excess strength may be due to rescattering in the final state. At lower recoil momentum the effect FSI has been shown both experimentally and theoretically to be smaller than $\approx$ 10 % in the 2 bbu channel. With the availability of continuum Faddeev calculations and new data for the electro disintegration of ^{3}He (and ^{3}H) in a large kinematic domain penetrating investigations of 3-body dynamics will become possible.

This work is part of the research program of the National Institute for Nuclear Physics and High-Energy Physics (NIKHEF, section K), made possible by financial support from the Foundation for Fundamental Research on Matter (FOM) and the Netherlands Organization for the advancement of Pure Research (ZWO).

References

1. Sauer, P.U., Proceedings of the Miniconference on the study of few-body systems with electromagnetic probes, NIKHEF-K (Amsterdam, 1981), p. 90
2. Day, D. et al., Phys. Rev. Lett. **43** (1979) 1143
3. Sick, I., Helvetica Physica Acta **58** (1985) 746
4. Jans, E., et al., Phys. Rev. Lett. **49** (1982) 974
5. Keizer, P.H.M. et al., Phys. Lett. B, **157B** (1985) 255
6. Keizer, P.H.M., Ph.D. Thesis (University of Amsterdam, 1985)
7. De Vries, C. et al., Nucl. Instr. Meth. **223** (1984) 1
8. Postma, H., et al., Nucl. Instr. Meth. **219** (1984) 292
9. De Forest, T., Jr., Nucl. Phys. **A392** (1983) 232
10. Meier-Hajduk, H. et al., Nucl. Phys. **A395** (1983) 332
11. Malfliet, R. and Tjon, J.A., Nucl. Phys. **A127** (1969) 161 and Tjon, J.A., private communication (1986)
12. Laget, J.M., Phys. Lett. **151B** (1985) 325
13. Meygaard, E. van and Tjon, J.A., to be published

PHOTODISINTEGRATION OF LIGHT NUCLEI
WITH THE LADON PHOTON BEAM

LADON collaboration*, presented by
S. d' Angelo
Dipartimento di Fisica – II Università di Roma
Istituto Nazionale di Fisica Nucleare – Sez. di Roma

ABSTRACT. – In this paper we present a short review of experimental data on the photodisintegration of deuteron and ^{4}He obtained with the monochromatic LADON photon beam.

INTRODUCTION

Nuclear forces and nuclear structure can be usefully explored by means of photonuclear reactions. The LADON beam obtained at the Frascati National Laboratories is a good source of monochromatic and polarized photons (see Table I). In particular it has been used to study two few body physics problems:

1) to examine the limits of validity of the current theories on nucleon-nucleon interaction by means of a high precision measurement of the total cross section for the

---

* D. Babusci, P. Belli, R. Bernabei, L. Casano, S.d' Angelo, M. P. De Pascale, G. Giordano, B. Girolami, A. Incicchitti, M. Mattioli, P. Picozza, D. Prosperi, C. Schaerf.

deuteron photodisintegration;

2) to test the charge symmetry in light nuclei by means of a new measurement of the $^4He(\gamma,p)^3H$ total cross section.

TABLE I – Main features of the LADON γ-ray beam.	
Energy range available	$\sim (10 \div 80)$ MeV
Energy resolution tipically:	3% at 15 MeV
	9% at 80 MeV
Bremsstrahlung background	$\sim 5\%$ on the whole spectrum
Intensity	$\sim 10^5 \gamma/s$ with 7.8 mm collimator
Polarization	$\geq 99\%$ linear

EXPERIMENTAL SET-UP

The same experimental set-up was used for both experiments. It is an integrated system composed of a gas container, surrounded by a 6.75-cm-thick NE213 liquid scintillator, viewed by two photomultipliers. The gas container is a 0.16-mm-thick aluminum tube (15 cm in lenght and 1.5 cm in diameter) with end caps of Lexan; a mechanical device maintains both gas and scintillator at the same pressure (typical operating pressure is about 30 bar). A schematic diagram of the apparatus is shown in Fig. 1.

The main trigger was provided by a threefold coincidence: a signal indicating the passage of the electron bunch through the laser cavity and the two pulses from the photomultipliers of the proton detector. A pulse shape discrimination was realized recording the total integrated current pulses from the two photomultipliers, their sum and the corresponding tail integrals. A pulse shape analyzer (PSA) jointly with a Time to a Digital Converter allowed also a further off-line p/γ discrimination.

A NaI(Tl) crystal (10 in. in lenght and 10 in. in diameter), whose threshold stability was periodically checked, was used to monitor the beam; its efficiency was calculated by means of a standard MonteCarlo program for electromagnetic showers. The photon beam energy profile, measured by a magnetic pair spectrometer, was continuously recorded.

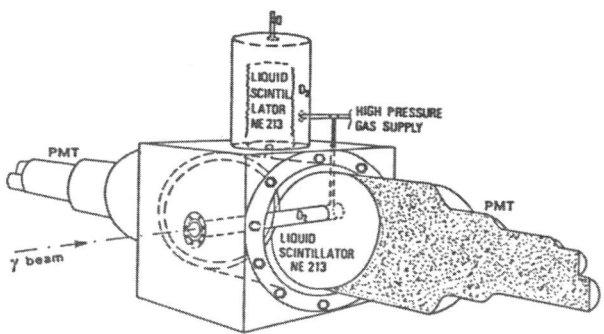

Fig. 1 – Schematic diagram of the target and detector system.

DEUTERON PHOTODISINTEGRATION

A recent critical review by Hadjimichael and Saylor [1] on a selected set of low energy experimental data on deuteron photodisintegration, suggests that these data do not completely fit with traditional theory. To explain the observed discrepancies they invoke the presence of quark degrees of freedom. But, studies made by Arenhövel [2] on unselected experimental data show that the standard theory including MEC and IC contributions works satisfactorily. To solve this dilemma one needs a measurement of the total cross section for deuteron photodisintegration with higher statistical and systematic accuracy. For this purpose we performed an high precision measurement using the LADON monochromatic photon beam [3], the described experimental apparatus and a data analysis procedure which gives us an high proton detection efficiency.

The electromagnetic background was rejected by using PSA and head/tail information. Moreover, measurements with an hydrogen filled target and a bremsstrahlung beam were performed to subtract the events coming, respectively, from spurious countings due to photon interactions with the target structure and those coming from the bremsstrahlung contamination.

TABLE II – Our experimental results for the ^{2}H(γ,n)p total cross section σ_{exp}, theoretical data σ_{th} of Cambi et al., ratio σ_{exp}/σ_{th} and systematic errors $\Delta\sigma_{exp}^{sys}$ as a function of the laboratory γ-ray energy (MeV).				
E_γ (MeV)	σ_{exp} (μb)	σ_{th} (μb)	σ_{exp}/σ_{th}	$\Delta\sigma_{exp}^{sys}$ (μb)
14.7 ± 0.1	925 ± 20	900.3	1.027 ± 0.022	44
19.3 ± 0.1	617 ± 9	627.3	0.984 ± 0.014	31
28.9 ± 0.1	361 ± 6	356.5	1.013 ± 0.017	12
38.2 ± 0.1	249 ± 3	239.1	1.041 ± 0.013	9
47.5 ± 0.2	177 ± 3	174.7	1.013 ± 0.017	6
57.5 ± 0.4	139 ± 3	133.1	1.044 ± 0.023	5
74.0 ± 0.5	97.6 ± 5.3	94.0	1.038 ± 0.056	3

The efficiency of the proton detector was estimated by a standard MonteCarlo calculation, assuming for the photoproton angular distribution the theoretical estimates of Cambi et al. [4]. This efficiency is about 80% at photon energy of 14.7 MeV and higher than 95% at photon energies equal or greater than 20 MeV.

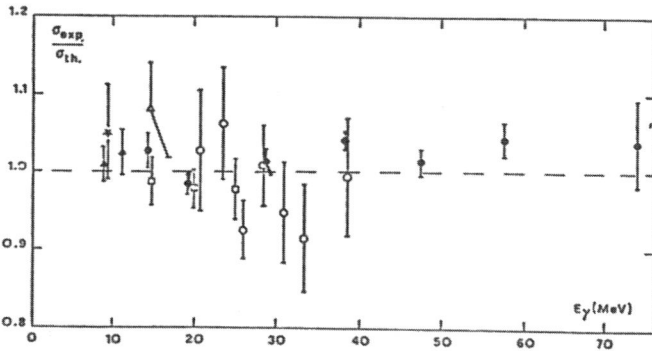

Fig. 2 – σ_{exp}/σ_{th} as a function of the photon energy. Filled circles, our experiment; open squares Ahrens [5]; filled triangles, Birenbaum et al. [6]; open circles, Bosman et al. [7]; open triangles, Stiehler et al. [8]; stars, Tudoric-Ghemo [9].

Our experimental results are shown in Table II. In Fig. 2 is also reported the ratio between our experimental cross section, σ_{exp}, and the corresponding theoretical data, σ_{th}, together with recent experimental results obtained with monochromatic beams.

Our measurements confirm, within the obtained accuracy, the substantial validity of current theory between 14.7 and 74.0 MeV.

$^4\text{He}(\gamma,\text{p})^3\text{H}$

Recently, some authors as Calarco et al. [10], made a crytical review of the available experimental data on the total cross section for the ^{4}He two body disintegration, suggesting a possible violation of the charge symmetry that cannot be justified only taking into account electromagnetic effects.

In order to make an estimate of this violation, it is very useful to consider the ratio, R_γ, of the photoproton and photoneutron cross sections for ^{4}He, which is very sensitive to the degree of isospin mixing in the nucleus. Current theories give for this ratio a value close to unity [11,12,13]. Up to now a lot of experiments [14] have been done but the results are not conclusive, since they are not in complete agreement. The controversy was opened by Berman, Fulz and Kelly [15] in 1970. They measured the photoneutron cross section with monochromatic photons and, below 30 MeV, they found values lower by about 50% than the most reliable data on the photoproton cross section. On the other side, different data of the photoproton cross section are in substantial disagreement, with variations of about 50 % at energies between threshold and 35 - 40 MeV. In this framework, Calarco et al. [10], following their critical analysis, evaluate a behaviour for the photoproton and the photoneutron cross sections that gives $R_\gamma \simeq 1.7$ around 25 MeV photon energy.

Bearing in mind this situation, we have planned our experiment to obtain a reliable value for the total photoproton cross section as little affected by systematic errors as possible. We used the same experimental apparatus and analysis procedure as described above. To eliminate background contaminations due to interactions with collimators and Lexan caps, we performed in this case further measurements with a ^{4}He target at about 1 atm. Moreover, spurious contributions were eliminated by subtracting data obtained with bremsstrahlung background.

The software analysis selects events using head/tail and PSA discrimination and subtracting events coming from bremsstrahlung and interactions with the structure of the target. The detector efficiency is calculated using a MonteCarlo procedure, that also allows us to study the most suitable threshold on the proton energy spectrum to eliminate contributions from other channels.

In Fig. 3 we report some very preliminary results † together with the Calarco et al. [10] evaluation of ^{4}He$(\gamma,p)^3$H and ^{4}He$(\gamma,n)^3$He cross sections.

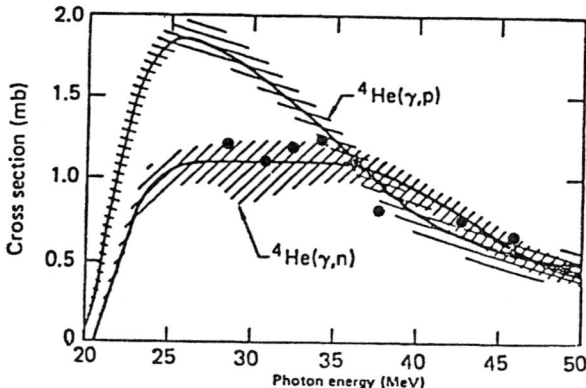

Fig. 3 – Evaluated ^{4}He$(\gamma,p)^3$H and ^{4}He$(\gamma,n)^3$He cross sections from Calarco et al. [10]; superposed filled circles are our preliminary results on ^{4}He$(\gamma,p)^3$H cross section.

In Fig. 4 we show the values of R_γ from our experiment ‡ together with the data of Phillips et al. [16], Dodge et al. [17] and F. Balestra et al. [18] (the existing direct measurements of R_γ) superposed on Calarco et al. evaluation.

Furthermore, a recent measurement [19] on π-^{4}He inelastic cross sections at $T_\pi = 180$ MeV with $\theta_{lab} = 30°$, gives a value $R_\pi = \sigma(\pi^+)/\sigma(\pi^-) = 1.05 \pm 0.08$ — averaged on the region of 1^- states ($E_x = 23 \div 30$ MeV) — while, if R_γ assumes the Calarco value, R_π is expected to be about 2.9. Therefore, this result indirectly supports our preliminar results.

In conclusion, the arguments favouring a large symmetry-breaking nuclear force

† Therein only statistical errors are reported. Additional systematic errors of $\approx 7\%$ are to be considered.

‡ For this purpose, we use the ^{4}He$(\gamma,n)^3$He cross section as evaluated by Calarco et al. [10].

and based on the previous ^{4}He photoproduction experiments need to be reexamined.

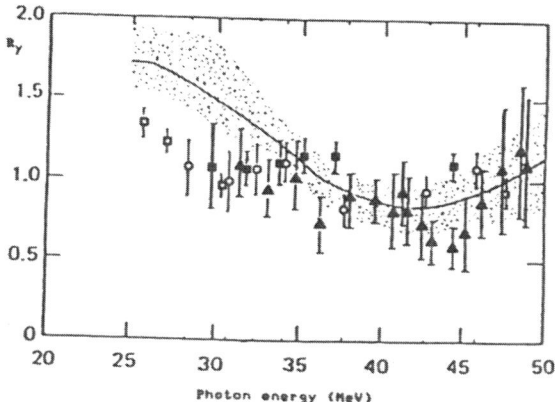

Photon energy (MeV)

Fig. 4 – R_γ as a function of the photon energy. Open circles, indicate this experiment; filled triangles, Phillips and al. [16]; filled squares, Dodge and Murphy [17]; open squares, Balestra et al. [18].

REFERENCES

[1] – E. Hadjimichael and D.P. Saylor, Phys. Rev. Lett. 45, 1776 (1980).

[2] – H. Arenhövel, Phys. Rev. Lett. 47, 749 (1981).

[3] – R. Bernabei, A. Incicchitti, M. Mattioli, P. Picozza, D. Prosperi, L. Casano, S. d'Angelo, M.P. De Pascale, C. Schaerf, G. Giordano, G. Matone, S. Frullani and B. Girolami, Phys. Rev. Lett. 57, 1542 (1986).

[4] – A. Cambi, B. Mosconi and P. Ricci, Phys. Rev. C26, 2358 (1982); J. Phys. G.: Nucl. Phys. 10, L11 (1984); private comunication.

[5] – J. Ahrens, H.B. Eppler, H. Gimm, M. Kröning, P. Riehn, H. Wäffler, A. Zieger and B. Ziegler, Phys Lett. 52B, 49 (1974).

[6] – Y. Birenbaum, S. Kahane and R. Moreh, Phys. Rev. C32, 1825 (1985).

[7] – M. Bosman, A. Boll, J.F. Gilot, P. Leleux, P. Lipnik and P. Macq, Phys. Lett. 82B, 212 (1979).

[8] – T. Stiehler, B. Kühn, K. Möller, J. Mösner, W. Neubert, W. Pilz and G. Schmidt, Phys. Lett. 151B, 185 (1985).

[9] – Tudoric-Ghemo, Nucl. Phys. A92, 233 (1967).

[10] – J.R. Calarco, B.L. Berman and T.W. Donnelly, Phys. Rev. C27, 1866 (1983).

[11] – A.M. Chung, R.J. Johnson and T.W. Donnelly, Nucl. Phys. A235, 1 (1974).

[12] – J.T. Londergan and C.M. Shakin, Phys. Rev. Lett. 28, 1729 (1972).

[13] – D. Halderson and R.J. Philpott, Phys. Rev. C28, 1000 (1983); Nucl. Phys. A359, 365 (1981).

[14] – For the $^4He(\gamma, p)$ cross section from photoabsorption and from capture reaction:

   – Yu.M. Arkatov, P.I. Vatset, V.I. Volshchuk, V.V. Kirichenko, I.M. Prokhorets and A.F. Khodyachikn, Sov. J. Nucl. Phys. 12, 123 (1971);

   – A.N. Gorbunov, Phys. Lett. 27B, 436 (1968); Proc. P.N. Lebedev Phys. Inst. [Acad. Sci. USSR] 71, 1 (1976);

   – F. Balestra, E. Bollini, L. Busso, R. Garfagnini, C. Guaraldo, G. Piragino, R. Scrimaglio and A. Zanini, Nuovo Cim. 38A, 145 (1977);

   – H.G. Clerc, R.J. Stewart and R.C. Morrison, Phys. Lett. 18, 316 (1965);

   – V.P. Denisov and L.A. Kul'chitskii, Sov. J. Nucl. Phys. 6, 318 (1968);

   – G.D. Wait, S.K. Kundu, Y.M. Shin and W.F. Stubbins, Phys. Lett. 33B, 163 (1970);

   – J. Sanada, M. Yamanouchi, N. Sakai and S. Seki, J. Phys. Soc. Jpn. 26, 850 (1969);

   – R. Mundhenke, R. Kosiek and G. Kraft, Z. Phys. 216, 232 (1968);

   – J.E. Perry, Jr. and S.J. Bame,Jr., Phys. Rev. Lett. 99, 1368 (1955);

   – W.E. Meyerhof, M. Suffert and W. Feldman, Nucl. Phys. A148, 211 (1970);

   – D.S. Gemmell and G.A. Jones, Nucl. Phys. 33, 102 (1962);

   – C.C. Gardner and J.D. Anderson, Phys. Rev. 125, 626 (1961);

   – R.C. McBroom, H.R. Weller, S. Manglos, N.R. Roberson, S.A. Wender, D.R. Tilley, D.M. Scopik, L.G. Arnold and R.G. Seyler, Phys. Rev. Lett. 45, 243 (1980).

   – see also ref. 9 and ref. 16.

   – For the $^4He(\gamma, n)$ cross section from photoabsorption and from capture reaction:

   – B.L. Berman, D.D. Faul, P. Meyer and D.L. Olson, Phys. Rev. C22, 2273 (1980);

   – L. Ward , D.R. Tilley, D.M. Skopik, N.R. Roberson and H.R. Weller, Phys. Rev. C24, 317 (1981).

[15] – B.L. Berman, S.C. Fultz and M.A. Kelly, Phys. Rev. C4, 723 (1971).

[16] – T.W. Phillips, B.L. Berman, D.D. Faull, J.R. Calarco and J.R. Hall, Phys. Rev. C19, 2091 (1979).

[17] – W.R. Dodge and J.J. Murphy II, Phys. Rev. Lett. 28, 839 (1972).

[18] – F. Balestra, E. Bollini, L. Busso, R. Garfagnini, C. Guaraldo, G. Piragino, R. Scrimaglio and A. Zanini, Nuovo Cim. 38A, 145 (1977).

[19] – C.L. Blilie, D. Dehnhard, D.B. Holtkamp, S.J. Seestrom-Morris, S.K. Nanda, W.B. Cottingame, D.Halderson, C.L. Morris, C. Fred Moore, P.A. Seidl, H. Ohnuma and K. Maeda, Phys. Rev. Lett. 57, 543 (1986).

FUTURE EXPERIMENTAL DEVELOPMENTS FOR FEW BODY PHYSICS AT NIKHEF-K

C. de Vries

National Institute for Nuclear Physics and High Energy Physics
P.O. BOX 41882, 1009 DB Amsterdam

Abstract

An outline is given of the future research program with the 500 MeV electron scattering facility at NIKHEF-K. Exclusive experiments in which recoil particles are detected require new technical developments involving a thin-window target and a special purpose detector. Triple-coincidence work, neutron knock-out and out-of-plane reaction studies need a large solid angle hadron detector and a neutron time-of-flight channel. This program will be further expanded after the proposed energy and duty factor increase of the present accelerator.

1. Introduction

At NIKHEF-K with its high figure of merit instrumentation for - in particular - coincidence experiments, future research will be focussed to a large extent on exclusive experiments. These experiments will be made possible not only through refinement of the present tools but also by implementation of new detection systems and (if financed by the government) by the planned energy and duty factor improvements of the accelerator (MEA).

The research program - in as much few-body systems are involved - will be elaborated upon in this paper. After a brief summary of the NIKHEF-K facility its present figure of merit for exclusive experiments is discussed. Some recent improvements of the present detection systems are indicated. Then a presentation is given of the importance of two-spectrometer experiments in which - in coincidence with the scattered electron - the recoil particle rather than the directly knocked out particle is detected . The recoil detector scheduled to be installed shortly is discussed. Next a description is given of i) a proposed large solid angle hadron detector which in conjunction with the existing spectrometers will allow triple-coincidence experiments and out-of-plane measurements and ii) a proposed neutron time-of-flight detector enabling to study for instance the (e,e'n) reaction in comparison with the (e,e'p) reaction and the virtual photon-neutron coupling in nuclear matter.

The physics requiring the above mentioned improvements in experimental techniques is envisaged still with a beam duty factor of 1-2%. A major step forward will be made if this duty factor is increased to 100 %. This talk will therefore be concluded with mentioning the 1984 proposal for a pulse-stretching device and, simultaneously, the increase of the maximum energy of the accelerator to 700 MeV.

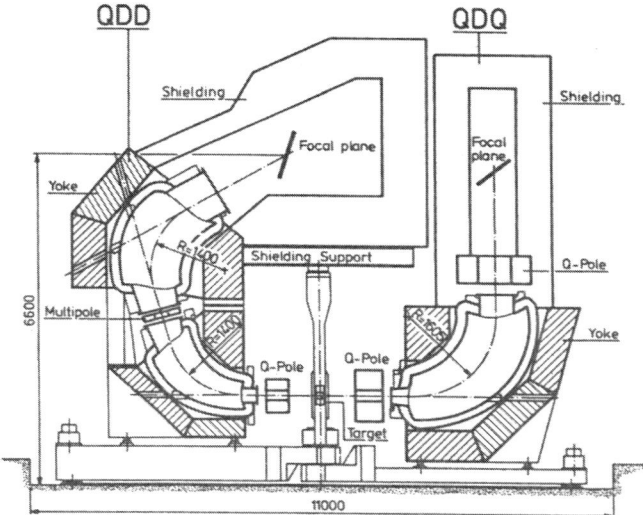

Fig. 1 Lay-out of the two-spectrometer setup, showing the rotatable platforms, magnet configuration and shielding. Also indicated are the scattering chamber and slit systems. The beam enters perpendicularly to the plane of the drawing. Measures are in mm.

2. Figure-of-merit for coincidence experiments

The presently available equipment for exclusive experiments consists of a pair of high-resolution spectrometers (configurations QDD and QDQ) in conjunction with the linear electron accelerator MEA (E_{max}= 500 MeV, duty factor 1-2%)

The characteristics of the facility are fully described in [1]. The two spectrometers and the lay-out of the detector telescopes in the focal planes are depicted in Fig. 1 and 2, respectively. The quality factor R, the ratio of true and accidental coincidences, for (e,e'p) experiments leading to discrete final states - is governed by the missing energy resolution (ΔE_m), the coincidence timing resolution (Δt), and the beam duty factor (d.f.). Fig. 3 displays the corresponding figure of merit of the NIKHEF-K facility in comparison with those of facilities elsewhere. A large number of exclusive (e,e'p) experiments fully employing the possibilities presented by this figure of merit has been performed. Referring to the literature for a survey of those experiments on ^{12}C, ^{27}Al, ^{51}V, ^{90}Zr and $^{206,208}Pb$ [2], I merely show here (Fig. 4) the missing energy (E_m) spectrum of the reaction $^{27}Al(e,e'p)^{26}Mg$ which underlines the quality of the data.

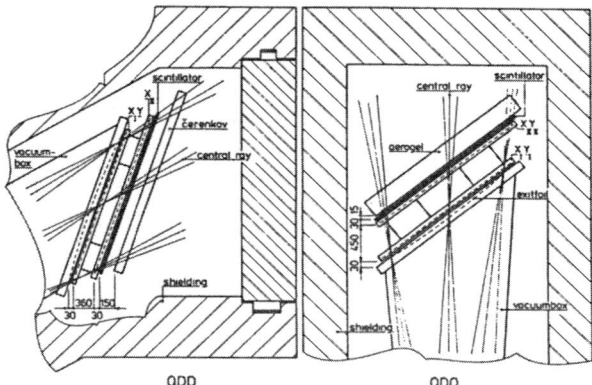

Fig. 2   Lay-out of the focal plane detection systems, showing the location of wire chambers (X_1, X_2, Y_1, Y_2), sinctillators and Cherenkov detectors. The focal plane coincides with the X_1 plane. Central $(p = p_0)$ and extreme $(p = p_0 \pm 5\ \%)$ rays in the spectrometers have been indicated.

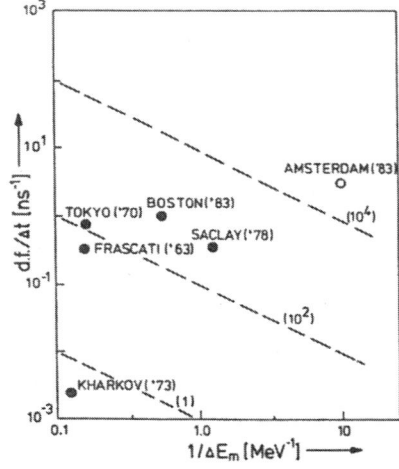

Fig. 3   Figures of merit of intermediate energy (e,e'p) facilities.

----- Lines of constant R: R is defined as the ratio between true and accidental coincidences

$$(\approx \frac{\text{beam duty factor}}{\Delta E_m . \Delta \tau})$$

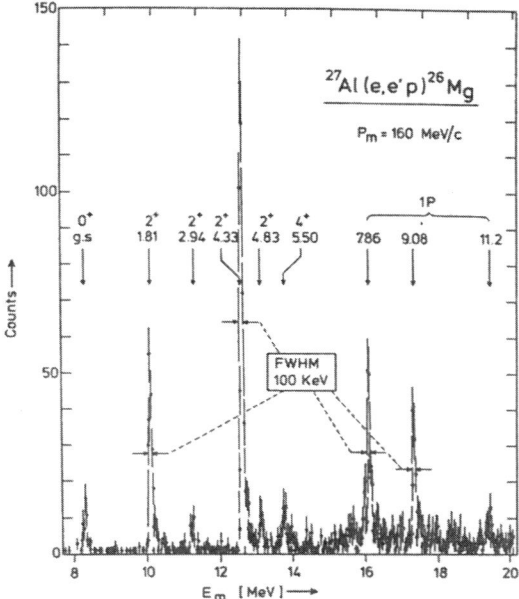

Fig. 4 Missing-energy (E_m) spectrum of the reaction ^{27}Al(e,e'p)^{26}Mg.

It should be mentioned that the extremely good missing energy resolution shown (100 keV) can only be obtained with thin (10 - 20 mg/cm^2) targets. As will be apparent below an important part of the future program requires the use of cryogenic gas targets with extensions in the beam direction of several cm. To suppress detoriation of the missing energy resolution, a fourth wire chamber has been installed to improve the angular resolution of the QDD spectrometer from 8 to 3 mrad. Hence, the contribution of the effect of kinematical broadening($E/M \sin \theta \Delta \phi$) to the widths of the peaks in the energy spectrum measured with the QDD spectrometer can be much better corrected for.

For coincidence experiments with extended targets the main advantage of this improvement is that the rejection rate for accidental coincidences will be highly improved. Results of tests in which - with a thin target and controlled displacements of the beam - target extension in the beam direction has been simulated demonstrate this (Fig. 5). All coincidence events for which the target position of both particles (obtained through backtracing the two position-direction focal plane vectors) do not match well within 2.5 mm can be unambiguously assigned to be accidental events.

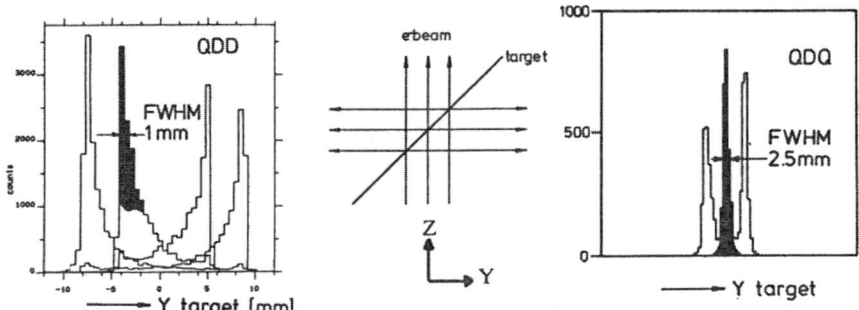

Fig. 5   Sensitivity of QDD and QDQ response as function of position of the (simulated) Z-position of the particles at the target.

3. Developments for recoil particle detection

We elaborate now on the importance of experiments where one detects the scattered electron in coincidence with the recoil particle rather than with the knock-out proton. One of those experiments, namely ^{3}He(e,e'd)^{1}H, has been executed already. It is discussed by De Witt Huberts [3] at this conference. Due to the more favorable true over accidental ratio, direct detection of the deuteron instead of the proton enabled to increase the region of the measured proton momentum distribution from 300 MeV/c to 500 MeV/c. On the basis of this succesfull experiment NIKHEF-K has been encouraged to focus part of its near future program on the study of few-body problems requiring the detection of recoil particles. Some experiments of this kind will be elucidated .

Due to its closed-shell character ^{4}He is of particular interest for the study of short-range correlations. It is tightly bound (E_{sep} = 20 MeV, rms radius = 1.672 ± 0.025 fm) and it features a pronounced central depression in its point proton density [4]. The latter effect may find its origin in the repulsive NN interaction at short distances [5] causing significant modifications of the standard non-correlated momentum distributions above, say , 300 MeV/c.

The proton momentum distribution has been investigated with the (p,2p) reaction [6]. Calculations using Eckart parametrizations of the wave function systematically underestimate the (p,2p) data at higher momenta, even if MEC effects and spin-orbit terms in the optical potential are taken into account. Electron-induced proton-knockout experiments [7] have been performed mapping out the momentum density distribution to 250 MeV/c, albeit with moderate resolution (10 - 15 MeV in the missing energy).

At NIKHEF-K (e,e'p) data have been taken with 350 keV missing energy resolution. Fig. 6 shows the spectral function for p_m = 40 MeV/c. A remarkable feature is the small cross section observed where the three- and four body breakup continua start. Although data have been collected up to 350 MeV/c the real-to-random coincidence ratio seriously hampers

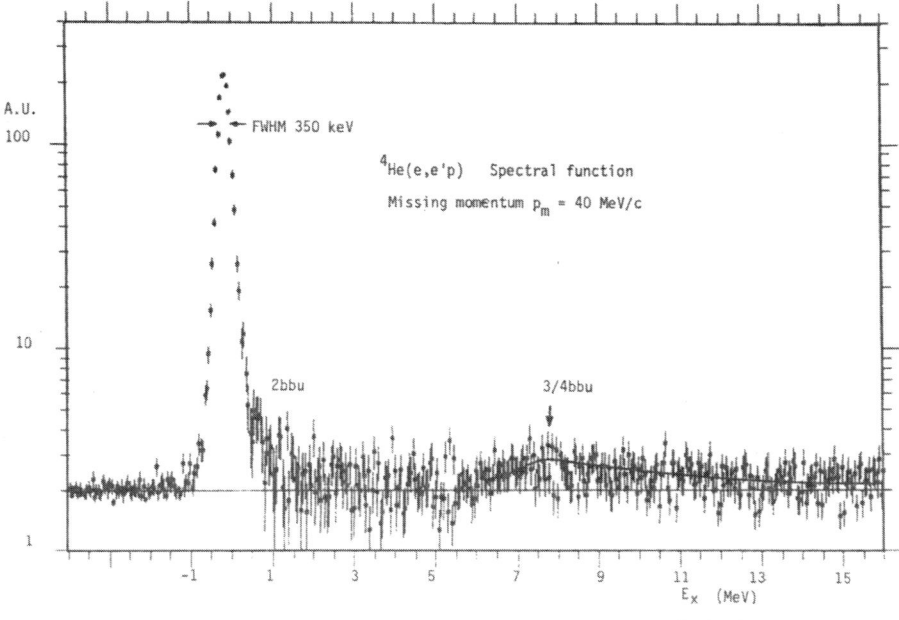

Fig. 6 ^{4}He(e,e'p) spectral function.

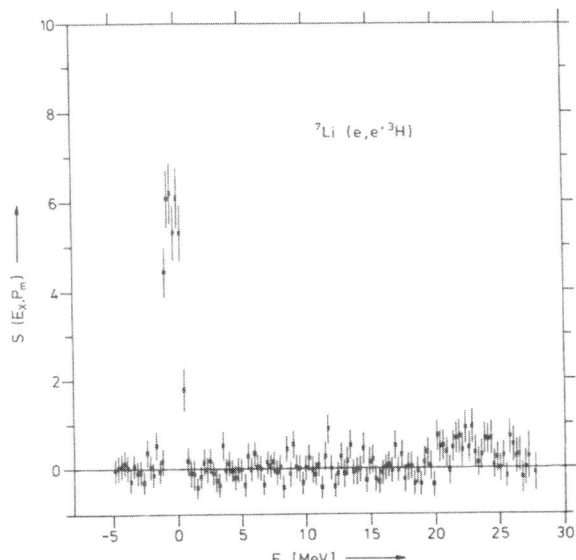

Fig. 7 ^{7}Li(e,e'^{3}H) excitation energy spectrum.

the (e,e'p) experiment. Therefore the ^{4}He(e,e'^{3}H)p reaction has to be studied instead of the ^{4}He(e,e'p)^{3}H reaction because it features a much better real-to-random ratio, like for the ^{3}He(e,e'd) experiment mentioned before.

Other interesting reactions, the study of which require the recoil detection technique, are i) ^{4}He(e,e'^{3}He)n and ^{4}He(e,e'd)d, providing important information about the neutron and the deuteron momentum distribution in ^{4}He and ii) ^{3}He(e,e'^{3}H)π^+ and ^{3}He(e,e'^{3}He)π° providing knowledge about pion production processes.

From the above given examples it is obvious that essential new information can be expected about ingredients of few-body systems through high figure-of-merit electron scattering coincidence experiments in which the hadrons ^{2}H, ^{3}H and ^{3}He are detected. As a first step to tune the facility for recoil detection, the air between the different multiwire proportional counters of the QDQ detection assembly has been replaced by helium. This reduces the multiple scattering effect enabling to detect tritons with much lower kinetic energy (12 MeV corresponding to 260 MeV/c recoil momentum) under accurate time-of-flight correction conditions. Fig. 7 shows an excitation energy spectrum of the ^{7}Li(e,e'^{3}H)^{4}He reaction.

In the analysis required to obtain this spectrum the tritons have been filtered out of a majority of protons and deuterons by means of pulse height discrimination in the two layers of scintillators which form part of the detection telescope. The spectrum shows at $E_m = 0$ MeV the peak corresponding with the ^{7}Li(e,e'^{3}H)^{4}He reaction for 19 MeV triton kinetic energy and $p_m = 90$ MeV/c and strength due to triton knock-out of the ^{4}He core of ^{7}Li above $E_x = 19$MeV.

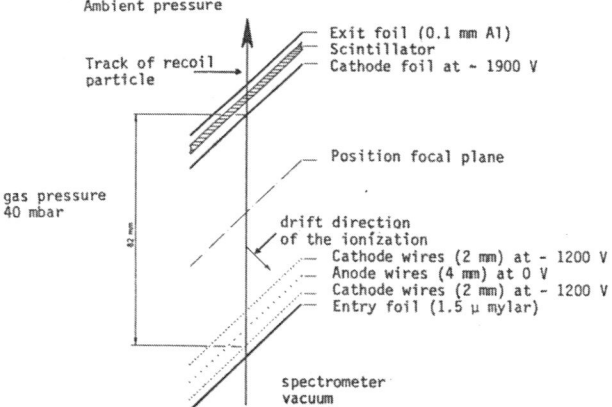

Fig. 8   The low-pressure recoil detector. In the driftspace, formed by the cathode foil and the wire planes, charge drifts from the track towards the MWPC structure at the bottom. The anode signals and the light output of the scintillator are analyzed to obtain the position and direction of the track, and to distinguish between different particles.

Detection of the recoil particles, ^{3}H and ^{3}He, with much lower kinetic energies requires a special purpose detector. A thin-window (4 μ mylar), low pressure (40 mbar) detector will be installed inside the (extended) vacuum chamber of the QDQ (Fig. 8). Multiple straggling, pulse heigth signals from the recoil detector and light output from the scintillator used for total energy determination have been investigated. The results show that, under good particle discrimination and angular resolution ($\approx$ 1°) properties, recoil particles (^{3}H, ^{3}He and ^{4}He) with extremely low kinetic energies (from 2 - 4 MeV) can be observed with this detector. The detector is under construction and will be tested on-line early 1987.

Obviously the effort to build the above described detector for recoil particles should be matched by the construction of a ^{4}He(^{3}He) target allowing also low kinetic energy recoil particles to emerge into the spectrometer without unduly disturbance of their properties (energy and direction). Fig. 9 shows the cryogenic gas target constructed for this purpose.

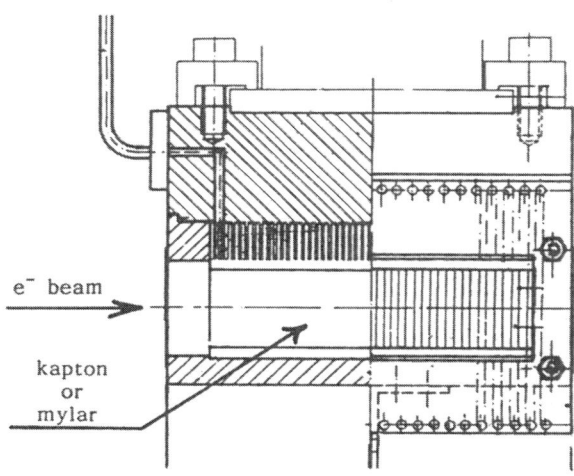

e$^-$ beam

kapton
or
mylar

Fig. 9 The cryogenic gas target for recoil particle experiments.

A mini-cooler maintains its temperature at T = 20 K allowing a heat dissipation of 10W. Its extension in the beam direction is about 10 cm (thus avoiding contributions from the copper end foils which are not "seen" by the spectrometers). The dimension perpendicular to the beam is 25 mm to limit straggling of the recoil particles in the ^{4}He gas itself. For the same reason a very thin target foil is required. Kapton foil (7 μ) supported by stainless steel wires showed to withstand gas pressures of 4 bar. For low kinetic energy ^{3}He particles, requiring even a thinner window, a 1.5 μ mylar window also supported by wires has shown to work reliable at a working pressure of 0.25 bar. With the numbers quoted experiments involving tritons with ≥ 2 MeV and recoil ^{4}He with ≥ 4 MeV will be feasible. The extension of the target in the beam direction will not influence unfavorably the real-to-random ratio, as has been pointed out before. First in-beam tests will take place shortly.

Succesful implementation of both the recoil detector and the cryogenic gas target will strongly influence the research at NIKHEF-K within the coming years. Essential information will then become available about momentum distributions of nucleons up to 500 MeV/c through the investigation of ^{4}He(e,e'^{3}H) and ^{4}H(e,e'^{3}He) reactions and about the fundamental pion production processes from ^{3}He(e,e'^{3}H)π^+ and ^{3}He(e,e'^{3}He)π° reaction studies [*].

4. Proposed hadron and neutron detection channels

In addition to the investigation of few-body systems discussed sofar NIKHEF-K will embark on programmes involving

- the investigation of N-N correlations and the role of meson exchange and isobar currents via (e,e'p) measurements in an out-of-plane configuration
- triple coincidence reactions of the type (e,e'pp), (e,e'pn) and (e,e'pπ) to study correlations and the excitation of delta's in nuclei
- the (e,e'n) reaction to study the quasi-free neutron knockout by direct detection of the neutrons.

These programmes require the installation of two new detection channels in conjunction with the QDD and QDQ spectrometers: a large solid angle scintillator telescope and a neutron time-of-flight detector. The use of such non-magnetic field particle detection channels have been hampered so far by the notoriously high background in electron scattering halls for intermediate energy physics. Encouraged, however, by the background conditions of our spectrometer hall - due to the careful design of the beam handling system as well as the beam dump area - tests have been performed establishing the feasibility to use such channels for the purposes mentioned above. Further tests on small-scale prototypes of the detectors will be carried out shortly.

---

[*] *Obviously these experiments will benefit from the availability of higher beam energies. For this purpose an upgrading program involving the increase of the peak r.f. power of the accelerator klystrons from 4 MW to 5MW is underway. It is foreseen that at the end of 1987 the maximum energy will be raised to 600 MeV.*

The two full-scale detector telescopes will probably be constructed as schematically indicated in fig. 10. The scintillator detector (fig. 10 a) consists of different slices of plastic scintillators in a telescopic arrangement. Ten layers with a thickness of 1 cm each will allow to stop 120 MeV protons. The layered construction assures a maximum separation for the amount of light produced by pions and protons as compared to electrons. The dimensions of the full-scale detector will be 40 cm x 10 cm which at a distance to the target of 50 cm amounts to an opening angle of 150 msr. In order to diminish the count rate in the first two scintillators, they consist of a large number of strips forming a hodoscope, the angular resolution of which is in the order of 10 mrad. ADC light output measurement will determine the particle energy. Electron background will be suppressed using two independent discriminator systems. One will be triggered on electrons, the other on particles with higher dE/dx. The energy resolution is anticipated to be better than 3 MeV for 100 MeV protons. The whole assembly (its background shielding included) will be constructed to enable also out-of-plane application.

The neutron time-of-flight channel (fig.10b) will be placed at forward angles at a distance between 5 and 10 meter from the target. Fast neutrons from the target area will be shielded against by a heavy-metal collimator. The shielding around the detector consists of a layer of water (or paraffin) to absorb relatively slow neutrons, a layer of (iron, copper or tungsten) heavy-metal shielding to slow down fast neutrons, a layer of lead to absorb neutron capture gamma-rays and finally a layer of lithium-carbonate to absorb the slowed down neutrons. The neutron detector itself is thought to contain (in a telescopic arrangement) the following elements: a thin scintillator as an anti-coincidence counter for charged particles, and, for reasons of counter efficiency, four liquid scintillators (NE 213). These 5 cm thick scintillators have excellent pulse shape discrimination properties for gamma-ray background suppression. Each scintillator will probably be followed by a two-stage low-pressure multi-wire proportional counter equipped with a suitable photo-cathode instead of the conventional photo-multiplier tube (see Fig. 11). This will provide position sensitivity in order to reject multiple hits. The timing resolution expected is in the order of 1 ns, which for 70 MeV neutrons and a flight path of 6.5 m results in an energy resolution of 1 MeV. The 12" diameter of the neutron detector determines the solid angle to be in the order of 1msr. Also this detector assembly will be constructed for out-of-plane application.

5. Proposed acccelerator improvement

The above given considerations show that NIKHEF-K intends to explore in the near future intriguing aspects of nuclear matter such as: high individual nucleon momentum components, virtual photon-nucleon coupling in the nuclear environment, meson exchange and isobar currents, electro-production processes for π^+ and π^o. As pointed out such an

a

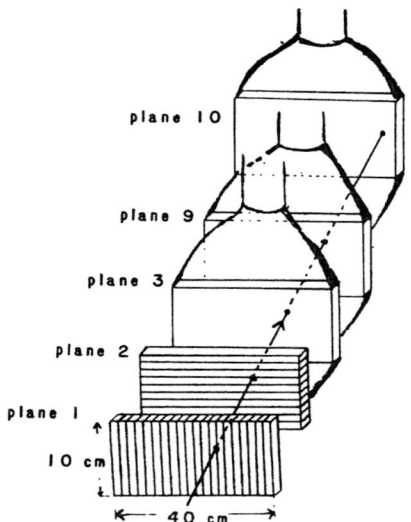

plane 10

plane 9

plane 3

plane 2

plane 1

10 cm

40 cm

b

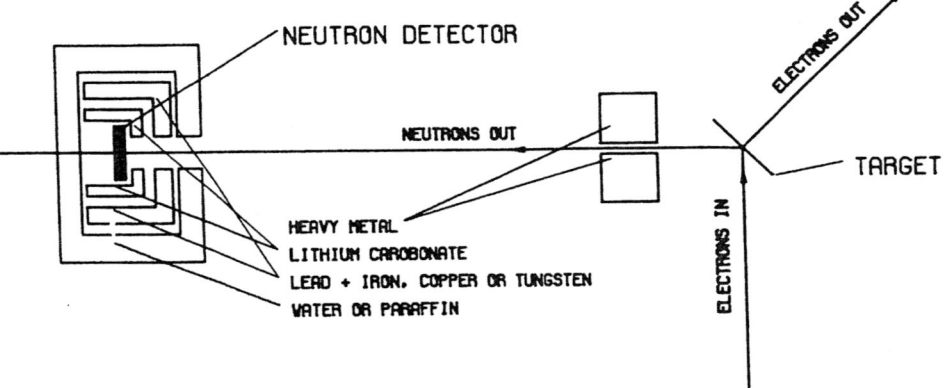

NEUTRON DETECTOR

NEUTRONS OUT

ELECTRONS OUT

TARGET

ELECTRONS IN

HEAVY METAL
LITHIUM CARBONATE
LEAD + IRON, COPPER OR TUNGSTEN
WATER OR PARAFFIN

Fig. 10   Schematic out-line of proposed detection channels
a) Scintillator hodoscope for large solid angle hadron detection
b) Neutron time-of-flight detector.

ambitious program requires large instrumental efforts. We anticipate that the program will cover at least the next five years. NIKHEF-K is confident that the major part of the program outlined can be carried out at relatively low beam duty factor (1 - 2 %), mainly due to the high-quality of the already existing equipment. Improving the beam duty factor will be ultimateley required to further extent this program as well as to open up new research lines. It is widely recognized that in the next decade intermediate energy physics will be dominated by facilities provided with high duty factor beams. Also NIKHEF-K intends to contribute to challenging programs which then become feasible. Therefore a proposal (called UPDATE) has been presented to the Dutch government to extend the 500 MeV accelerator with a pulse-stretching magnetic device to increase the duty factor up to 90% while simultaneously increasing the maximum energy of the accelerator to 700 MeV.

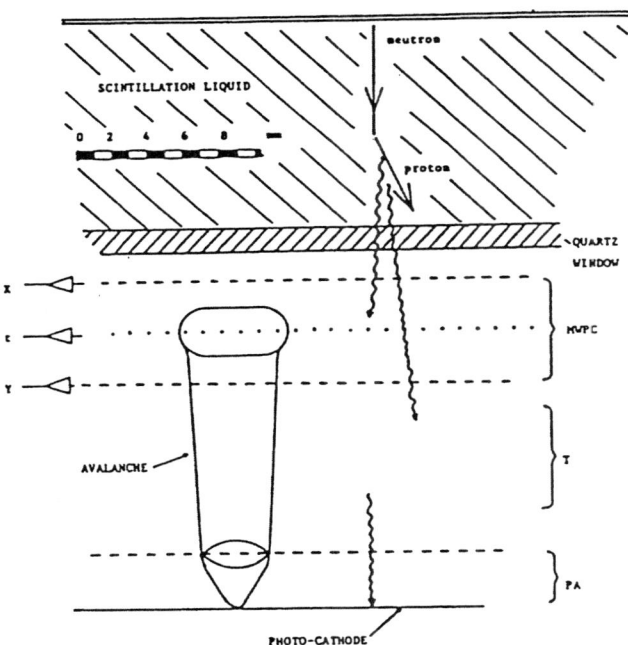

Fig. 11   Schematic representation of one scintillation module coupled to a wire chamber. First amplification of the photo-electrons occurs in region PA of the wire chamber after which these electrons are guided through a transfer region T to a multi-wire proportional chamber (MWPC) that gives a two-dimensional position and timing signal.

I wish to thank all my colleagues, physicists and technicians, who have brought forward the ideas underlying the research program.

This work is part of the research program of the National Institute for Nuclear Physics and High-Energy Physics (NIKHEF), made possible by financial support from the Foundation for Fundamental Research on Matter (FOM) and the Netherlands Organization for the Advancement of Pure Research (ZWO).

6. References

1. C. de Vries et al., Nucl. Instr. Meth. **223** (1984) 1
2. L. Lapikàs, Nucl. Phys. **A434** (1985) 85c
   G. van der Steenhoven et al., Phys. Lett. **156B** (1985) 151
   J.W.A. den Herder et al., Phys. Lett. **161B** (1985) 65
   P.K.A. de Witt Huberts, Proc. Workshop on Perspectives in Nuclear Physics., Trieste, Italy (1985), World Scientific
   G. van der Steenhoven et al., Phys. Lett. **156B** (1985) 146
   E.N.M. Quint et al., Phys. Rev. Lett. **57** (1986) 186
   G. van der Steenhoven et al., Phys. Rev. Lett. **57** (1986) 182
3. P.K.A. de Witt Huberts, ibid
   P.H.M. Keizer, The electro disintegration of ^{3}He studied with the ^{3}He(e,e'p)^{2}H and ^{3}He(e,e'd)^{1}H reactions, Ph.D. thesis, University of Amsterdam (1986)
4. J.S. McCarthy et al., Phys. Rev. **C15** (1977) 1396
5. A. Malechi et al., Phys. Lett. **36B** (1981) 61
6. W.T.H. van Oers et al., Phys. Rev. **C25** (1982) 390
7. V.A. Goldshtein et al., Nucl. Phys. **A355** (1981) 333

FUTURE EXPERIMENTAL DEVELOPMENTS FOR FEW-BODY PHYSICS AT MAINZ[*]

Reiner Neuhausen

Institut für Kernphysik

Johannes Gutenberg-Universität

D-6500 Mainz, Federal Republic of Germany

1. Introduction

The racetrack microtron MAMI B [1] now under construction at Mainz will provide a high-intensity electron beam with energies up to 840 MeV. The 100% duty factor, high-quality beam is considered to be ideal for the investigation of the nucleus in the medium energy region. Coincidence experiments with the detection of the scattered electron and the hadronic reaction products will become possible in a wide kinematical range yielding the information about the response of the nucleus to the transferred energy and momentum. By the method of bremsstrahlung tagging intense secondary photon beams with high monochromacity and well known fluxes will be available for photonuclear research work. Further developments will provide longitudinally polarized electron beams which will make extremely small interaction amplitudes accessible to the measurement via the interference terms.

To gain the optimal profit out of the new electron and photon beams highly sophisticated detection systems for the coincidence setups have to be designed in a proper way. The demands of large acceptances in solid angle and momentum without neglecting the overall resolution in energy and scattering angle lead to contradictory solutions in some cases and ask for flexible designs.

---

[*] Supported by Deutsche Forschungsgemeinschaft (SFB 201)

In the following report the present status of the construction work
for MAMI B is described, followed by a dicussion of the nuclear physics
research program which sets the requirements for the special features of
the future experimental facilities at MAMI B. These facilities are dis--
cussed to such an extent as they become apparent in the present design
considerations.

2. The status of the Mainz microtron MAMI B

The c.w. electron accelerator MAMI B [1] consists of a cascade of
three racetrack microtrons (Fig. 1) using a normal conducting, c.w. operated r.f. structure. The first stage with an output energy of 14 MeV was
first operated in 1979. The preliminary 180 MeV version MAMI A consisting
of the two first microtrons and a van de Graaff accelerator as an injector
is routinely in operation since 1983. The upgrade to the final version
MAMI B will consist in adding the third racetrack microtron and in replacing the present van de Graaff accelerator by an injection linac.
MAMI B with a maximum output energy of 840 MeV should be ready for operation by early 1989.

The 180 MeV version of MAMI housed in one of the existing experimental halls has been operated up to now for a total of about 9000 hours,
70 % of which were available for users experiments. The maximum energy
reached so far is 192 MeV, the maximum intensity 65 μA, limited by control
problems of the van de Graaff accelerator and of the r.f. supply which
both will be eliminated in the final version MAMI B. The energy width of
the beam of the properly operated machine is around 30 keV, compatible
with the design value of $\pm$18 keV. The beam emittance proved to be around
0.01 $\pi \cdot$mm$\cdot$mrad in either plane at full energy and compares favorably with
the somewhat conservative design values (Table 1).

The final 840 MeV version of MAMI will be housed in a new building
which is presently under construction and close to completion. The building consists of two accelerator halls, one for the 180 MeV injector, the
other for the third racetrack microtron (Fig. 2). The accelerator building
is completed by a beam guidance tunnel to the existing experimental halls,
an area for parasitic tagging experiments, technical facilities like a
power station, a control room and some offices for the accelerator staff.
Furthermore, a new experimental hall with an area of 30 m x 20 m and an
height of 16 m between the floor and the crane hook is in an early planning state. Construction work of this hall which should house the new
spectrometer setup (see below) is expected to be finished in spring 1989.

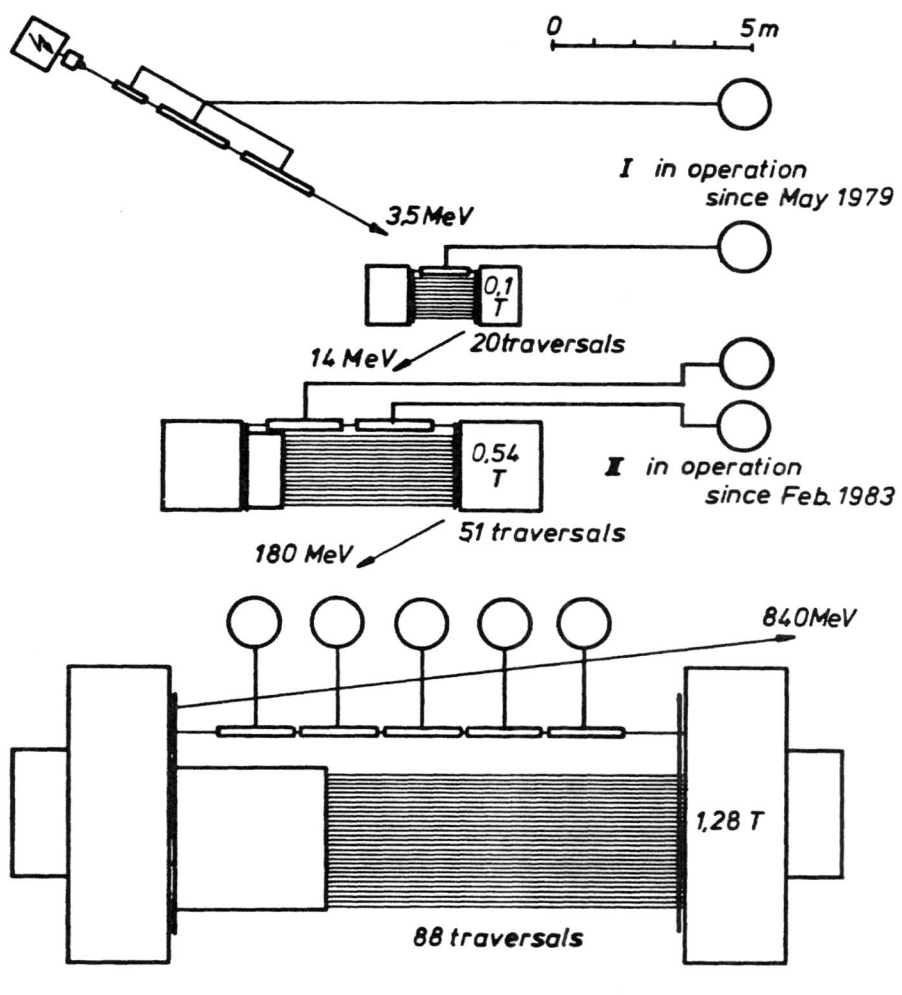

0   5m

I   in operation
    since May 1979

3,5 MeV

0,1 T

20 traversals

14 MeV

0,54 T

II in operation
    since Feb. 1983

180 MeV   51 traversals

840MeV

1,28 T

88 traversals

III under construction

scaled scheme of MAMI , Nov. 1985

Fig. 1: The cascade of three racetrack microtrons

Table 1: The essential parameters of MAMI B

	stage	I	II	III
General data:				
input energy	[MeV]	3.5	14	180
output energy	[MeV]	14	180	840
linac traversals		20	51	88
total power	[kW]		280	900
Magnet system:				
distance of magnets	[m]	1.66	5.59	12.85
flux density	[T]	0.1	0.56	1.28
max. orbit diameter	[m]	0.97	2.17	4.36
weight of one magnet	[t]	1.3	43	450
gap width	[cm]	6	7	10
R.F. system:				
number of klystrons		1	2	5
linac length (el.)	[m]	0.80	3.55	8.87
total r.f. power	[kW]	9	65	168
beam power	[kW]	1.1	17	66
energy gain per pass	[MeV]	0.59	3.25	7.5
Beam:				
energy width	[keV]	±9	±18	±60
emittance vert.	[π·mm·mrad]	0.17	0.04	0.01
hor.	[π·mm·mrad]	0.17	0.09	0.14

Injector: 100 keV gun, injector linac
Klystrons: Thomson-CSF TH 2075, 50 kW c.w., η=60%
Frequency: 2449.3 MHz

The design of MAMI B has been changed recently after detailed compu-
tations using the three-dimensional code PROFI had shown that the original
design value of 1.54 Tesla for the magnets of the third stage would not be
compatible with the field homogeneity requirements. The actual parameters
of MAMI B are compiled in Table 1. We expect the delivery of the two magnets of the third stage during the last quarter of 1986. It will take,
however, additional ten months for field measurements and corrections,
until the magnets will be ready for operation.

The injector linac which we have designed using the code PARMELA is
presently under construction and should be ready for test operation in
summer 1987. We expect that it may take almost a year to get sufficient
operation experience for routine operation within the close tolerances
required. The first two microtrons now forming MAMI A will have to be modified somewhat to be used as an injector for the third stage. Therefore,
MAMI A will be shut down for users experiments in late summer 1987. Its
reconstruction in the new hall will take place in the period between May
1988 and January 1989.

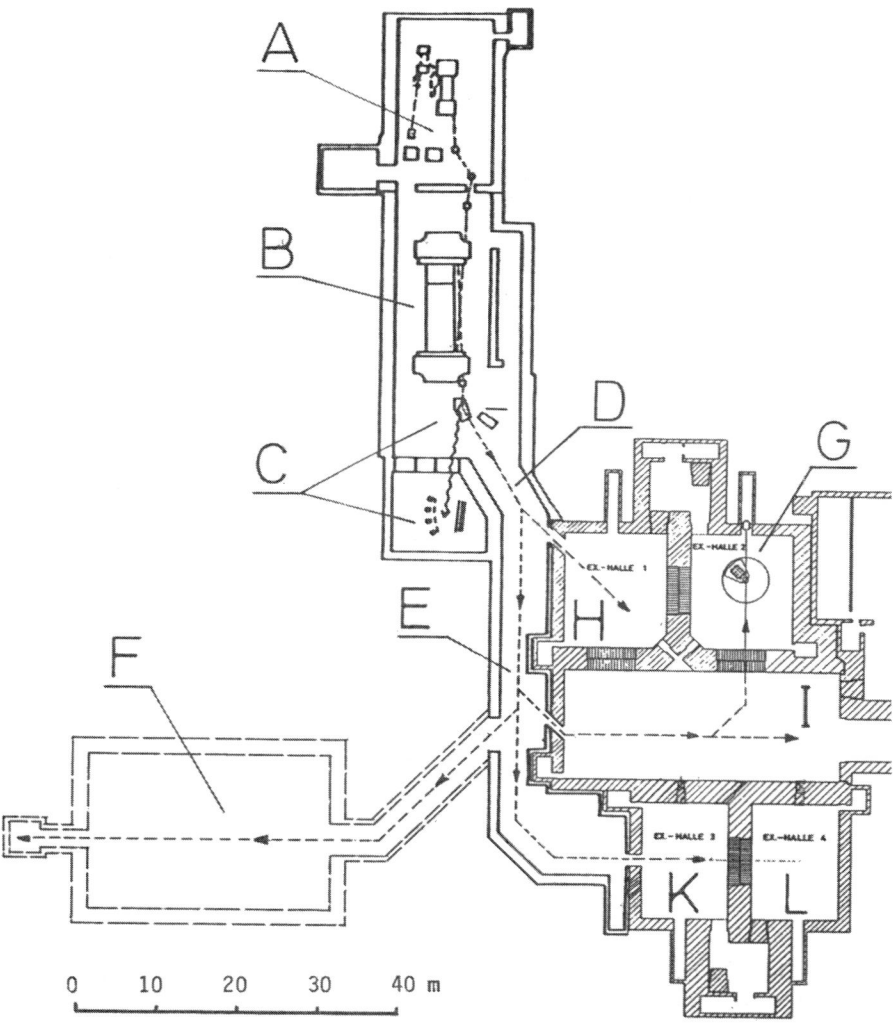

Fig. 2: Arrangement of the accelerator building and the experimental
halls. A: MAMI A, B: MAMI B, C: Area for parasitic tagging,
D: Area for future beam splitter, E: Beam guidance tunnel,
F: New experimental hall for three-magnetic-spectrometer
setup, G: 400 MeV station, H and I: Experimental areas for
various purposes, K and L: Low-intensity area for tagged
photon experiments

3. (e,e'x) and (e,e'xy) coincidence experiments

The comparison of the characteristics of the free with the bound nucleon will play an essential role in the nuclear physics research program with MAMI B. In the past contradictory experimental results, e.g. the quenching of the longitudinal structure function in quasielastic (e,e') scattering and the EMC effect, have led to speculations that the size of the nucleon may be increased drastically when bound in a nucleus. Limited data on the (e,e'p) reaction from Saclay and NIKHEF at quasifree kinematics reinforce the expectation that a systematic study of the nucleon knockout reactions can considerably contribute to the understanding of the structure of the nucleon inside a nucleus. The accurate separation of the four structure functions involved requires a large variation of the polarization parameter ϵ and the possibility of out-of-plane measurements.

A major part of the future experimental program will be devoted to the investigation of the excitation, propagation and decay of the Δ resonance in the nuclear medium. As an example, the Δ° resonance is generated by the (e,e'Δ°) reaction and detected by the measurement of the decay $\Delta^\circ \rightarrow \pi^-p$. The reaction can be completely measured in the (e,e'π^-p) triple-coincidence experiment where the two charged hadrons are detected in coincidence with the scattered electron. A striking feature of the study of the interplay of the Δ in a nucleus is that the well defined nuclear states serve as a "spin-isospin-filter" in the transitions. Since the wavefunctions are best known for light nuclei, these nuclei will be of special interest for (e,e'π^-p) experiments. Similar as in the (e,e'p) reaction studies an overall energy resolution of 100 to 200 keV is required to separate the final states of the residual nucleus. Only magnetic spectrometers with good imaging conditions can guarantee such a resolution.

Three magnetic spectrometers on a common pivot are proposed as a general setup for exclusive (e,e'x) and (e,e'xy) experiments. The choice of the design specifications (Table 2) are based on the kinematics investigated in detail for the (e,e'p) and (e,e'π^-p) coincidence experiments. A striking result is that in cases of large energy transfer the virtual photon is strongly directed to forward angles. Therefore, the electron and the hadron spectrometer should be capable to be set simultaneously at small angles. Large solid angles and large momentum acceptances are required to measure the small coincidence cross sections. Since the cross sections are proportinal to the flux of virtual photons which peaks at forward direction and decreases by several order of magnitudes with increasing electron scattering angle, we propose two electron spectrometers for the separation of the longidunal and transverse parts of the cross

Table 2: Specifications for the three-magnetic-spectrometer setup

A: Hadron spectrometer,   B: Forward electron spectrometer
C: Backward electron spectrometer or second hadron spectrometer

		A	B	C
Particles	(e,e'p)	p	e'	e'
	(e,e'pπ)	p	e'	π
Maximum momentum	[MeV/c]	650	840	500
Proton kinetic energy	[MeV]	200		
Pion kinetic energy	[MeV]			380
Solid angle	[msr]	40	5	40
Momentum acceptance	[%]	20	15	20-30
Electron angular range			5°-50°	40°-160°
Hadron angular range		10°-160°		40°-160°
Momentum resolution $\delta p/p$		< 10^{-4}	< 10^{-4}	< 10^{-4}
Angular resolution	[mrad]	$\approx$ 1	$\approx$ 1	$\approx$ 1

sections in (e,e'p) experiments. The first one (B) can reach the maximum momentum (840 MeV/c), but possesses a relatively small solid angle and a moderate momentum bite, and will be used exclusively at forward angles. The second one (C) is equipped with a large solid angle and a large momentum bite, but limited in the maximum momentum (500 MeV/c), and will be used at medium and backward angles. Furthermore, it will serve as the pion spectrometer in the (e,e'π^-p) experiments. The hadron spectrometer (A) for the detection of the proton is specified with a large solid angle and a large momentum bite, and limited to a maximum momentum of 650 MeV/c.

Magnetoptical designs of the spectrometers based on a QQDD configuration are in progress. As an important feature, QQDD designs yield favorable dispersion-to-magnification ratios, necessary for high momentum resolution, with moderate values of the dispersion, which help to keep the extension of the focal plane within reasonable limits. The projected arrangement of the three spectrometers in the experimental hall is shown in Fig. 3.

By measuring on the right and left side of the momentum transfer vector azimuthal particle emission angles Φ_x=0° and 180° are accessible. For out-of-plane measurements we plan to rotate the incoming electron beam by fixed angles between 20° and 45°. Depending on the specific kinematics the particle emission angle Φ_x reaches values between 30° and 90° in realistic cases.

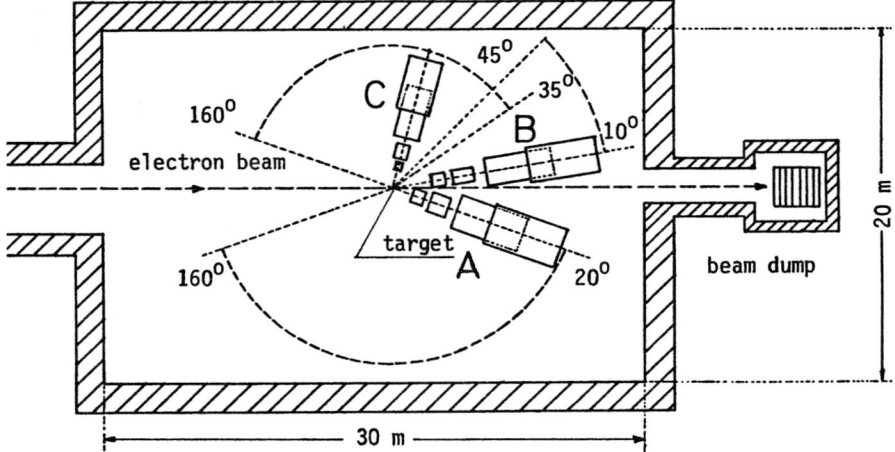

Fig. 3: The three-magnetic-spectrometer setup. A: Hadron spectrometer,
B: Forward electron spectrometer, C: Backward electron spectrometer or second hadron spectrometer.

4. Experiments with tagged photons

The crucial point of the research program with real photons are again
the coupling of the nuclear and subnuclear degrees of freedom. The main
goal of elastic photon scattering experiments will be the determination of
the strength distribution as a function of excitation energy, separated
into the multipolarities involved. The comparison of the shifts in the
strength distributions measured for the proton, few-body systems and heavier nuclei allows to draw direct conclusions on the coupling of the degrees of freedom.

The photoproduction of neutral pions is especially suited to study
the Δ resonance, because i) the π° production on the nucleus is almost
exclusively determined by the M1 amplitude, and ii) the transverse coupling of the photon to the Δ suppresses the elastic rescattering effects
known from pion scattering experiments. The coherent π° production leaves
the residual nucleus in the ground state, a process not possible in
charged pion production. A particular focus of the study is the comparison
of the π° production with the coherent η production which couples to the
$P_{11}(1440)$ and $S_{11}(1535)$ resonances.

The energy range of MAMI B is sufficient to excite the nucleon reso-
nances $P_{33}(1232)$, $P_{11}(1440)$, $D_{13}(1520)$ and $S_{11}(1535)$. The spectroscopy of
the decay of these resonances in different nuclear media from the few-body

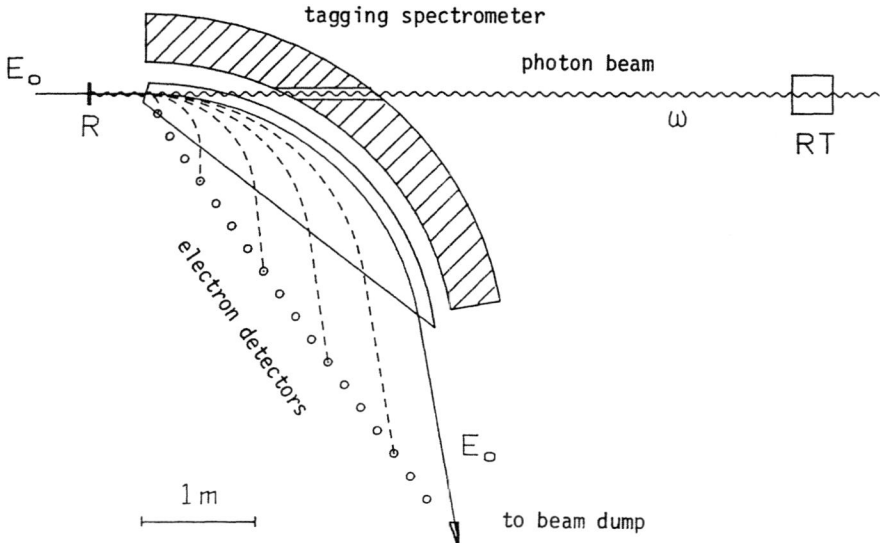

Fig. 4: The layout of the tagging system. E_o: Energy of primary
electron beam, ω: Energy of tagged photon, R: Radiator,
RT: Reaction target.

systems to heavier nuclei will open a direct access to the interaction of
the excited nucleon with nuclear systems.

The knockout of a nucleon from a deeply bound state by recoilless
production leaves the residual nucleus in a highly excited state, e.g. in
the Δ resonance. Again, the various decay channels can be studied.

Almost all photon experiments will make use of an intense tagged pho-
ton beam ($\approx 5 \cdot 10^7$/sec). The proposed tagging system (Fig. 4) provides photons in an energy range of 80 to 830 MeV with a primary electron energy of
840 MeV. The energy resolution of 100 keV is matched to the energy width
of the incoming electron beam. Linear polarized photons will be generated
by coherent bremsstrahlung at energies below 400 MeV and by suitable col-
limation of the primary electron and/or the secondary photon beam at
higher energies. Since the detection of the π^o and η mesons and of excited
states in the residual nuclei requires the measurement of several photons
per event, a 4π modular detection system (crystall ball) will be the central apparatus. Investigations of the most suitable scintillation material, the possibility of using drift chambers, the background conditions and
the appropriate geometries are under way. The tagged photon equipment will
be completed by a 0° magnetic spectrometer to measure charged particles
(p, π^+, π^-) emitted into the forward direction.

5. Experiments with longitudinally polarized electrons

The polarized electron research program addresses important questions in few-body systems. When using longitudinally polarized electrons, a small scattering amplitude in the presence of a large amplitude can be determined by the measurement of the interference term. These investigations are extremely difficult, but of fundamental importance. The program includes the measurement of the electric form factor of the neutron by the the quasifree scattering on the bound neutron in the reaction $^2H(\vec{e},e'\vec{n})p$, the determination of the small quadrupole contribution in the excitation of the Δ resonance, dominated by M1, and the measurement of the fifth structure function in the reaction $^2H(\vec{e},e'p)n$.

The technical experiences gained in Mainz while developing a polarized electron source for the 350 MeV linac will be transferred to the development of a new source for the c.w. operated linac of MAMI B. The electron pulses generated by photo-ionization in the GaAs cathode with a mode-locked laser will be matched to the micro-structure of the injection linac. The possible single pulse mode with variable frequency will allow time-of-flight measurements and, therefore, the suppression of the prompt background in the hadron detectors. Since the polarization vector will change its direction during the traversals in the microtron, the measurement of the polarization at the target with a Moeller polarimeter will deliver the information for a correct adjustment of the polarization vector at the source.

6. Summary

The construction of the Mainz microtron MAMI B and the development of a three-magnetic-spectrometer setup for (e,e'x) and (e,e'xy) coincidence experiments, of the equipment for tagged photon experiments and of a polarized electron source will keep the significance of nuclear physics with electromagnetic interactions at a very competitive level at Mainz. We believe that modern experiments with c.w. electron beams will considerably contribute to our understanding of nuclei and will be of special interest in investigating few-body systems.

References

1. Herminghaus, H. et al.: Nucl. Instr. and Meth. 138, 1 (1976)

FORWARD AND BACKWARD DEUTERON PHOTODISINTEGRATION CROSS SECTION AND A TAGGED PHOTON BEAM FROM BREMSSTRAHLUNG ON AN ARGON JET TARGET*

E. De Sanctis[1],M. Anghinolfi[2], A. Bertocchi[3],N. Bianchi[1], G.P. Capitani[1], P. Corvisiero[2], C. Guaraldo[1], P. Levi Sandri[1], V. Lucherini[1], L. Mattera[2], V. Muccifora[1], E. Polli[1], A.R. Reolon[1], G. Ricco[2], P. Rossi[1], M. Sanzone[2], M. Taiuti [2], G. Urciuoli[3], U. Valbusa[2] and A. Zucchiatti[2]

(1)   I.N.F.N. - Laboratori Nazionali di Frascati , P.O. Box 13, I-00044 Frascati , Italy.

(2)   Dipartimento di Fisica dell'Università di Genova e I.N.F.N.- Sezione di Genova, I-16146 Genova,Italy.

(3)   Laboratorio di Fisica dell'Istituto Superiore di Sanità and I.N.F.N.- Sezione Sanità, I-00185 Roma, Italy

Abstract. The cross section for the photodisintegration of the deuteron was measured simultaneously at 0°, 90° and 180° c.m. angles for the outgoing protons and at 220 MeV lab photon energy. A quasi-monochromatic photon beam, obtained by positron annihilation on a liquid hydrogen target, was used and the photon spectrum measured on-line by a pair spectrometer.The experimental apparatus consisted of a liquid deuterium target, a magnet sistem and three E-ΔE telescopes. The results agree with a recent fit to all modern experimental data.

A proposed monochromatic photon beam produced by the tagging technique is described. The radiator is a condensed molecular beam of argon installed in a straight section of the Adone storage ring. The recoil electron counters are placed in the magnetic field of the next dipole ring.

---

* presented by E. De Sanctis

1. Forward and backward deuteron photodisintegration cross section

The deuteron photodisintegration reaction, $^2H(\gamma,p)n$, is one of the most fundamental of the few-nucleon reactions and, therefore, it has been fairly extensively studied, both by experimentalists and theoretists. Nevertheless, in spite of the considerable effort spent so far on these studies, the knowledge of the cross section of the process is still unsatisfactory.

This is in particular true in the energy region between the pion emission threshold and the $\Delta(1232)$ resonance, where the spread of experimental values cover a factor of 2 in the absolute normalization, well outside the quoted error limits. Most of the observed disagreements should probably be ascribed to the use of continuous bremsstrahlung photon beams.

Recently the development of new techniques for producing monochromatic photon beam and of advanced computational capabilities has pushed the $^2H(\gamma,p)n$ reaction into the forefront of renewed experimental and theoretical interest.

In Fig. 1 are summarized the result of the most recent experiments all performed using monochromatic photons; specifically the experiment with photons produced by inverse Compton scattering by M. P. De Pascale et al. [1], for E_γ= 15-75 MeV; the positron annihilation photon beam measurement by E. De Sanctis et al. [2], for E_γ= 100-255 MeV, and the tagged photon studies by J. Arends et al. [3], and by K. Baba et al. [4], respectively for E_γ= 200-400 MeV and 180-600 MeV.

In the figure the cross section of the process has been expressed according to the usual Legendre polynomial expansion

$$\left(\frac{d\sigma}{d\Omega} \right)_{c.m.} = \sum_{L=0}^{3} A_L(E_\gamma)\ P_L(\cos\theta),$$

where θ is the angle between the incoming photon and outgoing proton momentum, and the $A_L(E_\gamma)$ coefficients, up to L= 3, has been fitted, for each photon energy E_γ, to the angular distributions provided by the above quoted experiments. As show in Fig. 1, the experimental situation appears considerably improved in the last years towards a fairly consistent set of experimental data with accuracy levels of the order of ±5% on the differential cross section values. Moreover it must be mentioned that there is also agreement with recent measurements of the inverse reaction[5,6], and consequently now a reasonable basis of experimental values is provided for comparison with the theory.

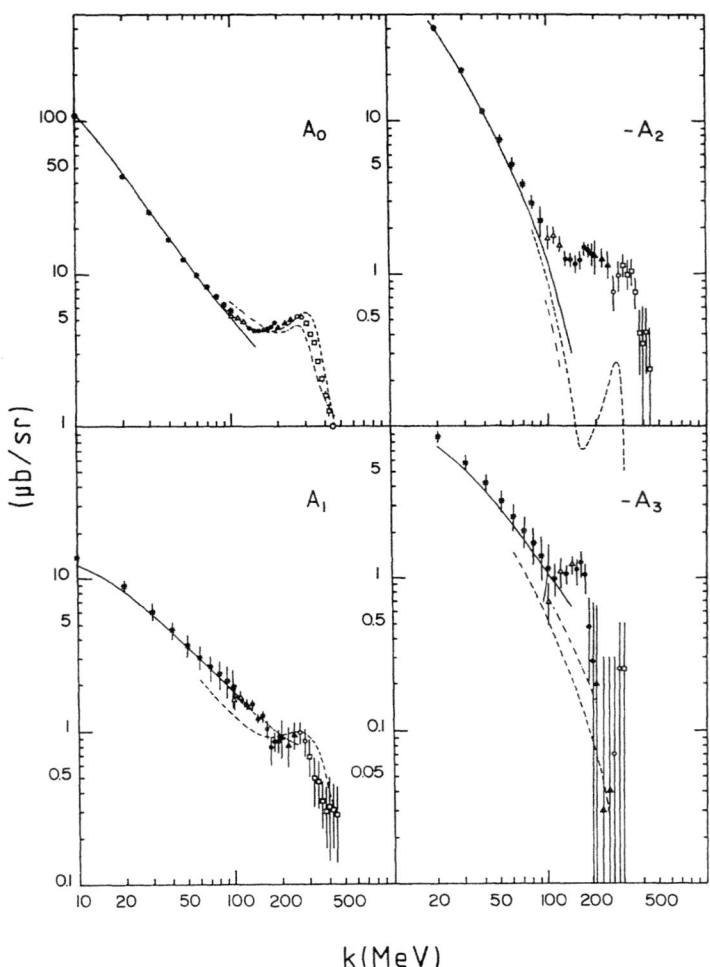

k(MeV)

Fig. 1 - Obtained results for the Legendre coefficients A_i (i= 0,.., 3) as a function of photon energy: asterisks, data from Ref. 1; open triangles, data from Refs. 2 , 9 and 5 (only at k= 100 MeV); solid circles data from Ref. 2 only; solid triangles, data from Refs. 2 and 3; open circles, data from Ref. 3 only; open squares, data from Refs. 3 and 4; dashed, dotted-dashed, and solid lines are from Refs. 7, 8 and 9, respectively

Fig. 1 also shows the values of the A_L coefficients deduced by fitting recent theoretical angular distribution calculations. The dashed curve is a calculation by Laget[7] performed using an expansion of the photodisintegration amplitude in terms of dominant diagrams; the dotted-dashed curve is a recent coupled channel calculation by Leidemann and Arenhövel[8], the solid line curve is a calculation by Cambi, Mosconi and Ricci[9] including second order relativistic corrections to the charge density. Despite the different theoretical approaches the results are qualitatively similar: the integrated cross section $4\pi A_0$, as well as the interference coefficients A_1 and (considering the error bars) A_3 are reasonably well reproduced, while $-A_2$ is strongly underestimated at energies greater than 100 MeV. Since A_2 determines the curvature of the angular distribution around 90°, the experimental data indicate that, above 100 MeV, the angular distributions are less isotropic than predicted by the theory.

We may therefore conclude that a better understanding of the two-body deuteron photodisintegration, at least above π production threshold, has still to be achieved.

From the theoretical side we may expect that the inclusion of ultrarelativistic corrections and/or a weaker tensor term, which already improved the agreement at forward angles below 140 MeV, may also reduce the observed discrepancy in the Δ excitation region.

From the experimental side one should provide a more complete set of data with the inclusion of the cross section values at the extreme angles . In order to accomplish this goal at Frascati it is presently going on a new measurement of the process in the photon energy range 100-260 MeV detecting simultaneously the protons ejected at 0°,90° and 180°. Therefore it will be possible to determine the forward to backward ratio of the cross section with reduced systematic errors and to check the absolute values by means of the 90° detector.

The measurements are carried out using the LEALE quasi-monochromatic photon beam produced by in-flight positron annihilation[11] on a liquid hydrogen target. A rectangular flat pole C-type magnet is used as an on-line pair spectrometer and the integrated photon flux on the deuterium target is measured by a gaussian quantameter.

The experimental apparatus is different from that we used in our previous experiment[2]: protons ejected from the target are deflected by a magnet, cylindrical in shape (120 cm diameter, 20 cm gap) having a central hole (Φ =48 cm) where the deuterium target can be inserted. The target consists of a vertical mylar cylinder (40 mm diameter, wall thickness 0.08 mm) , filled with liquid deuterium. Three E, ΔE telescopes, positioned respectively at 0°, 90° and 180°, measure the proton energy spectra. The apparatus was installed and tested.

Recently collected data at 220 MeV photon energy were preliminarly analysed. The 90° cross section value is in excellent agreement with our previous one[2], the 0° and 180° cross section values result very close to those given by the Thorlacius and Fearing[12] fit to all more modern data (data group I in ref. 12).

2. A tagged photon beam from bremsstrahlung on an Argon jet-target

In view of the great interest in nuclear physics studies with electromagnetically interacting probes, at Frascati it is foreseen to install an internal jet target on the electron storage ring Adone for producing a monochromatic high energy (up to 1200 MeV) photon beam through the tagging technique.

The use of internal targets in circulating beams antedates the availability of external beams from circular machines. In recent years, with improved understanding of beam dinamics and the construction of high energy synchrotron and storage rings, there have been a renewed interests in this option and growing activity in the development of suitable targets. The target which gives the largest luminosity is a type of condensed molecular beam[13] which provides a flow of gas at supersonic speed (hence the name of gas "jet" target) due to the expansion of gas from a vessel at high pressure and low temperature into the vacuum through an injector of very small aperture ($10 \div 150$ μm) and special geometry. The molecular jet flies forward along the axis of expansion and it is absorbed after having traversed the accelerator vacuum pipe. Only the core of the jet reaches the ultra-high vacuum of the ring via differential pumping stages where pratically all the uncondensed residual gas is pumped off.

Fig. 2 shows the schematic view of the jet target proposed for Adone. The Argon jet is produced in the chamber 1 (installed on top the Adone vacuum pipe) where the gas expansion takes place. The injector is a converging-diverging nozzle with special trumpet-shaped end part. Then the jet move across the machine vacuum pipe to the sink system, installed below the ring. We have interposed three differential pumping stages (each equipped with a 350 l/sec turbo-pump) to separate both the expansion and the sink chambers from the vacuum pipe in order to minimize the pressure rise in the interaction region ($\leq 10^{-8}$ torr). An additional pumping system (two 1000 l/sec turbo-pumps) is acting on the straight section of the ring where the jet target will be mounted, in order to reduce this rise pressure and limit the length of the region where the pressure is $\approx 10^{-8}$

torr. Two fast acting UHV valves separate the production and sink chambers from the Adone vacuum pipe to easy the jet on/off operations and to prevent the possible contamination of the ring in case of a large pressure bump due to breakdown of the target system.

The operating conditions are inlet pressure and temperature 6 bar and 150 °K, respectively, nozzle throat diameter 87 μm and semiaperture 3.5°. From a total flux of $\approx 10^{20}$ Ar- atoms/sec expanding from the nozzle the collimator system selects $\approx 5 \cdot 10^{18}$ atoms/sec, which corresponds to a target thickness of $\approx 10^{-8}$ g/cm^2 (Φ=1cm) on the path of the electron beam (that is at a distance of ≈ 25 cm from the nozzle).

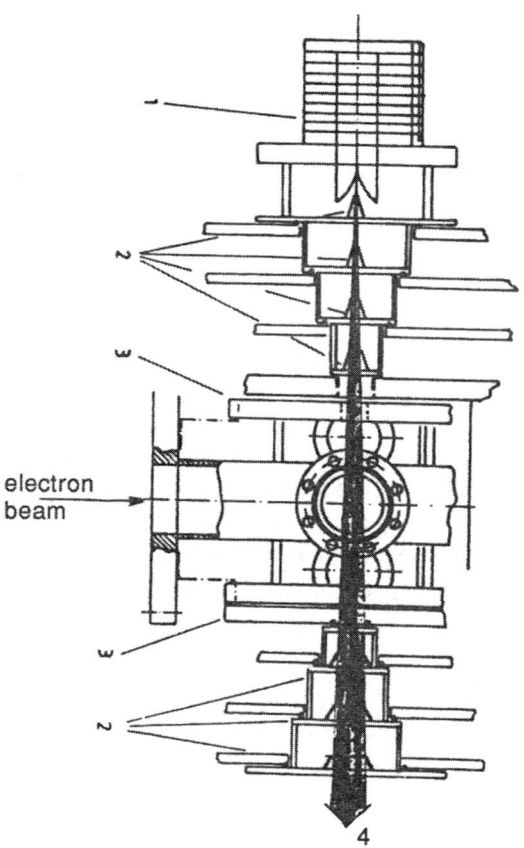

electron beam

Fig. 2 - Side view of the Argon jet target proposed for Adone: 1 gas expansion chamber; 2 collimators; 3 valves; 4 sink chamber.

The circumference of the Adone ring is approximately 105 m, so that a bunch of ultrarelativistic electrons takes about T_0 =351 nsec to make a round. The ring is divided in twelve identical lattice elements each consisting of a (n=1/2) bending magnet with a quadrupole doublet on each side.

Electrons are injected into the storage ring at an energy of 300 MeV (a few-turn injection will result in about 100 mA current circulating in the ring) and then accelerated to the desired energy by rising the magnetic field of the guiding magnets (this operation requires about 20 sec). The 51.4 MHz RF-cavity groups the circulating electrons into 18 buches, each ≈1nsec wide and ≈20 nsec apart.

After the rise in energy the Argon jet will be fired into the vacuum pipe and the electron beam lifetime $\tau=T_0/(\sigma_\varepsilon x)$ cut down to about 130 sec (T_0 is the revolution period, σ_ε the removal cross section and x the jet target thickness). Then the cycle is ended by lowering the field of the magnets to the injection value. The removal cross section involves only the process of bremsstrahlung in which the electron loses sufficient energy to place it outside the acceptance band-width of the ring ($\varepsilon= 0.01E_0$, E_0 being the machine energy). In fact the target thickness is so small that neither the multiple scattering nor the ionization losses contribute to the lifetime, being the RF- cavity able to compensate for both the growth in divergency and the mean energy losses.

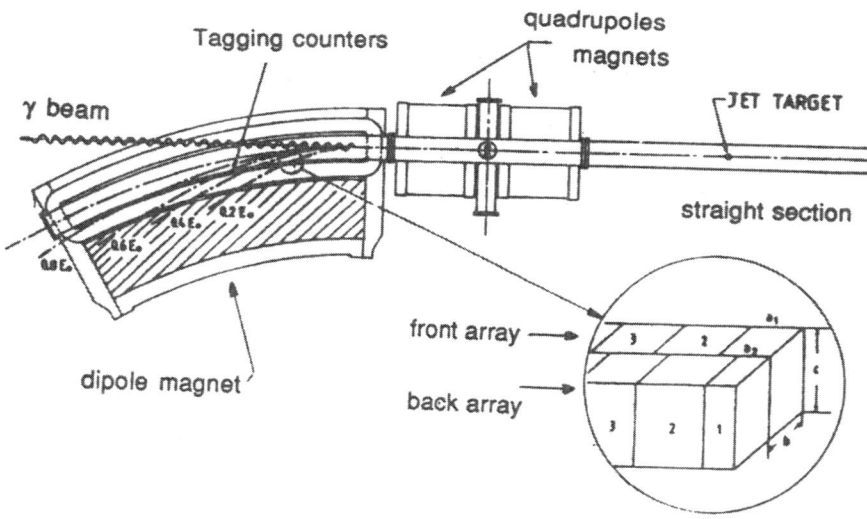

Fig.3 Schematic view of the tagging system

In Fig. 3 is sketched a lay-out of the apparatus: the Argon jet will be placed in a straight section (2.58 m long) between consecutive lattice elements and the recoil electrons will be momentum analyzed by the next dipole magnet and detected by a two-array scintillation counter hodoscope. This hodoscope will be placed between the ring vacuum pipe and the dipole magnet flux return joke. The scintillators have various sizes to give equal photon energy resolution ($\approx$1% at E_0 =1500 MeV) over the whole tagging range. The complete tagging system defines 80 energy channels covering the photon energy range k= (0.4÷08) E_0. This implies an extensive array of tagger detectors covering a side about 1m long of the bending magnet pole.

Since the determination of the photon energy relies on a coincidence between the tagging counters and the detector for the photoejected particles, the tagging method is subject to usual limitation due to random coincidences. In the normal operating mode the facility produces $\approx 10^8$ photons/sec in the whole tagging range. To make the operation of the tagged photon beam possible at the maximum intensity it is foreseen the installation of a new 350 MHz RF-cavity which makes the beam almost continuous in time (126 bunches 2.86 nsec apart).

References

1. M.P. De Pascale, G. Giordano, G. Matone, D. Babusci, R. Bernabei, O.M. Bilaniuk, S. d'Angelo, M. Mattioli, P.Picozza, D. Prosperi, C. Schaerf, S. Frullani and B. Girolami, Phys. Rev. C32, 1830 (1985); R. Bernabei, A. Incicchitti, M. Mattioli, P.Picozza, D. Prosperi, L. Casano, S. d'Angelo, M.P. De Pascale, C. Schaerf, G. Giordano, G. Matone, S. Frullani and B. Girolami, Phys. Rev. Lett. 57,1542 (1986).

2. E. De Sanctis, M. Anghinolfi, G.P. Capitani, P. Corvisiero, P. Di Giacomo, C. Guaraldo, V. Lucherini, E. Polli, A.R. Reolon, G. Ricco, M. Sanzone and A. Zucchiatti, Phys. Rev.C34, 413 (1986).

3. J. Arends, H.J. Gassen, A. Hegerath, B. Mecking, G. Noldeke, P. Prenzel, T. Reichelt, A. Voswinkel and W.W. Sapp, Nucl. Phys. A412, 509 (1981).

4. K. Baba, I. Endo, H. Fukuma, K. Inoue, T. Kawamoto, T. Ohsugi, Y. Sumi, T. Takeshita, S. Uehara, Y. Yano and T. Maki, Phys. Rev.C28, 286 (1983).

5. H.O. Meyer, I.R. Hall, M. Hugi, H.J. Karwowski, R.E. Pollock and P. Schwandt, Phys. Rev. C31, 309 (1985).

6. J.M. Cameron, C.A. Davis, H. Fielding, P. Kitching, J. Pasos,J. Soukup, J. Uegaki, J. Wesick, H.S. Wilson, R. Abegg, D.A. Hutcheon, C.A. Miller, A.W.

Stetz and I.J. Van Heerden, Nucl. Phys.A458, 637 (1986).

7. J.M. Laget, Nucl. Phys.A312, 265, (1978) and Can. J. Phys.62 , 1046 (1984).

8. W. Leidemann and H. Arenhovel, Phys. Lett., 139B, 22 (1984) and Can. J. Phys.62, 1036 (1984).

9. A. Cambi, B. Mosconi snd M. Ricci, Phys. Rev. C26, 2358(1982) and J. Phys. G10, L11 (1984).

10. R.J. Hughes, Z. Zieger, H. Waffler and B. Ziegler, Nucl. Phys.A267,329 (1976).

11. G.P. Capitani, E. De Sanctis, P. Di Giacomo, C. Guaraldo, V. Lucherini, E. Polli, A.R. Reolon, . Scrimaglio, M. Anghinolfi, P. Corvisiero, G. Ricco, M. Sanzone and A. Zucchiatti, Nucl. Instr. and Meth. 216, 307 (1983).

12. A.E. Thorlacius and H. Fearing, Phys. Rev. C33, 1830(1986).

13. O.F. Hagena and W. Obert, J. Chem. Phys. 56, 1793 (1972).

FUTURE EXPERIMENTAL DEVELOPMENTS IN FEW BODY PHYSICS AT BONN

Gisela Anton
Physikalisches Institut der Universität Bonn
Nussallee 12,D-5300 Bonn, Fed. Rep. Germany

Abstract:
Future experiments at the new electron stretcher ring ELSA in Bonn are
described. Especially the deuteron and the three nucleon system $t/^3$He
will be investigated using real and virtual photons. The improved
experimental equipment includes an energy tagged photon beam, linear
polarized photons and vector- and tensor polarized deuterons.

1. INTRODUCTION

In the following year the electron stretcher ring ELSA (Electron Stretcher
and Accelerator) at the Bonn University will go into operation. This new
machine allows a straight continuation of nearly thirty years of research
activities on light nuclei with electromagnetic probes.

During the last years the study of few nucleon systems in the intermediate
energy region was intensified. There is a rising interest not only in inclusive reactions but especially in exclusive reactions where two or more particles in the final state have to be detected. When Looking for detailed
effects there is an increasing need for measurements with good energy resolution and low errors.
In particular, for photon experiments an excellent method to achieve those
high quality data is the usage of a tagged bremsstrahlung beam in connection
with a detector of large solid angle and momentum acceptance. But such experiments are seriously affected by accidental event rate problems. For a process under investigation the ratio of the accidental event rate to the
prompt event rate is given by

$$\frac{N_{acc}}{N_{pr}} = const. \; I_{ave} \frac{1}{d}$$

I_{ave} : averaged beam intensity

d : duty factor of the accelerator

Obviously, a continous external beam at relatively low intensity is needed.
The duty cycle of the 2.5 GeV- synchrotron is only d=0.05. So, ELSA was constructed as a high duty cycle machine.

Fig. 1 shows a floor map of the electron accelerator facility at Bonn. The
electrons are preaccelerated by a Linac,injected to the 2.5 GeV-synchrotron
and then transfered to ELSA. ELSA has mainly two modes of operation (1). One

is a pure stretcher mode where the electrons - after a relatively short time
of injection into ELSA - are continously extracted so that a duty cycle of
d=0.95 is achieved. In the post accelerator mode a maximum energy of 3.5 GeV
is possible. Then the duty cycle is reduced to about d=0.5 .
ELSA will have two external beam lines: one provides for the PHOENICS experiment, the other is splitted and can alternatively deliver electrons to the
experiment SAPHIR or to the electron scattering facility.

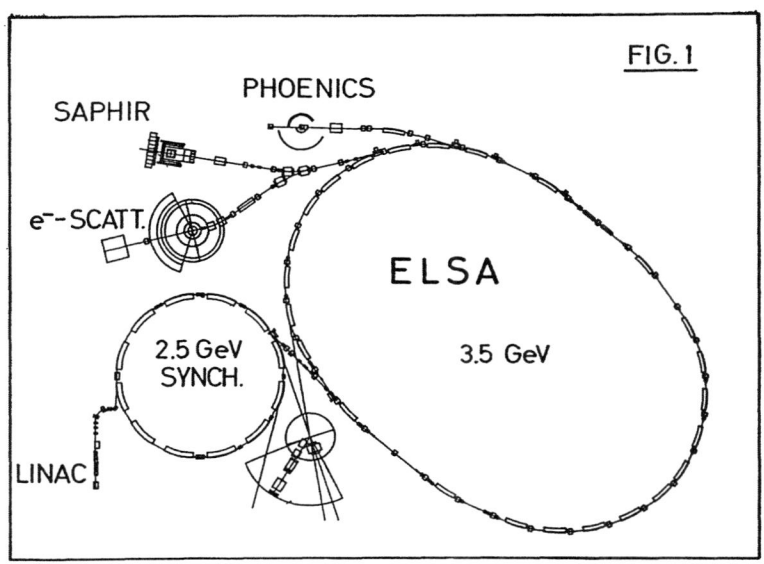

2. FEW BODY PHYSICS AT ELSA

The experimental program at ELSA includes the investigation of baryon resonances, nucleon form factors, vector meson excitation, deuteron reactions,
electromagnetic interaction of light nuclei etc. (17). In this talk I will
restrict myself to experiments with the deuteron and the three nucleon
system (tritium and ^{3}helium) as target nuclei.

The physics problems in the A=2,3- nuclei overlap to a great amount; so I
summarize the field of investigation.
The deuteron or t/^{3}He seem to be simple systems from kinematic arguments as
only 2 resp. 3 nucleons are involved. But the nucleons inside the nucleus
can change their identity by turning into Δ , N* or other baryon resonances.
They can exchange a number of bosons like π ,"σ", ϱ , ω ... These various phenomena are due to the fact that the nucleon is not elementary but is compo-
sed of quarks. Now the straight forward strategy would be to describe the
system on the quark level. But then there are many interacting particles
which undergo a complicated interaction.
Today the usual way to describe the two- and three nucleon system is based
on conventional nuclear physics in terms of baryons and mesons.

A first step in a calculation is the impulse approximation where only one
nucleon inside the nucleus interacts with the incoming particle (fig.2 a).

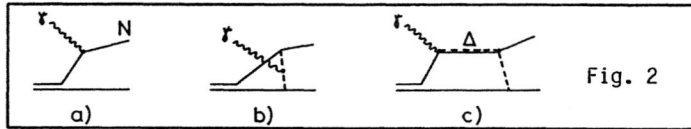

Fig. 2

But in contrast to a free nucleon, the fermi motion and off-mass shell kinematics have to be taken into account for the bound nucleon. A further step is the inclusion of meson exchange currents where the photon couples to an exchange meson (fig. 2 b). In addition the excitation of isobars like the Δ may be considered (fig. 3 c). Further ingredients of a calculation are the wave function of the nucleus, final state interaction, relativistic corrections, six-quark contributions etc..

2.1 Elastic Electron Scattering from Deuterons

The cross section for the elastic electron scattering off deuterons is given by (2)

$$\frac{d\sigma}{d\Omega} = \frac{d\sigma}{d\Omega}_{Mott} (A(q^2) + B(q^2) \tan^2 \frac{\theta_e}{2})$$

$$A(q^2) = G_C^2(q^2) + \frac{8}{9} \eta^2 G_Q^2(q^2) + \frac{2}{3} \eta G_M^2(q^2)$$

$$B(q^2) = \frac{4}{3} \eta (1 + \eta) G_M^2(q^2) ; \eta = q^2/4 M_d^2$$

This cross-section is proportional to the cross-section on point-like particles (Mott cross section). The term in brackets describe the influence of the deuteron as an extended object. This term contains combinations of the deuteron form factors (ff): the electric charge ff $G_C(q^2)$, the electric quadrupole ff $G_Q(q^2)$ and the magnetic dipol ff $G_M(q^2)$.

From a theoretical point of view the charge ff seems to be an interesting observable. Fig. 3 shows the result of a calculation (3) using the impulse approximation and with additional meson exchange currents. Obviously the position of the first diffraction minimum is very sensitive. In a model calculation (4) which includes a 6-quark component in the deutron wave function this dip will even vanish.

The structure functions $A(q^2)$ and $B(q^2)$ can be separated by varying the electron scattering angle θ_e.

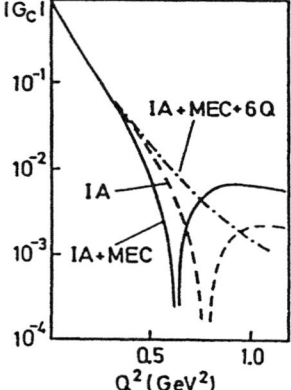

Fig. 3: Sensitivity of the deuteron charge form factor; dashed curve: IA from ref.3, full curve: IA + MEC from ref.3, dashed-dotted curve: 6q-component in the deuteron wave function, ref.4.

In this way the magnetic ff is determined. But the charge ff can only be separated from the quadrupole ff in a polarization experiment. One possibility is given by measuring the tensor polarization of the outgoing deuteron : $t_{20}(e\,d \rightarrow e'\,\vec{d})$ (5). Another possibility is to make use of a tensor polarized deuteron target : $T_{20}(e\,\vec{d} \rightarrow e'\,d)$. The two methods are equivalent via time reversal symmetry. A first measurement with a tensor polarized target was

performed at the 2.5 GeV- synchrotron at $q^2 = 0.5$ (GeV/c)^2 .The tensor polarization was 0.17 which is a relatively high value (6). The data are still
under analysis (7). This measurement will be continued under improved conditions at ELSA. Then, the interesting region up to $q^2 = 1.5$ (GeV/c)^2 will be
covered.

2.2 Photodisintegration of the Deuteron

The disintegration reaction $\gamma d \rightarrow p\ n$ is of basic importance in photo nuclear physics. Unfortunately,due to the spin of the particles involved,there
is a relatively high number(23) of independent observables for this reaction.
Among these observables the differential cross-section is not very sensitive
because it is proportional to the sum over squared helicity amplitudes. The
influence of a small amplitude can be recognized more easily in polarization
observables. As an example, fig.4 shows the result of a relativistic calculation from a Bonn group (6). One curve includes final state interaction,the
other does not.Obviously, the polarization observable is much more sensitive
to the different ingredients of the calculation then the differential cross-
section.
At ELSA we will have a linearly polarized photon beam. In connection with
the existing vector-and tensor polarized deuteron target, we will measure
single and double polarization observables at the experiment PHOENICS. We
hope that these data will substantially improve the experimental information
available on the deuteron photodisintegration.

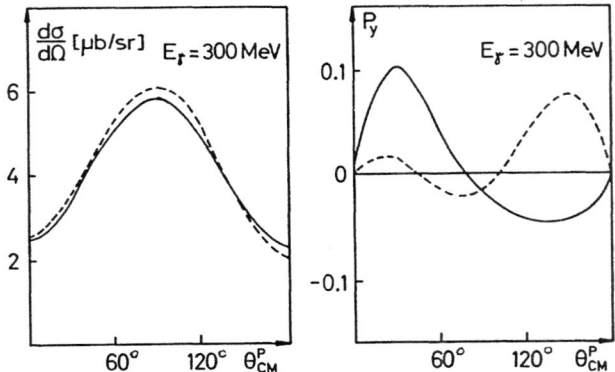

Fig. 4: Results of a relativistic calculation (6) for the
reaction $\gamma d \rightarrow p\ n$. Dashed line: including final state
interaction, full line: without final state interaction.

2.3 Pion Production from Deuterons

Above pion threshold the main contribution to the total γd- cross-section
comes from $\gamma d \rightarrow d\ \pi^o$ and $\gamma d \rightarrow N\ N\ \pi$ reactions. They form important
tests for three-body calculations . Both reactions can be measured with the
PHOENICS detector. In the case of three particles in the final state it is
not possible to cover the entire kinematic region. Here, one has to choose
special kinematic constellations.

Another interesting process is the $\Delta\Delta$ production via $\gamma d \rightarrow \Delta\Delta \rightarrow N\ N\ \pi\ \pi$.
In this way the interaction of a $\Delta\Delta$ -pair can be studied. A proper recon-
struction of this process is possible if three of the four final state particles are detected. This will be done with the SAPHIR detector.

A further point will be the search for narrow dibaryon resonances. Such an experiment was already performed at the 2.5 GeV- synchrotron (9). In the reaction $\gamma d \rightarrow p\, p\, \pi^-$ the invariant mass of the two protons showes a bump at 2014 MeV with a statistical significance of 4.7 standard deviations, fig. 5. The investigation of this and other narrow resonances will be continued with the SAPHIR detector.

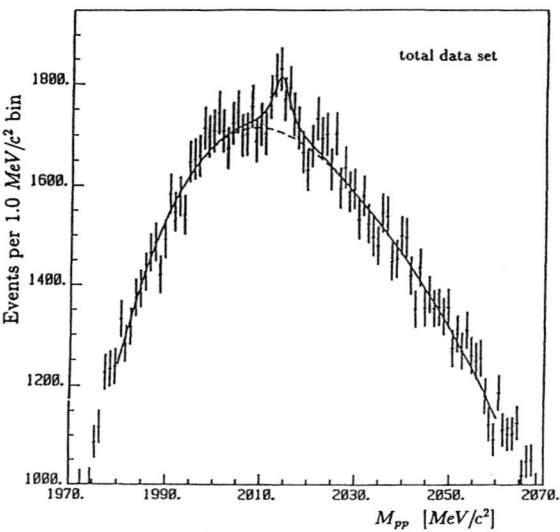

Fig. 5: Invariant mass distribution of the pp- pair in the reaction $\gamma d \rightarrow p\, p\, \pi^-$.

2.4 Photoreactions on t/^{3}He

The coherent pion production and the disintegration reactions

$$\gamma\, t \Big\langle \begin{array}{l} t\ \pi^o \\ ^3He\ \pi^- \end{array} \qquad\qquad \gamma\ ^3He \Big\langle \begin{array}{l} ^3He\ \pi^o \\ t\ \pi^+ \end{array}$$

$$\gamma\, t \Big\langle \begin{array}{l} d\ n \\ p\ n\ n \end{array} \qquad\qquad \gamma\ ^3He \Big\langle \begin{array}{l} d\ p \\ p\ n\ p \end{array}$$

will be studied with the PHOENICS detector.
Tritium and helium build an isospin doublet. So the isospin symmetric reactions mentioned above form good tests for calculations, especially for isospin symmetry breaking effects. As an example, fig. 6 shows data for the reactions $t(\gamma,\pi^-)^3He$ and $^3He(\gamma,\pi^+)t$ which were measured at the 500 MeV-synchrotron in Bonn. Obviously, the cross-section on the tritium target is larger than on the helium target. This is intuitively expected from the fact that in this energy region $\sigma(\gamma n \rightarrow \pi^- p)$ is larger then $\sigma(\gamma p \rightarrow \pi^+ n)$. A relatively old calculation (10), done in a simple impulse approximation, reflects this difference but it overestimates the cross-section. A calculation from Tiator et al. (11) where the fermi motion and off shell kinematics are taken into account is in agreement with the data concerning the energetic position of the resonance excitation. But this calculation cannot describe the

decrease of the cross-section at higher photon energies. In conclusion, there are many open problems in the field, for theorists as well as for experimentalists.

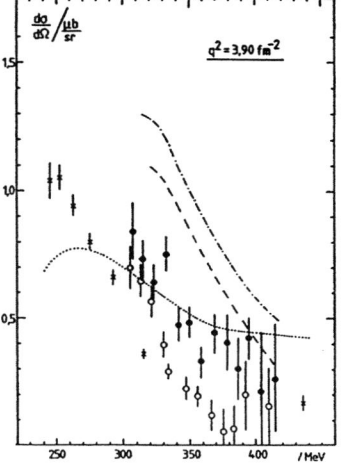

Fig. 6:
Results of pion production from the 3- nucleon system.
Data from Bonn (12):
● : t (γ,π^-) ^{3}He
o : ^{3}He(γ,π^+) t
Data from ref. 13 :
x : ^{3}He(γ,π^+) t
theoretical curves:
---: ^{3}He(γ,π^+)t   ref.10
-··-: t $(\gamma,\pi^-)^3$He ref.10
····: ^{3}He(γ,π^+)t   ref.11

3 EXPERIMENTAL FACILITIES AT ELSA

In this section, I will restrict myself to the description of the experiments PHOENICS and SAPHIR which are equipped with a tagged photon beam. A detailed description of the electron facility is given in ref.5
Most of the existing data on photonuclear reactions were obtained with untagged bremsstrahlung photon beams. The energy of the photons in such a beam varies between zero and the energy of the primary electron beam. So,for each event the actual photon energy has to be calculated backwards from the detected hadron kinematics. This method of photon energy determination is affected by large systematic and statictical errors. The huge discrepancies between data-sets from different laboratories emphasize this problem.
One possibility to improve the situation is to use an energy tagged bremsstrahlung beam. By this method the electron which has radiated the photon is detected in a magnetic spectrometer. The photon energy is simply given by the difference between the electron energy before and after the radiation process. I want to stress that the important advantage is the decoupling of the photon energy determination from the hadron detection.
Our photon tagging system covers a relatively large energy range :
E_γ = (0.2 - 0.95) E_o ,where E_o is the electron beam energy. The energy resolution varies between 5 MeV and 10 MeV.
In conjuction with a tagging system it is relatively easy to obtain a linear polarized photon beam. If the bremsstrahlung radiation target is replaced by a single crystal the according bremsstrahlung spectrum - which is detected via the tagging spectrometer - contains peaks which are due to a component of linearly polarized photons. This peak regions can be selected via the tagging counters.

3.1 The PHOENICS detector

A top view of the PHOENICS detector is given in fig.7 . This detector is composed of rings of scintillation counters which are centered around the tar-

430

get. These rings are formed by vertical scintillator bars. Each end of a
bar is connected to a photomultiplier tube. All counters together contain
174 multiplier tubes.
There are several hadron targets at our disposal: liquid hydrogen and deuterium as well as gaseous hydrogen,deuterium,tritium(!),^{3}Helium, ^{4}Helium etc..
Additionally, we can use vector polarized proton targets and vector- or tensor polarized deuteron targets (15). A frozen spin target which allows for
the orientation of the spin axis to any desired direction is under constuc-
tion.

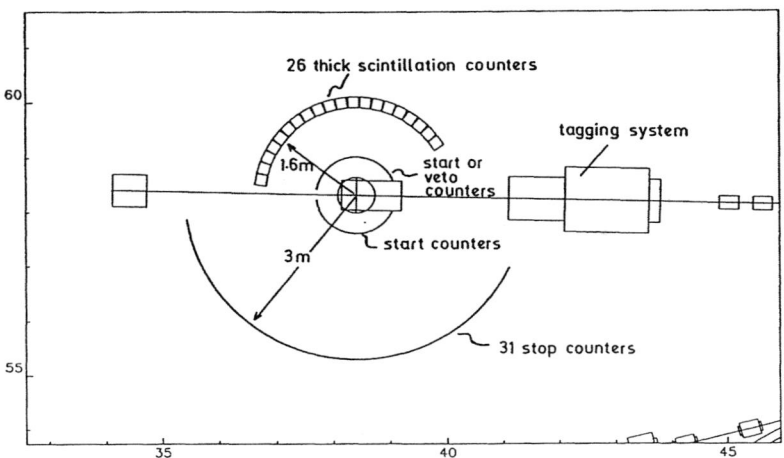

Fig. 7:   The PHOENICS detector

The PHOENICS detector has a large solid angle and momentum acceptance. In
connection with the energy acceptance of the photon tagging system a large
kinematic region of a process can be measured simultaneously. For most processes the detector gives a kinematic overdetermination. This is important
for the supression of competing processes.
PHOENICS is composed of scintillation counters only. This is a fast operating and simple detection system which can tolerate high data rates during
run-time as well as during the off-line analysis.
The mentioned features of PHOENICS make possible high quality data with low
statistic and systematic errors.

3.2 The SAPHIR detector

Fig.8 shows the SAPHIR detector in a stereo view and in a top view. This detector is especially designed for multiparticle final states. In this sense
it works complementary to the PHOENICS system.
SAPHIR contains a relatively large magnetic volume of 1 m^3 and a field of
0.6 T   The target which is placed in the center of the field is surrounded
by a cylindrical drift chamber.This drift chamber together with outer planar
drift chambers serves for the particle track identification. An arrangement
of scintillation counters is used for particle identification via time of
flight measurement and for trigger purposes. Photons can be identified in a
sandwich-type shower counter.The acceptance of the SAPHIR is 60% of the full
solid angle and the momentum resolution varies between 2% and 3% (14).

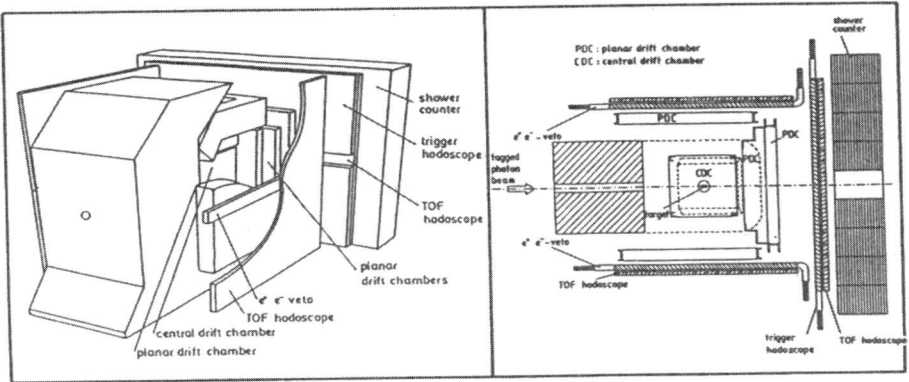

Fig. 8: The SAPHIR detector

References:

1. D.Husmann: Lecture Notes in Physics 234 (1985) 381
2. M.Goudin,C.A.Piketty: Nuov.Cim. 32 (1964) 1137
3. M.Gari, H.Hyuga: Nucl.Phys. A264 (1976) 409
4. A.P.Kobushkin: Sov.J.Nucl.Phys. 28 (1978) 252
5. J.M.Cameron et al.: Phys.Lett. 137B (1984) 315
6. W.Meyer et al.: Nucl.Instr. and Meth. A224 (1986) 574
7. D.Schablitzki: Ph.D. Thesis, in preparation
8. H.Huneke et al.: BONN-IR-80-24
9. B.Bock et al.: Nucl.Phys. A459 (1986) 573
10. J.L.Sanchez-Gomez,P.Pascual: Nucl.Phys B9 (1969) 153
11. L.Tiator et al.: Nucl.Phys. A333 (1980) 343
12. B.Bellinghausen et al.,to be published
13. D.Bachelier et al.: Nucl.Phys. A251 (1975) 433
14. W.J.Schwille: BONN-IR-86-26
15. B.Mecking, Lecture Notes in Physics 234 (1985) 353
16. W.Meyer : Nucl.Phys. A446 (1985) 381c
17. D.Menze, .Pfeil,W.Schwille: Proceedings of the 1984 Workshop on Photon
    and Electron Interaction at Intermediate Energies, Bad Honnef 1984,
    Lecture Notes in Physics 234 (1985)

RELATIVISTIC EQUATIONS

Franz Gross
Continuous Electron Beam Accelerator Facility
12070 Jefferson Avenue
Newport News, Virginia 23606 USA

and

College of William and Mary
Department of Physics
Williamsburg, Virginia 23185 USA

In the last 3 years I have given several talks on relativistic equations for nuclear physics, and on relativistic effects in few nucleon systems[1]. In preparing for these talks, and in responding to the discussion which followed them, my thinking about relativistic equations and relativistic effects has gradually evolved. I want to begin this talk by focusing on several issues which naturally arise when discussing this subject, and then turn to a brief discussion of relativistic three nucleon equations. The latter topic has not been reviewed recently, and is particularly of current interest because of the beautiful form factor measurements of the ^{3}He $-$ ^{3}H system recently completed at Saclay[2] and Bates[3], and the accompanying speculation that relativistic effects are important for understanding the three-nucleon system.

I. Issues

Relativistic equations for the two body scattering amplitude take the following very general form

$$M = V + V G M \qquad (1)$$

where M is the 2 body scattering matrix, V the relativistic kernel or potential and G the two body propagator. Equation (1) can be regarded as a shorthand for the infinite sum

$$M = V + V \ G \ V + V \ G \ V \ G \ V + V \ G \ V \ G \ V \ G \ V + \ldots.$$

$$= [1 - V \ G]^{-1} V \tag{2}$$

where the nth term in the sum is the nth Born approximation to the amplitude. Solving Eq. (1) is a method of summing the series (2) which is essential when many terms contribute to the series (which usually is the case for low energy scattering) or when the series diverges (near the bound state poles of M).

1.1 Lorentz invariance

The first issue concerns the meaning of "relativistic" in the content of this subject. What I shall assume is that the Eq. (1) and the corresponding series (2) are Lorentz invariant in the sense that (a) it is clear from the equations themselves how to calculate M in any frame, that (b) a law telling how to transform M from one frame to another can be deduced, and that this law makes it possible to write the components of M in any frame in terms of those in, say, the rest frame, so that M need be calculated originally only in one frame, and (c) matrix elements involving M can be explicitly shown to be Lorentz invariant using the transformation law for M.

Many equations which use relativistic kinematics or are derived from relativistic objects (such as Dirac spinors) do not satisfy these stringent requirements. For example, Bag models which are based on a relativistic lagrangian density are not Lorentz invariant because we do not yet know how to boost bag states, although some progress on this topic has been made recently. For the same reason, wave equations based on time-ordered perturbation theory which use relativistic kinematics may also not satisfy these criteria, even though they may provide an excellent dynamical description of the nucleon-nucleon interactions, and include some effects of relativistic origin[4].

It is important to compare intermediate energy data with Lorentz invariant calculations -- only in this way can we eliminate uncertainties arising from the breaking of Lorentz invariance and assure ourselves that we are really testing the underlying dynamics. This is particularly important for programs which study the few body system at intermediate energies, such as those which we expect to carry out at CEBAF.

1.2 Choice of propagator

The requirement of Lorentz invariance does not uniquely define the equation; it is necessary to specify the propagator. A large number of choices are possible, but three which have received the most attention are the Bethe-Salpeter (BS)[5], the one-particle-on-shell equation (G_1)[6], and the Light Front (LF)[7]. If two particles have total 4-momentum P and relative 4-momentum p, so that

$$P = p_1 + p_2$$

$$p = \frac{1}{2} (p_1 - p_2)$$

(3)

then in all cases the total momentum is conserved, and the propagator depends on p only. In the BS case it depends on all 4 components of p, so that for spin zero particles the BS propagator is

$$\int d^4p \; G_{BS}(p) = \int d^4p \; [m_1^2 - (\tfrac{1}{2}P + p)^2]^{-1}[m_2^2 - (\tfrac{1}{2}P - p)^2]^{-1}$$

(4)

where m_1, and m_2 are the masses of the two particles. For the one particle on shell equation (particle 2 if $m_2 > m_1$), the propagator depends only on three components of p, the 4th being constrained in a Lorentz invariant manner by the mass-shell condition, so that

$$\int d^4p \; \delta_+[m_2^2 - (\tfrac{1}{2}P - p)^2] \; G_1(p) = \int d^4p_2 \delta_+[m_2^2 - p_2^2] \; [m_1^2 - (P - p_2)^2]$$

$$= \int \frac{d^3p}{2E_2} \; [E_1^2 - (W-E_2)^2]^{-1}$$

(5)

where the last expression holds in the enter of mass of the pair, with $P = (W,\vec{0})$ and $E_{1,2} = (m_{1,2}^2 + \vec{p}^2)^{1/2}$. (It is important to identify the δ function with the volume integral -- particle 2 in this description is removed from the equation from the start and does not propagate.) One may also specify particle one to be on shell. Finally, in the LF formalism, the light front variables $p^{\pm} = p_o \pm p_3$ and $p_\perp$ are used, so that $p^2 = p^+p^- - p_\perp^2$ and the propagator can be written

$$\int d^4p \; \delta[m_2^2 - (\tfrac{1}{2}P - p)^2] \; G_{LF} = \int d^4p_2 \; \delta[m_2^2 - p_2^2] \; [m_1^2 - (P - p_2)^2]^{-1}$$

$$= \int_o^1 \frac{dx \; d^2p_\perp}{2x(1-x)} \; \left[\frac{m_1^2 + p_\perp^2}{1-x} + \frac{m_2^2 + p_\perp^2}{x} - W^2 \right]^{-1}$$

(6)

where, in the last expression, $P = (P^+, P^-, P_\perp) = (P, W^2/P, \vec{0})$ and x is the famous longitudinal momentum fraction

$$x \equiv \frac{P_2^+}{P^+} , \ 1-x = \frac{P_1^+}{P^+} . \tag{7}$$

Note that the final answer shows that this propagator is completely symmetric under the interchange of particles 1 and 2; we would have obtained the same result if we had begun with $\delta[m_1^2 - (1/2 \ P + p)^2]$. Furthermore, if both particles 1 $\underline{and}$ 2 are on the mass shell, then

$$P_1^- = \frac{m_1^2 + P_\perp^2}{(1-x)P^+} \ ; \ P_2^- = \frac{m_2^2 + P_\perp^2}{xP^+} \tag{8}$$

so that the LF propagator can also be written

$$\int_0^{P^+} \frac{dp^+ \ d^2 P_\perp}{2 \ P_2^+ \ P_1^+} \ [P_1^- + P_2^- - P^-]^{-1} \tag{9}$$

showing that the propagation is off the p^- shell. Hence, the LF formalism shows some resemblance to time ordered perturbation theory, with p^- playing the role of energy. However, the LF formalism is invariant under boosts in the z direction and rotations in the 1,2 plane, but is not invariant under rotations which mix the 3rd axis with the 1,2 plane, while the time ordered formalism is invariant under rotations, but not under boosts.

The LF formalism is very suitable for high energy problems where there is a preferred direction[8]. It has become a standard tool for the analysis of high energy quark interactions, but has seen less application to nuclear physics, where explicitly angular momentum conservation is a very useful tool[9].

It is amusing to note that the G_1 formalism and the LF formalism bear certain formal similarities[10]. If we introduce

$$x' = \frac{E_2 + P_z}{W} \tag{10}$$

we can transform (5) into (6), the only difference being that x' replaces x and the limits of integration on x' are from 0 to ∞ instead of 0 to 1.

1.3 Relationship between equations

It is important to realize that these equations are equivalent if the kernel V is chosen correctly. In proving such relations, it must be recognized that some equations depend on more variables, and equivalence can only be proved in a region where both amplitudes are defined. I will illustrate this by comparing the BS and G_1 equations. To do this introduce an operator P, which fixes the variables so that particle 2 is on shell. Suppose that M_{BS} is the BS amplitude, M_1' is its projection $M_{BS}P$, and M_1 is its double projection $PM_{BS}P$. Then if

$$M_{BS} = V_{BS} + V_{BS} \, G_{BS} \, M_{BS} \tag{11}$$

and V_1' satisfies the equation

$$V_1' = V_{BS}P + V_{BS} \, (G_{BS} - PG_1P) \, V_1' \tag{12}$$

it can be readily shown that

$$M_1' = V_1' + V_1' \, G_1 \, M_1 \quad . \tag{13}$$

Hence, to determine the BS amplitude when particle 2 is on shell in either the initial <u>or</u> final state, it is sufficient to solve the G_1 equation

$$M_1 = V_1 + V_1 G_1 M_1 \tag{14}$$

where $V_1 = PV_1'$, and use Eq. (13). This theorem is well known, but usually the role of Eq. (13) is not emphasized.

The significance of these observations is that one can just as well start with the G_1 equation as with the BS equation. Since the BS kernel V_{BS} is an infinite series, use of either of these equations must inevitably involve an approximation in which this series is truncated after a finite number of terms, usually the first corresponding to one boson exchange (OBE) or one gluon exchange (OGE). I have argued elsewhere that the series for V_1 is probably more convergent than that for V_{BE}, and will not develop this further here[1,11].

1.4 Singularities

The last issue I will discuss in this section is the presence of singularities in relativistic equations. These can readily arise because of the indefinite nature of the Lorentz metric; the squared 4-momentum is not positive definite as it is in non-relativistic physics, and this leads to singularities in propagators which are not always physical.

The singularities in the BS equation are usually eliminated by performing a Wick rotation of the energy variables to complex values. For the G_1 equation, singularities occur which must be eliminated in other ways. I wish to discuss one of these, the so called dissolution singularity, and show how it is dealt with[12].

The propagator (5) can be factorized:

$$G_1(p) = (E_1 + E_2 - W)^{-1} \ (E_1 + W - E_2)^{-1} \tag{15}$$

Note that this has singularities when $W = E_1 + E_2$ and when $W = E_2 - E_1$. The former is the usual elastic cut, and is physical and should be present. The singularity at $W = E_2 - E_1$ is spurious; it would imply that the interaction is strong at small W, regardless of the dynamics.

This singularity can be removed if it is desired to use the G_1 equation for small W. To understand its origin and to see how to remove it, return to the BS propagator

$$G_{BS} = [E_1^2 - (\tfrac{W}{2} + p_o)^2 - i\epsilon]^{-1} \ [E_2^2 - (\tfrac{W}{2} - p_o)^2 - i\epsilon]^{-1} \tag{16}$$

which has four poles in the p_o complex plane, as shown in Fig. 1. When $\vec{p}$ is small and $W \simeq m_1 + m_2$, the negative energy poles are widely separated from the positive energy ones, and integrands will be dominated by either the positive energy pole of particle 1 (if the p_o integration contour is closed in the lower half plane) or by the positive energy pole of particle 2 (if the upper half plane is used). However, when W is small, the positive energy pole of particle 2 is pinched by the negative energy pole of particle 1, and a singularity will arise unless both are retained. If both are kept this would lead to a generalization of (5):

$$\int d^4p_2 \ \delta_+[m_2^2 - p_2^2] \ [m_1^2 - (P - p_2)^2]^{-1} + \int d^4p_1 \ \delta \ [m_1^2 - p_1^2] \ [m_2^2 - (P - p_1)^2]^{-1}$$

$$= \int dp_o \ d^3p \ \frac{\delta \ (E_2 - \tfrac{1}{2} W + p_o)}{2 \ E_2 \ [E_1^2 - (W - E_2)^2]} + \frac{\delta (E_1 + \tfrac{1}{2} W + p_o)}{2 \ E_1 \ [E_2^2 - (W + E_1)^2]} \tag{17}$$

This leads to a coupled set of equations with potentials with different values of the relative energy p_o corresponding to different retardation.

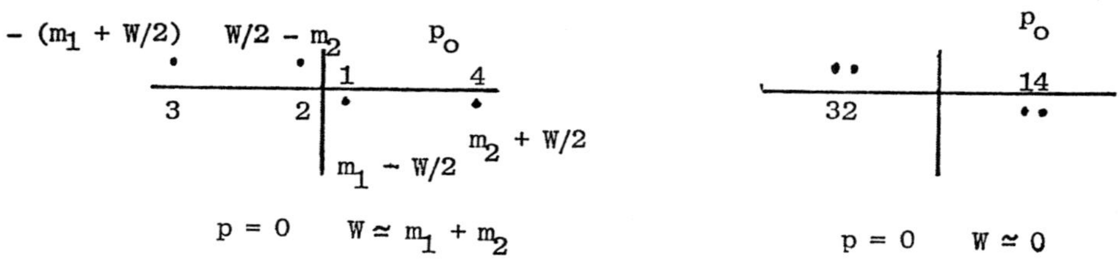

$-(m_1 + W/2) \qquad W/2 - m_2 \qquad p_o$

3 2 1 4

$m_2 + W/2$

$m_1 - W/2$

p_o

32 14

$p = 0 \qquad W \simeq m_1 + m_2$

$p = 0 \qquad W \simeq 0$

Figure 1

When $W \to E_2 - E_1$ the two δ functions become identical so that the different potentials become equal and the singularity at $W = E_2 - E_1$ cancels. At this point the propagator reduces to

$$\int dp_o \; d^3p \; \frac{(E_1 + E_2) \; \delta(\tfrac{1}{2}(E_2 - E_1) + p_o)}{2 \, E_2 E_1 \, [(E_1 + E_2)^2 - W^2]} \quad . \tag{18}$$

As long as one is interested in solutions far away from $W = E_2 - E_1$, it is an excellent approximation to neglect the negative energy channel. This is because the retardation factor in the potential which couples this channel to the positive energy channel is very large (making the potential small) unless W is small.

II. Relativistic Three Body Equations

Applications of relativistic two-body equations have been discussed widely in the past, and will be discussed by Tjon at this conference[18]. In the space remaining to me, I would like to discuss some recent work on a generalization of the G_1 equation to three particles[14].

The equations naturally have a Faddeev structure, with the spectator and one of the two interacting particles on shell. That the spectator must be on shell follows from a consideration of all ladder and crossed ladder exchanges between three particles. An example of a sequence of interactions between particle 1 and 2 followed by 2 and 3 is shown in Fig. 2; it can be seen in this special case that particle 1 and 3

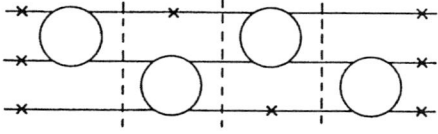

Figure 2

alternate as spectators, and the topology of the diagram requires that
the propagator correspond to particles 1 and 3 on shell, and particle 2
off shell. Consideration of other interactions leads one quickly to the
observation that all three propagators corresponding to the three
combinations of two particles on shell are needed. The equations are
shown diagrammatically in Fig. 3 in the case where all three particles
are different and three body forces are neglected. Note that a symmetric
treatment requires 6 Faddeev amplitudes instead of the three required in
non-relativistic physics, and that the two-body driving amplitudes are
identical to those obtained from the G_1 equation, except that all four
two-body amplitudes corresponding to the four possible choices of which
of the two particles in the initial or final state is to be on shell, are
needed. For identical particles this complication disappears, since the
Pauli-principle leads to the observation that the four possible two-body
amplitudes are all equal, and that there is really only one Faddeev
amplitude.

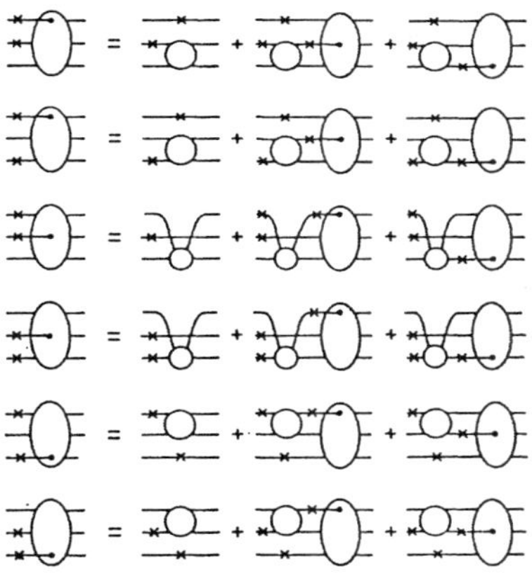

Figure 3

This approach has several advantages. First, note that the BS
treatment of the relativistic 3 body problem would have 8 internal
variables corresponding to the two independent 4 vector momenta. Using
the G_1 equation, the internal relative energies (or relative times) are
fixed in a covariant way by the two mass shell conditions, leaving two
internal relative three momenta, just as in the non-relativistic case.
Hence the G_1 approach has considerably simplified the problem. It is not
very much harder to to solve the three body problem using the G_1 equation
than it is to solve the non-relativistic three-body problem; the
equations can be reduced to coupled two dimensional integral equations.

The G_1 equation also satisfies the cluster property -- in the limit when one of the three particles is removed to infinity, the interaction between the remaining two particles is independent of the presence of the third, except for the requirement of energy conservation which constrains the energy W_{12} of the pair

$$W_{12} = M_T - E_3 \tag{19}$$

where M_T is the total (CM) energy of the three body system, and E_3 is the physical, on shell energy of particle 3.

The explicit form of the bound state equation for three identical spinless particles[15] is:

$$\Gamma(p_3, \tfrac{1}{2}(p_1 - p_2)) = -2 \int \frac{d^3k_1}{(2\pi)^3 2E_1} \frac{M(\tfrac{1}{2}(p_1 - p_2), \tfrac{1}{2}(k_k - k_2); P - p_3)}{m_2^2 - k_2^2}$$

$$\times \Gamma(k_1, \tfrac{1}{2}(p_3 - k_2)) \tag{20}$$

where $k_2 = P - p_3 - k_1$, and $M(p,k;P_{12})$ is the two body scattering amplitude which solves Eq. (5) and $\Gamma(p_3,p)$ is the Faddeev amplitude with the spectator momentum equal to p_3 and the relative momentum of the interacting 1-2 pair equal to p. All of these quantities are manifestly covariant, so the M matrix is a world scalar.

The Lorentz invariance of the helicity formalism makes it convenient for carrying out the partial wave decomposition. Using the paper by Wick[16] the algebra can be carried out, giving the result:

$$\Gamma(P\ p\ q\ j\ m) = - \sum_{j'm'} \int \frac{dp'}{E_{p'}} \frac{q'dq'}{E_{q'}} C \frac{M^j(q, q_o; W_{12})}{W_{12}(2E_o - W_{12})} \tag{21}$$

$$\times D(\chi, \theta, \theta')\ \Gamma(P\ p'q'j'\ m')$$

where

$$C = \frac{W_{23}'((2j + 1)\ (2j' + 1))^{1/2}}{8(2\pi)^3} \tag{22}$$

$$D(\chi, \theta, \theta') = \theta[1 - |\cos \theta'|]\ d^J_{m'm}(\chi)\ d^j_{mo}(\theta)\ d^{j'}_{m'o}(\theta')$$

where $\Gamma(P, p\ q\ j\ m)$ is the projected Faddeev amplitude with p as the magnitude of the spectator 3-momentum, q the magnitude of the relative 3 momentum of the interacting pair <u>in the CM of the pair</u>, j the angular momentum of the pair (defined in its CM), and m is the projection of this angular momentum in the direction of the spectator momentum $\vec{p}$ (equal to the negative of the 3 momentum of the pair). Similarly, M^j is the jth partial wave of the pair, with energy (in the CM of the pair) W_{12} and relative momenta q and q_o. The total angular momentum of the bound state is J. In terms of the variables p, q, p′ and q′ the kinematic quantities are:

$$W_{12}^2 = M_T^2 + m^2 - 2M_T E_p$$

$$W_{32}^{′2} = M_T^2 + m^2 - 2M_T E_{p'}$$

$$E_o = (m^2 + q_o^2)^{1/2} = \frac{1}{2W_{12}} [W_{12}^2 - W_{23}^{′2} + 2W_{23}^{′} E_{q'}]$$

$$p'q' \cos \theta' = (M_T - E_{p'})E_{q'} - W_{23}' E_p$$

$$p q_o \cos \theta = (M_T - E_p) E_o - W_{12} E_{p'}$$

$$pp' \cos \chi = W_{12} E_o - E_{p'}(M_T - E_p) \tag{23}$$

The d's are the usual Wigner rotation functions. The relations between p, p′, q′, and q_o, and the angles θ, θ', and χ are summarized in Figure 4.

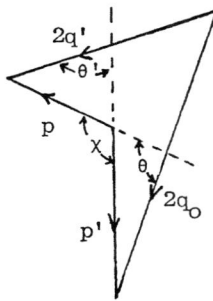

Figure 4

I have presented Eq. (21) in its entirety to show that an explicit formula exists, and to underline the fact that it is no more complicated than the usual non-relativistic formula. Eq. (21) uses the partial wave amplitudes of the two body driving terms derived in the two body rest system. The somewhat complicated relations between q_o and the integration variables p′ and q′ occurs naturally as a result of the

442

covariance.

I will close this talk by noting an unusual feature of the G_1 three-body equations. As the spectator momentum increases, the energy of the interacting pair, Eq. (19), decreases. Eventually the momentum and energy become equal, which occurs at

$$P_{crit} = \frac{4}{3} m \qquad (24)$$

if $M_T = 3m$. At this point the mass of the interacting pair is zero, and it is traveling at the speed of light; the Lorentz transformation effects are enormous. It looks at first as if the singularity discussed in Section 1.4 will arise here, but as it turns out the Lorentz effects over-compensate, and the amplitude is <u>zero</u> at this point, not singular! This happens because the propagator actually becomes (cf. Eq. 23)

$$W_{12}(2E_o - W_{12}) \underset{p \to p_{crit}}{\to} W_{23}'(2E_{q'} - W_{23}') \qquad (25)$$

and any singularity which might arise from $W_{23}' \to 0$ is cancelled by a similar factor in the numerator. Hence the propagator is finite, but because $q_o \to \infty$, the amplitudes $M^j \to 0$, making $\Gamma = 0$ at the critical point. Hence the 3 body equations can be truncated to $p \leq p_{crit}$; they are zero at the boundary and above this region are driven by two body amplitudes with space-like 4-momenta, a region of small effects safely ignorable.

The fully relativistic three body problem is therefore ready to be solved, but given experience with the non-relativistic problem we can anticipate that good numerical results are not likely to be ready for several years!

Acknowledgments

This work was supported by the U.S. Department of Energy through CEBAF and the National Science Foundation.

References

1.   F. Gross in Nuc. Phys. <u>A416</u> (1984) 387C; in "Electron and Photon Interations at Intermediate Energies", Ed. by D. Menze <u>et. al.</u>, Springer lecture notes in physics #234, p. 292, in "Few-Body Problems in Physics", Ed. by L.D. Faddeev and T.I. Kopaleishvili, World Scientific, p. 344; in "Proceedings of the LAMPF Workshop on Dirac Approaches to Nuclear Physics; LA-10438-C, p. 351; in the "Second Workshop on Prospectives in Nuclear Physics at Intermediate Energies", Trieste, March 1985 Ed. by S. Boffi, C. Ciofi degli Atti, and M.M. Giannini, World Scientific, p. 72.

2. F.P. Juster, et. al., Phys. Rev. Lett. 55 (1985) 2261.

3. W. Turchinetz, Proceedings of the CEBAF 1986 Summer Workshop, to be published.

4. See K. Holinde and R. Machleidt, Nucl. Phys. A327 (1981) 349; R. Machleidt, Trieste Workshop op. crit. p. 282.

5. See E.F. Van Faassen and J.A. Tjon, Phys. Rev. C24 (1981) 736; 28 (1983) 234; 30 (1984) 285; 33 (1986) 2105. J.A. Tjon to be published in the proceedings of the Tokyo Few Body Conference, August 1986, and contributions to this conference.

6. See F. Gross, Phys. Rev. 186 (1969) 1448; D10 (1974) 233.

7. See J. M. Namyslowski, Proceedings of the Graz Conference (1978) Lecture Notes in Physics #82, Ed. by Zingl et. al., p. 41; M. Chemtob, Nuc. Phys. A336 (1980) 299; L.L. Frankfurt and M.I. Strickman, Nucl. Phys. B148 (1979) 107; Physics Reports 76 (1981) 215.

8. G.P. Lepage and S.J. Brodsky, Phys. Rev. D22 (1980) 2157.

9. See, however, work by Frankfurt and Strickman, ref. 7.

10. F. Gross and B.D. Keister, Phys. Rev. C28, (1983) 823.

11. F. Gross, Phys. Rev. C26 (1982) 2203.

12. I am indebted to S.J. Wallace for calling my attention to this issue and for helpful discussions. See also V.B. Mandelzweig and S.J. Wallace, "QED Based Two-Body Dirac Equation", submitted for publication.

13. J. Tjon, contribution to this conference.

14. F. Gross, Phys. Rev. C26 (1982) 2226.

15. F. Gross, notes for lectures given at the Universitat Hannover, 1983, unpublished.

16. Wick, Ann. of Phys. (NY) 18 (1962) 65.

RELATIVISTIC CALCULATIONS IN FEW BODY SYSTEMS.

J.A. Tjon

Institute for Theoretical Physics
Princetonplein 5
P.O. Box 80.006
3508 TA Utrecht, The Netherlands

Calculations in the two and three nucleon systems are described using the field theoretical Bethe-Salpeter equation and the relativistic quasi potential approximations to it. Relativistic effects are discussed both for bound states and scattering solutions. Including the isobar degrees of freedom in a relativistic OBE model leads to a reasonable description of the dynamics of the NN system at intermediate energies.

I. Introduction.

In the dynamical description of composite systems at medium energies and large momentum transfers the effects of special relativity are expected to play a significant role. Apart from the inclusion of "minimal relativity" i.e. accounting for the proper relativistic kinematics, also the relativistic spin structure of the particles may have an important effect. Known examples are the so-called pair excitation contribution to the electromagnetic (em) processes in nuclei and the essential role of the Z-graph in the successful phenomenological Dirac description of elastic proton nucleus scattering at intermediate energies [1]. Here we give a brief review of a particular Lorentz covariant approach we have adopted in the past years in the study of the few nucleon system. It is based on a field theoretical framework with an underlying meson-nucleon dynamics.

II The deuteron problem.

One of the simplest systems to consider is the two nucleon system. To describe the scattering process we may use the Bethe-Salpeter (BS) equation as a starting point [2]. As a model for the nuclear force the kernel of the BS equation is approximated by the sum of one boson exchanges, given by π, ρ, ω, ε, η and δ mesons. The relativistic spin complication of the nucleons is fully accounted for. To make the driving force well behaved at high momenta strong form factors of the monopole type are introduced in the meson-nucleon vertices. The value of the cutoff parameter in these form factors is taken to be the same for all the mesons and is about 1.2 GeV. In the partial wave reduction of the BS equation the helicity formalism is used. The resulting equations are a coupled set of singular integral equations in two continuous variables and eight discrete spin channels. In addition to the spin states which have corresponding nonrelativistic ones also negative spinor states occur. Most singularities can be removed by applying a Wick rotation on the partial wave decomposed equations and which are then solved by Padé approximant techniques. To obtain a good description to the scattering phases pair suppression is needed in the intermediate NN states. This was achieved by employing a pseudo vector coupling theory for the pion [3].

The above relativistic one boson exchange (OBE) model for the nucleon-nucleon interaction has been used to study the em properties of the deuteron [4]. For a recent review of this work see Ref 5. One of the problems in describing the em interaction is that it is not independent of the dynamics of the composite system, since Ward identities have to be satisfied. In the usual estimates of mesonic exchange currents it is assumed that it can be determined perturbationally. Only information of the positive energy spinor components of the deuteron wavefunction is needed. These are obtained from a nonrelativistic calculation with realistic potentials like the Reid soft core. Such an analysis may however be inconsistent. In the relativistic model as discussed above a consistent calculation can be carried out. The validity of the Ward identity is equivalent to having the proper normalisation condition on the boundstate wave function.

To get some insight in the consistency problem we have studied the elastic em form factors of the deuteron in the above model. This can be done in a fully consistent way. It was found that the difference between a nonrelativistic calculation and the relativistic one was much smaller than expected from the usual perturbational estimate of the pair excitation graph contribution. Within a quasi potential (QP) approach the smallness of the relativistic effect can be understood in some detail [6]. Let us make a QP approximation to the BS equation in order to have a closer similarity to a nonrelativistic potential description. For this purpose let us use the Gross prescription i.e. put the spectator nucleon on the mass shell.

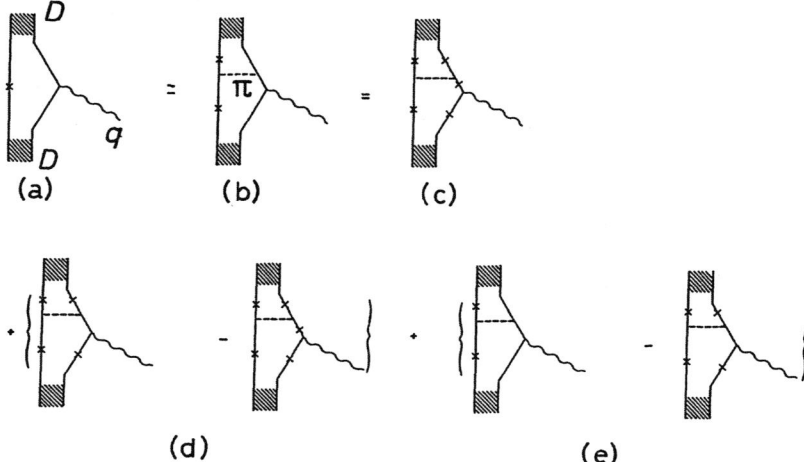

Fig. 1. The em deuteron vertex function in the Gross approximation: x means that the given nucleon is on mass shell. (c) corresponds to the nonrelativistic impulse approximation, while (d) and (e) are the pair term and dynamical correction respectively.

Similar results for the em form factors are then obtained as in the BS case. The various contributions can be analyzed perturbationally in such a quasi potential model. This is illustrated schematically in Fig. 1 for the em deuteron current. Since the deuteron wave function ψ satisfies the homogeneous QP equation

$$\psi = G_o V \psi \tag{1}$$

with G_o the Gross two-nucleon propagator and V the potential, we may replace it in the current by the right hand side of Eq. (1). It was verified numerically that the largest contribution of the once iterated wave function in the em deuteron current comes from the one pion exchange part in V and consequently graph (1a) can well be approximated by graph (1b). The latter can be rewritten as a sum of three contributions, corresponding to the nonrelativistic result (1c), the pair excitation contribution (1d) and a dynamical correction (1e). The major contribution in the latter correction comes from the deformation of the deuteron due to the presence of the negative energy spinor components in the intermediate states. It gives rise to a very specific off shell dependence in the positive energy spinor state components of the deuteron which is not

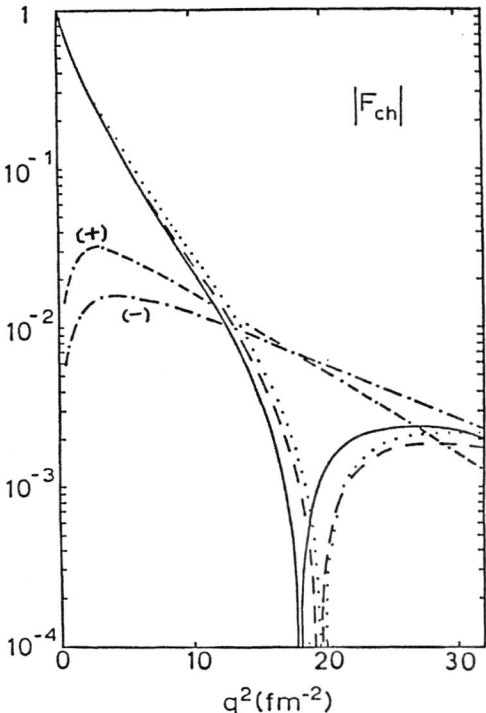

Fig. 2. The calculated charge form factor of ^{2}H for the relativistic OBE model with the Gross prescription. Complete calculation (solid line) and the static approximation (dashed line). Pair term (dash-dotted line) and dynamical correction (dash-dash-dotted line) have opposite sign. For comparison the Reid result is also shown (dotted line). Pseudo vector coupling for the pion is used.

present in a nonrelativistic calculation. In Fig. 2 are shown the various contributions to the em charge form factor of the deuteron. The static approximation which is close to the nonrelativistic one consists of neglecting the negative energy state contributions and the boost effects on the arguments of the wave function. Since the dynamical correction is opposite in sign to the pair term they tend to cancel each other at moderately large momentum transfer resulting into much smaller corrections than expected from the conventional estimates. In particular, in the magnetic form factor $B(q^{2})$ as measured recently at Saclay [7] some disagreement with theory remains if this dynamical correction is also included.

III. NN Scattering at intermediate energies.

At energies $T_{lab} > 330$ MeV pion production becomes possible. Since this takes place predominantly through the production of the P_{33} resonance, it is natural to extend the above relativistic OBE model to also allow for isobars in the intermediate states [8]. For the transition interaction π and ρ meson exchanges have been used. The diagrammatic representation of the extended model is shown in Fig. 3. Using the Rarita-Schwinger formalism

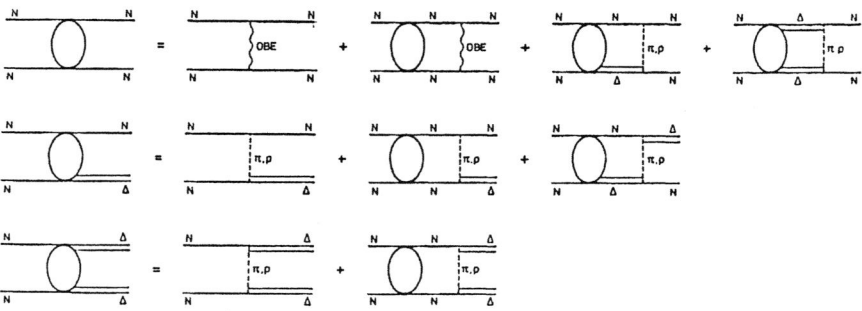

Fig. 3. Diagrammatic representation of the relativistic OBE model with isobars.

the propagator is described in terms of an energy dependent complex mass $m_\Delta = m_o - i \, \Gamma(q)/2$ with $m_o = 1236$ MeV. For the width Γ the Bransden-Moorhouse parameterisation [9] is used. The πN three-momentum q can readily be expressed in the πN invariant mass square $s_{\pi N}$. Following VerWest [10] we may assume that in the $N\Delta$ cm system $s_{\pi N}$ can be identified with the maximally allowed energy for the isobar i.e.

$$s_{\pi N} = (\sqrt{s} - m_N)^2 \tag{2}$$

with $\sqrt{s}$ the total energy of the two-nucleon system.

Since in the simpler model with only NN intermediate states the influence of the negative energy spinor states has been shown to be compensated by changes in the coupling constants, most calculations have been carried out neglecting these states for both the N and Δ in the equations. In Table 1 we show the effect of also including the negative energy states in the NN sector using the quasi potential approximation of the Blankenbecler-Sugar type. It leads in general to an additional repulsion in the various waves. Although the effect is sizable, by changing the coupling constant of the ε meson much of the difference can be removed. Using for $f^2_{N\Delta\pi}/4\pi = 0.35$, which is consistent with the isobar width, the gross features are in agreement with the experimental phase shift analysis. To obtain a reasonable state dependence of the p-waves the ρ exchange in the transition potential is needed. The resonantlike structures found experimentally in the $^1D_2$ and $^3F_3$ channels are also present in the considered model.

The inclusion of the Δ degrees of freedom is sometimes done by only modifying the NN force through addition of the NΔ and $\Delta\Delta$ box diagrams to

Table 1. The phase shift δ and inelastic parameter η for various waves with and without inclusion of the negative energy spinor states in the NN sector in the QP model.

Elab:	200 MeV		500 MeV		800 MeV	
			Negative energy spinor states included			
	δ	η	δ	η	δ	η
$^1S_0$	9.59	1.000	−25.33	0.969	−45.01	0.890
$^3S_1$	8.19	1.000	−25.39	1.000	−43.27	0.942
$^1P_1$	−26.53	1.000	−40.48	1.000	−46.01	0.961
$^3P_0$	−4.98	1.000	−33.77	0.970	−55.25	0.879
$^3P_1$	−15.87	1.000	−27.30	0.908	−38.55	0.614
$^3P_2$	15.21	1.000	11.89	0.942	5.23	0.731
$^1D_2$	10.56	1.000	15.79	0.900	2.49	0.703
$^3D_1$	−16.34	1.000	−13.94	1.000	−11.94	0.964
$^3D_2$	23.95	1.000	14.55	1.000	3.49(−1)	0.990
			Negative energy spinor states not included			
	δ	η	δ	η	δ	η
$^1S_0$	12.91	1.000	−21.52	0.971	−40.36	0.894
$^3S_1$	9.06	1.000	−24.41	1.000	−42.07	0.941
$^1P_1$	−23.19	1.000	−35.52	1.000	−38.91	0.953
$^3P_0$	−1.69	1.000	−27.61	0.972	−46.21	0.895
$^3P_1$	−12.64	1.000	−18.02	0.883	−23.43	0.528
$^3P_2$	16.58	1.000	15.12	0.937	10.14	0.710
$^1D_2$	10.88	1.000	16.97	0.898	4.13	0.698
$^3D_1$	−15.76	1.000	−10.05	1.000	−4.29	0.953
$^3D_2$	25.69	1.000	18.98	1.000	6.74	0.989

the potential instead of solving the complete coupled channel equations [11]. The above approximation corresponds to dropping the direct interaction between the NΔ states in our model. In Table 2 are given the calculated phase parameters in this approximation. Comparing the results with the corresponding ones in Table 1, we see that the effect is in general small.

Table 2. The calculated phase parameters in the QP model for the case that the direct interaction between the NΔ states is neglected. Negative energy spinor states are not included.

Elab:	200 MeV		500 MeV		800 MeV	
	δ	η	δ	η	δ	η
$^1S_0$	12.62	1.000	−21.69	0.972	−40.48	0.895
$^3S_1$	9.06	1.000	−24.42	1.000	−42.07	0.941
$^1P_1$	−23.20	1.000	−35.52	1.000	−38.91	0.953
$^3P_0$	−1.92	1.000	−28.07	0.978	−46.25	0.914
$^3P_1$	−12.64	1.000	−18.32	0.885	−24.08	0.542
$^3P_2$	16.53	1.000	14.93	0.941	10.42	0.722
$^1D_2$	10.74	1.000	16.59	0.900	3.54	0.714
$^3D_1$	−15.76	1.000	−10.05	1.000	−4.29	0.953
$^3D_2$	25.69	1.000	18.98	1.000	6.74	0.989

As is well known two particle unitarity is satisfied exactly by the BS equation below the one-pion production threshold. In the inelastic region the solution of the ladder BS equation may violate the unitarity condition sothat the inelasticity parameter η for a given partial wave may exceed unity [12]. This is shown in Fig. 4 for the case of the relativistic OBE model without isobars [13]. In the region up to the second inelastic threshold unitarity can be restored exactly by renormalizing the nucleon propagator by the lowest order self energy bubble diagram, which can readily be seen using the Cutkowski cutting rules. Going beyond the two-pion production threshold the unitarity condition breaks down, but in actual calculations the violations appear to be small and we may expect that the associated two-pion cuts are of minor importance. Another part of the relativistic OBE model with isobars can be improved [13]. Instead of using a fixed mass model for the Δ we may describe it as a πN scattering process using a separable interaction. As the NΔπ vertex function a product is used of the usual OBE form factor and a scattering function, which

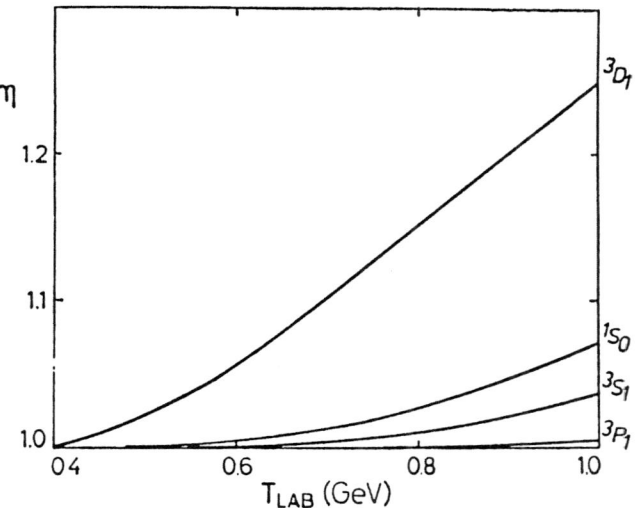

Fig. 4. Violation of the unitarity condition in the relativistic OBE without isobars using the BS equations.

depends on the relative four–momentum square of the πN system. In order to get a good effective range a relatively low cutoff mass is needed in the scattering function. An excellent fit to the phaseshifts can be found in this way for the P_{33} channel.

In Fig. 5 are shown the calculated phase parameters for the NN channels which exhibits clearly resonantlike behaviour. The two curves B

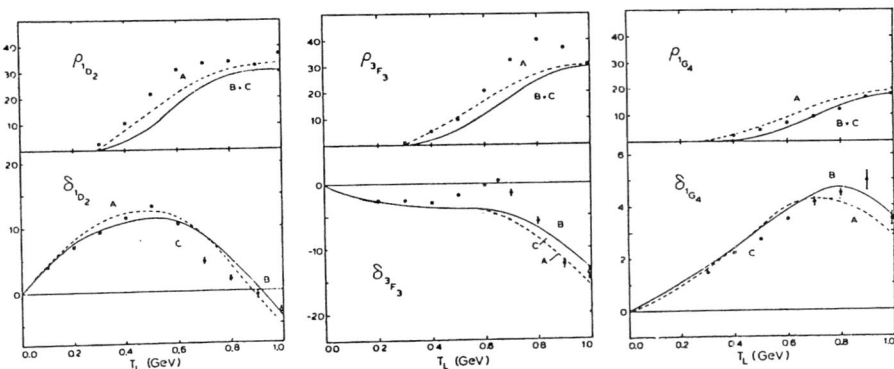

Fig. 5. The calculated phase shift δ and inelastic parameter ρ for the fixed mass model (curve A) and for the unitary model (curves B and C). The experimental data points are from Ref 14.

and C correspond to somewhat different choices for the ε and ω coupling constants. For comparison the results are shown for the fixed mass model (curve A). We see that the inelasticity parameter at lower energies is considerably smaller for the unitary model. The major reason for this is that the assumption (2) neglects the kinetic motion of the nucleon. Taking the physically more acceptable prescription [13,15]

$$s_{\pi N} = (\sqrt{s} - \sqrt{p_1^2 + m_N^2})^2 \quad , \tag{3}$$

with p_1 the three-momentum of the nucleon, the inelasticity indeed reduces substantially near the one-pion production threshold. More recently we have studied the effect of introducing the πd channel in the quasi potential equations [16], showing that for at least the $^1D_2$ channel most of the missing strength in the inelasticity can be ascribed to the coupling to the πd channel. For the $^3F_3$ channel there still appears to be a persistent lack of strength in the enelasticity and the pronounced structure in the phaseshift around 800 MeV is not well reproduced.

IV. The tri-nucleon system.

With regard to the tri-nucleon system actual calculations have been rather rudimentary and on the level of only including special relativity in a minimal way. The groundstate problem has been studied using various versions of QP equations [17-22]. Calculations indicate that the relativistic effect on binding is small and of the order of 0.3 MeV. They differ however in the predictions of its sign. For the relativistic OBE model using the quasi potential equations from Ref 17 the ^3H binding energy has been determined in the two-channel approximation using the $^1S_0$ and $^3S_1$ two-nucleon T-matrix as input [23]. A unitary pole expansion has been used. The calculated binding energy of $E_{3_H} = -7.5$ MeV is larger in magnitude than found for the Reid soft core potential in the same approximation. This additional binding can be ascribed to the lower D-state probability of 4.7 % for this model and a small part to relativity. In Fig. 6 is shown the resulting ^3H_e charge form factor for the case that 8 terms have been used in the separable expansion of the two-body T-matrix. Similarly as in the Reid soft core case the secondary maximum is much too low as compared to the experiment. To study off mass shell effects we have recently solved the relativistic field theoretical Bethe-Salpeter-Faddeev (BSF) equations for the three nucleon system under certain simplifying assumptions [24]. The relativistic spin structure of the nucleons is neglected and it is assumed

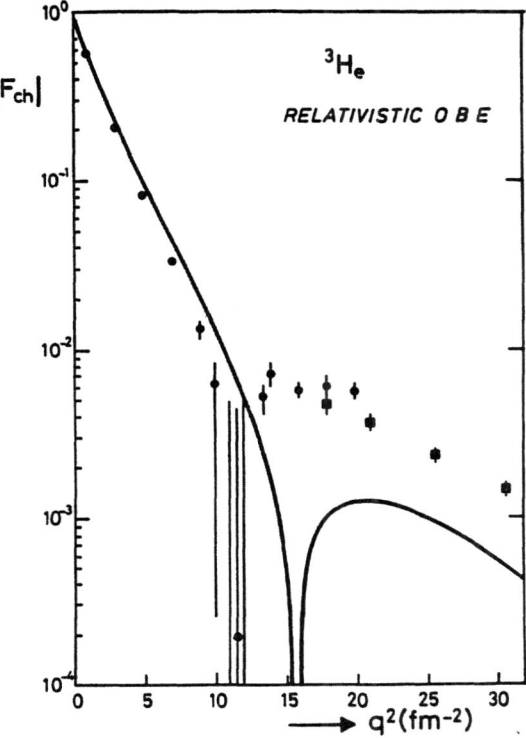

Fig. 6. The calculated charge form factor of ^3H_e for the relativistic OBE model in the QP approximation.

that the NN interaction is given by separable interactions. For such interactions the BS equation for two particles can be solved in a closed form. Two kinds of form factors have been used, the Yamaguchi type

$$g_\gamma(p_o p) = \frac{1}{-p_o^2 + p^2 + \beta^2 - i\epsilon} \qquad (4)$$

and the Tabakin type

$$(5) \qquad g_{T_n}(p_o, p) = \frac{-p_o^2 + p^2 + \alpha^2}{-p_o^2 + p^2 + \gamma^2 - i\epsilon} \; \frac{p_c^2 + p_o^2 - p^2}{(-p_o^2 + p^2 + \beta^2 - i\epsilon)^2}$$

with $n = \frac{3}{2}$ and 2. Here p and p_o are the relative three-momentum and energy between the two nucleons. With the separable ansatz for the two-nucleon T-matrix the BSF equations reduce to two-dimensional integral equations. After Wick rotation these equations can be solved using the standard ratio and Padé approximant methods. Some results are shown in Table 3 for various combinations of form factors used in the $^1S_0$ and $^3S_1$ channels. For comparison are given the results using the nonrelativistic and QP approximation of Ref 17. These calculations indicate that the binding

Table 3. The calculated triton binding energies in MeV for the BS, QP and nonrelativistic equations with various choices of the two-nucleon separable interactions.

$^3S_1$	$^1S_0$	NR	QP	BS
Y	Y	10.65	11.00	11.09
Y	$T_{3/2}$	7.95	8.22	8.26
$T_{3/2}$	Y	8.17	8.29	8.43

energy corrections are small and attractive in accordance with some of the previously found results. The em charge form factors of the tri-nucleon system has also been calculated for this model in the static approximation. One interesting result is that in the field theoretical case although the binding energy increases more correlation than in the nonrelativistic case is found in the em form factor. Work is in progress to employ more realistic interactions in the form of a multi-term separable representation of the nuclear force.

References

1. S.J. Wallace Lecture Notes of the summerschool at Tokyo (1985); J.A. Tjon, Proceedings of the fourth MiniConference Amsterdam, p75 (1985).

2. J. Fleischer and J.A. Tjon, Nucl. Phys. B84, 375 (1975).

3. J. Fleischer and J.A. Tjon, Phys. Rev. D21, 87 (1980).

4. M.J. Zuilhof and J.A. Tjon, Phys. Rev. C22, 2369 (1980).

5. J.A. Tjon, Proc. XI-th International Conference on Few Body Systems, Tokyo-Sendai august 1986.

6. M.J. Zuilhof and J.A. Tjon, Phys. Lett. 84B, 31 (1979); Phys. Rev. C24, 736 (1981).

7. S. Auffret et al., Phys. Rev. Lett. 54, 649 (1985).

8. E. van Faassen and J.A. Tjon, Phys. Rev. C30, 285 (1984).

9. R.H. Bransden and R.G. Moorhouse, The Pion Nucleon System (Princeton University Press, Princeton).

10. B.J. VerWest, Phys. Rev. C25, 482 (1982).

11. R. Machleidt, K. Holinde and Ch. Elster, Phys. Repts., to be published.

12. M. Levine, J. Wright and J.A. Tjon, Phys. Rev. <u>157</u>, 1416 (1967).

13. E. van Faassen and J.A. Tjon, Phys. Rev. <u>C33</u>, 2105 (1986).

14. R.A. Arndt et al., Phys. Rev. <u>D28</u>, 482 (1983).

15. W.M. Kloet and J.A. Tjon, Phys. Rev. <u>C30</u>, 1653 (1984).

16. E. van Faassen and J.A. Tjon, to be published.

17. A.D. Jackson and J.A. Tjon, Phys. Lett. <u>32B</u>, 9 (1970).

18. V.S. Bhasin, H.J. Facob and A.N. Mitra, Phys. Lett. <u>32B</u>, 15 (1970).

19. E. Hammel, H. Baier and A.S. Rinat, Phys. Lett. <u>85B</u>, 193 (1979).

20. L.A. Kondratyuk, J. Vogelzang and M.S. Fanchenko, Phys. Lett. <u>98B</u>, 405 (1981).

21. W. Glöckle, T-S. H. Lee and F. Coester, Phys. Rev. <u>C33</u>, 709 (1986).

22. G. Rupp, L. Streit and J.A. Tjon, Proc. IX European Conf. on Few Body Problems in Physics, Tbilisi, 1984 (World Scientific, Singapore 1985).

23. J.A. Tjon, Proc. VII International Conf. on Few Body Problems in Nuclear and Particle Physics, New Delhi 1975, p567 (North-Holland, Amsterdam 1976).

24. G. Rupp and J.A. Tjon, to be published.

RELATIVISTIC FADDEEV CALCULATIONS OF πNN SYSTEMS

Humberto Garcilazo

Kernforschungszentrum Karlsruhe, Institut für Kernphysik,
D-7500 Karlsruhe, Federal Republic of Germany

The five reactions of the πNN system πd → πd, πd → πNN, NN → NN, NN → πd, and NN → πNN are calculated simultaneously using a relativistic Faddeev theory in which the spectator particles are on the mass shell and the exchanged particles are off the mass shell. This theory includes also a direct nucleon-nucleon interaction as represented by the Paris potential.

The amplitudes T_{dd}, $T_{\Delta d}$, T_{NN}, T_{dN}, and $T_{\Delta N}$ corresponding to the processes πd → πd, πd → πNN, NN → NN, NN → πd, and NN → πNN, obey integral equations of the form[1]

$$T_{ij} = V_{ij} + \sum_{k=d,\Delta,N} V_{ik}\, \tau_k\, T_{kj} \; ; \quad i,j = d,\Delta,N \, , \tag{1}$$

$$V_{dd} = 0 \, , \tag{2}$$

$$V_{NN} = V_{OPE} + V_{HM} \, , \tag{3}$$

where in our particular case d represents the nucleon-nucleon isobars $^1S_0$ and $^3S_1$-$^3D_1$, Δ represents the pion-nucleon isobars S_{11}, S_{13}, P_{13}, P_{31}, and P_{33}, N represents the pole part of the P_{11} isobar, V_{OPE} is the nucleon-nucleon one-pion-exchange potential, and V_{HM} is the contribution of heavy meson exchanges to the nucleon-nucleon interaction. These equations satisfy two and three-body unitarity. Relativistic invariance is fulfilled in our model,[2] by putting the spectator particles on the mass shell and the exchanged particles off the mass shell.[3] Thus, the transition potentials V_{ij} in Eq. (1), are standard Feynmann diagrams corresponding to a one-particle-exchange mechanism, and the isobars are four-component spinors for particles of spin j and mass $\sqrt{s}$, where j is the total angular momentum of the two-body channel and $\sqrt{s}$ its invariant mass. The functions τ_d, τ_Δ, and τ_N, are proportional to the reduced on-shell amplitudes of the two-body subsystems $e^{i\delta}\sin\delta/k^{2\ell+1}$. Since the transition potentials proceed either by pion or by nucleon exchange, we multiply every vertex by the form factor

$$f_i(t) = \frac{\Lambda_i^2 - m_i^2}{\Lambda_i^2 - t} \; ; \quad i = \pi, N \; , \tag{4}$$

where t is the invariant mass squared of the exchanged particle (which is off the mass shell). We took Λ_N = 1600 MeV/c, since we have found that the results are quite insensitive to its value. The results on the other hand, are very sensitive to the value of Λ_π and we have found that using pseudo-vector coupling for the πNN vertex requires $\Lambda_\pi \sim 650$ MeV/c.

If we use for the direct nucleon-nucleon interaction only the one-pion-exchange part $V_{NN} = V_{OPE}$, this theory works very well for the πd elastic and breakup reactions $\pi d \to \pi d$ and $\pi d \to \pi NN$, but it works only poorly for the reactions involving the NN channel, NN → NN, NN → πd and NN → πNN. This is due to the fact that the one-pion-exchange potential V_{OPE} contains only the long-range part of the nucleon-nucleon force. Thus, in order to include a more realistic treatment of the direct nucleon-nucleon interaction V_{NN}, we use for it the Paris potential.[4] The Paris potential however, contains an uncorrelated two-pion exchange piece which preceeds by the excitation of intermediate Δ states, so that we have to modify the transition potentials $V_{\Delta N}$ so as to insure that these contributions are not included twice below the pion production threshold. This modification can be done without violating any of the unitarity conditions.[5]

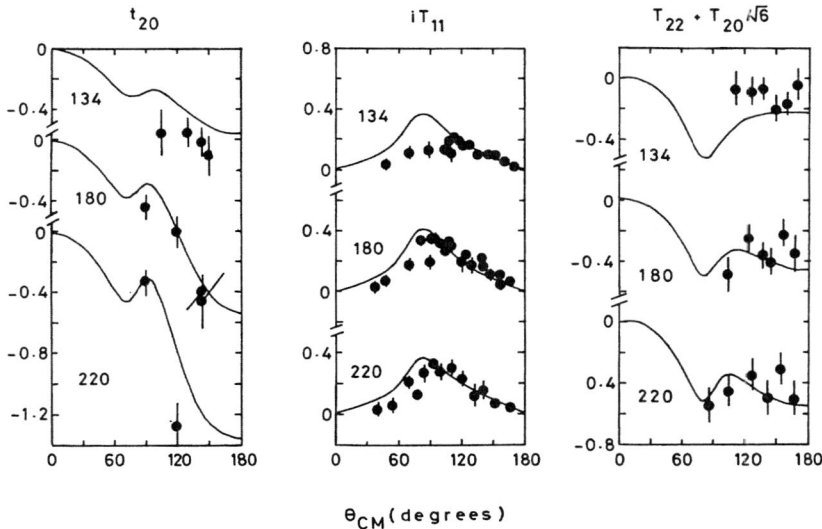

Fig. 1. Polarization observables of the reaction $\pi d \to \pi d$ at three energies.

458

We show in Fig. 1 some results of this theory for the reaction $\pi d \to \pi d$ at incident pion energies of $T_\pi = 134$, 180, and 220 MeV, where we compare with three different polarization observables. The tensor polarization t_{20} was until recently a very controversial observable since there were contradictory measurements of it by different experimental groups[6] as well as contradictory predictions by several theoretical groups[7]. This controversial situation has now been settled and no serious dissagreement remains any more. The data for the linear combination of obsrvables $T_{22} + T_{20}/\sqrt{6}$ is preliminary data taken recently by the Karlsruhe-SIN group[8] but already one can see that it represents no serious problem for the theory. Similar situation exists for the vector analyzing power iT_{11}. The results of Fig. 2

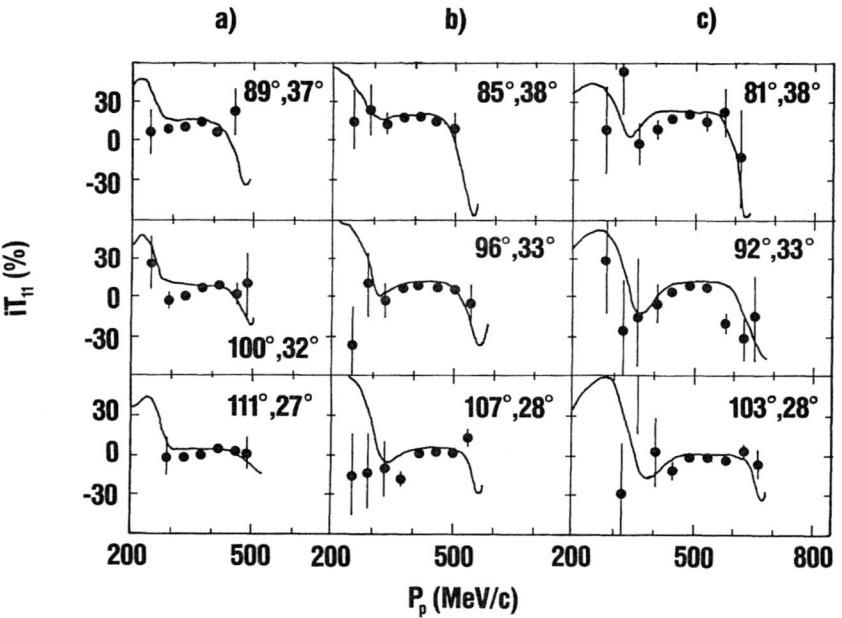

Fig. 2. Vector analyzing power iT_{11} of the reaction $\pi^+ d \to \pi^+ pn$ at a) $T_\pi = 180$ MeV, b) $T_\pi = 228$ MeV, and c) $T_\pi = 294$ MeV, as a function of the proton momentum. The first angle is that of the pion and the second one that of the proton.

correspond to the reaction $\pi^+ d \to \pi^+ pn$ at three different incident pion energies. The vector analyzing power iT_{11} of this reaction has been measured recently by the Karlsruhe-SIN colaboration[9]. Here the theory is in excellent agreement with the data not only for the vector analyzing power, but also for the differential cross section[9]. We show in Fig. 3

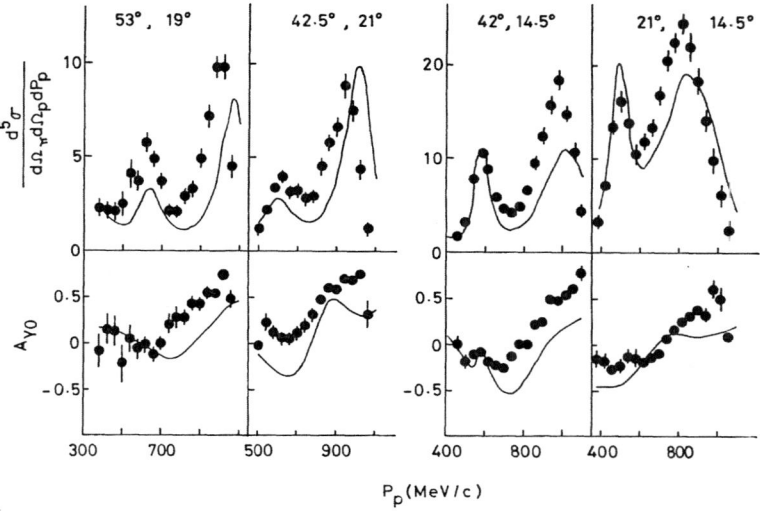

Fig. 3. Differential cross section and vector analyzing power A_{Y0} of the reaction $pp \to \pi^+ pn$ at 800 MeV, as a function of the final proton momentum. The first angle is that of the pion and the second one that of the proton. The experimental data is from Ref. 10.

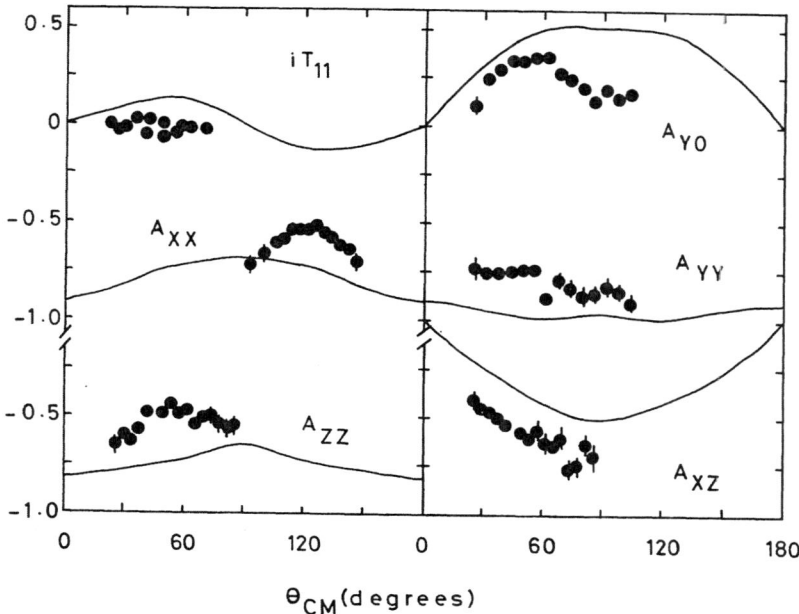

Fig. 4. Polarization observables of the reaction $pp \to \pi^+ d$ at 538 MeV.[11]

some results for the reaction pp → π⁺pn at an incident proton energy of
800 MeV, where we compare both the differential cross section and the vector analyzing power A_{Y0}. The peak of the differential cross section at the
lower proton momentum is due to the nucleon-nucleon final-state interaction
while the peak at the higher momentum is due to the formation of an intermediate Δ^{++} resonance. Both peaks are reproduced by the theory, although
the agreement with the data for this reaction is not as good as in the case
of the previously discussed π⁺d → π⁺pn reaction. This is due to the fact
that while the reaction π⁺d → π⁺pn proceeds by nucleon exchange so that the
pnd vertex is completely determined by the deuteron wave function, the
reaction pp → π⁺pn on the other hand proceeds by pion exchange so that in
this case we have in addition the possibility of exchanging a ρ meson between the proton and the Δ^{++} (which we have not taken into account), while
there are also uncertainties due to the unknown nature of the πNN vertex
(whether it is pseudovector or pseudoscalar or what is more likely a linear
combination of both). The results in Fig. 4 correspond to the reaction pp →

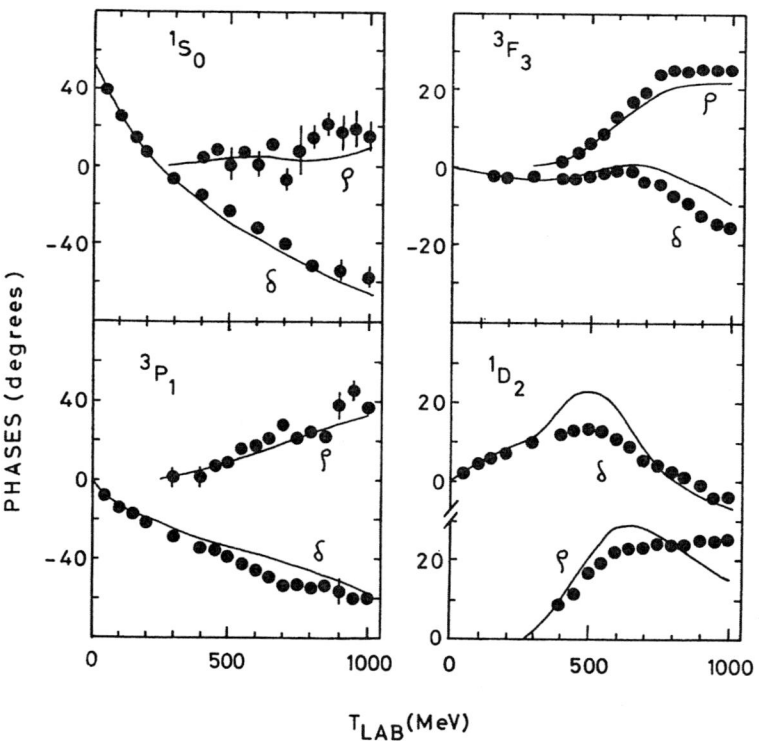

Fig. 5. Several nucleon-nucleon partial-wave amplitudes between 0 - 1 GeV.

π^+d at an incident proton energy of 538 MeV, where we show a comparison with six different polarization observables. In this case as we see, there is agreement between theory and data only to the level of 0.2 which corresponds to 10 % of their allowed range of variation. Finally, we show in Fig. 5 as an example of the reaction NN $\rightarrow$ NN, some phase shifts and inelasticities between 0 and 1 GeV. In the case of the $^3F_3$ and $^1D_2$ amplitudes, the corresponding Argand diagrams show a resonant-like behavior in agreement with the data.[12]

To conclude, we have shown that the five reactions of the πNN system can be described simultaneously within a single theoretical model based in a relativistic Faddeev theory which includes a realistic treatment of the direct nucleon-nucleon interaction.

REFERENCES

1. Y. Avishai and T. Mizutani, Nucl. Phys. A326, 352 (1979); B. Blankleider and I. R. Afnan, Phys. Rev. C24, 1572 (1981); A. S. Rinat, Y. Starkand, and E. Hammel, Nucl. Phys. A364, 486 (1981).

2. H. Garcilazo, KFK preprint 86-1, to be published,

3. H. Garcilazo, Phys. Rev. Lett. 45, 780 (1980); Phys. Lett. 99B, 195 (1981); F. Gross, Phys. Rev. C26, 2226 (1982).

4. M. Lacombe et al., Phys. Rev. C21, 861 (1980)

5. H. Garcilazo, to be published.

6. W. Grüebler et al., Phys. Rev. Lett. 49, 444 (1982); Phys. Rev. Lett. 52, 333 (1984); E. Ungricht et al., Phys. Rev. Lett. 52, 333 (1984); Phys. Rev. C31, 934 (1985). Y. M. Shin et al., Phys. Rev. Lett. 55, 2672 (1985); G. R. Smith et al., Phys. Rev. Lett. 57, 803 (1986).

7. I. R. Afnan and R. J. McLeod, Phys. Rev. C31, 1821 (1985); H. Garcilazo , Phys. Rev. Lett. 53, 652 (1984).

8. C. R. Ottermann, talk presented at the "Workshop on Spin Phenomena in Hadronic Interactions", Sept. 1986, Wuppertal.

9. W. Gyles et al., Phys. Rev. C33, 583 (1986); C33, 595 (1986).

10. A. D. Hancock et al., Phys. Rev. C27, 2742, (1983).

11. A. B. Laptev and I. I. Strakovsky; A Collection of Experimental Data for the Reaction pp $\rightarrow$ dπ^+ Process. Leningrad 1985.

12. R. A. Arndt et al., Phys. Rev. D28, 97 (1983).

QUARKS IN NUCLEONS AND NUCLEI

C. W. Wong
Department of Physics, University of California
Los Angeles, CA 90024, U.S.A.

A gluon-exchange quark model is found to give baryon-nucleon spin-orbit forces rather similar to those of traditional meson-exchange models. Hidden-color basis states of multiquark systems are expressed in terms of states involving color-singlet hadrons. The size of various nucleon form factors is used to separate the interior perturbative region of quarks from the exterior nonperturbative region of hadrons. The distinction between the baryon interior and its exterior appears to be relevant in baryon spectroscopy and in nuclear forces.

1. Introduction

By quarks in nuclei [1], we usually do not refer to the virtual mesons in nuclei, even though they are also made up of quarks, i.e., $q\bar{q}$ pairs. We refer rather to the nucleon core of quarks in nuclei. This intuitive picture of baryons gains conceptual support from the two-phase picture of baryons, as modeled for example by the MIT bag [2]. In this picture, there is an outside nonperturbative phase of pions and other virtual mesons of broken symmetries, and an inside perturbative phase involving quarks and gluons interacting more weakly [3-5].

At low energies, explicit quark degrees of freedom are not expected to come into play. This expectation is supported by the empirical successes in low-energy phenomenology of the Skyrme topological soliton model [6], which does not involve quarks explicitly. The rather similar predictions on low-energy nucleon properties given by a variety of models — big bag, little bag or no bag at all — gives rise to the "Cheshire cat" picture that at low energies the quarks could well disappear into thin air, provided that they leave behind their topological "grin" [7].

Explicit quark effects are expected to show up [8] at sufficiently high energies, when one is probing the small distances of perturbative QCD. It is not clear, however, where this inside region begins. One can probably say in a very rough way that the quark description might be preferable in an energy regime where the number and width of hadron resonances cause them to overlap badly. Past experience in high energy scattering and reactions suggests that while this is qualitatively to be expected, it is nevertheless difficult to specify the point beyond which the quark description will become more effective [9]. For example, we do not know if this point begins at 2 GeV where tensor glueball are expected [10].

In this talk, I would like to describe a few anecdoctal results which on the one hand appear to illustrate the difficulty of distinguishing between quark and hadron degrees of freedom in nucleons and nuclei, and on the other hand gives a guess on the location of the ill-defined boundary between the quark and the hadron degrees of freedom in nucleons.

I would like to cover the following four topics: (1) quark models of baryon-nucleon spin-orbit forces, (2) the hadronic representation of hidden color basis states, (3) nucleon core size, and (4) the quark model of baryons.

2. Quark Models of Baryon-Nucleon Spin-Orbit Forces

Baryon-nucleon (BN) spin-orbit (SO) forces are of interest because they are responsible for part of the experimentally observed B-nucleus SO splittings. Since these forces have rather short ranges, they might involve quarks in some way. Their spin dependence could provide a particularly specific and perhaps clearer window into quark effects in nuclei.

Recent studies of simple quark models of BN SO forces show that they are surprisingly similar to the familiar one-boson-exchange (OBE) [11-13]. Figure 1 from He, Wang, and Wong [13] gives a simple comparison of the

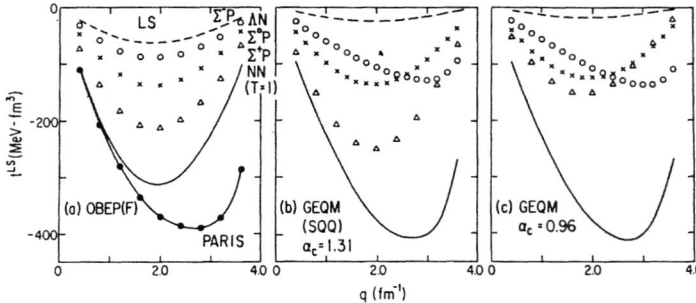

Fig. 1. Baryon-nucleon symmetric spin-orbit t matrices
in the Born approximation.

symmetric SO forces in different BN channels by showing the resulting t matrices calculated in the Born approximation. Figure 1a gives results from OBE potentials: the NN T=1 results from the Paris potential [14] are shown as dots connected by a smooth curve, while the rest are calculated from the Nijmegen OBEP model F [15] which has a hard core. Figure 1b shows the corresponding results from a simple gluon-exchange quark model (GEQM) with only symmetric quark-quark (SQQ) SO interactions. The model parameters have been adjusted to fit the Paris NN T=1 results. If the anti-symmetric qq (AQQ) SO interactions are also included, with model parameters readjusted to fit the same Paris NN T=1 results, one gets the results of Fig. 1c.

Comparison among these figures shows that the $\Sigma^0 p$ t matrices are practically model independent. The weak $\Sigma^- p$ t matrices become weaker in the GEQM's, while the $\Sigma^+ p$ t matrix is about 17% stronger in GEQM (with SQQ interaction only) than in the OBEP(F), but 30% weaker when the AQQ terms are also included. In the ΛN channel, the larger GEQM results at large values of q come from contributions absent in the OBEP(F) because they come from inside the hard-core radius.

These results appear to suggest that measurements of BN polarization to better than 10% might be of help in distinguishing a GEQM from a meson-exchange (ME) model. This is only partly true, because both the OBEP and the GEQM have many sources of uncertainties when absolute comparisons are made. Until these uncertainties are understood and brought under control, it is hard to interpret experimental data with confidence. The situation with the hypernuclear SO splittings is even less clear cut because of the complexities of many-body systems.

At this time, we prefer to concentrate instead on the similarities seen in different models at the 20-30% level in all BN channels. This similarity appears surprising at first sight: ME nuclear forces appear primarily as the folding of ME contributions between quarks into the quark distributions in nucleons. Gluon exchange, on the other hand, vanishes in the folding approximation because each nucleon is separately a color singlet. Quarks must be exchanged between the nucleons before any nuclear force appears.

The similarity between these models has a simple explanation, however [12]. The GEQM results turn out to be dominated by the interaction between the pair of quarks (say 3 and 4) which are Pauli exchanged between the nucleons. The dominant folding contribution of the ME process involving the same quarks simulates this Pauli exchange process in the following way: It contains the spin-isospin factor $(\vec{s}_3+\vec{s}_4)(v_0^\omega+v_0^\rho\vec{\tau}_3\cdot\vec{\tau}_4)$, which simplifies to $2v_0(\vec{s}_3+\vec{s}_4)P_{34}^\sigma P_{34}^\tau$ for the S=1 quark pair when $v_0^\rho = v_0^\omega$ is SU(6) symmetric. The result still differs from the quark exchange operator in GEQM in color and spatial structures. Both are unimportant, but for different reasons. The difference in color structure simply gives rise to an overall factor which is absorbed into the unknown coupling constant α_s of the GEQM. The spatial structures are short-ranged in both models, and are quite similar to each other at the 20-30% level. Analogous arguments hold also for the exchange of K* mesons between a strange and a nonstrange quark.

3. Hadronic Representation of Hidden-Color Basis States

It is sometimes convenient to use basis states made up of two color-octet clusters, i.e., the so-called hidden-color basis states. These hidden-color states are not confined, but can be rearranged into color-singlet hadronic clusters. As a result, they can break up asymptotically into the physical hadronic channels and cannot readily be distinguished from other hadronic processes. In the following, we review known arguments showing why the recoupling into hadronic states is always possible.

The situation is simpler for hidden-color dimeson states (see, for example, [16]). If 1,2 are quarks and $\bar{3},\bar{4}$ are antiquarks, a color-singlet state can be recoupled to

$$| (2\bar{3})_1 (1\bar{4})_1 > = P_{12} | (1\bar{3})_1 (2\bar{4})_1 >$$

$$= (\frac{1}{3} + \frac{1}{2}\lambda_1 \cdot \lambda_2) \; | (1\bar{3})_1 (2\bar{4})_1 >$$

$$= \frac{1}{3} | (1\bar{3})_1 (2\bar{4})_1 > + \frac{1}{3}\sqrt{8} \; | (1\bar{3})_8 (2\bar{4})_8 > \; , \qquad (1)$$

where we have used the result $<(\lambda_1 \cdot \lambda_2)^2> = 32/9$.

The situation for dibaryons is a little more complicated [17]. The 6q color w.f. has permutation symmetry [222]. This is a 5-dimensional irreducible representation which has equivalent cluster decompositions $S_6 \supset S_3 \times S_3$ of the form [21]x[21] and [111]x[111]. The first is the hidden-color basis of Harvey [18], while the second is a color-singlet basis. The latter is made up of product color singlet w.f. $\phi_0 (y_j) \phi_0 (z_j)$. The five basis states needed can be constructed either by using Schmidt orthogonalization starting with $y_1 z_1 = (123)(456)$, and then going on to the remaining four states labeled by $y_2 z_2 = (124)(356)$, etc., or more elegantly by using matrix elements of suitable permutation operators. The color-singlet and the hidden-color bases can be related to each other through the five young tableaux Y_i of [222].

The 6q w.f. is totally antisymmetric when the space, spin and isospin (XST) function is also included:

$$\Psi(6q) = \sum_{i=1}^{5} a_i \Psi_C([222]Y_i) \Psi_{XST}([33]\widetilde{Y}_i)$$

$$= \sum_{i,j} a_i b_{ij} \phi_0(y_j) \phi_0(z_j) \Psi_{XST}([33]\widetilde{Y}_i) \, , \tag{2}$$

where $\widetilde{Y}_i$ is conjugate to Y_i. A familiar form emerges if the already anti-symmetrized w.f. is antisymmetrized once more

$$\Psi(6q) = \mathcal{A} \sum_{j=1}^{5} c_j \phi_0(y_j) \phi_0(z_j) \Psi_{XST}(j)$$

$$= \sum_j c_j \, \mathcal{A}_{y_j z_j} [\mathcal{A}_{y_j} \mathcal{A}_{z_j} \phi_0(y_j) \phi_0(z_j) \Psi_{XST}(j)] \, , \tag{3}$$

where a cluster decomposition of the antisymmetrizer $\mathcal{A}$ ensures that the factor [ ] is made up of a linear combination of products of baryon w.f. The final antisymmetrizer $\mathcal{A}_{y_j z_j}$ generates the expected $6!/(3!3!)=20$ terms for each product baryon configuration in [ ]. This baryonization of 6q w.f.'s includes as special cases Harvey's NN, NΔ and $\Delta\Delta$ physical-basis states.

Thus all 6q states can be expressed in terms of these $\mathcal{A}(B_1 B_2)$ states, where B_i is a baryon state. This is the complete physical basis, since only hadrons can normally be liberated. Indeed the correct asymptotic boundary conditions can be applied only in these physical channels.

For example, dibaryon resonances could appear as loosely bound systems of excited baryons. Since excited baryons will decay, primarily by pion emission, such systems will be indistinguishable from $B^2 \pi^n$ structures [19]. The known B^2 resonances might well be of this type [20].

At low energies, the hadronic description appears to have an advantage, since it is closer to the physical observables. It is well known, however, that the hadronic description becomes very messy as the energy increases. Hence one would like to switch over from a hadronic description to a quark/gluon one at some point. Examples where the quark/gluon description might be more convenient come readily to mind: (1) short-range nuclear force and nuclear structure [1], (2) multiquark systems such as multiquark bags [21], quark/gluon plasmas [22] and nuggets [23], (3) glueballs [24], (4) deep inelastic structure functions [25], and (5) Drell-Yan processes of direct $q\bar{q}$ reactions [26]. Indeed, we expect that quarks will provide a more elegant description at sufficiently high energies or sufficiently short distances.

The question then is where does nonperturbative QCD in the baryon interior give place to perturbative QCD in the baryon exterior. In nuclear physics, this question can be recast into a simple form: where does the meson cloud end and the quark core begin in a nucleon?

4. Nucleon Core Size

The most detailed information on hadron sizes has been obtained from electromagnetic (e.m.) form factors. The simplest of these is probably that of the pion, which because of its zero spin has only one e.m. form factor $G_\pi(t)$. Experimentally, it is found to be $[1-(t/\mu^2)]^{-1}$, where $\mu \simeq$ 0.73 GeV [27]. The resulting pion rms radius is

$$r_\pi = [6(d/dt)G_\pi(t)|_{t=0}]^{1/2} = \sqrt{6}/\mu = 0.66 \pm 0.02 \text{ fm.} \tag{4}$$

This result can be understood very simply in the vector dominance model (VDM) [28]. In this model the (real or virtual) photon interacts with a hadronic target H only indirectly through the intermediary of neutral vector mesons ρ_0, ω_0, ϕ_0, etc., as shown in Fig. 2. The most important of

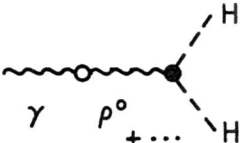

Fig. 2. Vector dominance of electromagnetic form factors.

these vector mesons is ρ_0 so that

$$G_H(t) \simeq [1-(t/m_\rho^2)]^{-1}[1-(t/\Lambda_H^2)]^{-1}, \tag{5}$$

where m_ρ = 0.77 GeV and the second factor on the right is the ρHH form factor. The experimental result of Eq. (4) shows that the $\rho\pi\pi$ vertex is essentially a point. In this VDM picture, the pion e.m. form factor comes primarily from the ρ propagator which describes the combined effect of the virtual ρ_0 meson cloud around the pion and the apparent size of the photon when it is near hadrons. That this entanglement between the probe and the target is real and impossible to separate can be seen from the description of high-energy photon-hadron reactions in terms not only of a ρ_0 component of the photon, but also a dipion component, which is nothing but the entire Fig. 2 itself (with H = π) [29]. An apparent size of the photon can also be seen in the width of the elastic diffraction peak in the Compton scattering of photons from protons [30].

In the case of the proton charge form factor $G_E(t)$, the experimental data have been fitted to eq. (5) to yield $\Lambda_N \simeq 0.93 \pm 0.03$ fm [31]. The resulting ρNN vertex has a size of

$$r_{\rho NN} = \sqrt{6}/\Lambda_N \simeq 0.52 \text{ fm.} \qquad (6)$$

A recent analysis in the context of the Skyrmion model yields comparable results [32]. This gives an estimate of the nucleon core size r_N.

The value of r_N is of special interest in nuclear physics because in Yukawa's meson exchange theory of nuclear forces one distinguishes between the virtual mesons which carry the interaction and the meson-NN vertices which are the sources of virtual mesons. The nucleon source size has a rather complicated field-theoretical description [33], and is usually parametrized empirically by a cutoff mass Λ_N:

$$v(t) = \sum_i v_i \left[1-(t/m_i^2)\right]^{-1} \left[1-(t/\Lambda_N^2)\right]^{-2}. \qquad (7)$$

In the Bonn potential [34], which includes explicit two-pion contributions, a value of $\Lambda_N \simeq 1.2$ to 1.6 GeV is needed. This gives $r_N \simeq 0.3$ to 0.4 fm.

In the naive quark model, r_N arises from the spatial extension of the quark w.f. Its value is not easy to estimate reliably from experiment, but it is possible to obtain some upper bounds under various assumptions concerning how nuclear forces are generated [35]. For example, the long-range potential $\left[1-(t/m^2)\right]^{-1}$ due to the exchange of a meson of mass m can be reproduced by a naive quark model with colorless contact qq interactions if the nucleon core size r_N were $\sqrt{3}/m$. This works out to be 2.5 fm if m = m_π and 0.44 fm if m = m_ρ. The result must be smaller if there are meson exchanges between quarks. Analysis of the empirical NN spin-orbit forces using a quark model with single vector (V) meson exchanges gives a result of only 0.35 fm. On the other hand, if the qq interaction is pure one-gluon exchange, the resulting nuclear forces have no direct or folding contribution, but only quark exchange contributions which appear with an overall negative sign. A larger nucleon core size ($\simeq 0.46$ fm) is then needed.

We should note that this nucleon core size is the same as that of the VNN vertex seen in the e.m. form factor of Fig. 2. The result from nuclear forces has the advantage that the nucleon form factor appears twice, once for each nucleon, and that the uncertainty concerning the apparent photon size is absent. Of course, we are not counting the exchanged virtual meson as part of the core, but the core might still be dressed to some extent, so that the bare core size is smaller.

To summarize, the proton could be as large as the radius (0.8 fm) of its

Dirac e.m. form factor if one includes the virtual vector meson cloud and considers the virtual photon as a point particle. If the vector meson cloud is excluded, the remaining nucleon core radius could be less than half of its "dressed" value.

5. Quark Model of Baryons

Simple constituent quark models of baryons appear to work surprisingly well. One exception is the spin-orbit mass splittings in P-wave baryons which are still poorly described in most quark models including that of [36]. Such problems in baryon spectroscopy raise questions concerning the reliability of naive quark descriptions of nuclear forces in systems which are at least twice as complicated.

Most naive quark models have difficulty fitting P-wave spin-orbit (SO) mass splittings in nucleons because the color coupling constant α_s obtained by fitting the Δ-N mass difference is too strong. Some of these well-known problems can be seen in the following theoretical expressions for certain mass splittings in the single $(1s)^2 1p$ shell model:

$$\delta \overline{N} = N(\tfrac{3}{2}) - N(\tfrac{1}{2}) = \tfrac{3}{2} (I_G - I_c) + \tfrac{9}{20} I_G \simeq 20 \text{ MeV} ,$$

$$\delta \Delta = \Delta(\tfrac{3}{2}) - \Delta(\tfrac{1}{2}) = \tfrac{1}{2} I_G + \tfrac{3}{2} I_c \qquad \simeq 80 \text{ MeV} ,$$

$$\text{meson } (^3P_2 - {}^3P_0) = 6(I_G^M - I_c^M) \qquad \simeq 240 \text{ MeV} ,$$

and

$$\Lambda(\tfrac{3}{2}) - \Lambda(\tfrac{1}{2}) + \delta \Delta \simeq 3(I_G - I_c) \simeq 200 \text{ MeV} . \tag{8}$$

where the baryon mass are denoted by the symbol N, Δ or Λ. I_G and I_c are certain radial integrals from the qq SO force arising from the one-gluon-exchange (G or OGE) and confinement (c) potentials, with $I_G = (\Delta-N)_G/2$, i.e., half of the OGE contribution to Δ-N. The integral I_G^M for mesons is slightly larger if meson size parameters are smaller. $\overline{N}$ is the center of mass of the two N(j) states, the position being unchanged by configuration mixing in the single-shell model.

We can see that large values of I_G or I_c are inconsistent with the experimental mass splittings. The Capstick-Isgur model [36] does well for $\delta \Delta$ and mesons by using small values of $I_G \simeq 40$ MeV and $I_c \simeq 20$ MeV. It is only fair in $\delta \overline{N}$, and poor in $\delta \Lambda$. The small I_G requires a small value of α_s (<0.6), while the large Δ-N needed is generated by a large wave distortion effect in the strongly attractive hyperfine interaction.

On the other hand, a small α_s is hard to reconcile with the naive quark model of nuclear forces either without $q\bar{q}$ excitations ($\alpha_s \simeq 5$ is needed in [12,13] or with $q\bar{q}$ excitations ($\alpha_s \simeq 3$ is needed in [37]). This inconsistency in the values of α_s used in different models suggests that the strong interaction dynamics might be more complicated. We were thus led to consider the effects of additional physical processes.

To do this, Wang and I [38] introduce a pomeron contribution ($I_c \rightarrow I_c - I_p$) for the repulsive scalar component arising from various closed inelastic channels, and flavor-symmetric sea contributions proportional to $\frac{2}{3} + \vec{\lambda}_i^f \cdot \vec{\lambda}_j^f$ separated into vector, pseudoscalar and scalar terms. These four additional parameters are then fitted to nonstrange baryon SO mass splittings:

$$I_p = 40 - 100 \text{ MeV}, \quad I_{sea}(V) = -(40 - 60) \text{ MeV},$$

$$I_{sea}(PS) = -70 \text{ MeV}, \quad I_{sea}(S) = 15 - 10 \text{ MeV}. \tag{9}$$

When these terms are added to $\delta\Lambda$ of Eq. (8), its value comes out quite well.

It is interesting to note that the empirical I_p compares well with a theoretical value of very roughly 60 MeV deduced from the Regge theory [38, 39]. The signs of $I_{sea}(V)$ and $I_{sea}(PS)$ are both opposite to those of meson exchanges. This suggests that sea excitations carries a color-symmetric factor of $<1 + \frac{3}{2} \vec{\lambda}_i^c \cdot \vec{\lambda}_j^c>_B = -3$. However, the magnitudes of all these sea integrals are much weaker than those obtained from meson-exchange nuclear forces when they are brought down to the quark level. These features appear to be consistent with their possible origin in the perturbative regime of the baryon interior. Even the relative weakness of $I_{sea}(S)$ could be related to the lesser importance of P-wave scalar $q\bar{q}$ pairs or higher-order $(q\bar{q})^2$ excitations in perturbative QCD.

6. Conclusions

We find that at the 20-30% level, the gluon-exchange quark model gives essentially the same baryon-nucleon spin-obit forces as the traditional meson-exchange models in a variety of baryon-nucleon channels. The hadronic and the hidden-color representations give mathematically equivalent bases for the description of multiquark systems such as nuclei, thanks to the antisymmetrization of quark wave functions. We expect that the hadronic description is more effective at low energies or large distances, while the quark description might be more economical at high energies or small distances. The separation between these regimes can be characterized by a nucleon core size of quarks of less than 0.3 - 0.5 fm. There are indications from both baryon spectroscopy and nuclear forces that the dynamics of the perturbative regime of the baryon interior is different from the nonperturbative regime of the baryon exterior.

If this picture looks familiar, this is because it is close to the Little Bag model of Brown and Rho [4].

This work was supported in part by NSF grant PHY84-11473. I am indebted to F. Wang for collaboration in deriving eq. (3) before we found out the prior result of Oka and Yazaki. I also want to thank M. Rho for stimulating discussions. This is an expanded version of a talk first given in the Stony Brook Conference and Symposia (in honor of G. E. Brown), September 4-6, 1986.

References

1. Wong, C.W.: Phys. Rep. 136, 1 (1986).

2. Chodos, A., et al.: Phys. Rev. D 9, 3471 (1974);
   DeGrand, T., et al.: Phys. Rev. D 12, 2060 (1975).

3. Marciano, W., Pagels, H.: Phys. Rep. 36C, 137 (1978);
   Bander, M.: Phys. Rep. 75, 205 (1981).

4. Brown, G.E., Rho, M.: Phys. Lett. 82B, 177 (1979).

5. Callan, C.G., Dashen, R.F., Gross, D.J.: Phys. Rev. D 19, 1826 (1979).

6. Zahed, I., Brown, G.E.: Phys. Rep. 142, 1 (1986);
   Holzwarth, G., Schwesinger, B.: Rep. Prog. Phys.

7. Nadkarni, S., Nielsen, H.B., Zahed, I.: Nucl. Phys. B253, 308 (1985).

8. Mueller, A.H.: Phys. Rep. 73, 237 (1981); Comments Nucl. Part. Phys.
   14, 205 (1985).

9. Collins, P.D.B., Martin, A.D.: Rep. Prog. Phys. 45, 335 (1982);
   Fiałkowski, K., Kittel, W.: Rep. Prog. Phys. 46, 1283 (1983).

10. Fishbane, P.M., Meshkov, S.: Comments Nucl. Part. Phys. 13, 325 (1984);
    Ward, B.F.L.: Phys. Rev. D 33, 1900 (1986).

11. Suzuki, Y., Hecht, K.T.: Nucl. Phys. A420, 525 (1984);
    Morimatsu, O., et al.: Nucl. Phys. A420, 573 (1984);
    Morimatsu, O., Yazaki, K., Oka, M.: Nucl. Phys. A424, 412 (1984);
    Wang, F., Wong, C.W.: Nucl. Phys. A438, 620 (1985).

12. He, Y., Wang, F., Wong, C.W.: Nucl. Phys. A451, 653 (1986).

13. He, Y., Wang, F., Wong, C.W.: Nucl. Phys. A454, 541 (1986).

14. Lacombe, M., et al.: Phys. Rev. C 21, 861 (1980).

15. Nagels, M.M., Rijken< t.A., de Swart, J.J.: Phys. Rev. D 20, 1633 (1979).

16. Robson, D.: Prog. Part. Nucl. Phys. 8, 257 (1982).

17. Oka, M., Yazaki, K.: Intl. Rev. Nucl. Phys. 1, 489 (1984).

18. Harvey, M.: Nucl. Phys. A352, 301 (1981).

19. MacGregor, M.: Phys. Rev. Lett. 42, 1724 (1979).

20. Tatischeff, B.: Phys. Lett. 154B, 107 (1985).

21. Jaffe, R.L.: In: Proc. Topical Conf. on Baryon Resonances, Oxford, 1976. Ross, R.T., Saxon, D.H. (eds.). Chilton, Didcot, England 1976, p. 455

22. Baym, G.: invited talk given in Intl. Conf. and Symposia on Unified Concepts of Many-Body Problems, Stony Brook, NY, September 4-6, 1986.

23. De Rújula, A., Glashow, S.L.: Nature 312, 734 (1984).

24. Robson, D.: Nucl. Phys. B130, 328 (1977).

25. Thomas, A.W.: Prog. Part. Nucl. Phys. 11, 325 (1984).

26. Drell, S.D., Yan, T.M.: Phys. Rev. Lett. 25, 316 (1970); Ann. of Phys. 66, 578 (1971).

27. Amendola, S.R., et al.: Phys. Lett. 146B, 116 (1984).

28. Sakurai, J.J.: Currents and Mesons U. of Chicago: 1969, chapt. 3; In: Properties of the Fundamental Interactions, Part A. Zichichi, A. (ed.). Bologna: Compositori 1973, p. 243.

29. Yennie, D.R.: Rev. Mod. Phys. 47, 311 (1975).

30. Sakurai, J.J.: In: Laws of Hadronic Matter. Zichichi, A. (ed.). New York-London: Academic Press 1975, p. 252; Perl, M.L.: High Energy Hadron Physics. New York: Wiley 1974, chap. 19.

31. Massam, T., Zichichi, A.: Nuovo Cim. 43, 1137 1966); 44, 309 (1966).

32. Brown, G.E., Rho, M., Weise, W.: Nucl. Phys. A454, 669 (1986).

33. Brown, G.E., Jackson, A.D.: The Nucleon-Nucleon Interaction. Amsterdam: North Holland 1976, chap. 6.

34. Machleidt, R.: Lecture Notes in Phys. 197, 352 (1984); Holinde, K.: Phys. Lett. 157B, 123 (1985).

35. He, Y., Wang, F., Wong, C.W.: Phys. Lett. 168B, 177 (1986); Nucl. Phys. A448, 652 (1986).

36. Capstick, S., Isgur, N.: Phys. Rev. D (1986).

37. Fujiwara, Y., Hecht, K.T.: Nucl. Phys. A444, 541 (1985); A451, 625 (1986); A456, 669 (1986); Phys. Lett. 171B, 17 (1986).

38. Wang, F., Wong, C.W.: Quark potential model of baryon spin-orbit mass splittings, UCLA preprint (1986).

39. Rijken, T.A.: In: Proc. Intl. Conf. on Few Body Problems in Nuclear and Particle Physics, Quebec, 1974. Slobodrian, R.J., Cuec, B., Ramavataram, R. (eds.). Quebec: Presses Univ. Laval 1975, p. 136.

THE NUCLEON FORM FACTOR IN QCD

Carl E. Carlson
Physics Department
College of William and Mary
Williamsburg, VA 23185, USA

Abstract. We review the results obtained for nucleon form factors using QCD and perturbation theory. We also make some comments upon the data, and indicate what momentum transfer can suffice for seeing quark effects that are calculable in simple ways.

I. Introduction

The application of Quantum Chromodynamics (QCD) to calculation of nucleon form factors[1,2], which I take to mean calculations using perturbation theory with QCD, has been a source of interest and controversy[3]. PQCD at a simple level gives predictions about the high Q^2 behavior of the nucleon form factors. A brief catalog may be in order already: the nucleon electromagnetic form factors G_M and F_2 fall asymptotically like $1/Q^4$ and $1/Q^6$, respectively[1,2]; the axial vector form factors g_A and g_P also fall like $1/Q^4$ and $1/Q^6$, respectively[4]; and the N-Δ transition form factors behave like the elastic nucleon form factors if the definitions are made in parallel[5]. In general, the leading amplitude is the one that preserves the hadrons' helicity[6]. This statement is often striking if expressed in terms of multipole amplitudes. For the N-Δ transition the ratio F_{E2}/F_{M1} should be $\sqrt{3}$ at high Q^2[5], and for high Q^2 electron-deuteron elastic scattering one has $G_C/G_Q = Q^2/6M_d^2$ [7]. With some special knowledge of the nucleon wave functions one can also calculate the absolute normalization of the leading nucleon form factors[1,2]. We will indicate how these calculations are done and elaborate some of the results.

An important reason for the interest in the nucleon form factor in QCD is to learn when, i.e., at what momentum transfer, the quark model will have something to say about the nucleus that could not be said using traditional nuclear physics. We shall comment about when PQCD seems to be giving qualitatively correct answers for the nucleon form factors.

We begin by making some comments on the data in Section II, and go on to the calculations and results in Section III.

II. Comments on the data.

Since one of the predictions of perturbative QCD is that G_{Mp} falls like $1/Q^4$ at high Q^2, we start in Fig. 1 by showing $Q^4 G_{Mp}$ plotted vs. Q^2[8]. The data does flatten out for Q^2 above 5 $(GeV)^2$ and this is some benchmark of where perturbative QCD works. There is some falloff of $Q^4 G_{Mp}$ above 5

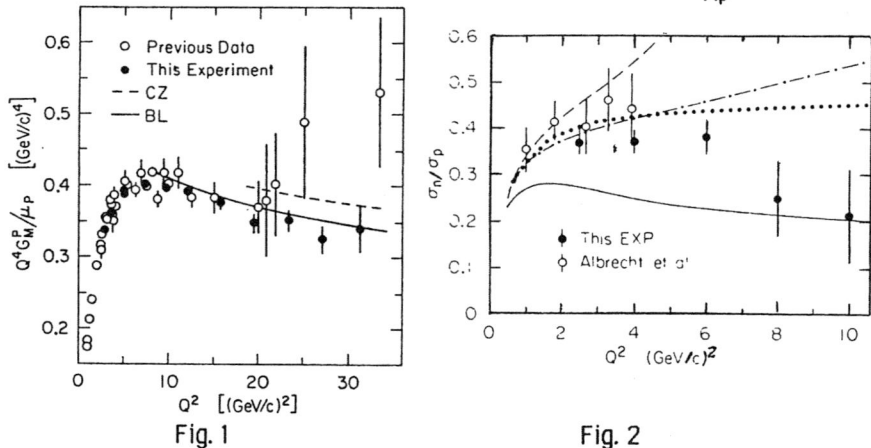

Fig. 1   Fig. 2

(GeV)^2 which may be a residual of a rapidly falling nonperturbative contribution to G_{Mp} or may be an additional expected QCD falloff with $\log Q^2$.

Now consider G_{Mn}. There is data on σ_n, the electron-neutron differential cross section, at Q^2 = 2.5, 4, 6, 8, and 10 (GeV)^2 but only at one scattering angle[9], so that a separation of G_{Mn} and G_{En} is impossible. The most salient feature of the data, shown in Fig. 2, is that σ_n/σ_p falls roughly like $1/Q^2$ between (say) 5 and 10 (GeV)^2 and the ratio is roughly 1/4 at the higher Q^2. Recalling that

$$\sigma_N \sim G_{EN}^2 + \tau\, G_{MN}^2 [\, 1 + 2(1+\tau)\tan^2\theta/2 \,] \tag{1}$$

for N = n or p, $\tau = Q^2/4m_N^2$, and θ the lab scattering angle, and also recalling that the contibution of G_{Ep} to σ_p is small at high Q^2 leads to

$$\left| G_{Mn}/G_{Mp} \right| \leq 1/2 \tag{2}$$

at Q^2 of 10 GeV^2. Possibly $Q^4 G_{Mn}$ flattens out at 10 GeV^2 and is (in magnitude) about 1/2 of $Q^4 G_{Mp}$ thereafter. However the falling of σ_n/σ_p suggests another possibility, namely that G_{Mn} is very small[10]. Then σ_n must then be dominated by G_{En} (which need not be very big, but of about the size of G_{Mp} or F_{1p}) and the $1/Q^2$ falloff of σ_n/σ_p is natural. If G_{Mn} is small at high Q^2, it connects on to the low Q^2 data with $F_{1n} \approx 0$ at all Q^2 [10].

We will below consider all the possibilities for G_{Mn}. Data on g_A and the leading N-Δ transition amplitude will be mentioned at the proper times.

III. Calculations and results.

We will work in a special frame where the initial nucleon is moving fast in (say) the z-direction. We view the nucleon as three quarks moving almost parallel to each other. The high Q^2 results depend only on the wave function with transverse momenta integrated, giving the "distribution amplitude,"

$$\phi(x) = \int [d^2 k_T] \, \psi(x, k_T). \tag{3}$$

The variables x_i for quark i are the light cone momentum fractions

$$x_i = k_i^+/p^+ = (k_i^0 + k_i^3)/(p^0 + p^3) \tag{4}$$

where k_i is the momentum of quark i and p is the momentum of the nucleon. The argument x means the collection (x_1, x_2, x_3), and similarly for the transverse momenta $k_{iT} = (k_i^1, k_i^2)$. The incoming photon (or vector boson for g_A) enters from the transverse direction, which means $q^+ = 0$ and $q_T \neq 0$.

All the processes of interest can be diagrammed as in Fig. 3a, and given algebraically like the formula for G_{MN},

$$G_{MN} = \int [dx][dy] \, \phi(y) \, T_H(x, y, Q) \, \phi(x) \tag{5}$$

where T_H is called the "hard scattering amplitude" generally and in this case is the magnetic form factor for taking three parallel moving quarks into three parallel moving quarks.

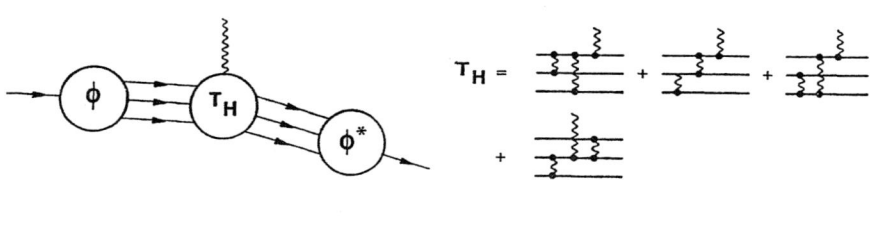

Fig. 3a   Fig. 3b

T_H is calculated using perturbation theory. There are 42 Feynman diagrams in lowest order, but only 14 of them are nonzero, and only the four shown in Fig. 3b need to be calculated independently; the other ten are obtained from them by symmetries. We will not display T_H [1], but will make two remarks on the powers of Q^2 it contains.

(1) Neglecting the quark mass and the binding of the quarks, the Q^2

dependence of T_H can be got by dimensional analysis. Each quark line gives a factor of Q from the initial and final quark wave functions [recall that $u(q)\gamma_\mu u(q) \sim q_\mu$ so with no mass or binding energy scale, the wave functions either give a factor Q or zero], a factor $1/Q$ from each fermion propagator, and a factor $1/Q^2$ for each gluon propagator. Hence for a system of n quarks, we have for the Q^2 dependence of the Feynman diagram

$$Q^n \ (1/Q^{n-1}) \ (1/(Q^2)^{n-1}) = Q/(Q^2)^{n-1}.$$

or $1/Q^3$ for three quarks. There is one other point, which is that we must look up the standard definition of G_{MN} and relate it to the Feynman diagram. This is easily done, and gives another factor of Q that leads to the advertized $G_{MN} \sim 1/Q^4$.

(2) Comment (1) applies without modification only to the leading form factor, which is the one that requires no quark helicity flips. If a quark must flip helicity, there is an extra factor of $O(m/Q)$ giving faster falloffs and smaller contributions to cross sections from nonleading form factors.

While T_H can be calculated explicitly, the distribution amplitude ϕ cannot, at least not at present. One can try to make plausible or flexible choices for ϕ, for starters a simple symmetric form

$$\phi(x) = N \ (x_1 x_2 x_3)^\eta \tag{6}$$

which has one parameter, the power η, and as an example with more flexibility an expansion in terms of orthogonal polynomials of two variables, the Appell polynomials $\tilde{\phi}_i$,

$$\phi(x) = x_1 x_2 x_3 \sum N_i \ \tilde{\phi}_i(x) , \tag{7}$$

where the $x_1 x_2 x_3$ factor damps singular contributions from T_H at the edges of the integration region. There is some slow, dependent on $\log Q^2$, evolution of the distribution amplitude with Q^2 which we mainly ignore, but it is worth commenting that there is an exact statement possible about the nucleon distribution amplitude as Q^2 becomes superhigh(i.e., log log $Q^2 \to \infty$). Then the distribution amplitude is proportional to just $x_1 x_2 x_3$, which in terms of the above forms means η(log log $Q^2 \to \infty$) = 1 or only $N_0(\log\log Q^2 \to \infty) \neq 0$.

The form factors have been worked out using both of the above distribution amplitudes and the results can be found in the references [4,5,11]. There is an additional constraint, which is the normalization

$$\int [dx] [d^2k_T] \ |\psi(x,k_T)|^2 = P_{3q} \tag{8}$$

where P_{3q} is the probability of finding the three quark Fock component in the nucleon. The consequences of the normalization condition can be studied by taking a factorized approximation

$$\psi(x,k_T) = \phi(x) \times (768\pi^4/\beta^4) \ \exp \ (-\Sigma \ k_{iT}^2/2\beta^2) \tag{9}$$

where $(2/3)^{1/2}\beta$ is the RMS transverse momentum of the quarks. If one has a $\phi(x)$ that is good for the form factor data, the normalization condition determines what RMS quark tranverse momentum is implied: a few hundred MeV is good[3,11].

What sort of results do we get?

(1) Proton magnetic form factor. The $1/Q^4$ falloff of G_{Mp} has been shown, so we consider the normalization of the calculated G_{Mp} compared to the data. We can give an important "go" theorem[11] (as opposed to a "no go" theorem, which is a statement that something cannot be done): we can obtain the observed normalization of G_{Mp} using perturbation theory and QCD. To do so, the distribution amplitude must be broader than the superasymptotic one. In the simple symmetric form, for example, the power η must be less than one. This is also required for another reason. The sign of G_{Mp} is not guaranteed positive in perturbative QCD since not all the terms in T_H and not all the quarks' charges have the same sign. If $\eta > 1$,

G_{Mp} comes out negative and $\eta < 1$ is needed for G_{Mp} to be positive.

(2) Neutron magnetic form factor. The conclusions depend on what we think the data is. If G_{Mn} is intermediate between zero and $(-1/2)G_{Mp}$ one can use the simple symmetric form or some symmetric combination of Appell polynomials for the distribution amplitude. If $G_{Mn} \approx -1/2\ G_{Mp}$, then there is a serious problem involving the normalization condition. There is (it seems) no distribution amplitude that gives $G_{Mn} \approx -1/2\ G_{Mp}$ with an acceptable RMS quark transverse momentum and is symmetric among the two parallel helicity quarks. An asymmetric distribution amplitude is necessary. The same conclusion holds although more weakly if $G_{Mn} \approx 0$.

Chernyak and Zhitnitsky[2] use a QCD sum rule technique to obtain some moments of the distribution amplitude, and find a strong asymmetry. Their results do not fix the distribution amplitude definitively and do not fix the magnitude of G_{Mp} but remarkably enough all the distribution amplitudes which are consistent with their moments give $G_{Mn} \approx -1/2\ G_{Mp}$. Gari and Stefanis[12], allowed a violation of one of the (independent) Chernyak and Zhitnitsky moment constraints, and found a suitable and asymmetric distribution amplitude that gives $G_{Mn} \approx 0$ with a good G_{Mp}.

(3) Axial vector form factor. That g_A falls like $1/Q^4$ asymptotically is assumed by the experimenters in fitting their data and seems all right. The question then concerns the normalization of g_A at high Q^2. The distribution amplitudes suggested for the nucleon magnetic form factor data all give g_A/G_{Mp} somewhat above unity[4], and this is in accord with the data. The data for g_A however only extend to $Q^2 = 3$ GeV^2 , which is a bit low.

(4) The leading N-Δ transition form factor. This means the non-helicity flip form factor; in multipole amplitudes it is a linear combination of the

M1 and E2. The exact definition parallels G_{Mp}; we call the form factor $G_{MN\Delta}$ to emphasize the parallelism even if it is not purely magnetic, and $G_{MN\Delta} \sim 1/Q^4$. This seems in accord with the data. (A pre-QCD definition of an N-Δ form factor which is widely noted to fall faster than G_{Mp} is predicted to fall like $1/Q^5$. See Ref. [5])

Interestingly, the ratio $G_{MN\Delta}/G_{Mp}$ is quite different for different distribution amplitudes. In particular, those that give $G_{Mn} \approx -1/2\, G_{Mp}$ tend to give $G_{MN\Delta} \approx 0$, and the data suggests $G_{Mp\Delta^+} \approx G_{Mp}$. This supports the idea that G_{Mn} is small, or at least that G_{Mn} is not $-1/2\, G_{Mp}$[13].

In conclusion, perturbative QCD makes a number of predictions for nucleon form factors, makes us think about the quark wave functions, and may be valid at Q^2 of not too many GeV^2.

I wish to thank my friends and collaborators for all their work and to thank the National Science Foundation for support.

References

1. G.P. Lepage and S.J. Brodsky, Phys. Rev. D 22, 2157(1980).
2. V.L. Chernyak and I.R. Zhitnitsky, Nucl. Phys. B246, 52(1984); V.L. Chernyak and A.R. Zhitnitsky, Phys. Rep. 112, 173(1984).
3. N. Isgur and C.H. Llewellyn-Smith, Phys. Rev. Lett. 52, 1080(1984).
4. C.E. Carlson and J.L. Poor, Phys. Rev. D 34, 1478 (1986).
5. C.E. Carlson, Phys. Rev. D (to be published).
6. S.J. Brodsky and G.P. Lepage, Phys. Rev. D 24, 2848(1981).
7. C.E. Carlson and F. Gross, Phys. Rev. Lett. 53, 127 (1984).
8. R.G. Arnold et al, Report SLAC-PUB-3810 (October 1985).
9. S. Rock et al, Phys. Rev. Lett. 49, 1139(1982).
10. M. Gari and W. Krümpelmann, Z. Phys. A 322, 689 (1985) and Phys. Lett. B173, 10 (1986).
11. C.E. Carlson and F. Gross, Report LUTP85-12 & CEBAF PR-85-005(1986).
12. M. Gari and N.G. Stefanis, Phys. Lett. B 175, 462 (1986).
13. C.E. Carlson, M. Gari, and N. Stefánis, in preparation.

THE NN INTERACTION AND THE STRUCTURE OF FEW NUCLEON SYSTEMS WITHIN QCB

B.L.G. Bakker

Natuurkundig Laboratorium der Vrije Universiteit,

Amsterdam, the Netherlands.

Abstract

The description of few-nucleon systems, keeping account of explicit quark
degrees of freedom, is considered. One particular model, the Quark Compound
Bag model, is reviewed. Applications of this model to nucleon-nucleon scattering, the deuteron and the three-nucleon bound state are discussed.

1. Introduction

One of the goals of nuclear physics is to understand the nature of the forces that hold nuclei together. In particular the few-body community has
been very active in constructing, applying and testing realistic interactions. Nowadays it is widely believed that the quark-structure of hadrons
must have consequences for the description of the interactions between the
hadrons. The successes of the meson theory of the nuclear interaction suggest that at medium and long range mesonic degrees of freedom are well suited to describe the interhadronic forces in the most economical way. However,
at short distances the situation is, at best, unclear.

Realistic nucleon-nucleon forces rely on phenomenological parameters in order to describe the region of interparticle distances less than about 1.0-
1.5 fm. So a natural place for quark degrees of freedom to come into play
is this central or core-region.

Different approaches to the incorporation of quark degrees of freedom in
the hadronic interactions have been put forward. For reviews of some recent
work we refer to ref. [1]. The variation in the different models reflects
a basic problem in this field: our inability to solve the equations of QCD

at nuclear distances. Here we present recent results for a model that was suggested a few years ago by the ITEP-group: the Quark Compound Bag (QCB) model [2]. Reports on some earlier developments of this model can be found in refs. [3,4].

There is a close connection between the QCB-model and the Jaffe-Low P-matrix [5]. The latter has been introduced to account for the elusiveness of multiquark states. The P-matrix is a real and symmetric object whose poles correspond to the (artificially) confined multiquark states. It can be computed from the experimental data if only the long range part of the hadronic interaction is known. In cases where the tail of the potential is well understood, the known properties of the P-matrix [6] can be used to detect deficiencies in the data. For an extended application of P- matrix methods to nucleon-nucleon scattering one may consult ref. [7]. Because of its important rôle in the construction of the QCB-model, we will discuss in sect. 2 the most important properties of the P-matrix. Sect. 3 contains the basic formalism of the QCB model. An important field of applications is the two-nucleon system. In sect. 4 we present results for $^3S_1$ - $^3D_1$ scattering states. Sect. 5 is devoted to our results for the deuteron and in the last section we quote some results for the three-nucleon systems.

2. The P-matrix

The P-matrix is defined as the logarithmic derivative of the wave-function matrix at a certain radius b:

$$P_\alpha(k;b) = \dot{u}_\alpha^{(+)}(k;b) [u_\alpha^{(+)}(k;b)]^{-1} \tag{1}$$

Here $u_\alpha^{(+)}(k;r)/r$ is the wave function matrix at radius r and CMS momentum k. The dot denotes a partial derivative with respect to the radius and α denotes the quantum numbers of the partial wave considered. (For coupled channels, e.g. $\alpha = {}^3S_1 - {}^3D_1$, u_α, $\dot{u}_\alpha$ and P_α are 2 x 2 matrices).

An important property of P, that is useful in the analysis of experimental data, is that the diagonal matrix elements of P are non-increasing functions of k. In those cases where the interaction beyond r = b is accurately known, e.g. in the nucleon-nucleon case for b $\geq$ 1.0 - 1.5 fm, the link between P and the experimental data is very strong and a violation of $(\partial P/\partial k)_{diag} \leq 0$ indicates that something must be wrong with the data [7]. For our purposes another property of P is important: it has simple poles at those energies where det $[u_\alpha^{(+)}(k;b)]$ vanishes. This property connects

the P-matrix poles to the confined multiquark states [5,6]. One may exploit a boundary-condition model like formalism for the hadronic interaction, the P-matrix playing the rôle of a boundary-condition matrix [6,8].
It connects the outside, hadronic, world with the inside, multiquark,
region,
The value of b, the position of the boundary, is not easily determined.
Estimates for the case of nucleon-nucleon scattering give b = 1.0 - 1.5 fm
[2,5]. If we identify the P-matrix poles found in nucleon-nucleon scattering with the energies of the multiquark states calculated by the Nijmegen
group [9], a value of b close to 1.2 fm is found. This is the value we
adopt in what follows.

3. The Quark Compound Bag model

The QCB-model is a semiphenomenological model that incorporates the important rôle that the vacuum plays in QCD. According to this theory, only
colourless objects exist in the vacuum and, consequently, non-overlapping
hadrons can only exchange white objects. But, at sufficiently small interparticle distances the hadrons overlap and gluon exchanges become important. This decisive difference between the short-range and the peripheral
regime is accounted for in the QCB-model by the feature that at a certain
distance between the hadrons, b, the transition between the purely hadronic phase and the multiquark bag state takes place.
From now on we restrict our discussion tot the case of two-nucleons. Then
the two-nucleon interaction for $r \geq b$ is described by a realistic meson-
exchange interaction. (We have adopted the Paris potential [10], but this
is of no great consequence as all realistic NN interactions pretty much
coincide beyond 1.2 fm).
In the inner region, $r \leq b$, the total wave function can be split into two
parts: a piece built as a linear combination of confined six-quark states,
ψ_ν, which we call QCB-states, and a hadronic part ψ_h which has the form

$$\psi_h = A \{\psi_N(123)\psi_N(456)\psi_{NN}(\vec{r})\} \tag{2}$$

where A is the six-quark antisymmetrizer. The full wave function $\psi = \sum_\nu a_\nu \psi_\nu$
+ ψ_h is a solution of the Schrödinger-equation with the complete six-quark
Hamiltonian. The states ψ_ν are confined states that vanish outside b. Presently, our inability to solve QCD at distances of the order of 1 fm makes
a calculation of ψ impossible. However, we can deduce the form of the

effective interaction for ψ_h. If we write the hamiltonian H as a matrix:

$$H = \begin{bmatrix} H_h & H_{hq} \\ H_{qh} & H_{qq} \end{bmatrix} \tag{3}$$

where H_{qq} acts only on the confined part $\sum_\nu a_\nu \psi_\nu$ and the parts H_{hq} and H_{qh} connect the hadronic and the confined part of ψ, one can eliminate the confined part in the standard way to obtain the effective hadronic hamiltonian

$$H_{eff} = H_h + V_{hqh} \tag{4}$$

where the energy-dependent and non-local interaction is given by

$$V_{hqh}(\vec{r}',\vec{r};E) = \sum_\nu \frac{\langle\vec{r}'|H_{hq}|\psi_\nu\rangle\langle\psi_\nu|H_{qh}|\vec{r}\rangle}{E - E_\nu} \tag{5}$$

The propagator of the confined states is a sum over poles as H_{qq} has a discrete spectrum only.

For the formfactors we make the following Ansatz:

$$f_\nu(\vec{r}) = \langle\vec{r}|H_{hq}|\psi_\nu\rangle = \frac{1}{r}\left[c_\nu\delta(r-b) + x_\nu(E-E_\nu)\eta_\nu(r)\right]Y_{\ell sj}(\hat{r}) \tag{6}$$

The first term corresponds to the interaction at the surface b. At present we do not need a more sophisticated form for the surface interaction than the simple δ-function. The second, volume, term is due to the overlap of the QCB-state ψ_ν with the hadronic part ψ_h. The parameter x_ν has the meaning of the probability for finding a two-nucleon component in the state ψ_ν, if the function $\eta_\nu(r)$ is properly normalized inside b. For convenience we take the Ansatz:

$$\eta_\nu(r) = N_\ell r j_\ell(\beta_\ell r) \tag{7}$$

The parameter β_ℓ is chosen such that $\beta_\ell b$ is the first positive zero of $j_\ell(z)$. The quantum number ℓ is the orbital angular momentum of the two nucleons.

Upon solving the Schrödinger-equation with the hamiltonian (4) one calculates the P-matrix at b, which we shall call the QCB P-matrix. It has poles at $E = E_\nu$, where E is connected to the Mandelstam-variable s by $s = 4m_N^2 + 4 m_N E$. Close to a pole E_ν (or s_ν) we can write:

$$P(E) \underset{E \to E_\nu}{\sim} \frac{\tau_\nu}{E - E_\nu} \tag{8}$$

The residue-matrix τ_ν is in the coupled-channels case given by

$$\tau_\nu^{\ell'\ell} = m_N \, c_\nu^{\ell'} \cdot c_\nu^{\ell} \tag{9}$$

The parameters c_ν^{ℓ} and E_ν can be determined by a P-matrix analysis of the data [7].

The other parameters x_ν^{ℓ} can be found by a fit to the scattering data at all energies we want to consider.

Finally we give the form of the QCB P-matrix for the case that only one QCB-state is taken into account.

$$P^{\ell'\ell}(k;b) = k \left\{ \frac{d}{dx}[x j_\ell(x)] / x j_\ell(x) \right\}_{x=kb} \cdot \delta_{\ell'\ell} \tag{10}$$

$$- \frac{m_N(E-E_\nu) a_\nu^{\ell'}(k) a_\nu^{\ell}(k)}{D_\nu(E)}$$

The form factors $a_\nu^{\ell}(k)$ are

$$a_\nu^{\ell}(k) = \frac{c_\nu^{\ell}}{E-E_\nu} - x_\nu^{\ell} \sqrt{\frac{2}{b}} \, \beta_\ell \, \frac{1}{k^2 - \beta_\ell^2} \tag{11}$$

and the denominator function $D_\nu(E)$ is

$$D_\nu(E) = 1 - m_N(E-E_\nu) \sum_\ell \frac{(x_\nu^{\ell})^2}{k^2 - \beta_\ell^2} \tag{12}$$

The variable k is again the CMS momentum: $s = 4(m_N^2 + k^2)$.

4. Results for nucleon-nucleon scattering

For the value of b that we have adopted, 1.2 fm, only one QCB-state (P-matrix pole) is clearly visible below 1 GeV laboratory energy. However, higher QCB-states will also have some infuence. Below $T_{lab} = 500$ MeV we found it possible to approximate their effect by adding to the form (10) a constant term $-\sqrt{\xi^{\ell'}.\xi^{\ell}}$. The values of the parameters ξ^{ℓ} are determined by the requirement that the deuteron wave function is a solution of the Schrö-dinger-equation with the hamiltonian (4) at the correct binding energy. In table 1 we give the parameters of the QCB-potential determined in this way. Fig. 1 shows the corresponding phase shifts.

Table 1. QCB-potential parameters

b = 1.20 fm	E_{ν} = 0.34595 GeV
$c_{\nu}{}^{o}$ = 0.30173 GeV$^{\frac{1}{2}}$	$c_{\nu}{}^{2}$ = 0.10667 GeV$^{\frac{1}{2}}$
$x_{\nu}{}^{o}$ = 1.04577	$x_{\nu}{}^{2}$ = 0
ξ^{o} = 0.10236 GeV	ξ^{2} = 0.05876 GeV

Clearly the QCB-model can fit the data very well up to 500 MeV.

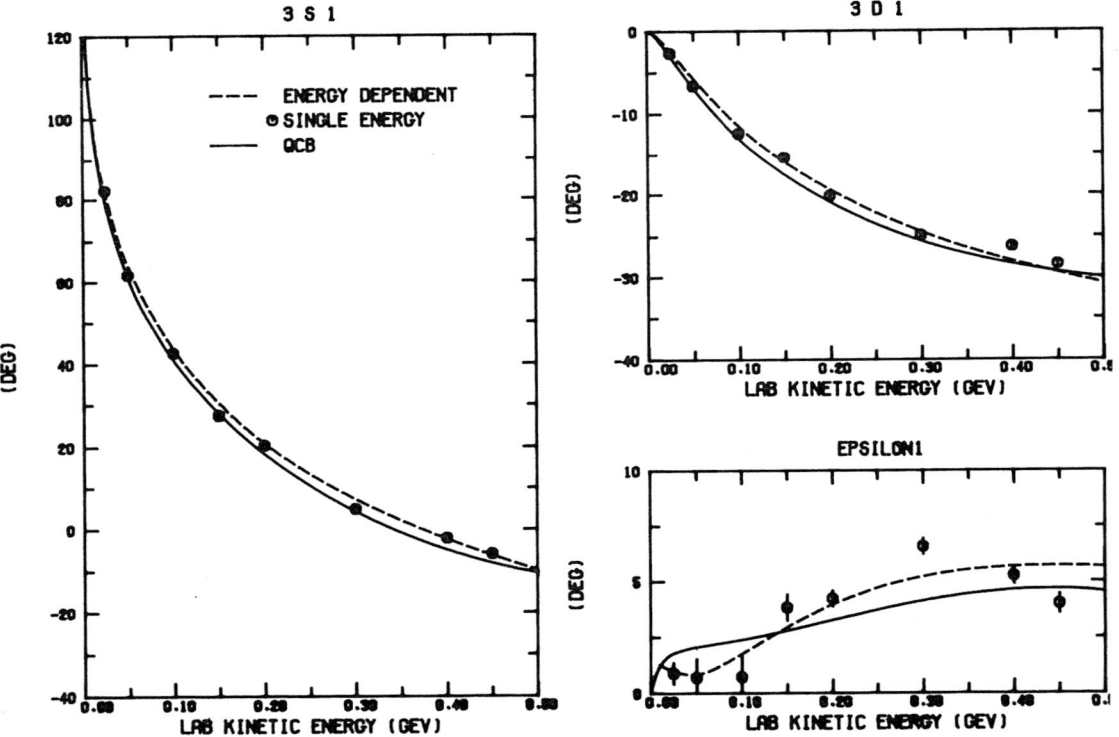

Fig. 1. Bar-phase shifts $\bar{\delta}_{3S_1}$, $\bar{\delta}_{3D_1}$ and $\bar{\epsilon}_1$ for the QCB-potential with the parameters given in table 1. The scattering length and effective range are a = 5.49 fm, r_o = 1.79 fm.

5. The deuteron

When the QCB-model is used for the deuteron one can calculate the probability that the constituents are in the core region:

$$P_{6q} = 1 - \int_b^\infty dr \ [u^2(r) + w^2(r)] \ / \ < \psi | \psi > \tag{13}$$

The functions $u(r)$ and $w(r)$ are the S- and D-wave components of the deuteron. The probability for finding the deuteron in the QCB-state ψ_ν is

$$P_{QCB} = |a_\nu|^2 \ / \ <\psi|\psi> \tag{14}$$

The coefficient a_ν is given by

$$a_\nu = \sum_\ell \{ \frac{\dot{c}_\nu^\ell u_\ell(b)}{E - E_\nu} + x_{\nu 0}^\ell \int_0^b dr \ \eta_\nu^\ell(r) u_\ell(r) \} \tag{15}$$
$$= a_\nu^S + a_\nu^V$$

In principle the QCB-states can affect all quantities of the deuteron. For an arbitrary operator O we write, for one state ψ_ν:

$$<O> \ = \ <\psi|O|\psi> \ / \ <\psi|\psi>$$

$$<\psi|O|\psi> \ = \ a_\nu^2 <\psi_\nu|O|\psi_\nu> + 2a_\nu <\psi_\nu|O|\psi_h> + <\psi_h|O|\psi_h>$$

Thus we can write:

$$<O> \ = \ P_{QCB} \ O_{QCB} + \frac{2a_\nu O_{int} + O_{conv}}{<\psi|\psi>} \tag{16}$$

We take only the lowest QCB-state into account, which we suppose to be a s^6 state. It has a quadrupole moment equal to zero. Then we find

$$Q = (2a_\nu \ Q_{int} + Q_{conv}) \ / \ <\psi|\psi> \tag{17}$$

$$Q_{int} = \frac{1}{\sqrt{50}} \int_0^b dr \ r^2 \left[\frac{x_\nu^0 \eta_\nu^0(r) w(r) + x_\nu^2 \eta_\nu^2(r) u(r)}{2} - \frac{x_\nu^2 \eta_\nu^2(r) w(r)}{\sqrt{8}} \right]$$

$$Q_{conv} = \frac{1}{\sqrt{50}} \int_0^\infty dr \ r^2 \left[u(r)w(r) - \frac{w^2(r)}{\sqrt{8}} \right]$$

For the mass-distribution we find the following correction to the expectation value of the radius:

$$\langle r^2_{int} \rangle = \int_0^b dr \; r^2 [x_\nu {}^0 \eta_\nu {}^0 (r) u(r) + x_\nu {}^2 \eta_\nu {}^2 (r) w(r)] \tag{18}$$

$$\langle r^2_{QCB} \rangle = \langle \psi_\nu | r^2 | \psi_\nu \rangle$$

The corrections to the magnetic moment are

$$\langle \mu_{int} \rangle = a_\nu {}^V (\mu_p + \mu_n) - \frac{3}{2} (\mu_p + \mu_n - \frac{1}{2}) a_{\nu,2} {}^V \tag{19}$$

$$\langle \mu_{QCB} \rangle = \langle \psi_\nu | \mu | \psi_\nu \rangle$$

In table 2 we give the numerical values of these corrections. For $\langle \mu_{QCB} \rangle$ we have taken the estimate by Kondratyuk $\langle \mu_{QCB} \rangle = 1.2$ nm. For $\langle r^2_{QCB} \rangle$ no estimate as yet exists. (We give the values of the experimental data on these parameters in parentheses).

Table 2. Deuteron properties for the QCB-model of table 1

$a_{\nu,0} {}^V = -0.030294$	$a_{\nu,2} {}^V = 0$
$a_\nu {}^S = 0.068031$	$a_\nu = 0.037736$
$P_D = 5.51\%$ $\quad P_{6q} = 10.4\%$	$P_{QCB} = 1.4\%$
$Q = 0.281497$ fm^2	$Q_{int} < 0.2\%$
$(0.28590(30)$ fm$^2)$	
$\mu_{conv} = 0.818138 \quad \mu_{int} = 0.019400$	$\mu_{QCB} = 0.016477$ nm
$(0.857406(1)$ nm$)$	
$\langle r^2_{conv} \rangle = (2.0167$ fm$)^2$	$1.9650(45)$ fm
	$1.9576(68)$
$a_t = 5.49$ fm	$r_{ot} = 1.79$ fm
$(5.419(7)$ fm$)$	$(1.759(5)$ fm$)$

The wave functions of our model are shown in fig. 2. Beyond 1.2 fm they coincide with the wave functions of the Paris deuteron; at r = 1.2 fm the occurence of the δ-function in V_{hqh} leads to a discontinuity in the derivative which is characteristic for the QCB-model.

Finally we note that the relation between the asymptotic normalization A_s and the effective-range parameter extrapolated to the deuteron-pole

$\rho_1(-\alpha^2, -\alpha^2)$ is also changed. In the notation of Sprung et al. [11]one has

$$A_S^2 = \frac{2\alpha}{1 - \alpha\rho_1(-\alpha^2, -\alpha^2)}$$

(20)

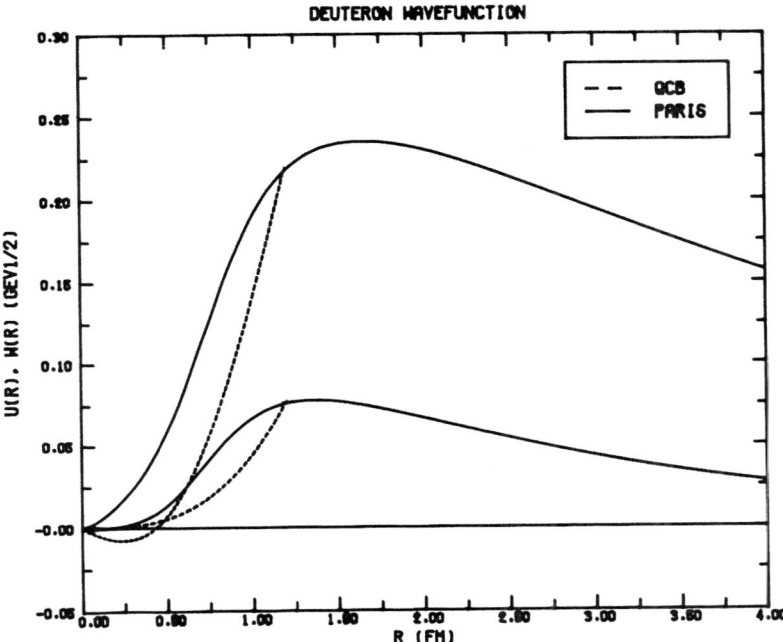

Fig. 2. The deuteron wave functions according to the QCB-model of table 1.

Here $\hbar^2\alpha^2/m_N$ is the binding energy of the deuteron and $\rho_1(-\alpha^2, -\alpha^2)$ is the effective range parameter extrapolated to the deuteron pole. In the presence of a QCB admixture this relation changes to

$$A_S^2\left\{1-\sum_{\ell;\ell}\left[\frac{c_\nu^\ell\tilde{u}_\ell(b)c_\nu^{\ell'}\tilde{u}_{\ell'}(b)}{(E_D - E_\nu)^2} - \tilde{\eta}_\nu^{\ell'}\tilde{\eta}_\nu^\ell\right]\right\} = \frac{2\alpha}{1 - \alpha\rho_1}$$

(21)

The quantity $\tilde{\eta}_\nu^\ell$ is defined as $\tilde{\eta}_\nu^\ell = x_\nu^\ell\int_0^b dr\tilde{u}_\ell(r)\eta_\nu^\ell(r)$ and $\tilde{u}_\ell(r)$ is the deuteron wave function component with the symptotic normalization A_S divided out.

6. Three nucleon properties

The QCB-interaction being non-local and energy dependent is expected to give results for the three-nucleon observables that can be different from those calculated with the help of more traditional realistic models. In addition the occurrence of QCB states modifies the off-shell two-body t-matrix in the three-body space. This effect has been studied by the ITEP-group.[12] and results for cases where the peripheral meson-exchange interaction was neglected have been reported by them [13]. The main result of that investigation is that it is possible to get the correct value (8.5 MeV) for the triton-binding energy. The modification due to the QCB-admixture is very important.
If the latter is neglected and the Faddeev equations are solved with the unmodified off-shell two-body t-matrices, the binding energy is only 4.3 MeV.

Acknowledgements

I want to thank H. Dijk, Yu.A. Simonov, Yu.S. Kalashnikova and I.M. Narodetskii for numerous comments and stimulating discussions. This work forms a part of a collaboration between the Physics Department of the Vrije Universiteit at Amsterdam and the Institute for Theoretical and Experimental Physics at Moscow.

References

1. De Tar, C., Ann. Rev. Nucl. Sci. 33, 235 (1983)
   Faessler, A., Prog. Part. Nucl. Phys. 11, 171 (1984)
2. Simonov, Yu.A., Phys. Lett. 107B, 1 (1981)
   Simonov, Yu.A., Sov. J. Nucl. Phys. 36, 422 (1982)
   Simonov, Yu.A., Nucl. Phys. A416, 109C (1984)
3. Narodetskii, I.M., Proc. IX Eur. Conf. Few Body Prob. in Phys., Tbilisi 1984
4. Bakker, B.L.G., Proc. X Eur. Symp. Dyn. Few Body Syst.,

Gy, Bencze, P. Doleschall , J. Revai (eds.) Balatonfüred 1985, p.37

5. Jaffe, R.L., Low, F.E., Phys. Rev. D 19, 2105 (1979)

6. Bakker, B.L.G., Mulders, P.J., Ann. Rev. Nucl. Sci. 37 (1986) to appear

7. Mulders, P.J., Nucl. Phys. A416, 99C (1984)
   Bakker, B.L.G., Grach, I.L., Narodetskii, I.M., Nucl. Phys. A424, 563 (1984)

8. Feshbach, H., Lomon, E.A., Ann. of Phys. 29. 19 (1964)

9. Mulders, P.J., Aerts, A.T., De Swart, J.J., Phys. Rev. D21, 2653 (1980)

10. Lacombe, M., Loiseau, B., Richard, J.M., Vinh Mau, R., Côté, J., Pirès, P., De Tourreil, R., Phys. Rev. C21, 861 (1980)

11. Sprung, D.W.L., Kermode, M.W., Klarsfeld, S., J. Phys. G8, 923 (1982)

12. Kalashnikova, Yu.S., Narodetskii, I.M., Sov. J. Nucl. Phys. 42, 203 (1985)

13. Veselov, A.I., Grach. I.L., Kalashnikova, Yu.S., Narodetskii, I.M., Sov. J. Nucl. Phys. 42, 347 (1985)

ON DESCRIPTION OF NUCLEON AND PION FORM FACTORS
IN QUANTUM CHROMODYNAMICS *

A.S. Gorsky

M.V. Terentyev

Institute of Theoretical and Experimental Physics

Moscow

For pion and nucleon electromagnetic form factors a unified picture of the sequence of theoretical regimes in the whole variation range of q^2 is described. The picture is in agreement with the available experimental data and provides with certain predictions in the domain where the data are absent.

The subject of this report is a concise exposition of an approach to the theoretical description of electromagnetic form factors (FF) of hadrons. Actually we deal with two examples: the pion FF and the nucleon FF. One can hardly be satisfied with the present dynamical description of FF in a wide region of the momentum transfer squared (q^2). There are some q^2 regions where the approaches are based upon quite different pictures of the physical process. Therefore people are sometimes in controversy as to an interpretation of the data on the pion and nucleon FF's and yield quite different predictions on the behaviour of the FF at large q^2 where no measurements have been performed yet.

Experimental investigation of exclusive FF is a difficult problem. This is a reason why the relevant data have been accumulated slowly. In the space-like region ($Q^2 = -q^2 > 0$) the proton FF have been measured for

---

*Paper submitted in absentia

$0 \leq Q^2 \stackrel{<}{\sim} 30$ [1], the neutron FF - for $0 \leq Q^2 \stackrel{<}{\sim} 10$ [1], the pion FF - for $0 \leq Q^2 \stackrel{<}{\sim}$ $\stackrel{<}{\sim} 3.5$ [2]. In the time-like region ($q^2 \geq 0$) the proton data are available for $3.5 \stackrel{<}{\sim} q^2 \stackrel{<}{\sim} 4.5$ [3], and the pion data - for $0 \leq q^2 \stackrel{<}{\sim} 6$ [4], and also at the points corresponding to the ψ - and ψ' - meson decays [5] (see also [27]). (Here and in the following we use GeV^2 units for q^2).

We outline here a possible physical picture in describing FF's; we believe it is consistent in various q^2 regions, unified for the pion and nucleon FF and the most probable. Note that the picture outlined here is not quite conventional. Only references to the works which are directly relevant to the approach we advocate are given in the report. The methods of theoretical description of FF at different q^2 which we advocate are given suitably in Fig. 1, which is common for pion and nucleon. The text below is, in fact, a commentary to the figure.

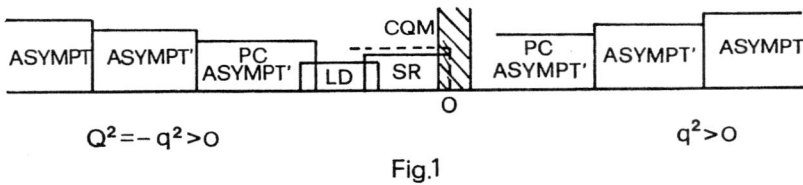

Fig.1

The shaded area in Fig. 1 is a resonance region. SR is the region where the QCD sum rules are applicable. CQM is the region of the constituent quark model. It is indicated by a dashed line, unlike the other regions, because CQM exists independently of QCD. The upper boundary of the region (in the Q^2 scale) is indefinite, as the model is not quite rigid, but this region is hardly extended beyond $Q^2 \geq 1$. The calculations in the framework of CQM employ the graphs of Fig. 2, which involve vertices determined by the wave function (WF) of the hadron composed of the constituent quarks.

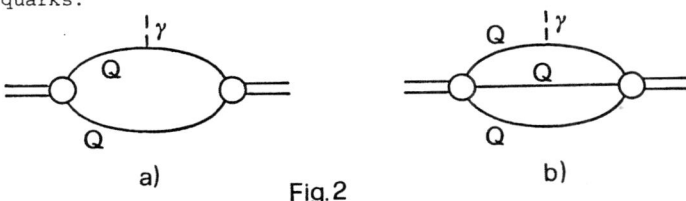

Fig.2

LD in Fig. 1 is the region of the local quark-hadron duality. Here the pion and nucleon FF's are calculated with the graphs in Fig. 3a and

494

3b respectively. The vertices, there, are given by the matrix elements $<0|\eta_H|H>$ of the local quark currents ($H = \pi$ or N). For pion we have $\eta_\pi \sim$ $\sim \bar{q}\gamma_\mu\gamma_5 q$, for nucleon: $\eta_N = (qC\gamma_\nu q)\gamma_5\gamma_\nu q$. The integrals in the invariant mass squared of the two (three) quarks incident to the current vertices are cut off at s_o, which specifies a duality interval.

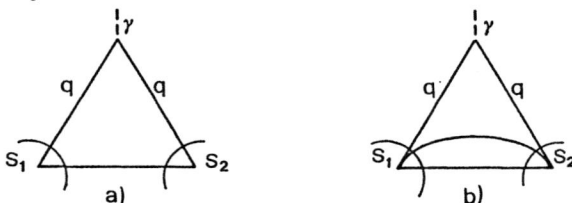

ASYMPT' in Fig. 1 is the region of the QCD asymptotics, where the hard quark rescattering due to the gluon exchange (Fig. 4) is dominating. As to the hadron WF, it may be non-asymptotical - the prime in the name of the region indicates this stipulation.

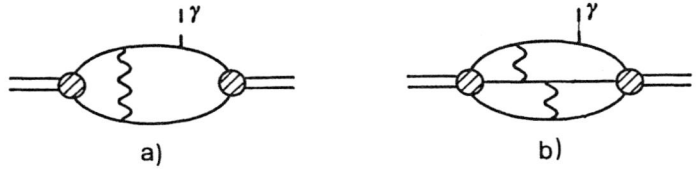

PC in Fig. 1 are the power corrections which exist within the hard rescattering mechanism. These corrections result from higher twists in the hadron WF. They can be due, in particular, to contributions from diagrams with an additional emission of gluons during the hard stage of the process.

ASYMPT in Fig. 1 is the genuine asymptotics of FF in the framework of perturbative QCD (Fig. 5). If the hadron WF at $Q^2 \sim 10$ deviates considerably from the asymptotics, the genuine asymptotics region is not available practically. In the other extreme case the pion WF is, probably, close to the asymptotics already for $Q^2 \sim 1$. In this situation the ASYMPT' mechanism is actually disappearing from the scheme, as it coincides with the ASYMPT mechanism.

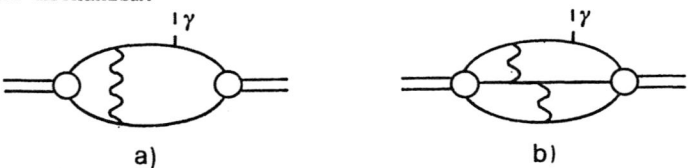

The boundaries between the regions in Fig. 1 are determined up to a factor of 2 ÷ 3, and they are appreciably different for pion and nucleon. In the leading terms the behaviour of FF at large positive q^2 must be the same as that for large negative q^2. Meanwhile the approach to the $q^2 > 0$ asymptotics is, in general, slower, and the boundary of the asymptotic region can hardly be estimated in the theory. Therefore we do not indicate the lower boundary of the PC+Asympt' region in the right hand side of Fig. 1.

Now let us turn to a more detailed characteristic of each region. For the pion FF the most elaborated analysis in the CQM has been performed in Ref. [6], where the authors used WF of the relativistic quark model [7,8] with parameters fitted to many low-energy properties of the pion physics [8,9]. Mention also a recent work [10] performed along the same line. The description of Ref. [6] has been extended up to $Q^2 \sim 6$ and a reasonable agreement with the experimental data is attained everywhere. Unfortunately, a similar procedure has not been performed yet for the nucleon FF. Nevertheless, as we have already mentioned, it is unlikely that CQM is reasonable for $Q^2 > 1$ and the agreement for higher Q^2 stems from the fact that the CQM formulae are matched to those of the LD method at $Q^2 \sim 1 \div 2$ and acquire a quite different meaning there. Actually, for $Q^2 \gtrsim 1$ the resulting FF in CQM depend weakly on details of the hadron WF and on the mass of the constituent quark, so the graphs of Fig. 2 and 3 do not differ in essence.

The SR provide also with a fair description of the data on FF. The calculations have been performed for pion [11-13] and nucleon [14]. The three-point correlator $\langle \eta_\pi(x) J_\mu(0) \eta_\pi(y) \rangle$ where J_μ is the electromagnetic current, was analyzed in [11-13], and the two-point correlator in an external field was considered in [14]. The pion FF in the region $0.5 \lesssim Q^2 \lesssim 4$ was described in the framework of the standard SR method [11,12]. (For $Q^2 \to \infty$ the method cannot be employed, as the relative contribution from the power corrections is increasing and the effect of continuum is large as compared with the leading FF term. For $Q^2 \to 0$ nonphysical singularities appear, and there are uncontrollable contributions from higher-twist operators). A modification of the usual SR method was proposed [13,14] which enabled the authors to describe the data on the pion and nucleon FF in the region $0 < Q^2 \leq 1$. In principle, one cannot exclude the possibility that the description of the nucleon FF in the SR method can be extended to $Q^2 \leq 3 \div 4$, i.e. as far as that for the pion FF.

The lower boundary of the domain of applicability of LD is determined with the values of Q^2 at which the asymptotical freedom of quarks is maintained in the graphs of Fig. 3, namely, for $Q^2 \gtrsim 1$. Formally, the method is valid up to energies at which the virtuality of the spectator quark falls to a few GeV^2; it is the case for $Q^2 \lesssim 7 \div 10$ for pion, $Q^2 \lesssim 15 \div 20$ for nucleon. However, already for $Q^2 \gtrsim 4$ for pion and $Q^2 \gtrsim 10$ for nucleon the calculations acquire an essential dependence on the quark distribution in the hadron, thus suppressing the predictive validity of the model and impeding its comparison with the experimental data.

The LD method has been formulated in [12,15] to be applied to the calculation of the pion and nucleon FF. The underlying physical assumption is that the integral in the invariant mass of two (three) quarks in a narrow interval $0 < s < s_o$, of the graphs of Fig. 3, equals the integral over the same region of the pion (nucleon) contribution to the double spectral function $\rho(s_1, s_2, q^2)$ for the three-point correlator $\langle \eta_H(x) J_\mu(0) \eta_H(y) \rangle$, where $H = \pi, N$. The contribution from the region $s > s_o$ cancels with that from many-particle states in $\rho(s_1, s_2, q^2)$. The parameter is s_o; one sets $s_o \approx 0.75$ for the pion FF and $s_o \approx 2.3$ for the nucleon FF. A fair agreement has been attained [12,15], in the whole, both for the FF behaviour as function of Q^2, and for relations between neutron and proton magnetic FF's.

In the region considered, the experimentally observed behaviour of FF is given by $F_\pi(Q^2) \sim (Q^2 + 0.55)^{-1}$ for pion, and $G_M(Q^2) \sim (Q^2 + 0.71)^{-2}$ for the nucleon magnetic FF. The LD yields a good numerical approximation to these functions with a more complicated expression, arising from integrals of the graphs in Fig. 3. Formally the LD method looks like setting certain WF in the graphs of Fig. 2, yet one should have in mind that in essence the LD approach does not assume that pion (nucleon) consists of two (three) quarks, as in CQM. It takes into account solely a contribution to FF from the corresponding component of the Fock column. Besides,

as we have already mentioned, the method deals with the massless current
quarks in the asymptotical freedom region. The fact that the LD method
reproduces correctly the Q^2 dependence of FF can be explained qualitatively from the following kinematical arguments. Let us firstly consider
the pion FF. The graph in Fig. 3a, when calculated in the infinite pion
momentum reference frame, leads to the integral of the same type as that
in CQM: $F_\pi(Q^2) \sim \int d^2k_\perp dx \ (1-x)^{-1} x^{-1} \phi_1(k_\perp,x) \phi_2(k_\perp,xq_\perp,x)$, where $q_\perp^2 = -Q^2$,
$k_\perp$ is the transverse momentum,

x is a fraction of longitudinal quark momentum. The "wave func-
tion", $\phi_1(k_\perp,x)$, for massless quarks is responsible for a cut-off of the
integral in the variable $s_1 = -k_\perp^2/x(1-x) \underset{\sim}{<} s_0$, where s_1 is the effective
mass of a pair of quarks incident to one of the vertices. Similarly, $s_2 =$
$= -(k_\perp+xq_\perp)^2/x(1-x) \underset{\sim}{<} s_0$, hence $x_{eff} \sim s_0/q_\perp^2$, and $F_\pi(Q^2) \sim \int dx \sim q_\perp^{-2}$.

The nucleon FF is effectively determined by the double integral
$\int dx_1 dx_2$ and again we have $x_j \sim s_0/q_\perp^2$, since the mass of any quark is limited, as well as the virtual mass of all three quarks. Therefore we have
$G_M(Q^2) \sim q_\perp^{-4}$.

As Q^2 is increasing further, we get into the region of the hard rescattering (PC + ASYMPT' and ASYMPT in Fig. 1). No experimental data are
available here. For the nucleon FF, the theoretical predictions are rather uncertain. (We have no reliable information on details of the nucleon
WF, and the power corrections have not been calculated). It was found,
meanwhile, that the ASYMPT mechanism, being extended to $Q^2 \sim 20$ (where one
would hope to match it to the LD mechanism), yields a wrong sign of the
proton FF, and its magnitude smaller by a factor of 100. (The above statement results from the asymptotical formulae for $G_M(Q^2)$ obtained in Ref.
[16,17] and from the normalization of the asymptotical nucleon WF, as
given in [18]). Probably, one can speculate on the non-asymptotical behaviour of the nucleon WF. In particular, the desired value of $G_M(Q^2)$ at
$Q^2 \sim 20$ has been obtained in this way in Ref. [19] in the ASYMPT' regime.
To this end, one should assume a very nonsymmetric quark distribution in
the nucleon; besides, the gluon virtualities in the hard rescattering
processes do not exceed 1 GeV^2 in this regime, because of kinematical
reasons. With all this in view, this description is close to being incon-

sistent. Therefore we feel it is more natural to assume that in the nucleon case the change of the regimes above the LD regions takes place, in general, in the manner as in the pion case (see below).

We shall discuss the pion FF in the hard rescattering region going from large Q^2. An expression for $F_\pi(Q^2)$ in the ASYMPT region has been derived in ref. [20] (unfortunately, with incorrect absolute normalization). A general method for calculation of FF asymptotics has been developed in Refs. [21-23]. However, the formulae in the ASYMPT region, when extended up to $Q^2 \sim 4 \div 6$, lead to a value of FF that is by a factor 4 less than the observed magnitude. Evidently there is an essential change of the regime in the interval between ASYMPT and LD.

This new mechanism is the power corrections to the hard rescattering mechanism. Calculations of the PC due to a contribution from the pseudo-scalar current (the twist-3 component in the quark WF of the pion) have been performed in [24]. In subsequent works[25,26] it was shown that corrections from some other operators, in particular, those containing the gluon field are also substantial. The PC are considerable throughout the region $10 \underset{\sim}{<} Q^2 \underset{\sim}{<} 10^2$ if the pion WF of the leading twist is close numerically to its asymptotics already for $Q^2 \sim 1$ (the arguments for that have been presented in [6,9]). On the other hand, if the intermediate ASYMPT' regime takes place in a wide region (as it was argued in [27]), then the PC have at least the same order of magnitude as the ASYMPT' contribution at $Q^2 = 4 \div 6$. Therefore the expected behaviour of the pion FF in the region of $Q^2 \underset{\sim}{>} 4$ is $F_\pi(Q^2) \sim A_\pi/Q^4 + B_\pi/Q^2$, where (up to logarithmic factors) A_π/Q^4 is a contribution from the PC, and B_π/Q^2 is a contribution from the asymptotical graphs in Figs. 4a, 5a. We expect that $A_\pi \underset{\sim}{\sim} 5B_\pi$ (in units of GeV^2) if the ASYMPT' regime from [27] takes place, and $A_\pi \underset{\sim}{\sim} 30B_\pi$, if the ASYMPT regime is established at $Q^2 \sim 10$, i.e. immediately after the LD region (as advocated in [9]).

Likewise, the expected behaviour of the nucleon magnetic FF in the region of $Q^2 \geq 20$ is: $G_M(Q^2) \sim A_N/Q^6 + B_N/Q^4$, where B_N/Q^4 is determined by the graph in Fig. 4b. (For the nucleon there are no reasons to believe that the ASYMPT' region is absent - see Ref. [9]), and A_N/Q^6 is a possible contribution from the PC, $A_N \sim (10 \div 100)B_N$.

Note that a good semiphenomenological description of the data on the pion FF can be obtained throughout the total region of q^2 if one adopts a model where the assumption of the dominance of the vector meson resonance contribution (ρ,ω,ρ') at $q^2 \sim 1$ is combined with restrictions imposed by the requirements of analyticity and the asymptotical behaviour (see the Ref. [28]). At least one arbitrary parameter specifying a matching of the asymptotics and the resonance region is always present in models of this type. A similar semiphenomenological description of the nucleon FF has not been developed until now. References to a number of works concerning a phenomenological description can be found in [29] (for pion case) and [30] (for nucleon case).

Mention also computer calculations of WF moments[31] and pion FF[32] performed in the lattice version of QCD. A discussion of these works is beyond our purpose here.

REFERENCES

1. Mestayer M.D. et al.: SLAC Report n. 214 (1978).
    Rock S. et al.: SLAC Publ. 2949 (1982).

2. Bebek C. et al.: Phys. Rev. D17, 1693 (1978).

3. Bisello D. et al.: Nucl. Phys. B224, 379 (1983).
    Delcourt B. et al.: Phys. Lett. B86, 395 (1979).
    Bassompiere G. et al.: Phys. Lett. B68, 477 (1977).

4. Esposito B. et al.: Lett. Nuovo Cim. 28, 337 (1980).
    Barbellini G. et al.: Lett. Nuovo Cim. 6, 557 (1973).
    Bukin A. et al.: Phys. Lett. B73, 226 (1977).

5. Vannucci F. et al.: Phys. Rev. D15, 1816 (1977).

6. Bagdasaryan A.S. et al.: Yad. Fiz. 38, 402 (1983).

7. Terentyev M.V.: Yad. Fiz. 24, 207 (1976).

8. Aznauryan I.G. et al.: Yad Fiz. 36, 1278 (1982).

9. Terentyev M.V. et al.: Yad. Fiz. 38, 213 (1983).

10. Jacob O., Kisslinger L.: Phys. Rev. Lett. 56, 225 (1986).

11. Ioffe B.L., Smilga A.V.: Nucl. Phys. B216, 373 (1983), Phys. Lett. B114, 353 (1982).

12. Nesterenko V., Radyushkin A.: Phys. Lett. B115, 410 (1982).

13. Nesterenko V., Radyushkin A.: Pis'ma ZETP 39, 576 (1984).

14. Belyaev V.M., Kogan Ya.I.: Preprint ITEP n. 29 (1984).

15. Nesterenko V., Radyushkin A.: Phys. Lett. B128, 439 (1983).

16. Avdeenko V.A. et al.: Yad. Fiz. 33, 481 (1981).

17. Brodsky S. et al.: Phys. Rev. D23, 1152 (1981).

18. Belyaev V.M., Ioffe B.L.: ZEPT 83, 876 (1982).

19. Chernyak V., Zhitnitsky A.: Nucl. Phys. B216, 52 (1984).

20. Farrar G., Jackson D.: Phys. Rev. Lett. 35, 1416 (1975).

21. Chernyak V.L. et al.: Pis'ma ZEPT 26, 760 (1977).

22. Efremov A.V., Radyushkin A.V.: TMF 42, 147 (1980).

23. Brodsky S., Lepage G.: Phys. Rev. D22, 2157 (1980).

24. Geshkenbeyn B., Terentyev M.: Yad. Fiz. 39, 873 (1984).

25. Gorsky A.S.: Preprint ITEP n. 168 (1984).

26. Chernyak V., Zhitnitsky A.: Phys. Rep. 112, 173 (1984).

27. Chernyak V., Zhitnitsky A.: Nucl. Phys. B201, 492 (1982).

28. Geshkenbein B., Terentyev M.: Yad. Fiz. 40, 758 (1984).

29. Amendolia S. et al.: Preprint CERN-EP/86-34 (1986).

30. Bisello D. et al.: Nucl. Phys. B224, 379 (1983).

31. Gottlieb S., Kronfeld A.: Phys. Rev. Lett. 55, 2553 (1985).

32. Wilox W., Woloshin R.: Phys. Rev. Lett. 54, 2653 (1985).

HADRONIC PROBES OF FEW NUCLEON SYSTEMS

N.E. Davison
Department of Physics, University of Manitoba,
Winnipeg, Manitoba, R3T 2N2, Canada
and
TRIUMF
4004 Wesbrook Mall, Vancouver, B.C., V6T 2A3, Canada

Abstract

Recent studies of few nucleon systems using hadronic probes are commented on. Emphasis is placed on nucleons and pions as probes. The sensitivities of these probes to reaction mechanisms and nuclear properties are discussed. Impressive advances have occurred in several areas, while success in others has been of an evolutionary nature.

Introduction

Many impressive advances have been made within the last year within the broad subject of hadronic probes of light nuclei. This applies not only to investigations of the properties of the nuclei themselves, but also to the study of reaction mechanisms. There is, of course, a continual interplay between these two aspects of light nuclear physics: only when mechanisms are understood to a certain degree, can more refined pictures of the structure be extracted, and only as the framework defined by the nuclear structure becomes clear, is it possible to judge what mechanisms are likely to be important.

It is important that we continue to make progress. The goals of studies of light nuclear systems must surely include the ability to undertake a confident assessment of the importance of genuine three-body interactions or more exotic phenomena such as signatures of the quark structure of nucleons. In this context, some of the most important progress in the last year centres around the indications that we are continuing to identify those problems that are most likely to maintain the momentum of progress. In this review, therefore, there will be an attempt to emphasize not only the clear successes, but also cases in which there has been an improvement due to the introduction of a new mechanism, or the identification of the need for a specific type of new data. There will also be an effort to put as much weight on the pion as a probe as on the nucleon. This is partly due to personal interest and partly because there have been many recent reviews that have stressed the nucleon with the result that the pion has been relatively neglected. In addition, from a practical point of view, nucleon and pion probes are beginning to address the same questions and many important amplitudes in nucleon-nucleon (NN) processes involve pions.

In the past year, there have been impressive advances in understanding reaction mechanisms for both pion and nucleon induced reactions in light nuclei. A personal choice that summarizes this review includes the following:

1) The discrepancy between data and theory for the $^2H(p,2p)n$ reaction at intermediate energies has been greatly reduced and possible reasons for the remaining missing strength have been identified.

2) The spectral function describing the momentum distribution of a proton or a deuteron in ^3He has been measured to much larger momentum transfers than were available before. Data are now available for $^3He(p,2p)d$, $^3He(p,pd)p$, $^3He(\pi^+,pp)p$ and $^3He(e,e'p)d$. While the data from all these reactions are not in perfect accord, the agreement is sufficient to permit evaluation of the source of the remaining difficulties.

3) Theory and experiment for πd elastic scattering and for π-induced breakup of the deuteron are in good agreement, so good in fact, that a new round of experiments is needed to permit observation of discrepancies.

4) The quality of fits to π absorption data on light nuclei especially π^- absorption on ^3He is greatly improved. While the quality of the fits is not yet nearly as good as for πd scattering, significant improvements have been obtained.

On the other hand, there are also problems. This is especially true for pion interactions. Although the $pp \rightarrow d\pi^+$ and $\pi^+d \rightarrow pp$ reactions have been studied for quite some time, and although qualitative and even quantitative fits have been obtained, there is a disturbing difficulty in obtaining really good fits to this fundamental process. In the opinion of this reviewer, it is still premature to start talking seriously about signatures of quark effects because there are several other more mundane effects to be taken into account before the admittedly more exciting possibilities of sub-nucleonic degrees of freedom are investigated. Even in the remaining difficulties, though, there is much to find encouraging. For the most part, difficulties seem to involve questions of detail that we may hope to resolve sufficiently well that the really interesting questions of new physics may be approached in the not too distant future.

Elastic Scattering

Care must be taken to distinguish the scattering of pions from the scattering of nucleons because the theories used in the two cases are quite different. In the case of nucleon scattering, the complexity of the computational problems and the limitations of potential models in representing the NN interaction mean that at high and intermediate energies, Faddeev techniques and extensions to more than 3 bodies are not practical. One is therefore forced to use either optical model or multiple scattering approaches, such as the Glauber theory, both of which contain major approximations the effects of which are not always clear. In the case of pions, on the other hand, the spin structure is sufficiently simple that a Faddeev representation is feasible. Unfortunately, one must treat at least the pion relativistically and a computationally viable relativistic three-body formalism free of severe approximations does not yet exist. In spite of this, however, the results give a good representation of the data.

Nucleons

For Nd elastic scattering the scattering matrix is a 6x6 matrix of complex amplitudes. Of these, 12 remain after application of rotational invariance, parity conservation and time-reversal invariance. Removing a common phase, one needs to measure 23 observables at each energy and scattering angle to completely determine the scattering matrix. To measure 23 independent observables is an immense undertaking, but already it is either complete or nearly so at two energies. Sperisen [1] has pointed out that at a proton laboratory energy of 10 MeV, there is available a set of 19 pd elastic scattering observables: the unpolarized differential cross section, 5 analyzing powers, 3 proton to proton and 10 proton to deuteron polarization transfer coefficients. These measurements are independent but, of course, do not constitute a complete set. At 800 MeV, a complete set of 24 observables now exists [1] and further measurements are being made to extend the range of momentum transfers for which data are available. The set at 800 MeV consists of the unpolarized differential cross section, 4 analyzing powers, 4 proton to proton spin transfer coefficients and 15 second and third order spin observables obtained with both the proton beam and the deuteron target polarized.

In the case of proton scattering from ^{3}He, the data set is much smaller, but an analysis [2] of data covering the range of incident proton energies from 100-1000 MeV was completed during the last year, using Glauber multiple scattering theory with both parameterized NN interactions and NN

scattering amplitudes obtained from phase shift analyses [3]. Somewhat surprisingly, the results obtained with the parameterized interactions were
better than those obtained from phase shift analyses. This implies that the
parameterized interactions are acting as effective interactions. The problem with the results obtained using the realistic NN amplitudes appears to
lie with excess interference between the single and double scattering terms.
This is consistent with the fact that the constant ratio of the real to
imaginary parts of the parameterized scattering amplitudes limits the degree
of interference between the first- and second-order scattering.

A number of potential improvements to the calculations were examined in
Ref. 2. Ascribing the problem to insufficiencies in the ^{3}He wave function
used was rejected on the basis of calculations carried out by Narboni [4].
These calculations suggested that using a ^{3}He wavefunction with S' and D
state terms did not remove the discrepancy between theory and experiment.
Improved off-shell corrections and the use of a relativistic treatment such
as has been successful in nucleon-nucleus scattering [5] was judged to be of
potential value. Unfortunately, a formulation for nucleon scattering from a
spin one-half target does not exist. Indications are that use of a ^{3}He form
factor explicitly dependent on the spin and isospin might also help [6], but
it would be necessary to verify independently the validity of the new form
factors. In summary, there is considerable room for improvement in the calculation of proton elastic scattering from ^{3}He, but a great deal of work
needs to be done both theoretically and experimentally before a satisfactory
description becomes available.

Pions

Before commenting on the elastic scattering of pions, it is necessary
to say something about the theories used. The scattering of pions from the
deuteron is usually treated within the context of the NN → NNπ system of
reactions, using a relativistic formulation. Namyslowsky [7] proposed in
1968 several criteria that a proper relativistic 3-body theory should
satisfy. In general, a satisfactory theory should:

1) be relativistically covariant,

2) preserve 2- and 3-body unitarity,

3) have correct 4-momentum conservation in intermediate states

4) have the proper off-shell continuation. In Namyslowsky's original
   formulation, this was achieved by requiring time reversal
   invariance,

5) have proper cluster decomposition and this decomposition should be
   unique for the 2- and 3-body clusters.

To this might be added the requirement that the theory should have the
correct low energy limit. The first two criteria are automatically

satisfied by the procedures developed by Aaron, Amado and Young (AAY) [8] which are a practical application of the Blankenbecler–Sugar reduction [9]. The Blankenbecler–Sugar reduction has the somewhat unwanted characteristic that it forces all the particles to be on the same "shell" which does not have to be the mass shell. With the proper manipulation, however, this shell can be chosen so as to satisfy criterion 3. Criteria 4 and 5 taken together lead to an incompatibility between the correct clustering (factoring of the S matrix into the product of S matrices for the two subsystems) and the correct low energy limit. An excellent discussion of these questions in given is Ref. 10. It is important to note that there is not at present any computationally viable theory that satisfies all these criteria.

Garcilazo has systematically developed a version of the AAY theory for treating πd elastic scattering and pion-induced breakup of the deuteron in a number of papers presented over several years [11]. The emphasis has been on fairly low pion energies, although this is not a constraint intrinsic to the theory. Generally, the fits to elastic scattering differential cross sections are excellent at lower energies, although significant deviations do show up at an incident pion energy of 256 MeV. That this is not a fundamental problem is illustrated by the work of Andrade et al [12] in which inclusion of additional πN interactions in the $^5S_2$ and $^5P_3$ waves gives a significant improvement both to the differential cross sections and vector analyzing powers.

Garcilazo's work on the πd system has found its most recent expression in the work of Gyles et al. on differential cross sections [13] and analyzing powers [14] for pion-induced breakup of the deuteron. The fits to the data, a sample of which is shown in Fig. 1, are excellent in most cases. For the analyzing power work, the statistical accuracy of the data is much reduced, but there is little or no indication of deviation of theory from experiment. Further experiments have been approved at TRIUMF to improve the statistical accuracy in the pion-induced breakup work.

Nucleon-Induced Breakup Reactions

Work on nucleon-induced breakup reactions seems to be following an evolutionary path suggesting that a great deal of work will be necessary before a detailed agreement between theory and experiment is achieved.

Breakup of the Deuteron

It has been known for some time that the Impulse Approximation (IA) predicts cross sections that are 10% to 20% too low for $^2H(p,2p)n$ or $^2H(e,e'p)n$. Many "small" effects could be at the origin of this discrepancy. For instance, half-off-shell NN or eN amplitudes are required in the IA, but

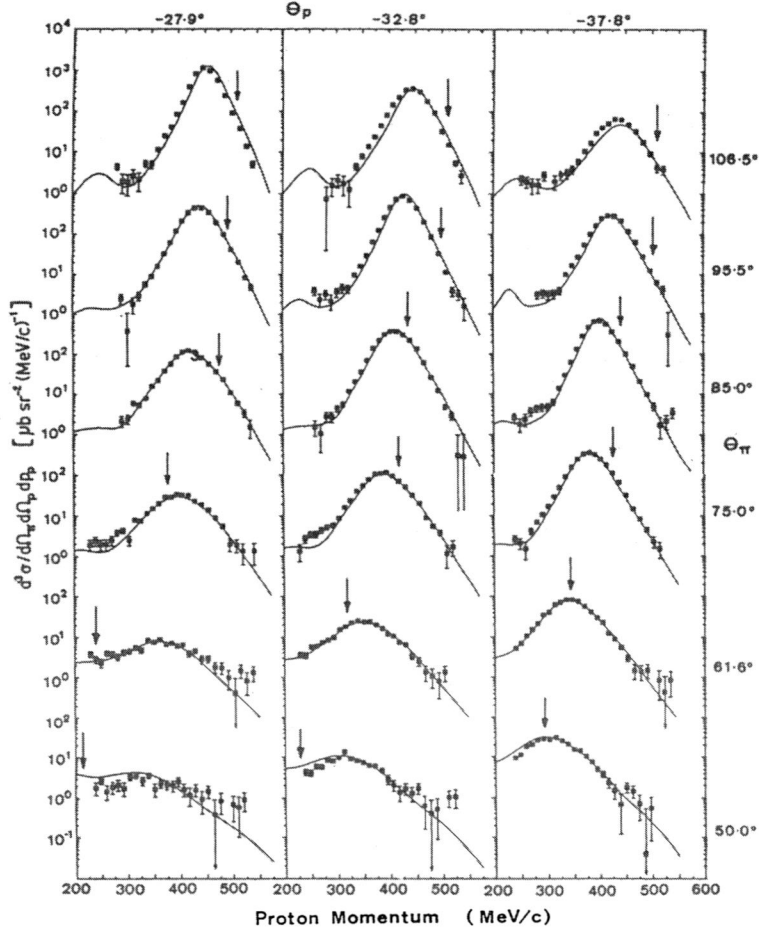

Fig. 1. Comparison of experimental data for pion-induced breakup of the deuteron with calculations based on the work of Garcilazo. The positions where the πp invariant mass is equal to the Δ mass are indicated by arrows.

almost universally, the on-shell amplitudes obtained from elastic scattering data are substituted. In addition, final state interactions, non-nucleonic components in the deuteron, projectile rescattering [for (p,2p)] and meson exchange corrections [for (e,e'p)] are expected to be important. The importance of non-nucleonic components has recently been demonstrated by data for ^{2}H(p,2p)n at large angles taken at an incident proton energy of 507 MeV [15]. These data were compared with calculations by Yano [16] in which the effects of the Δ were taken into account as a non-IA contribution to the reaction. Results are shown in Fig. 2. Final state interactions and rescattering corrections have still not been included in these calculations, and the quality of results even without inclusion of these effects strongly suggests that the Δ plays a major role in this reaction.

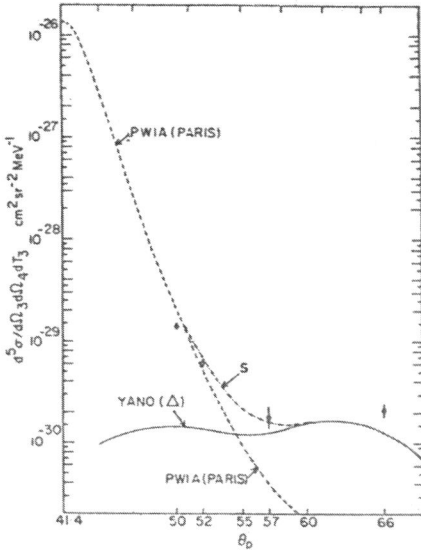

Fig. 2. The differential cross sections at four symmetric
angles for the detected protons. The dashed curve is a
PWIA prediction with the Paris potential, while the solid
curve is for virtual excitation of the Δ. The dot-dash
curve is the coherent sum of the two.

Breakup of ^{3}He

The study of the proton induced breakup of ^{3}He has also seen major
advances over the last year. This reaction has been studied previously at a
number of energies in the range 35 to 590 MeV [17], but the recoil momenta
were never sampled above 130 MeV/c. See Ref. 18 and references therein.
This is an important deficiency because it is the high momentum components
of the wavefunction that contain information that goes beyond the indepen-
dent-particle aspects of nuclear properties. They are also closely
associated with the short range behaviour of the nucleon-nucleon force. By
way of comparison, the ^{3}He wavefunction has been studied using the (e,e'p)
reaction out to approximately 800 MeV/c in the scaling variable $y = \vec{k}.\vec{q}/q$.
In the study by Epstein et al. [18], the ^{3}He wavefunction was measured out
to approximately 530 MeV/c by studying both the ^{3}He(p,pp)d and ^{3}He(p,pd)p
reactions at incident proton energies of 300 and 450 MeV. Analysis was
carried out within the framework of the Plane Wave Impulse Approximation
(PWIA). A comparison of the two reactions allows the factorization assumptions of the PWIA to be tested, and use of the PWIA itself allows a comparison with ^{3}He(e,e'p)d and ^{3}He(e,e'd)p.

Results of this study are shown in Fig. 3. The solid curve is that of
Ciofi degli Atti [19] obtained using the variational wavefunction of Nunberg
et al with the correct asymptotic behaviour of the harmonic oscillator basis

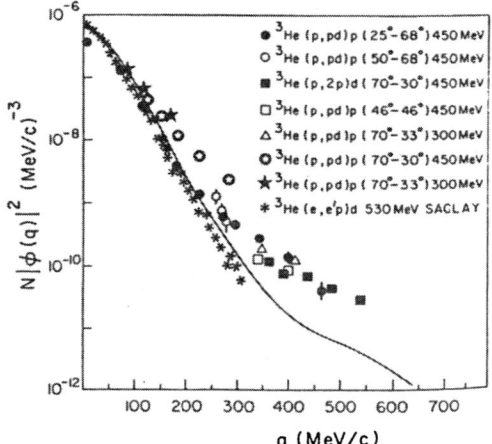

Fig. 3. Experimental spectral functions from the work of Ref. 18. The solid curve is from the work of Ciofi degli Atti [19].

expansion wave function. The (p,pp) and the (p,pd) results agree well suggesting that the factorization procedures used in the PWIA are adequate. For the central region of momenta, 50<q<200 MeV/c, the data agree fairly well both with the PWIA results and with the ^{3}He(e,e'p)d results. Above q=200 MeV/c, however, both the electron data and the theoretical curve fall well below the proton data and the discrepancy increases as the value of q increases. There thus seems to be a significant excess of high momentum components seen in the proton data as compared with what is seen in the electron data. It has been pointed out by Sick et al. [20] that strongly interacting probes may be scattered more than once and that the probability of two step processes becomes comparable with the probability of finding high momentum components in the wavefunction. These data present a definite challenge in working out the details of the reaction mechanism but also a great opportunity. Not only are there now data with electron and proton probes, but as will be pointed out below, with pion probes as well. With these three different probes, we may expect significant clues as to both the reaction mechanisms and details of the ^{3}He wavefunction.

Pion Production and Absorption

The study of pion production and absorption is closely associated with the studies of elastic pion scattering mentioned above. The theoretical approaches used differ greatly from those used for elastic scattering, but they show a complementary side of the study of the NN → NNπ system. In addition, it has been possible to obtain the momentum distribution of nucleons in ^{3}He from pion absorption studies and a comparison with results from ^{3}He(e,e'p) is of interest.

The study of pion absorption and production in light nuclei opens up a broad set of new questions. One of the most important is to understand why there seem to be difficulties in obtaining good fits to data for pp → dπ⁺ when the prediction of πd elastic scattering was so good. It is recognized that the fact that there is explicit pion production introduces a major perturbation into the formalism compared with pion elastic scattering, and the fully relativistic formalism that was so successful for πd scattering is no longer applicable. Nevertheless, it is, at least, disappointing that the results for pp → dπ⁺ are not better. Several groups have studied the formalism of pion production and absorption within the context of the NN → NNπ system of reactions. Several deserve specific mention. The work of the "Australian School" is particularly noteworthy as they have made major progress in the study of the pp → dπ⁺ reaction. The paper by Afnan and McLeod and references therein provide a good introduction to this work. For the study of 3-body final states, particularly for the reaction pp → pnπ⁺, the formalism of Dubach, Kloet, Silbar and others, described in Ref. 21, is frequently used.

In the following, a brief discussion of the current status of studies of pp → dπ⁺ is given. This is used as an introduction to the studies to be emphasized; namely, pion absorption on ³He.

The pp → dπ⁺ Reaction

This reaction has received extensive attention because it is one of the simplest of all production reactions governed by the strong interaction. A recent evaluation of studies of this reaction is given by Afnan and McLeod [22], and Fig. 4 is taken from that paper. It can be seen clearly that the theory reproduces the data qualitatively, but that a quantitative fit is elusive. Afnan and McLeod discuss several specific shortcomings of the theory. They point out that:

(i) A fully relativistic theory appears to be necessary. The formalism they have used to date treats the pion relativistically, but not the nucleons. This is done so that the correct clustering proper-ties of the three-body amplitudes are obtained [10]. Developing a fully relativistic formalism with explicit pion production and absorption is a difficult problem because the existing covariant formalisms do not have the proper clustering properties, and it is difficult to evaluate the importance of this deficiency. See also [10] for a discussion of this problem.

(ii) There may be problems with the use of separable Yamaguchi form factors for the NN → NNπ system. In particular, Afnan and McLeod point out that the required changes in the πNN coupling constant are significantly in excess of the limit set by partial

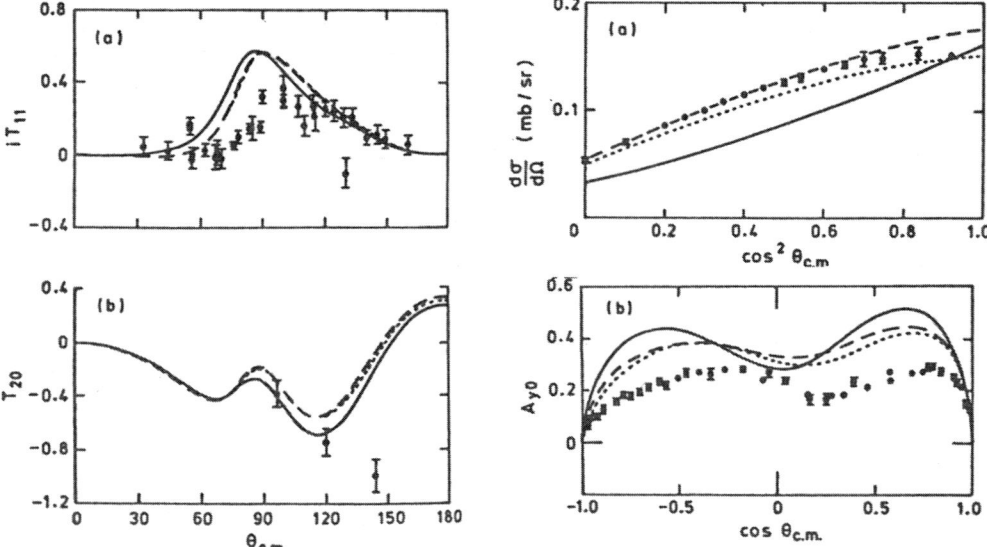

Fig. 4. Observables for πd elastic scattering at 256 MeV (left) and at 800 Mev (right). The theoretical results correspond to including all S- and P-wave πN amplitudes and NN amplitudes in (i) $^3S_1-^3D_1$, solid line, (ii) $^3S_1-^3D_1$, S- and P-wave, dashed line, and (iii) all S-, P- and D-wave channels.

conservation of axial-vector current (PCAC). This appears to suggest that a different form factor is needed.

(iii) There is presently a clear asymmetry in the theory in that the resonance is not treated on the same basis as the nucleon. This leads to undercounting. In particular, the Δ can only emit a forward pion since a backward pion would correspond to a two-pion intermediate state. Several attempts are under way to alleviate this problem [23,24], but to date no numerical results for a fully symmetric treatment of the nucleon and isobar are available.

To this should be added the question of whether or not it is sufficient to include only the P_{33} and P_{11} πN interactions. Especially below the resonance, it is likely that πN interactions in the S-waves will be important. There is evidence from pion absorption on ^{3}He that such waves may be important, but the best way to decide the question is likely to be detailed studies of the np → ppπ⁻ reaction. This reaction is much less dominated by the Δ resonances than is pp → pnπ⁺ and the "minor" amplitudes are expected to play a relatively more important role. There are approved experiments to study the np → ppπ⁻ reaction at both TRIUMF and Saclay.

Pion Absorption

Pion absorption on light nuclei, especially ^{3}He, is of great importance because one can study absorption on both T=0 and T=1 pairs. In the case of the T=0 pairs (pn), the process should be very similar to that of the $\pi d^+ \to$ pp reaction, but in an environment which causes th: pair to be closer together on the average than in the deuteron. Absorptic on T=1 pairs (pp) is the inverse of the np $\to$ ppπ^- reaction which can have both T=0 and T=1 components in the initial state unlike the pp $\to$ dπ^+ reaction. Of course, other mechanisms may appear in the case of ^{3}He, especially processes in which all three nucleons participate in the absorption. Although the majority of the absorption is expected to be on pairs of nucleons, there is significant evidence that in nuclei more than two nucleons are involved a significant fraction of the time. Finally, it is possible to measure the momentum distribution of the nucleons in ^{3}He by examining the width of the angular correlation between outgoing nucleons produced in a two-nucleon (2N) absorption.

There have been several recent studies of pion absorption on ^{3}He, but the study by Aniol et al. [25] is significant for the breadth of questions addressed. This work involved absorption of both π^- and π^+, each at 62.5 and 82.8 MeV. Firstly, they examined the relative strengths of the 2N and 3N absorption processes. For π^+ absorption, the total 3N contribution was small – typically about 5% of the 2N contribution. On the other hand, it contributes about 30% of the total T=0,J=1 absorption cross section. These results are not in good agreement with recent calculations by Oset et al. [26] which indicate that the 2N process should contribute between 90 and 95% of the total at these energies. Aniol et al. point out that this disagreement is surprising in the light of the success that the calculations have had in fitting the energy dependence of pion absorption on ^{12}C and the A dependence of the absorption cross section at an incident pion energy of 165 MeV.

In addition, Aniol et al. extracted the momentum distribution of the protons in ^{3}He by examining the angular correlation between proton pairs following π^+ absorption. Unfortunately, they do not express their results in the form of a spectral function as was done in the case of the ^{3}He(p,2p) and ^{3}He(p,pd)p results of Ref. 15. Rather they compare their angular correlation for π^+ absorption with the ^{3}He(e,e'p) data [27] taken at 600 MeV and expressed in the same manner as their pion absorption data. The results are shown in Fig. 5. One can see that the pion and electron data are essentially in agreement, but it is difficult to evaluate to what momentum transfer the agreement extends. Recall that in the comparison of proton and electron data, the real disagreement was found for momentum transfers in excess of 200 MeV/c. It would be interesting to see all three sets of data expressed in the same fashion.

512

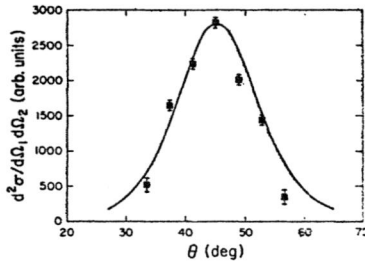

Fig. 5. Angular correlation taken at 165 MeV with one proton detected at 120°. The line is calculated using a spectator model using the (e,e'p) data of Ref. 27.

The absorption of a π^+ on ^{3}He is expected to be very similar to the reaction $\pi d \to pp$, and it is found that, indeed, the shape of the angular distribution for 2N absorption on ^{3}He resembles that on deuterium. In addition, when the cross sections are examined, it is found that the quasideuteron absorption in ^{3}He is enhanced by a factor of about 1.5 over that for π^+ absorption on the deuteron itself. This is interesting because in the simplest model [28], one finds that there should be 1.5 pn pairs in ^{3}He coupled to the deuteron quantum numbers. Such agreement suggests that the absorption process is insensitive to the radial wavefunction of the absorbing pair.

The situation is much less satisfactory for π^- absorption than it is for absorption. The experimental angular distribution is distinctly asymmetric about 90°, but the origin of this asymmetry is not clear. Two possible causes are presently recognized. Firstly, the asymmetry may be due to interference between the 2N and 3N processes. On the other hand, it may be due to interference between even and odd partial waves in the 2N amplitude. Adopting the latter interpretation, Aniol et al find that at pion bombarding energies between 60 and 80 MeV, π^- absorption on a proton pair is dominated (70%) by an angular momentum L=1 between the pion and the nucleon pair. This transition has a total spin-parity of $J^\pi=1^+$ and total isospin T=0 which does not allow formation of an intermediate Δ resonance. A P_{11} isobar can be formed, however, and may be making a significant contribution through a process in which the final nucleons are in a T=0 state. It is interesting to compare the magnitude of the T=0 cross section required by π^- absorption on ^{3}He as obtained by Aniol et al. with the upper limit of 200 µb found by Kleinschmidt et al. for the reaction np $\to$ ppπ^- [30]. Aniol et al. deduce that the T=0 cross sections should be 19±4 and 31±6 µb for 62.5 and 82.8 MeV pion energy respectively (401 and 442 MeV for the np $\to$ ppπ^- reaction). It is to be hoped that the np $\to$ ppπ^- experiments planned for TRIUMF and Saclay will be able to extract a value for the T=0 cross section as this may contribute to resolving the question of whether the asymmetry is due to interference between 2N and 3N processes, to interference between odd and even partial waves within the 2N cross section alone, or to a combination of these possibilities.

In the light of these uncertainties, no conclusions will be warranted until it becomes possible to calculate the differential cross section for $^3\text{He}(\pi^-,\text{Pn})\text{n}$. A significant step in this direction has been taken by Maxwell and Cheung [31], who have included a contact rescattering term constructed from the phenomenological $\pi\pi NN$ S-wave vertex of Koltun and Reitan [32] along with the usual P-wave interactions. The results they obtain are dependent on details of the model, and in the words of the authors, the "... results should be interpreted in a more semi-quantitative than quantitative fashion". Nevertheless, their results, (Fig. 6), suggest that it may be possible to obtain reliable predictions for both π^+ and π^- absorption in the near future.

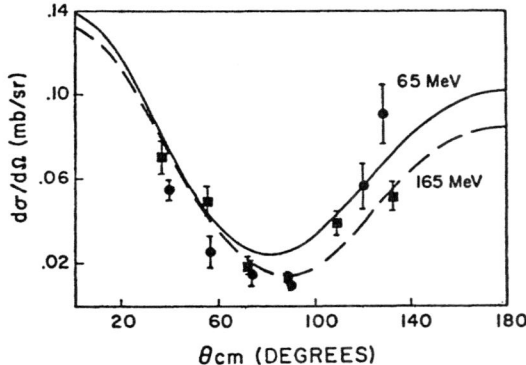

Fig. 6. Data for π absorption on ^3He at 65 MeV and 165 MeV. Each curve has been normalized so as to give the correct total cross section. The calculations are from Ref. 31.

Conclusions

A great deal of useful information has been obtained over the last year with hadronic probes of light nuclei, and there appears in most cases to be definite progress in understanding of the underlying physics. There is definitely a great deal to be accomplished both experimentally and theoretically before it will be possible to realistically address the questions of whether subnucleonic degrees of freedom are necessary. Nevertheless, the rate of progress permits hope that it may be possible to reach this goal in the near future.

This work was supported in part by the Natural Sciences and Research Council of Canada.

References

1. F. Sperisen, in Proceedings Sixth International Symposium on Polarization Phenomena in Nuclear Physics, Osaka, 1985: J. Phys. Soc. Japan, 55, 852 (1986).

2. D.K. Hasell, A. Bracco, H.P. Gubler, W.P. Lee, W.T.H. van Oers, R. Abegg, D.A. Hutcheon, C.A. Miller, J.M. Cameron, L.G. Greeniaus, G.A. Moss, M.B. Epstein and D.J. Margaziotis: Phys. Rev. C34, 236 (1986).

3. R.A. Arndt, L.D. Roper, R.A. Bryan, R.B. Clark, B.J. VerWest, and P. Signell, interactive dial-in program SAID.

4. Ph. Narboni: Nucl. Phys. A205, 481 (1973).

5. J.A. McNeil, J.R. Sheppard and S.J. Wallace: Phys. Rev. Lett. 50, 1439 (1983); J.R. Sheppard, J.A. McNeil and S.J. Wallace: ibid, 50, 1443 (1983).

6. M.J. Paez and R.H. Landau: Phys. Rev. C29, 2267 (1984).

7. J.M. Namyslowsky: Nuovo Cim. 57A, 355 (1968).

8. R. Aaron, R.D. Amado and J.E. Young: Phys. Rev. 174, 2022 (1968).

9. R. Blankenbecler and R. Sugar: Phys. Rev. 142, 1051 (1966).

10. S. Morioka and I.R. Afnan: Phys. Rev. C23, 852 (1981).

11. H. Garcilazo: Phys. Rev. Lett. 45, 780 (1980); Nucl. Phys. A360, 411 (1981); Phys. Rev. Lett. 48, 577 (1982); Phys. Rev. Lett. 53, 652 (1984).

12. Sarah C.B. Andrade, Erasmo Ferreira and H.G. Dosch: Phys. Rev. C34, 226 (1986).

13. W. Gyles, E.T. Boschitz, H. Garcilazo, W. List, E.L. Mathie, C.R. Ottermann, G.R. Smith, R. Tacik and R.R. Johnson: Phys. Rev. C33, 583 (1986).

14. W. Gyles, E.T. Boschitz, H. Garcilazo, E.L. Mathie, C.R. Ottermann, G.R. Smith, S. Mango, J.A. Konter and R.R. Johnson: Phys. Rev. C33, 595 (1986).

15. C.F. Perdrisat, V. Punjabi, M.B. Epstein, D.J. Margaziotis, A. Bracco, H.P. Gubler, W.P. Lee, P.R. Poffenberger, W.T.H. van Oers, Y.P. Zhang, H. Postma, H.J. Sebel and A.W. Stetz: Phys. Lett. 156B, 38 (1985).

16. A.F. Yano: Phys. Lett. 156B, 33 (1985).

17. I. Slaus, M.B. Epstein, G. Paic, J.R. Richardson, D.L. Shannon, J.W. Verba, H.H. Forster, C.C. Kim, D.Y. Park and L.C. Welch: Phys. Rev. Lett. 27, 751 (1971); S.N. Bunker, M. Jain, C.A. Miller, J.M. Nelson and W.T.H. van Oers: Phys. Rev. C12, 1396 (1975); A.A. Cowley, P.G. Roos, H.G. Pugh, V.K.C. Cheng and R. Woody III, Nucl. Phys. A220, 429 (1974); R. Frascaria, V. Comparat, N. Marty, M. Morlet, A. Willis and N. Willis: Nucl. Phys. A178, 307 (1971); P. Kitching, G.A. Moss, W.C. Olsen, W.J. Roberts, J.C. Alder, W. Dollhopf, W.J. Kossler, C.F. Perdrisat, D.R. Lehman and J.R. Priest: Phys. Rev. C6, 769 (1972).

18. M.B. Epstein, D.A. Krause, D.J. Margaziotis, A. Bracco, H.P. Gubler, D.K. Hasell, W.P. Lee, W.T.H. van Oers, R. Abegg, C.A. Miller and A.W. Stetz: Phys. Rev. C32, 967 (1985).

19. C. Ciofi degli Atti, E. Pace and G. Salme, in Proceedings of the Miniconference on the Study of Few-body Systems with Electromagnetic Probes, Amsterdam, 1981, p103; P. Nunberg, D. Prosperi and E. Pace: Nucl. Phys. A285, 58 (1977).

20. I. Sick, D. Day and J.S. McCarthy: Phys. Rev. Lett. 45, 871 (1980).

21. W.M. Kloet and R.R. Silbar: Nucl. Phys. A338, 281 (1980); ibid A364, 346 (1981).

22. I.R. Afnan and R.J. McLeod: Phys. Rev. C31, 1821 (1985).

23. I.R. Afnan and B. Blankleider; Phys. Rev. C32, 2006 (1985).

24. T.-S.H. Lee and A. Matsuyama: Phys. Rev C32, 516 (1985).

25. K.A. Aniol, A. Altman, R.R. Johnson, H.W. Roser, R. Tacik, U. Weinands, D. Ashery, J. Alster, M.A. Moinester, E. Piasetzky, D.R. Gill and J. Vincent: Phys. Rev. C86, 1714 (1986).

26. E. Oset, Y. Futami and H. Toki, Proceedings of the International Conference on Particles and Nuclei, Heidelberg, 1984.

27. E. Jans, P. Barreau, M. Bernheim, J.M. Finn, J. Morgenstern, J. Mougey, D. Tarnowski, S. Turck-Chieze, S. Frullani, F. Garibaldi, G.P. Capitani, E. de Sanctis, M.K. Brussel and I. Sick: Phys. Rev. Lett. 49, 974 (1982).

28. D. Ashery, R.J. Holt, H.E. Jackson, J.P. Schiffer, J.R. Specht, K.E. Stephenson, R.D. McKeown, J. Ungar, R.E. Segel and P. Zupranski: Phys. Rev. Lett. 47, 895 (1981).

29. R.R. Silbar and E. Piasetzky: Phys. Rev. C29, 1116 (1984).

30. M. Kleinschmidt, Th. Fischer, G. Hammel, W. Hurster, K. Kern, L. Lehman, E. Rossle and H. Schmitt: Z. Phys. A298, 253 (1980).

31. O.V. Maxwell and C.Y. Cheung: Nucl. Phys. A454, 606 (1986).

32. D.S. Koltun and S. Reitan: Phys. Rev. 141, 1413 (1966).

PION INTERACTION WITH FEW-BODY SYSTEMS
EXPERIMENTS AT SIN

G. Backenstoss
University of Basel, Klingelbergstr. 82
CH-4056 Basel, Switzerland

Abstract

Relevant experiments in progress or being accepted at SIN are
listed and those having produced at least preliminary results
pertaining to pion production and absorption are being discussed.

Introduction

A considerable number of experiments at SIN is dedicated to
study the interaction of pions with few body systems. This is
not surprising in view of the fundamental nature of these interactions. In table 1 a list of experiments in progress or in
preparation is shown where experiments with a single nucleon
and with nuclei with A > 4 have been omitted. For the very important π d scattering experiments it is referred to a paper
contributed directly by this group to this workshop. In this
paper therefore, we concentrate on π production, relevant pionic atoms and π absorption processes.

π Production

The Geneva-SIN group has studied in a systematic and complete
way the elastic $\vec{p}$ $\vec{p}$ interaction at various energies using a polarized $\vec{p}$ beam as well as polarized targets. In continuation of
these efforts the group has also studied inelastic processes.
The reactions $pp \rightarrow d\pi^+$, $pp \rightarrow pn\pi^+$ and $pp \rightarrow pp\pi^0$ have not been
described theoretically to a real satisfying level. The first reac-

T a b l e 1 : Few body experiments at SIN

Collaboration (Spokesman)	
Basel-Karlsruhe (G. Backenstoss, H. Ullrich)	Special reaction channels of π^+ and π^- absorption in light nuclei
Neuchâtel-Caltech (E.Bovet, J.Gimlett)	Experimental determination of the strong interaction shift in the 2p-1s transition of pionic H and D atoms
Geneva-SIN (R. Hess)	Complete experimental reconstruction of the scattering amplitudes $pp \to d\pi^+$
Karlsruhe-SIN-TRIUMF (C. Ottermann)	Measurement of t_{20} in π-d scattering with a tensor polarized d-target
Karlsruhe-SIN (W. Gyles)	Polarization effects in the π-d breakup
Basel-Karlsruhe-LANL-Wien-Zagreb (G.Backenstoss, H.Ullrich)	π^+/π^- absorption in tritium
Erlangen-Nürnberg-Karlsruhe (A. Hofmann, H.W. Ortner)	$\pi^+ p \to \pi^+ \pi^- n$ near threshold
Karlsruhe-SIN-TRIUMF-Regina (W. Gyles)	Measurement of the $\pi d \to \pi^\circ$ NN charge exchange reaction
Neuchâtel-SIN-IMP/ETHZ (E. Bovet, W. Beer)	A precision measurement of the 1s strong interaction shift and width in pionic hydrogen and pionic deuterium

tion $pp \to d\pi^+$ with a two body final state, being the most simple one, has been studied by the group. Taking into account as observables one amplitude, two polarization parameters, 4 spin correlation parameters and 20 deuteron asymmetries a 17 parameter fit was made to the data [1]. Certain symmetries could be applied but also unknown parameters of the Carbon polarimeter had to be accounted for. Consequently 6 amplitudes with 5 unarbitrary phases had to be fitted and have been compared to theoretical predictions by Locher [2]. They are in relatively good agreement with the experimental results obtained independently of any theoretical hypothesis.

π Atoms

Pionic atoms are an important tool to investigate the π-nucleus interaction at practically zero pion energy. The observables energy shift ΔE caused by the strong interaction as compared to the purely electromagnetic energy and the level width Γ can be directly related to the real Re(a) and imaginary part Im(a) respectively of the π-nucleus scattering length a. For the "few-body" nuclei ^{3}He ΔE and Γ have been measured previously [3]. But for πD with the K_α X-ray as low as 2.5 keV only recently a measurement had been performed at SIN with a curved graphite-crystal spectrometer. [4]. ΔE was determined to be 5.5 (±0.8) (stat) ±0.5 (syst) eV whereas Γ could not yet be measured with the atteined energy resolution of about 25 eV. From this Re(a) could be extracted consistent with that obtained from π D scattering data. In fig. 1 differences between isotopes (^{1}H, ^{2}H; ^{3}He ^{4}He) of Re(a) are shown which differ significantly from the free πn scattering length as a simple minded approach would suggest, particularly for the lightest isotopes.

It would be very desirable to make these comparisons also for Im(a). The Neuchâtel-SIN group aims at much more precise measurements by use of a bent Silicon crystal spectrometer operated in reflection from which a resolution of the order of 0.5 eV is expected.

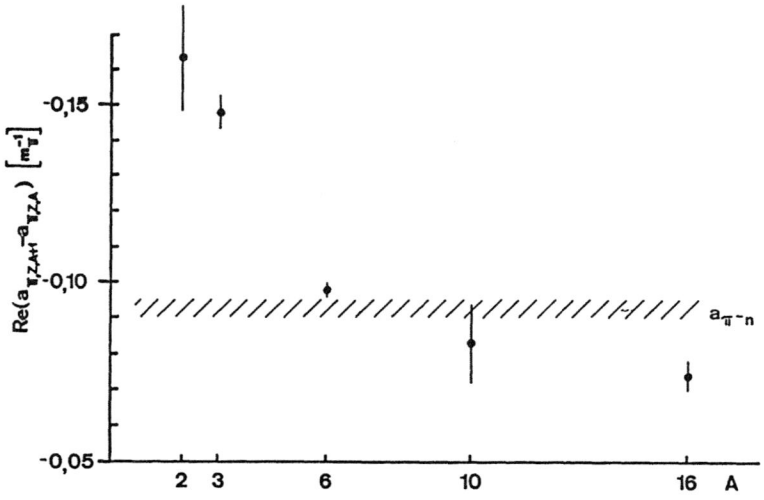

Fig. 1 Isospin differences between the real parts of the $π^-$-nucleus scattering lengths for several light nuclei. The hatched region marks the value of the elementary ($π^-$n) scattering length.

π Absorption

Pion absorption has been the subject of a research program at
SIN for a number of years. It is aiming at the identification
and understanding of the fundamental absorption processes. To
do this the study of the absorption on a 3N system is indicated.
Energy-momentum conservation requires a single nucleon to have
a momentum of 500 MeV/c in order to be able to absorb a pion,
a momentum scarcely available in nuclei. The π-deuteron absorption has been treated equally well theoretically and experimentally and offered the so far widely used model for pion absorption in heavier nuclei, i.e. the quasi-deuteron absorption. The
absorption process is believed to proceed predominantly by the
ΔN channel. However, the shortcomings are obvious. Absorption
on nucleon pairs with T=1 where the ΔN channel is strongly suppressed, cannot be studied with deuterons. The test of the real
predominance of the 2N absorption process cannot be done in the
absence of a further nucleon and a possible influence on the
nuclear density cannot be seen in d alone. Therefore, pion absorption on ^{3}He has been investigated with the aim to
- separate and study individually quasifree 2N absdorption channels, final state interaction with a 3^d nucleon including the
Nd channel and "true" 3N absorption
- study the dependence on quantum numbers of the initial state
such as isospin and angular momentum and on the pion energy.

With stopping pions (only π$^-$) less degrees of freedom are present, the dominant s-absorption provides isotropy of particle
emission facilitating measurements of branching ratios. In
flight absorption on the other hand is possible for π$^-$ and π$^+$,
energy and angular momentum can be varied and an angular distribution with respect to the pion beam direction may be observed.

Experimentally the absorption on a 3N system has the advantage
that 3 particles are emitted at the maximum. Therefore, the final state is determined kinematically complete if the momenta of
two particles are measured (the energies of two particles and
the angle between them for absorption at rest). This is provided
by a two arm apparatus consisting of a large TOF position sensitive spectrometer capable to detect charged particles and neutrons (~15% detection efficiency) [5] and a modular total energy spectrometer for charged particles equipped with track defining MWP-chambers.

Data reduction and evaluation is done by
- Mass separation of charged particles by energy and TOF measurements
- Reconstruction of the tracks originating in the target
- Reconstruction of missing mass
- Dalitz plot

As an example for the separation of the emitted quasifree pn
pair and the p and n peaks due to the FSI of nn and pn respectively the Dalitz plot of the reaction $\pi^- {}^3He \rightarrow pnn$ (fig. 3b) may be
considered.

Absorption at rest

The results on the stopping π^- in ${}^3He$ have been published [6],
the essentials are summarized shortly here:
- The branching ratios of pnn, dn and $\pi^0 t$ final states have been
  measured to be respectively 60.8, 11.7 and 13.5% leaving 14%
  for all radiative channels (γpnn, γdn, γt)
- The pure QFA amounts only to 2/3 of all non-radiative true absorption processes
- Reaction $\pi^- pn \rightarrow nn$ exceeds $\pi^- pp \rightarrow pn$ by a factor of about R=7
- 3N events are strongly correlated around 180°, very few statistically distributed events are present
- From K-X-ray measurements s-absorption is measured to be 64%
- The dominance of the absorption on T=0 pairs is not caused
  mainly by p wave absorption as derived from K-X-ray coincidence measurements.

The theoretical explanations of these findings are at present
still rather marginal.

Absorption in flight

The experiments to study absorption in flight were deviced to
single out the different absorption channels observed already
with stopping π^- to measure the energy dependence and angular
distribution of the differential and total cross sections. By
investigating also π^+ absorption a further degree of freedom
with respect to the isospin dependence of the absorption mechanism becomes available. Also in flight the final state is kinematically fully determined, hence the recoil momentum distribution
of the spectator nucleus can be derived. As can be seen from
fig. 2 it agrees up to 150 MeV/c very well with the one derived
from (e,e'p) data, prooving thus the quasifree 2N absorption me-

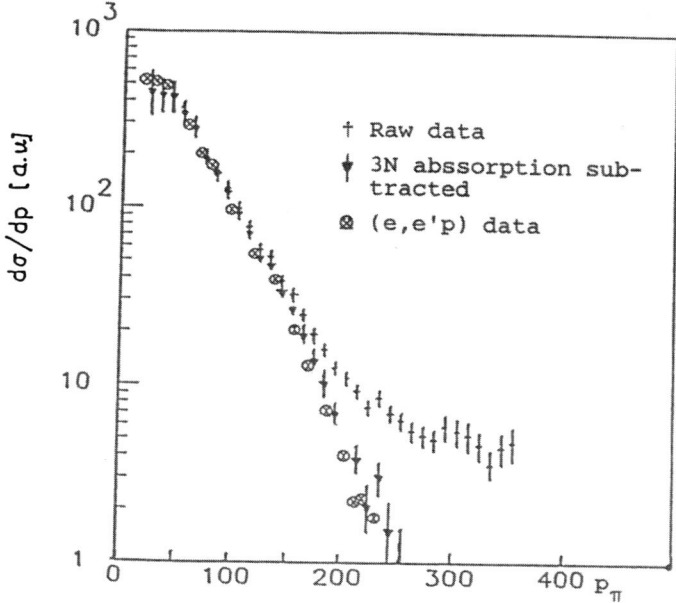

Fig. 2
Momentum dis-
tribution of
the recoil
(spectator)
nucleon for
the π^+ ^{3}He→(pp)p
reaction com-
pared with
(e,e'p) data
(normalized)
from ref. 7.

chanism. Beyond that another mechanism is present (3N absorption) which we will discuss later.

Quasifree 2N absorption. In Fig. 3 the three dimensional
Dalitz plot for the processes π^+pn → pp (a) and π^-pp → pn (b)
are shown. Process (b) is much harder to measure than process
(a) because as for pions at rest absorption on a T=1 pair (pp)
is strongly suppressed and the detection of the neutron is further reduced to ~15% due to the lower detection efficiency.
Therefore, other weaker channels such as the two FSI enhance-
ments become noticeable.

In Fig. 4 the measured energy dependence of the ratio

$$R = \frac{\sigma_{QFA}[\,^3He(\pi^+,pp)p]}{\sigma_{QFA}[\,^3He(\pi^-,pn)n]}$$ is shown together with calculations [8]

wich are far from being satisfying, particularly for stopping
pions. The π^- absorption on a T=1 pair proceeding without Δ
dominance seems to be a real theoretical test ground because
probably a number of channels (e.g. P_{11}) have to be taken into
account.

The angular dependence of the quasifree differential cross sec-

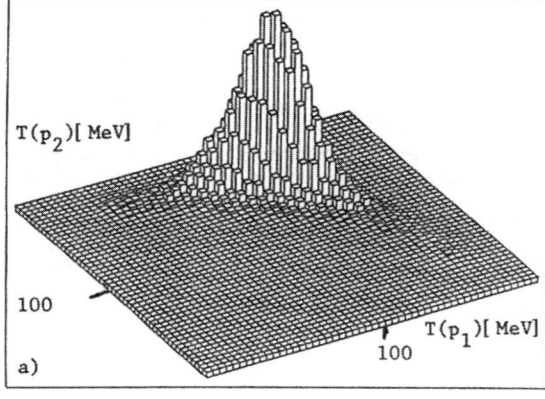

a)

Fig. 3
Dalitz plot for
220 MeV/c
a) $\pi^+ \, ^3He \rightarrow (pp)p$

and

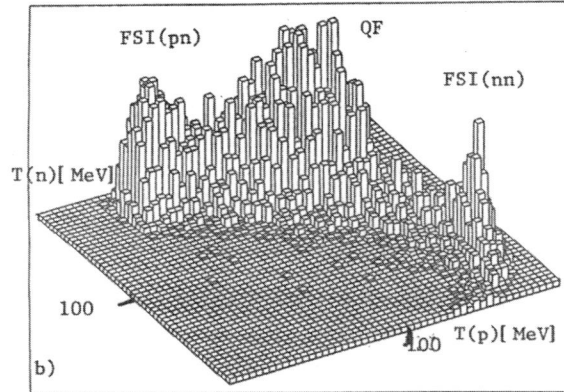

b)

b·) $\pi^- \, ^3He \rightarrow (pn)n$

where also the
FSI (nn) peak at
the right and FSI
(pn) peak on the
left side of the
quasifree
($\pi^- pp \rightarrow pn$) peak
are visible.

tion on the CM angle with respect to the beam axis has been measured and fitted by Legendre polynomials

$$d\sigma/d\Omega = \sum_{i=0}^{2} A_i P_i (\cos\theta)$$

In table 2 the coefficients A_i and the total cross section
$\sigma = 2\pi A_0$ are given for $T_\pi = 120$ MeV. For π^- a slight anisotro-
py around 90° ($A_1 \neq 0$) is observed originating most likely from
interference between different isospin amplitudes. In table 3
preliminary data of σ_{tot} for other p_π are presented.

A comparison with absorption on deuterons (σ_{tot} as well as A_2)
was made between the processes $^3He(\pi^+, pp)p$ and $^2H(\pi^+, pp)$. The
energy dependence of both σ_{tot} and A_2 are very similar, the ^3He
data being 1.1 ± 0.2 times larger, i.e. smaller than 1.5 the number of the T=0 pairs available in ^3He.

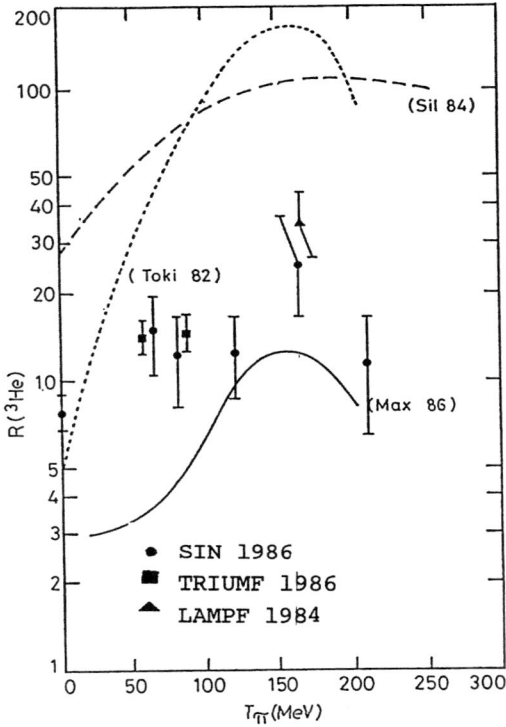

Fig. 4

Isospin ratio between
$\pi^+pn \to pp$ and $\pi^-pp \to pn$
absorption in dependence of the pion energy.
The curves show calculations as given in ref. [8].

Table 2:

π^+ and π^- Differential Cross
Sections for $p_\pi=220$ [MeV/c],
$d\sigma/d\Omega = \sum\limits_{i=0}^{2} A_i P_i (\cos\theta)$
$A_i [\mu b/sr]$, σ_{tot} [mb]

	π^+	π^-
A_o	1884 ± 42	76 ± 2
A_1	——	-37 ± 4
A_2	2104 ± 99	103 ± 4
σ_{tot}	12 ± 2	0.95 ± 0.25

Table 3:

π^+/π^- Absorption Cross Sections on ^{3}He for different
pion momenta p_π
QFA: quasifree π2N absorption
3N : (3N) absorption

	p_π[MeV/c]		σ[mb]	
QFA	150	π^+	7.0	± 2.0
	220	π^+	12.0	2.0
		π^-	0.95	0.25
	270	π^+	13.0	2.5
	320	π^+	7.0	1.5
3N	150	π^+	1.1	0.5
	220	π^+	3.9	0.5
		π^-	3.7	0.6
	270	π^+	4.4	0.8
	320	π^+	2.3	0.6
		π^-	3.6	0.6

3N Absorption. The width of the quasifree peak is given by
the Fermi momentum distribution in ³He. But outside this region
i.e. towards the centre of the Dalitz plot a considerable event
rate prevails for π^+ and π^- somewhat differently than for stopping π^-. By reconstruction of the mass of the target nucleus
background events could be excluded. As already published [9]
these 3N events are only proportional to the 3 body phase space.
This process has been investigated for more pion energies and
is displayed in fig. 5 exhibiting a significant energy dependence perhaps suggesting a double Δ process. The cross sections
are also shown in table 3. They are larger than those for the
quasifree process on T=1 pairs.

Absorption on ⁴He. Pion absorption has also been studied
recently on ⁴He. The motivation was to find out whether also
here the absorption can be described in terms of the mechanisms
observed in the 3N system or whether new processes become important. Here only the NNd final state is kinematically completely determined and hence was investigated primarily. The recoil
momentum distribution turns out to be very similar to that in
fig. 3 indicating that the 4th nucleon acts merely as spectator.
Since also the recoil distribution reconstructed from events
where two nucleons have been measured one is tempted to conclude that also there one nucleon is spectator, the absorption
hence taking place on a 3N subcluster. At the moment it is too
early to present absolute total cross sections for the various
reactions.

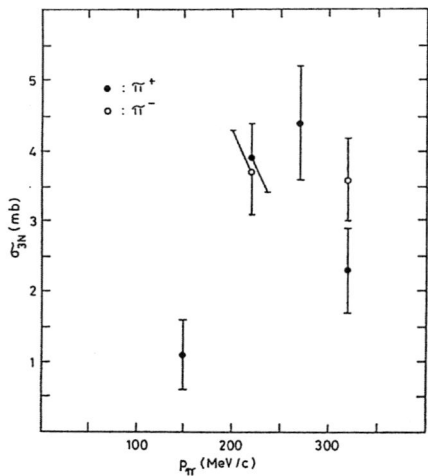

Fig. 5

Dependence of the 3N absorption cross section on the
pion momentum

But in table 4 some ratios are given. R_1 is very similar to that of the ^{3}He indicating also here the dominance of the absorption on a T=0 pair. R_2 proofs charge symmetry with the experimental error of 15%. R_4 suggests also an absorption involving 3 nucleons to be stronger than absorption on a T=1 pair. In general the pion absorption in ^{4}He follows the pattern seen in ^{3}He indicating that the additional neutron does not change the picture basically confirming the fact that the 3N system is the essential test ground to understand pion absorption in general.

T a b l e 4 : Relative Cross Sections for π^+/π^- Abssorption in ^{3}He at p_π = 220 MeV/c

$$R_1 = \frac{\sigma[^4He(\pi^+,pp)pn]}{\sigma[^4He(\pi^-,pn)nn]} = 11 \pm 2$$

$$R_2 = \frac{\sigma[^4He(\pi^+,dp)p]}{\sigma[^4He(\pi^-,dn)n]} = 1.1 \pm 0.2$$

$$R_3 = \frac{\sigma[^4He(\pi^+,dp)p]}{\sigma[^4He(\pi^+,pp)pn]} = 0.16 \pm 0.03$$

$$R_4 = \frac{\sigma[^4He(\pi^-,dn)n]}{\sigma[^4He(\pi^-,pn)nn]} = 1.6 \pm 0.3$$

Reference

[1] P. Bach, R. Hess et al., SIN Newsletter 18 (1986) 28

[2] M. Locher SIN Preprint PR 85-12

[3] I. Schwanner et al., Nucl. Phys. A412 (1984) 253

[4] E. Bovet et al., Phys. Lett. 153B (1985) 231

[5] S. Cierjacks et al., Nucl.Instr. and Meth.A 238 (1985)354

[6] G. Backenstoss et al., Nucl. Phys.A 448 (1986) 567

G. Backenstoss et al., Phys. Lett.B 115B (1982) 445

D. Gotta et al., Phys. Lett. B 112B (1982) 129

[7] E. Jans et al., Phys. Rev. Lett. 49 (1982) 974

[8] O.V. Maxwell and C.Y. Cheung, Nucl. Phys. A454 (1986) 606

R. Silbar and E. Piasetzki, Phys. Rev. C29 (1984) 1116

H. Toki and H. Sarafian, Phys. Lett. 119B (1982) 285

[9] G. Backenstoss et al., Phys. Rev. Lett. 55 (1985) 2782

ANTINUCLEON-NUCLEUS EXPERIMENTS

G. Bendiscioli and A. Rotondi

Dipartimento di Fisica Nucleare e Teorica, Università di Pavia

INFN Sezione di Pavia

Abstract. A short review of recent experimental results on the $\bar{p}p$ and $\bar{p}$-nucleus interaction with $\bar{p}$ at rest and with momentum up to 600 MeV/c is presented. Hints of some possible future measurements are given too.

1. Introduction

A relevant progress has been made in the antinucleon physics at low momenta (<1 GeV/c) in the last years, particularly after the advent, in 1983, of the Low Energy Antiproton Ring (LEAR)[3] at CERN. In this paper we review briefly some recent experimental results and give some hints on future measurements. Both the antiproton-nucleon and antiproton-nucleus interactions will be considered, with a particular attention to the few nucleon systems. In these notes we do not treat the interesting problems about the existence of $\overline{NN}$ resonances, baryonium and glueballs. For these topics, the reader is referred to more extensive reviews of the antinucleon physics, which can be found in the proceedings of the Telluride[1] and Tignes[2] Conferences.

2. The elementary antinucleon-nucleon ($\overline{NN}$) interaction

Information on the $\overline{NN}$ interaction is obtained usually from the study of the annihilation and scattering processes of antiprotons on free protons and on nuclei and from experiments on antiprotonic atoms. It results that annihilation is the most important process at low energies, as it dominates over elastic scattering and charge-exchange reactions. Its study is therefore essential to the understanding of the $\overline{N}N$ forces and, more generally, of the strong interactions between hadrons at a microscopic level and at small distances. The theoretical explanation of the number and variety of annihilation final states and of intermediate resonant states, and the dependence of their branching ratios or partial cross-sections on the quantum numbers of the initial state (energy, angular momentum, isospin, spin) is still an open problem[4]. Therefore, at present phenomenological potential approaches as well as new experimental studies are of great importance.

From $\overline{p}p$ scattering and charge-exchange data available before LEAR, several $\overline{N}N$ potential were deduced, among which we mention those of Dover and Richard[5] (DR) and of the Paris group[6]. These two potentials are similar in their medium and long range part (≥ 0.9 fm), deduced from the G-parity transformation upon a NN one boson exchange (OBE) potential. They differ in the short range (annihilation) part obtained phenomenologically from $\overline{p}p$ data. In the DR approach a local, spin and isospin independent and complex Woods-Saxon form is employed, characterized by a very deep imaginary part (~ 1 GeV). In the Paris potential the short range part has a more flexible form, isospin and (strongly) spin dependent, and no real annihilation potential is considered[7]. A variety of narrow resonances and bound states in the $\overline{N}N$ system was predicted by the potential approaches. For further details on these potentials see Ref. (8).

$\overline{p}p$ scattering experiments performed before[9] and after[10,11] the LEAR advent have shown also that the elastic scattering angular distribution is strongly forward peaked even at low energies, and that the annihilation cross section exceeds by a factor two the total elastic one. A $\overline{p}p$ elastic scattering amplitude at forward angles has been derived in the form

$$f_{\overline{p}p}(q) = \frac{k}{4\pi} \sigma_p (i+\rho_p) e^{-\beta_p^2 q^2/2} \tag{1}$$

where k and q are the incident and transferred momentum respectively, σ_p is the total $\overline{p}p$ cross section, $\rho_p = \mathrm{Re}\, f_{\overline{p}p}(0)/\mathrm{Im}\, f_{\overline{p}p}(0)$ and β_p^2 is the slope parameter.

Recent values of these parameters at low energy, measured at LEAR[10] are given in Tab. I and in fig. 1.1. Other values of these parameters, averaged from the world data up to 1984, are reported in Ref. (12). These data indicates a diffractive behaviour of the scattering amplitude, similar to the case of the pp interaction above 1 GeV.

Tab. I - p̄p scattering amplitude parameters at low energy. Ref. (10).

Momentum (MeV/c)	σ_p (mb)	β_p^2 $(GeV)^{-2}$	ρ_p
219	292.1±23.6	61.2±16.5	0.093±0.098
287	229.3± 2.0	32.3± 1.7	-0.103±0.018
590	140.0± 2.6	22.8± 0.94	0.203+0.11

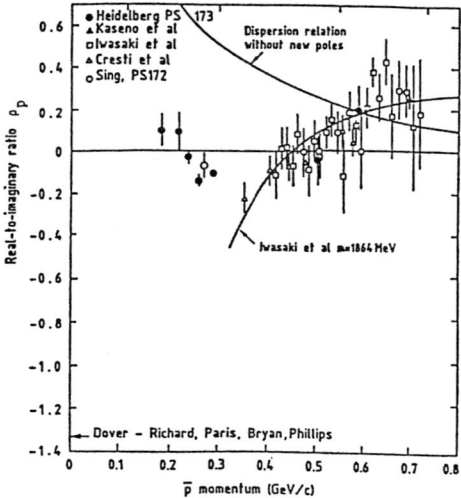

Fig. 1.1 - The real-to-immaginary part of the p̄p forward scattering amplitude as a function of the p̄ momentum. From Ref.(17).

The p̄ nucleus interaction is a combination of the elementary p̄N interaction and of nuclear matter effects. Its study allows one to obtain information on the dependence of the N̄N potential on spin and isospin and to filter out particular components of the N̄N force. For instance, in p̄ atoms with nuclei of isospin zero the pion exchange is absent[17].

Moreover, it has been shown[13] that the forward peaked nature of the $\bar{N}N$ amplitude, eq. (1), permits to apply to the low energy $\bar{p}$-nucleus case the Glauber approach, employed with success in the p-nucleus scattering analysis above 1 GeV (14,15,16). This makes possible to obtain information on the parameters σ_n, β_n, ρ_n of the $\bar{p}$-neutron scattering amplitude. In this context, in the following we will refer often to the elementary ratio between $\bar{p}$-n and $\bar{p}$-p total cross sections, denoted as $R = \sigma_n/\sigma_p$.

3. Antiprotonic atoms

Antiprotonic atoms give information on the $\bar{p}$-nucleus and the elementary $\bar{p}N$ interaction mainly through measurements of the hadronic shifts and widths of the energy levels in the last observable transition of the x ray spectrum.

Measurements of the hadronic energy shift and width of the 1s levels (spin and isospin multiplets) in the $\bar{p}p$ system (protonium) lead to the determination of the $\bar{p}p$ scattering length a($\bar{p}p$).

The ratio $\rho(\bar{p}p)$ = Re a($\bar{p}p$)/Im a($\bar{p}p$) is equivalent to the ratio of the real-to imaginary part of the $\bar{p}p$ forward scattering amplitude at zero energy (see eq. 1). Different potentials predict Re a($\bar{p}p$) $\approx$ -1 (within 10%) and $\rho(\bar{p}p) \approx$ - 1.3 (within 3%)[17].

Three experiments (PS 171, PS 174, PS 175) at LEAR are devoted to the study of the $\bar{p}p$ atom. Tentative experimental results indicate a negative Re a($\bar{p}p$)[17]. Moreover, the large value of the measured width Γ_{1s} seems to preclude the existence of narrow $\bar{N}N$-states very near the threshold[18].

Measurements have been performed also in deuterium, ^{3}He, ^{4}He and on heavier nuclei[17,18], including isotopic series. We mention three points.

First, changes of hadronic shifts and widths in a chain of isotopes indicate the strength of the $\bar{p}n$ interaction (pure I = 1 state). Clear isotope effects have been observed on $^{6/7}$Li (see fig. 3.1), $^{16/17/18}$O and $^{58/64}$Ni. The shift $\Delta\epsilon$ and the width $\Delta\Gamma$ of the energy levels between isotopes can be evaluated by a perturbative calculation which gives $\rho(\bar{p}n)$ = 2 $\Delta\epsilon\Delta\Gamma$ = Re V/Im V. The measurements on the oxigen isotopes[20] give for

530

this ratio a value around -1, close to the theoretical predictions for the p̄n and the p̄p interactions[17]. This is remarkable, because the values of ρ(p̄p) from Coulomb interference measurements in low energy scattering experiments are close to zero at the lowest p̄ momenta (see fig. 1.1).

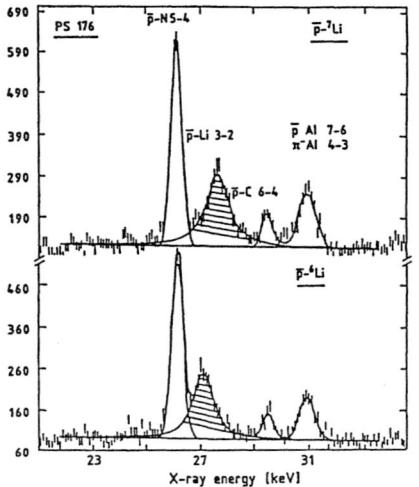

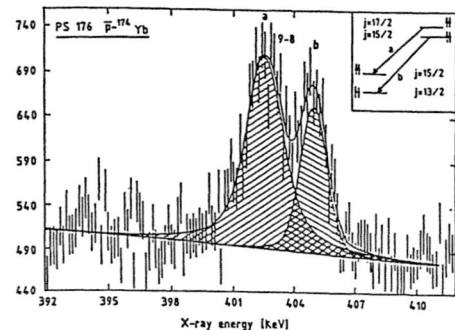

Fig. 3.1 Part of the X-ray spectrum of antiprotonic ⁶Li and ⁷Li showing the isotopic effect in the 3→2 transition. From Ref. (17)

Fig. 3.2 Part of the X-ray spectrum of antiprotonic ¹⁷⁴Yb, showing the fine structrure components of the 9-8 transition. Ref. (17).

Second, spin-orbit effects may be evidentiated in even-even nuclei (spin-zero), where the hadronic level width is large enough, but smaller than the fine structure splitting. Clear evidence of significant differences of spin orbit effects in different fine structure levels is given by the measurements on ¹⁷⁴Yb (see fig. 3.2)[17], which is surprising if compared to the very weak spin dependence shown by the elastic scattering data[31].

Finally, the results on the p̄ ⁴He system do not support the suggestion that antiproton bound states in the He nucleus exist with binding energies and widths of a few MeV[21].

Further news on recent antiprotonic atom measurements are given in Refs. (1,2,17 and 18).

4. $\bar{p}$-nucleus elastic and inelastic scattering

At present there is very few information on this kind of reaction on light nuclei between 50 MeV and 1 GeV. Old low statistics elastic scattering $\bar{p}$ ^{2}H data[24] obtained with bubble chambers, were analyzed in the framework of Glauber theory and a ratio $R = \sigma_n / \sigma_p \sim 1$ was deduced[13].

More recently, the PS 184 collaboration studied extensively at LEAR $\bar{p}$ scattering on C, O, Ca, Pb nuclei, furnishing high quality data[22,23], which have been extensively analyzed with microscopical[8,22,25,26] phenomenological[22,27,28] and Glauber[29,30] models.

Since angular distributions relative to nuclei between ^{2}H and ^{12}C are lacking, we report in fig. 3.1, as representative examples, some elastic and inelastic $\bar{p}$ ^{12}C data at 46 and 180 MeV. The diffractive behaviour due to the presence of the strong annihilation channel is evident.

We see also that the difference between microscopic calculation employing Dover-Richard and Paris potential is not very significant (fig. 4.1 a) and b)). The same ambiguity is reproduced also in a recent comparison between Paris and relativistic models[26]. It has been suggested that measurements of polarization and of angular distributions in reactions as ^{12}C $(\bar{p}\bar{p}')$ C^{12*} involving transitions to unnatural parity states could set same constraints to the two body spin dependent $\overline{NN}$ amplitudes[8] but the present LEAR data of thys type[23,31] are of too poor quality and not conclusive (see also fig. 4.1 c).

These facts indicate that, because of the annihilation, the elastic and inelastic $\overline{N}$-nucleus scattering reduces to a quasi free $\overline{N}$-N interaction on the nuclear surface. In these conditions the process is determined by the long range spin-isospin averaged part of the $\overline{NN}$ amplitude, which is well defined in all current models by fitting $\bar{p}p$ scattering data.

As a further experimental proof of this annihilation dominance, we mention the negative result of the experiment PS 184 in the search of $\bar{p}$ bound states in light nuclei via the reaction A$(\bar{p},p)$ X. Statistically significant results were obtained with ^{6}Li and with a scintillator target for outgoing proton energies between 120 and 290 MeV. No evidence for narrow peaks[32] corresponding to bound or resonant $\bar{p}$-nucleus states was

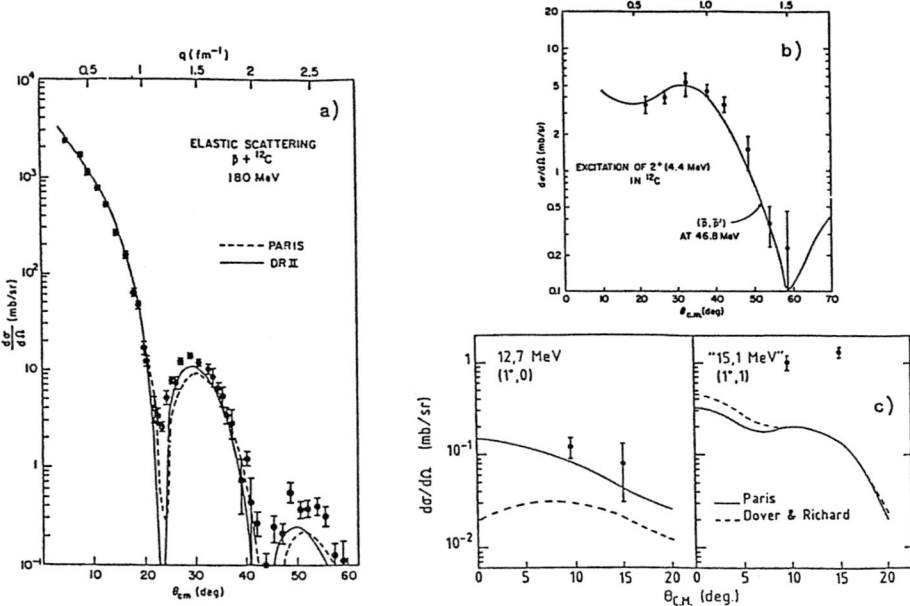

Fig. 4.1 - $\bar{p}^{12}C$ elastic and inelastic distributions measured at Lear. Microscopical calculations with Dover-Richard and Paris potentials are also shown. In fig. 1.b the DR and Paris potentials give similar results (solid line). Figs. a and b from Ref. (8), Fig. c from Ref. (23).

found in the proton spectra. Experimental limits for the production of $(\bar{p}\text{-}^5He)$ and $(\bar{p}\text{-}^{11}B)$ states on ^6Li and ^{12}C, respectively, were deduced for different outgoing proton energies and level widths. Considering proton energies close to the incident $\bar{p}$ energy, i.e. states in which the $\bar{p}$ binding energy is equal to the binding energy of the ejected proton, such limits (3 σ) are ~ 12 μb/sr in ^6Li and ~ 40 μb/sr in ^{12}C, about one order of magnitude lower than theoretically predicted[33].

5. $\bar{p}$-nucleus reaction cross section

Data on ^4He and Ne and preliminary measurements on ^3He have been recently obtained at LEAR[29,34]. These data are added to previous $\bar{p}^2H$ measurements made in bubble chambers[24,35,36]. In fig. 5.1 the $\bar{p}^4He$ charged particle multiplicity (charged meson and nuclear fragments) observed in the streamer chamber at three different lab. energies is shown.

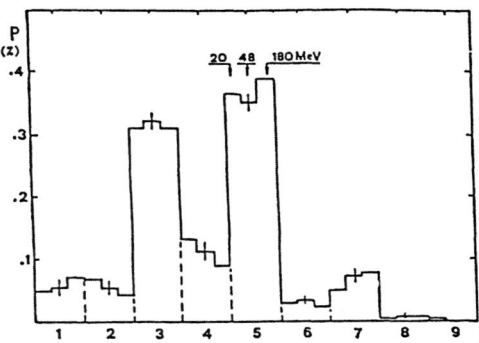

Fig. 5.1 - Charged prong multiplicity M of $\bar{p}^4$He reaction events at three lab. kinetic energies. From Ref. (37).

We see that at 20 MeV (where the kinematics allows only annihilations) the 2 and 3 prong event percentages are statistically the same as at the higher energies, where these two channels might contain also inelastic events (break-up, knock-out, etc.). This fact and the measurements (in progress) of 2 and 3 charged prong events, permit to set an upper limit of 5% to the inelastic processes in ^{4}He at these energies.

Another interesting feature of fig. 5.1 is given by the even prong channels, which, by charge conservation, include only $\bar{p}$ interactions on a neutron with a ^{3}He recoil[34]. The sum of these channel gives the $\bar{p}^4$He→ ^{3}He + X cross section, a datum relevant for the astrophysics[38]. This and other reaction cross sections on light nuclei are shown in fig. 5.2.

Perhaps, the most striking feature of this figure is the increase of the $\bar{p}^2$H reaction cross section, which becomes comparable with that for ^{4}He around 300 MeV/c (50 MeV). Although broad resonances in the $\bar{p}^2$H system have been predicted[43] before these recent $\bar{p}$-^{4}He measurements, at present there is no clear explanation of this effect. Since $\bar{p}^2$H data came from bubble chambers, where it is difficult to measure at low energy, and where different measurements of the $\bar{p}^2$H total elastic cross section are in disagreement (see fig. 5.3)[35,39], the need of new $\bar{p}^2$H experiments at low energy is evident. Extensive Glauber analyses of $\bar{p}^2$H reaction cross section data lead to contradictory results[39] because of the mentioned disagreement between the data.

More consistent results, obtained with $\bar{p}^4$H data, are shown in fig. 5.4. Here a neutron scattering amplitude with a ratio $R = \sigma_n/\sigma_p = \beta_n^2/\beta_p^2$

534

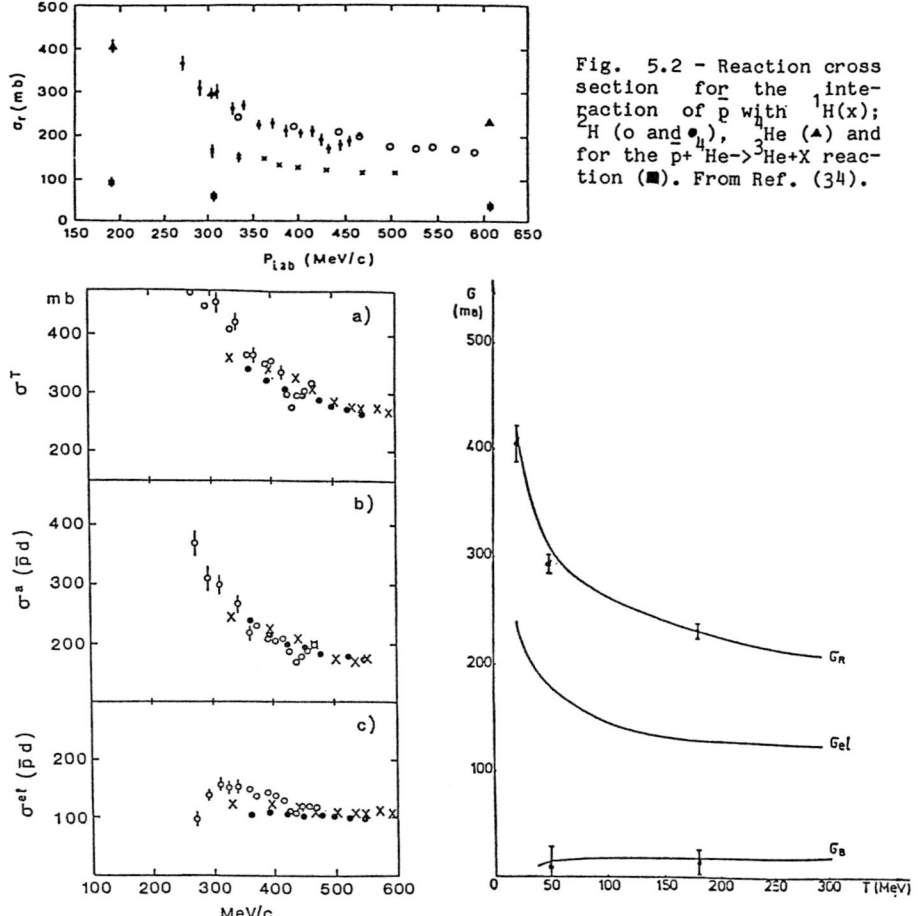

Fig. 5.2 - Reaction cross section for the interaction of $\bar{p}$ with ^{1}H(x); ^{2}H (o and $\bullet_4$), ^{4}He (▲) and for the p+^{4}He->^{3}He+X reaction (■). From Ref. (34).

Fig. 5.3 - Total, annihilation and elastic $\bar{p}^2$H cross sections vs $\bar{p}$ momentum. Comparison among the data of Ref.(24) (x) Ref.(35) (o) and Ref. (36) (●).

Fig. 5.4 - Glauber calculations for reaction σ_R, elastic σ_{el} and inelastic σ_B total cross sections of $\bar{p}^4$He scattering (solid lines). Note that $\sigma_{tot}=\sigma_R+\sigma_{el}$ and $\sigma_{ann}=\sigma_R-\sigma_B$. Data are from Ref.(34,37)

as a free parameter and $\rho_n = \rho_p$ has been employed (see eq. 1). The input parameters σ_p, β_p and ρ_p are taken from Tab. I. The results for R are reported in Tab. II, together with the values R_b from measurements on $\bar{p}^4$He annihilation events observed in the streamer chamber [39]. In these measurements, annihilations on a neutron or on a proton have been distinguished and the ratio $R_b = (\sigma_n/\sigma_p)_b$ for nucleon bound in ^{4}He has been obtained.

Tab. II – Ratio σ_n/σ_p for free nucleons (R) and for nucleons bound in ^{4}He (R_b)

Momentum	Energy MeV	R	R_b
at rest	at rest	–	0.42 ± 0.05
192.8	19.6	0.40 ± 0.12	0.64 ± 0.07
306.2	48.7	0.71 ± 0.07	0.69 ± 0.06
607.7	179.6	0.94 ± 0.6	0.65 ± 0.09

The 200 MeV/c case aside, where the comparison might be meaningless because the validity of Glauber theory is uncertain and the $\bar{p}p$ scattering parameters are measured with large errors (see Tab. I), it seems that $R = R_b$ at 300 MeV/c and $R > R_b$ at 600 MeV/c, indicating possible nuclear binding effects. We note that Glauber theory analyses of scattering distributions[30] and total cross sections[29] on the heavier nuclei confirm the values of R <1 reported in Tab. II.

Since, taking into account the isospin dependence of the $\bar{N}N$ amplitude, one can write

$$R = \frac{\sigma_n}{\sigma_p} = \frac{2\sigma(I=1)}{\sigma(I=0) + \sigma(I=1)} \tag{2}$$

a ratio R < 1 at low energy means that the amplitude in the isosinglet state is dominant. This fact is in agreement with the current $\bar{N}N$ potentials. Hence, we think that new theoretical investigations on the connection between potential and Glauber approaches to the $\bar{p}$-nucleus reactions could improve our knowledge of the $\bar{N}N$ interaction.

To obtain further information on the $\bar{p}$-light nucleus annihilation mechanisms, diagrams like those of fig. 5.5 on ^{4}He must be considered. For the $\bar{p}\,^2$H annihilation many studies of this kind have been performed,

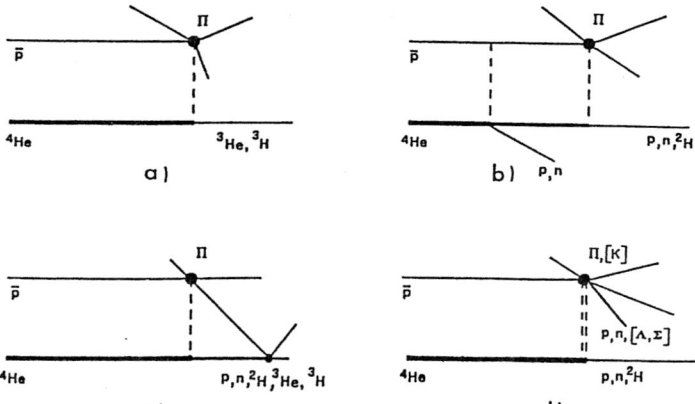

Fig. 5.5 - Annihilation mechanisms in ^{4}He. a): impulse approximation; b): initial state interaction of the incoming $\bar{p}$ with the nucleus followed by annihilation; c): interaction of the annihilation pions in the final state; d): annihilation on two or more nucleons.

clarifying the role of off-shell effects and of the structure of the momentum spectrum of the spectator proton[40,41]. For the $\bar{p}$ ^{4}He system the only attempt we know in this direction is that of Nazaruk[42]. Considering diagrams like those of fig. 5a) and 5c), he calculated the $\bar{p}$ ^{4}He → ^{3}He+ X and $\bar{p}$ ^{4}He → ^{2}H + X cross sections reported in fig. 5.6. We see that the experimental points for the ^{3}He production are in distinct disagreement with the theoretical predictions. It is clear that further studies of this type are necessary, also in view of a quantitative interpretation of the multiplicities[34] and momentum spectra of the pions from the $\bar{p}$-

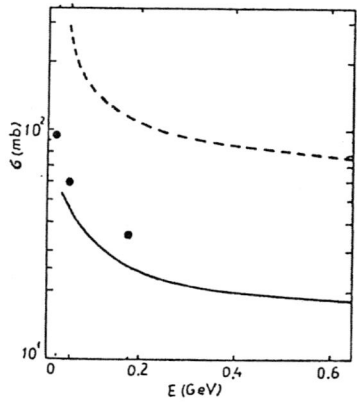

Fig. 5.6 - Theoretical cross sections for ^{3}He (solid line) and ^{2}H (dashed line) production in the $\bar{p}$- ^{4}He annihilation[42]. The experimental points for the ^{3}He production are taken from Ref. (34). See also fig. 5.2.

^{4}He annihilation.

6. Search for exotic effects

With exotic one usually means processes that are forbidden on a free nucleus while they may take place on nucleons bound in the nucleus. Here we mention briefly the annihilation on more than one nucleon (see diagram in fig. 5.5d) and the excitation of new degrees of freedom of the nuclear matter, such as the nucleon deconfinement with formation of a quark gluon plasma. Common signatures of these phenomena in the $\bar{p}$-nucleus annihilation should be an excess of strange praticles (Hyperons and Kaons) produced and anomalies in the angular and energy distribution of pions and nuclear fragments. Discussion on these topics are found, for instance, in Refs. (44,45,46,47).

Preliminary searches of exotic phenomena have been made by two experiments at LEAR[29,48]. Both experiments have found signals of $\bar{p}$ annihilating deeply inside medium-light nuclei, a favourable condition for the production of exotic reactions. The streamer chamber collaboration found also no increase in the yield of strange particle production in He and Ne with respect to the free nucleus annihilation (2-3%).

However an analysis of $\bar{p}$Ne data at 180 MeV has revealed an anomalous percentage of hyperon, not possible on a free nucleon at this energy. A ratio of the production cross section $\sigma(\Lambda^\circ)/\sigma(K_o) - 2$ has been measured[49]. This seems to be hardly explanable with Kaon-nucleon rescattering, because of the high number of Λ's produced and of the typical features of the rapidity and the angular distribution of the pions produced in the primary vertices in the presence of a Λ°.

Signatures of quark gluon plasma formation in the $\bar{p}$He annihilation with coagulation of the plasma on expansion into mesons and baryons have been studied theorically. A significant probability of annihilation on two nucleon, about 10%, has been predicted[47]. Moreover, fast outgoing protons ($\sim$ 100 MeV kinetic energy) and a shift to lower values of the pion momentum distributions are expected. Hence, two peaks should be observed in the energy spectrum of the pions emitted in the $\bar{p}$-^{3}He and $\bar{p}$-^{4}He annihilations[47].

Clearly, to detect unambiguously deconfinement signals and quark-gluon effects, exclusive measurements with high resolution and statistics are necessary. As an example of these measurements, we mention two reactions proposed recently[50,51]:

$$\bar{p} + {}^2H \rightarrow \bar{\pi} + p \qquad\qquad\qquad (3)$$

$$\bar{p} + {}^3He \rightarrow n + p \qquad\qquad\qquad (4)$$

Reaction (4) has not yet been observed. Reaction (3) has been experimentally observed at rest and a probability $(0.9 \pm 6.4)\ 10^{-5}$ on the total annihilation probability was found[52]. These annihilations can be explained in two ways: a) annihilation on several nucleons, that is on six or nine quark bags and b) a "standard mechanism" with annihilation on a single nucleon followed by absorption of pions in the final state (see fig. 5.5 c).

In a study of the standard mechanism, triangle diagrams like that of fig 5.5 c) for ${}^4He$ have been calculated with known techniques[51] for ${}^2H$ and ${}^3He$ and probabilities $(10^{-7}-10^{-10})$ for the reaction (3) and $< 10^{-11}$ for the reaction (4) have been found. We see that the datum on ${}^2H$ at rest is at least two orders of magnitude higher than this theoretical expectation.

Reactions (3) and (4) will be studied with the OBELIX spectrometer[50] at LEAR in the future. Such data will be of great help in the understanding of short range nucleon correlations and of the large quark bags in the light nuclei.

7. Conclusions

. Antinucleon-nucleus physics has been undertaken in a systematic way only in the last years and we have shown the first results of the exploratory studies. New high resolution and high statistics exclusive experiments are planned for the future to obtain deeper information on the nuclear forces and the nuclear structure. Some of them will be performed at CERN in the next year, when LEAR will be again operative.

References

1. Walker, G.E., Goodman, C.D., Olmer, C. (eds.): Antinucleon- and nucleon-nucleus interaction. New York and London: Plenum Press 1985

2. Gastaldi, U., Klapisch, R., Richard, J.M., Tran Thanh Van, J.,(eds.)
   Physics with antiprotons at LEAR in the ACOL era. Gif sur Yvette:
   Edition Frontières 1985

3. Gastaldi, U., and Klapisch, R.: Report CERN-EP81-06

4. Armenteros, R., and French, B.: NN interactions. In High Energy Physics. Burhop, E.H.S. (ed.) New York; Academic Press 1969.
   Montanet, L. et al.: Physics Reports 63, 149 (1980)

5. Dover, C.B. and Richard, J.M.: Phys. Rev. C21, 1466 (1980)
   Buck, W.W., Dover, C.B. and Richard, J.M.: Ann. of Phys.121,47(1979)
   Dover, C.B. and Richard, J.M.: Ann. of Phys. 121, 70 (1979)

6. Lacombe, M. et al.: Phys. Rev. C21, 861 (1980)

7. Côtè J. et al.: Phys. Rev. Lett. 48, 1319 (1982)

8. Dover, C.B. and Miller, C.J.: Ref. (1) pag. 25

9. Flaminio, V. et al.: Report CERN-HERA/84-01 Geneva 1984

10. Brückner, W. et al.: Phys. Lett. 158B, 180 (1985)
    Phys. Lett. 166B, 113 (1986)

11. Beard, C.I. et al.: Report CERN-EP/84-140, Geneva 1984

12. Iwasaki, H., et al.: Nucl. Phys. A433, 580 (1985)

13. Kondratyuk, L.A. et al.: Sov. J. Nucl. Phys. 33, 413 (1981)

14. Glauber, R. and Matthiae G.: Nucl. Phys. B21, 135 (1970)

15. Bassel, R.H. and Wilkin, C.: Phys. Rev. 174, 1179 (1968)

16. Alkhazov, G.D. et al.: Phys. Rep. 42, 89 (1978)

17. Poth, H.: Lecture notes in physics 243, 357 (1985)

18. Koch, H.: 2nd Conference on the intersections between particle and
    nuclear physics. Lake Louise, Canada 1986. Report CERN-EP/8-78(1986)

19. Guigas, R. et al.: Phys. Lett. 137B, 323 (1984)

20. Köhler, Th. et al.: Phys. Lett. 176B, 327 (1986)

21. Davies, J.D. et al.: Phys. Lett. 145B, 319 (1984)

22. Garreta, D. et al.: Phys. Lett. 135B, 266 (1984)
    Phys. Lett. 139B, 464 (1984)
    Phys. Lett. 149B, 64 (1984)
    Phys. Lett. 151B, 473 (1985)
    Phys. Lett. 169B, 14 (1986)

23. Garreta, D. et al.: Nucl. Phys. A456, 557 (1986)

24. Bizzarri, R. et al.: Nuovo Cimento 22A, 285 (1974)

25. Di Marzio, F. and Amos, K.: Nucl. Phys. A454, 453 (1986)

26. Picklesimer, A. et al.: Phys. Lett. 163B, 311 (1985)

27. Ingemarsson, A.: Nucl. Phys. A454, 475 (1986)

28. Friedman, E., and Lichtenstadt, J.: Nucl. Phys. A455, 573 (1986)

29. Balestra, F. et al.: Nucl. Phys. A452, 573 (1986)

30. Dalkarov, O.D., and Karmanov, V.A.: Nucl. Phys. A445, 579 (1985)

31. Birsa, R. et al.: Report CERN-EP/85-28, Geneva 1985

32. Garreta, D. et al.: Phys. Lett. 150B, 95 (1985)

33. Heiselberg et al.: Phys. Lett. 132B, 279 (1983)

34. Balestra, F. et al.: Phys. Lett. 149B, 69 (1984)
    Phys. Lett. 165B, 265 (1985)

35. Kalogeropoulos, T., and Tzanakos, G.S.: Phys. Rev. D22, 2585 (1980)

36. Hamilton, R.P.: Phys. Rev. Lett. 44, 1182 (1980)

37. Batusov, Yu.A. et al.: JINR Rapid Communications 12 (1985)

38. Batusov, Yu.A. et al.: Lett. Nuovo Cimento 41, 223 (1984)
    JINR Rapid Communications 6 (1985)

39. Balestra, F. et al.: Report CERN-EP/86-104 (1986)

40. Alberi, G. and Bajzer, Z.: Lett. Nuovo Cimento 32, 99 (1981)

41. Alberi, G. et al.: Nuovo Cimento 53A, 191 (1979)
    Kondratyuk, L.A.: Sov. J. Nucl. Phys. 24, 247 (1976)

42. Nazaruk, V.I.: Phys. Lett. 155B, 323 (1985)

43. Grach, I. et al.: Report ITEP-12, Moscow (1982)

44. Rafelski, J.: Phys. Lett. 91B, 281 (1980)

45. Rafelski, J. and Mueller, B.: Phys. Rev. Lett. 48, 1066 (1982)

46. Phatak, C.S. and Sarma, N.: Phys. Rev. C31, 2113 (1985)

47. Phatak, C.S. and Sarma, N.: Unpublished report

48. McGaughey, P.L. et al.: Phys. Rev. Lett. 56, 2156 (1986)

49. Piragino, G.: Report CERN-EP/86-75, Geneva 1986

50. Armenteros, R. et al.: Report CERN/PSCC/86-4 PSCC/P95, Geneva (1986)

51. Kondratyuk, L.A. and Sapozhnikov, M.G.: To be published in Yad.Fiz.

52. Bizzarri, R. et al.: Lett. Nuovo Cim. 2, 341 (1969).

HIGHLIGHTS OF THE INTERNATIONAL SYMPOSIUM ON THE

THREE-BODY FORCE IN THE THREE-NUCLEON SYSTEM

B.L. Berman

Department of Physics

The George Washington University

Washington, DC

The International Symposium on the Three-Body Force in the Three-Nucleon System was held at the George Washington University in Washington, DC on April 24-26, 1986. It was attended by 122 physicists from 58 universities and national laboratories in 14 countries.

The goals of the Symposium were to review the subject of three-body nuclear physics, to focus sharply on the three-body-force aspects of this subject, and to formulate a program of experiment and theory in this field for the next five years. The subject matter of the Symposium was divided into four regimes:

1) the bound-state properties,
2) the long-wavelength region,
3) the intermediate-energy region,
4) short-range phenomena.

For each of these four areas there was a Working Group composed of both experimentalists and theorists, jointly led by one of each. Each Working Group was charged with the responsibility of producing a program of recommended research in the three-body physics of its regime. The Working-Group Leaders were given complete freedom to invite speakers and otherwise to structure their sessions so as to make the most efficient use of the limited time of the Symposium. This resulted in a hectic and crowded but very productive two days. My presentation here is largely the record of how very well the Working-Group Leaders did their job.

No conference had ever before dealt expressly with the three-body force in nuclear physics, despite the fact that an understanding of such forces is essential for a quantitative assessment of the importance of quark substructures, relativistic effects, and other much-discussed exotic nuclear phenomena as well as of the traditional questions of the static and dynamic properties of nuclear few-body systems. The Washington Symposium was

unique not only in detailing our present state of knowledge
in this field but also in setting forward a research
program for building upon this knowledge to enhance our
understanding of three-body forces and with it our
comprehension of the basic underpinnings of nuclear
physics.

The Proceedings of the Symposium fill 530 pages.
Clearly, I cannot cover all the papers presented, and I
apologize in advance for those that I have omitted. My
selection is the result of a point of view: that those
experimental results which show discrepancies with
theoretical calculations based upon two-body forces might
yield some insight into the problem of where to look for or
to expect effects of three-body forces. Of course, one
must always be cautious that the two-body input is complete
and therefore not ascribe to a three-body force the result
of a deficiency or shortcoming in the two-body force. As
we shall see, this <u>caveat</u> has particular relevance to
continuum calculations.

The first, and arguably the most important question
approached at the Symposium is, what do we mean by a three-
body force? The opening paragraph of McKellar's paper[1]
states it well: "The three-body force between nucleons has
been around for a long time as a concept -- at least since
1938.[2] However, it is only in recent years that most
nuclear physicists have taken the concept seriously. This
delay has been due to the great difficulties that have been
in the way of deciding that the 3-nucleon force is
necessary for the understanding of nuclear properties.
Over the last five years or so it has become apparent that
2-nucleon forces, used in a non-relativistic framework, do
not quantitatively describe the properties of nuclei. The
fault could be in many places -- relativistic effects,
many-nucleon interactions, renormalization of the
interaction by the nuclear medium, quark effects, etc. Of
course the distinction between the various items on this
list is not clear cut, and we find people using different
labels to describe the same basic effects."

McKellar goes on to illustrate the various kinds of
terms, illustrated by the diagrams shown in Figs. 1-5, that
might well be the most important contributors to the
nuclear three-body force. Friar and Frois, in their
Working-Group 1 summary,[3] add, in Fig. 6, the diagram for
pion scattering from a virtual pion being exchanged between
nucleons, and point out as well that, like many other
features of nonrelativistic theories, certain apparently
three-body contributions can be explained alternately by or
arise naturally in a fully relativistic treatment.

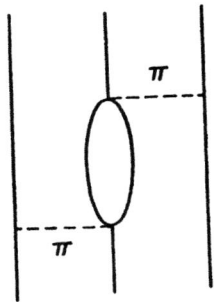

Fig. 1. The fundamental
π,π exchange 3-nucleon
potential. The blob
represents the $\pi N \rightarrow \pi N$
amplitude with the
forward propagating
Born term subtracted.

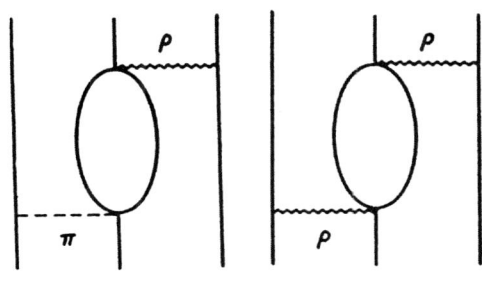

Fig. 2. π,ρ and ρ,ρ
contributions to the
3-nucleon potential.

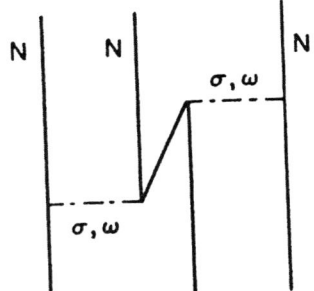

Fig. 3. σ and ω exchange
Z-graph contributions to
the 3-nucleon potential.

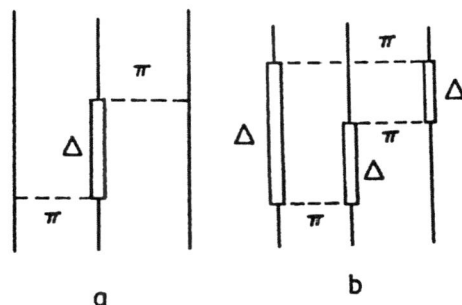

Fig. 4. a. The Δ contribution
to the π,π exchange potential.
b. The 3-Δ contribution to the
3π exchange potential.

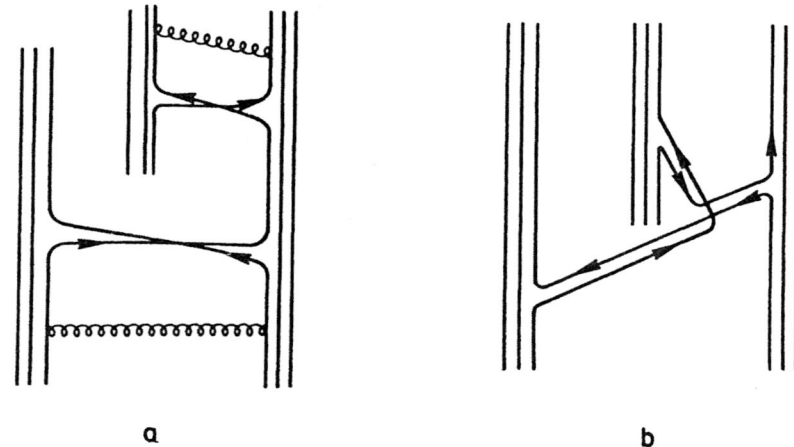

Fig. 5. a. Quark-gluon contributions to the 3-nucleon potential. b. A topologically equivalent distortion of part a, demonstrating its duality to t-channel meson exchange.

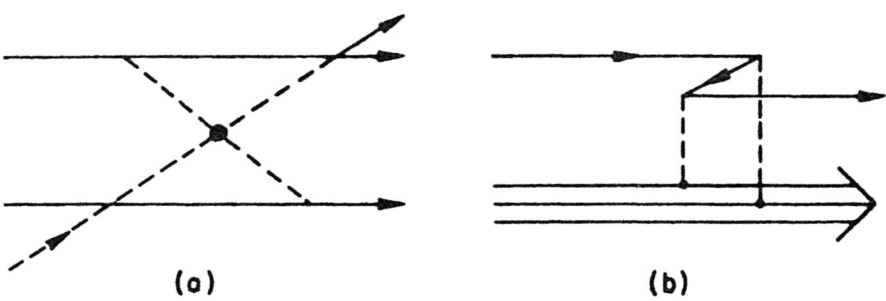

Fig. 6. The scattering of a pion from the (exchanged) pion cloud in the nucleus shown in part (a) is a true three-body force, while the Z-graph for proton-nucleus scattering depicted in part (b) is a three-body-force effect implicitly included in Dirac treatments of the proton projectile.

Further amplifications of the quark-cluster descriptions of the three-body force as given by Chang and Gross in their Working-Group 4 summary[4] is illustrated in Fig. 7, and a comparison of this description (with quark-gluon exchange) with the three-baryon descriptions (with meson exchange) is made in Fig. 8, from the American Physical Society summary paper of Gibson.[5]

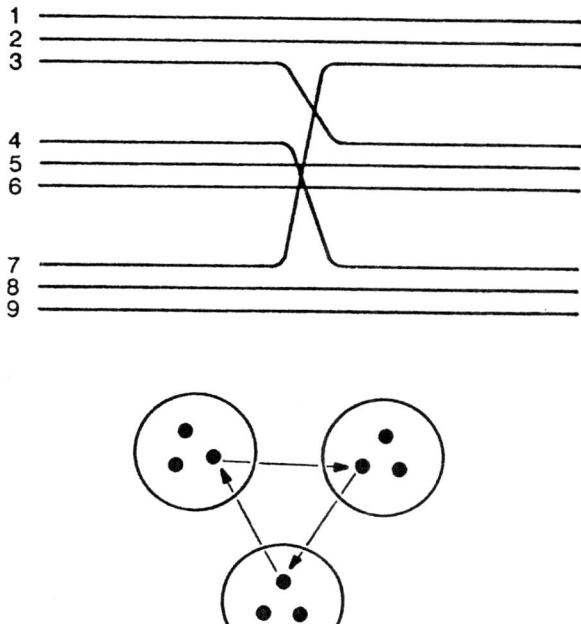

Fig. 7. Antisymmetrization of the nine-quark wave function introduces a term in which $r_3 \to r_4 \to r_7 \to r_3$. Matrix elements of this term introduce effective terms which depend on three coordinates, and hence play the role of three-body forces.

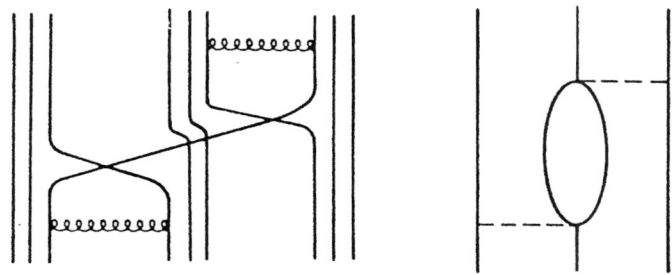

Fig. 8. Comparison of a three-body-force term in terms of three-quark clusters and quark-gluon exchange with a three-baryon description and meson exchange.

Turning now specifically to the three-nucleon system, Sick[6] focuses attention on the greater likelihood of finding three-body-force effects when the three nucleons are in a triangular configuration ("star geometry") than when they are aligned ("collinear geometry"), as illustrated in Fig. 9. This observation became one of the underlying themes of the Symposium, leading to an emphasis on higher partial waves and d-state observables, especially at short range or high momentum transfer, as can be seen from the large effect of the three-body force on the d-state density in ^{3}He at short range shown in Fig. 10.

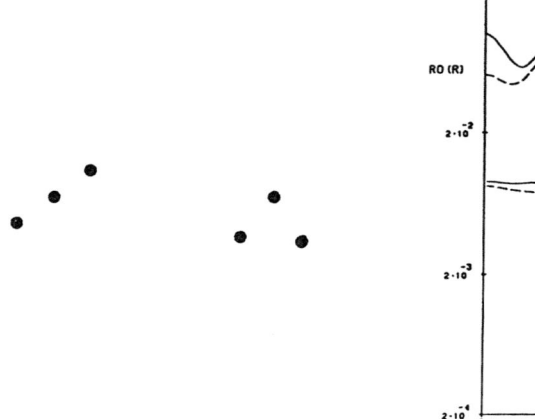

Fig. 9. Aligned and triangular 3-nucleon configurations.

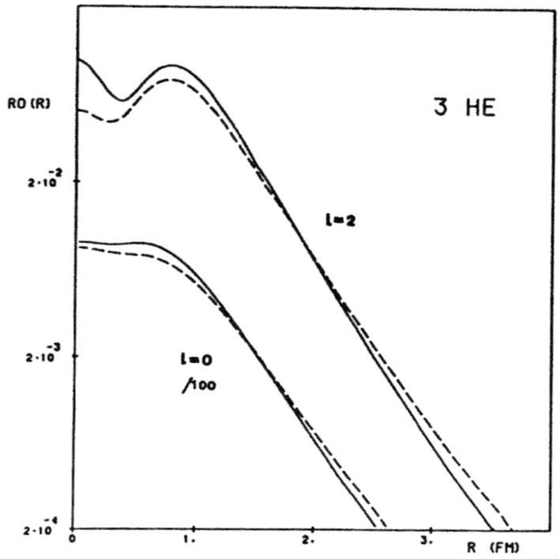

Fig. 10. The s- and d-state densities of ^{3}He calculated (Ref. 7) without (dashed) and with (solid) a three-body force.

Further examples where three-body forces clearly play an important role are summarized in Table I, from the summary presented at the Lake Louise meeting by Nefkens.[8] In each of these four cases, the inclusion of a three-body force in the calculation brings the calculated result into significantly better agreement with the experimental data: (i) the triton binding energy is increased by about 1.8 MeV (see, for example, Ref. 9); (ii) the trinucleon charge radii are reduced by about 0.1 fm (see, for example, Ref. 10) and are now in agreement with the data, as shown in Fig. 11 (from Ref. 9); (iii) the asymptotic d-state normalization constant is increased by about 10% (see Refs. 10 and 11); and (iv) the n-d doublet scattering length is reduced by no less than a factor of two, bringing it into excellent agreement with experiment, as shown in Fig. 12 (from Ref. 12).

Table I. Comparison of the results of two-body-force
calculations with experimental data in the A=3 system.

	Experiment	Two-body-force results	
1. Triton binding energy (MeV)	8.48	RSC (34ch.) 7.35	AV14 7.67
2. ^{3}He rms charge radius (fm)	1.93±0.03	2.04	2.02
3. Asymptotic normalization constant D_2(fm^2) for ^{3}H → nd	-0.279±0.012	-0.243	
4. n-d doublet scattering length $^2A_{nd}$ (fm)	0.65±0.04	1.31	

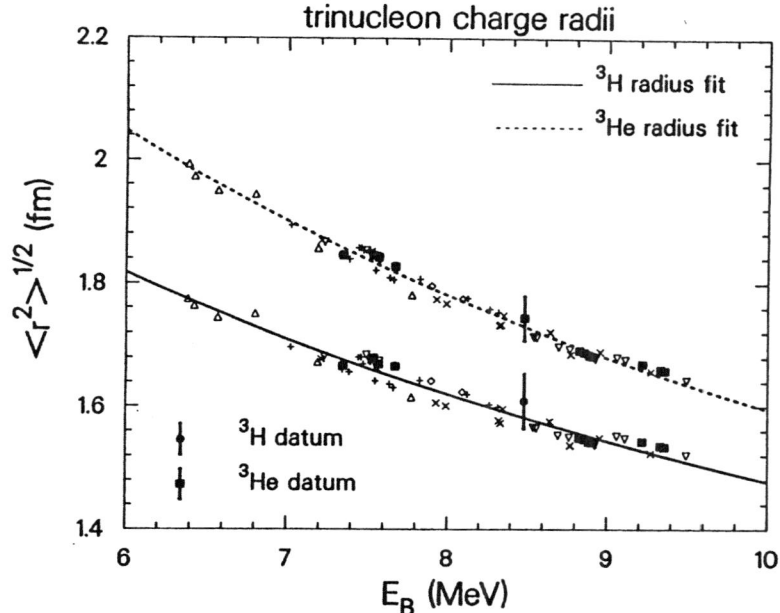

Fig. 11. A collection of model ^{3}He and ^{3}H radii (for point-like protons) for model Hamiltonians with a wide range of binding energies are plotted. Also plotted are least-squares-fitted curves to the theoretical values.

548

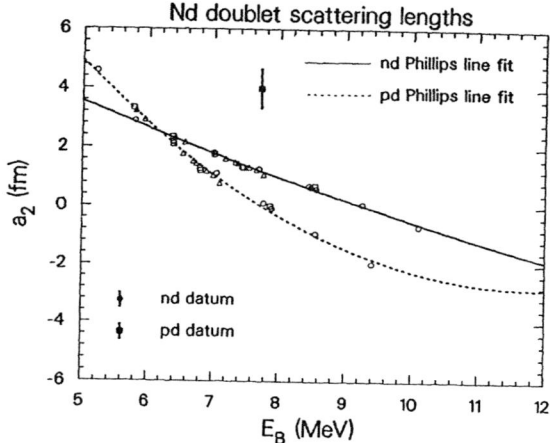

Fig. 12. The doublet nucleon-deuteron scattering length vs. the trinucleon binding energy. The results with three-nucleon forces included are denoted by triangles.

The new elastic electron-scattering data for ^{3}H and ^{3}He from Saclay[13] that made possible the detailed confrontation of theoretical predictions of the charge and magnetization radii and form factors and thus gave substance to many of the Symposium discussions are shown (in part) in Fig. 13 (from Ref. 3). A provocative point is that while the charge form factor for ^{3}H is reproduced satisfactorily by "conventional" nuclear theory based upon two-body forces, that for ^{3}He is not (see Fig. 13).

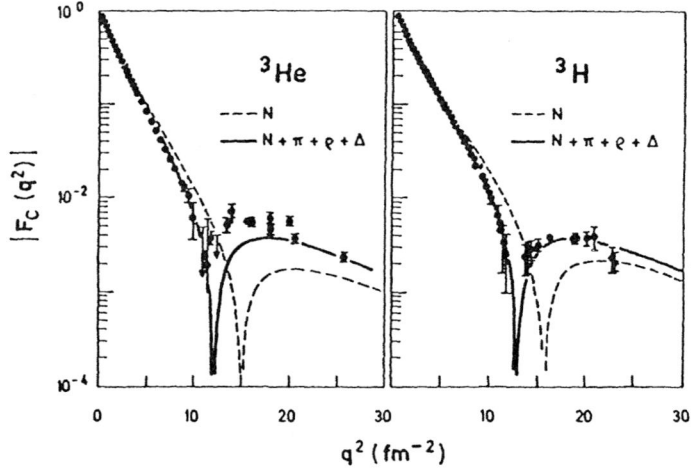

Fig. 13. Charge form factors for ^{3}He and ^{3}H (from Ref. 3).

A new method for determining the odd-nucleon matter form factors for ³H and ³He from pion elastic-scattering cross sections was presented by Nefkens.[14,15] The validity of this method follows from fundamental properties of the π-N interaction near the Δ resonance at scattering angles near the non-spin-flip dip, and thus can throw light on these matter distributions in the momentum-transfer window from about 2 to 4 fm⁻². The results to date agree with the magnetization form factors obtained from electron scattering; what is exciting is that if future measurements succeed in reducing the present experimental uncertainties by a factor of about 3 to 5 (which is, in fact, feasible at present), they will constitute a direct determination of the strength of the three-body force mechanism of Fig. 6(a).

The long-wavelength continuum region was much discussed at the Symposium, with the general consensus that the existing kinds of theoretical calculations are not adequate. The evidence for this consensus comes from several sources, including (i) the precise new n-d scattering measurements, (ii) certain three-body breakup measurements, with both hadronic and electromagnetic probes, and (iii) new measurements of spin-dependent observables, especially tensor polarizabilities.

Brandenburg, in his historical summary,[16] observes, "In what must have been one of nature's finest finesses, we were given the Coulomb force between charged particles to entice experimentalists to measure p-d reactions and theorists to do n-d calculations." Gibson comments,[5] "The good news is that the n-d data are now of precision comparable to that obtained in p-d scattering."

Tomusiak and Weller, in their Working-Group 2 summary,[17] point out the inadequacy of current continuum Faddeev calculations, as illustrated in Figs. 14 (TUNL data, from Ref. 18) and 15 (Karlsruhe data, from Ref. 19), which perhaps get worse with increasing energy (see Ref. 19). They also point out the importance of kinematic conditions in three-body breakup reactions, as illustrated by the star-geometry (Fig. 16) data of Bodek et al.,[20] shown in Fig. 17. However, the fact that these data are inadequately described by a Faddeev calculation (also shown in Fig. 17) might be attributable more to the neglect of higher partial waves than to three-body forces; in any case, a better continuum calculation is needed.

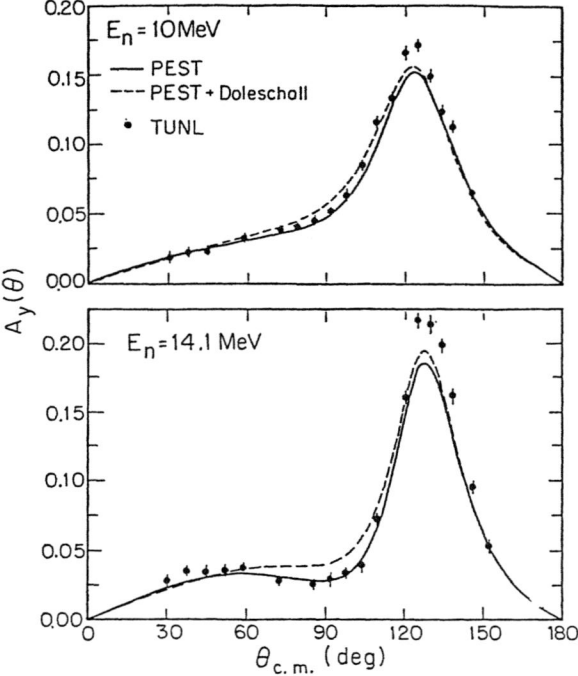

Fig. 14. Results of Faddeev calculations compared with n-d elastic $A_y(\theta)$ data (Ref. 18).

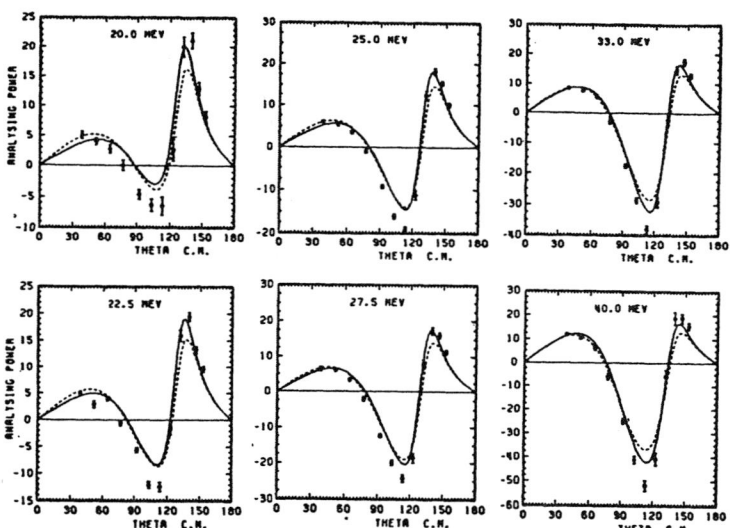

Fig. 15. The n-d elastic analyzing-power data of Klages et al.[19] for E_n from 20 to 40 MeV. The curves represent the results of Faddeev calculations. Solid line: Graz II potential; dashed line: Doleschall 4T 4B potential.

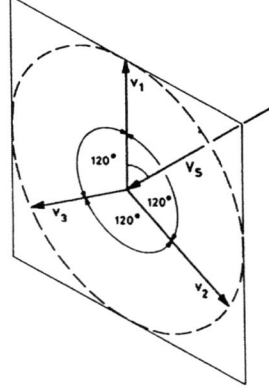

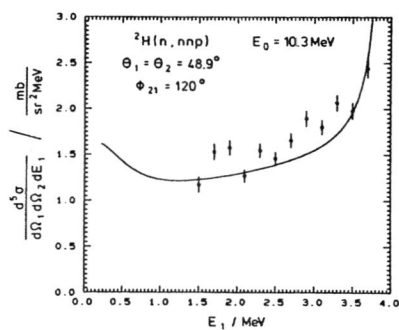

Fig. 16. Fully symmetric star geometry (V_S is the velocity of the center of mass).

Fig. 17. Star-geometry cross section (one pair), corrected, with SASA calculation (solid line).

In his status report on three-nucleon photonuclear reactions,[21] Lehman discusses the ^{3}H(γ,n) and (γ,2n) data of Faul et al.,[22] shown in Fig. 18. While the continuum Faddeev calculation of Gibson and Lehman[23] reproduces the two-body-breakup data reasonably well, it overpredicts the three-body photodisintegration results, again showing the need for more complete continuum calculations when three-body-force effects might be sizable.

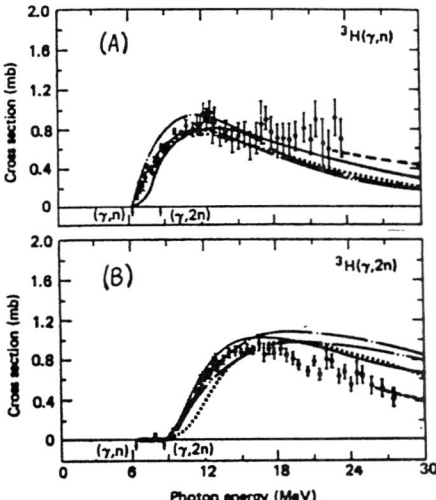

Fig. 18. Comparison of the results of continuum Faddeev calculations (solid curves, Ref. 23) with the ^{3}H photodisintegration data (Ref. 22): (A) two-body breakup; (B) three-body breakup.

Lehman also discusses the possible sensitivity of the tensor polarizability T_{20} in $\vec{p}$-d capture, where both E1 and E2 radiation are important, to three-body effects. The close connection of this quantity to the d-state component in ^{3}He is illustrated in Fig. 19. It should be noted, however, that even allowing the deuteron d-state component to vary until a best fit to the data of Ref. 24 is achieved, the calculated result is not a satisfactory representation of the data. Here once again, any meaningful test of a three-body-force effect awaits a better continuum calculation.

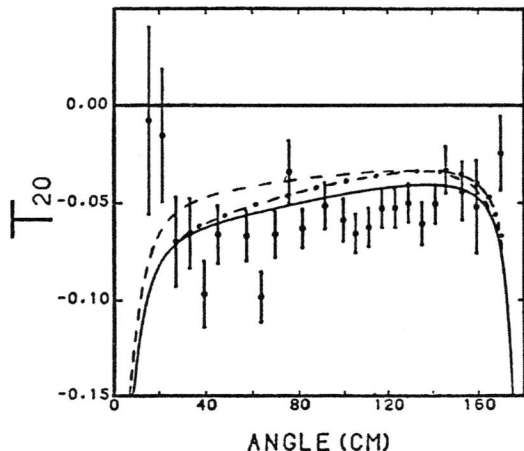

Fig. 19. T_{20} for ^{1}H($\vec{d},\gamma$)^{3}He, from Ref. 24. The solid line, the dashed line, and the dot-dashed line correspond to 7%, 2%, and 9% deuteron d-state, respectively.

As the excitation energy of the three-nucleon system is raised into the resonance region, the pion and the Δ isobar become explicit constituents of the nuclear system. Cameron and Sauer, in their Working-Group 3 summary,[25] show the importance of taking into account the Δ (in particular) in calculations of three-nucleon properties and reactions [see especially Sauer's report to Working Group 1 (Ref. 26)]. Even when this is done, however, the uncertainties surrounding the treatment of meson-exchange currents, final-state interaction, and relativistic effects in this medium-energy regime make any quantitative assessment of three-body-force effects unrealistic at the present time.

Still, several recent results reported at the Symposium may provide clues as to how to proceed in uncovering three-body effects. Cierjacks[27] reported results of pion-induced two- and three-body breakup measurements of ^{3}He done at SIN. These detailed studies demonstrate that three-nucleon absorption of the incident resonant pions is far more probable than heretofore thought, ranging from 20 to 30% in the resonant region and seeming to follow the resonance shape. Whether this large three-nucleon absorption probability is indicative of a three-body-force effect is open to conjecture at this time.

Cameron[28] reports strong anomalies in measurements of analyzing powers in both the radiative-capture and π^0-production channels for the $\vec{p}$-d reaction at medium energies at TRIUMF. The results for latter are shown in Fig. 20. One sees that the data are not even remotely reproduced by the calculation, thus showing the overriding importance at these energies of final-state-interaction and perhaps other effects not presently included in the theory.

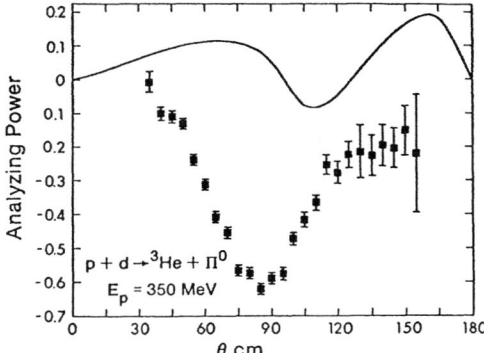

Fig. 20. Analyzing power for the reaction $\vec{p}d \to \pi^0$ ^{3}He at E_D=350 MeV (from Ref. 29). The solid curve is from Refs. 30 and 31.

At still higher energies, yet another piece of evidence points to the importance of spin observables as potential indicators of three-body effects. This evidence comes from the comprehensive series of measurements carried out by the UCLA group of Igo, Whitten, and collaborators at LAMPF of the spin observables in p-d scattering at 800 MeV, reported on at the Symposium by Adams[32] and by Bleszynski.[33] Figure 21 shows the vector and tensor analyzing-power data compared with the results of calculations with and without the inclusion of three-body-force effects. Clearly, the inclusion of three-body-force effects improves the agreement with the data markedly, for <u>both</u> kinds of spin-dependent observables.

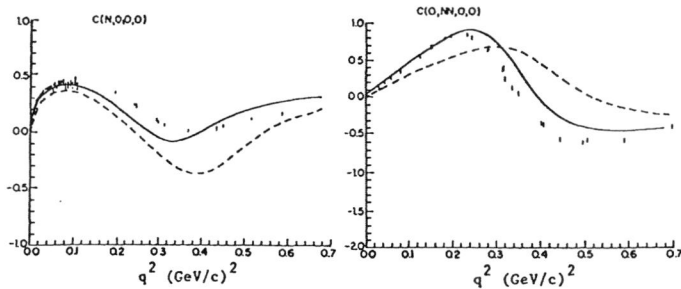

Fig. 21. Vector and tensor analyzing powers for p-d scattering at 800 MeV. The data are from Ref. 32; the dashed curves result from relativistic calculations using two-body forces only, while the solid curves result from the inclusion of three-body forces as well (from Ref. 33).

554

From inelastic electron-scattering measurements in the GeV energy range one finds that the data can be parametrized as a function of only one variable instead of two (energy and momentum transfer). In this so-called y-scaling regime, one can derive a function of this y parameter and compare it with theoretical calculations. Sick[6,34] has done this for the inclusive ^{3}He(e,e') cross sections of Day et al.,[35] shown in Fig. 22; his results are shown in Fig. 23. The result of a Faddeev calculation,[36] also shown in Fig. 23, fall increasingly below the data as $|y|$ increases, but the addition of a three-body-force component reduces the disagreement with the data noticeably. Sasakawa presented new results[10] that show how far this procedure can be carried out. It turns out that the addition of a (three-body) Tucson-Melbourne force component with a larger π-N vertex mass cutoff Λ improves the fit still further for large $|y|$, but the fit at $|y| \leq$ 300 MeV/c becomes worse, and the triton binding energy, when computed with $\Lambda \geq$ 700 MeV, becomes too large.

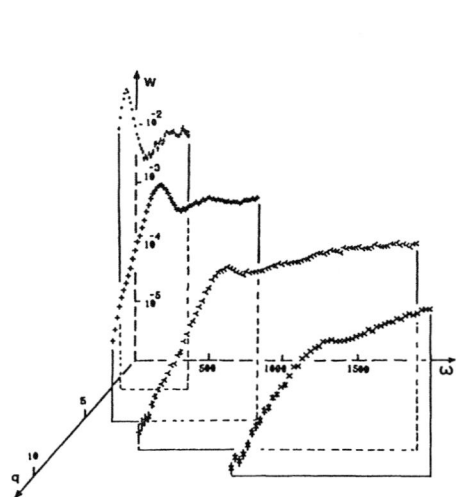

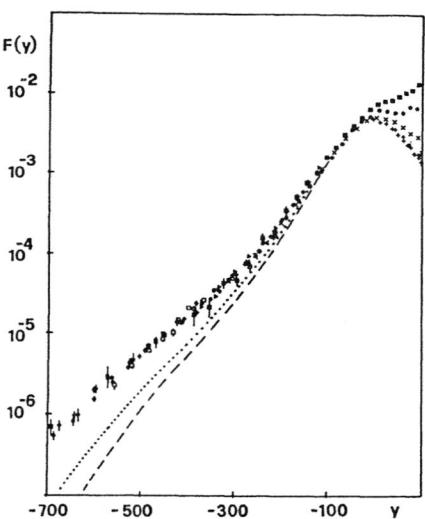

Fig. 22. Inclusive ^{3}He (e,e') cross sections as a function of energy loss and momentum transfer (from Ref. 35).

Fig. 23. Scaling function F(y) as a function of y derived[34] from the data of Fig. 22; the dashed curve gives the corresponding Faddeev result[36]; the dotted curve includes the change produced by including a three-body force.[10]

Lastly, we turn to the results presented by Kisslinger[37] based upon his hybrid quark-hadron model.[38,39] Figure 24 shows the fits to the trinucleon charge form factors obtained. Since in this model nonrelativistic wave functions are used outside of some critical radius and thus discontinuities occur at the boundary, it is not yet clear whether this model has any predictive power; still, the quality of the fits shown in Fig. 24 is impressive.

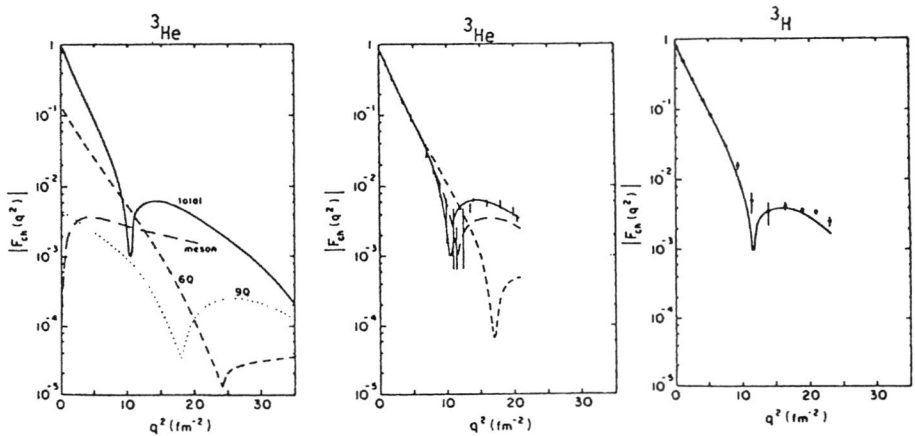

Fig. 24. Fits to the ^{3}He and ^{3}H charge form factors obtained from the hybrid quark-hadron model.[37] Also shown are the relative contributions of the 6- and 9-quark parts. Clearly the 9-quark part has something to do with the short-range part of the three-body force, and its relative size gives some indication of the importance of this contribution (from Ref. 4).

Chang and Gross[4] also discuss other theoretical approaches to short-range effects in the three-nucleon system, but it is clear that with nearly a total lack of data in this regime, at least until some are obtained at CEBAF and perhaps at other high-energy facilities, most of these theoretical efforts will remain untested. It is, of course, a formidable but vital challenge to the theorists to guide the planning of future experimental programs under these conditions.

I would like to conclude by quoting Gibson's concluding remarks[5]: "Physics progresses in cycles. At present the experimentalists have leaped over our computational resources and are ahead of the theorists. This has not always been so. We have taken giant strides in the last three to five years in our efforts to understand three-body forces in nuclear physics. (An indication of the interest in the topic is given by the fact that the organizers originally hoped for 60 participants and 120 came.) Unfortunately, we have reached only the end of the beginning of the investigation. However, that guarantees that much exciting research lies ahead!"

Acknowledgments

I am pleased to acknowledge the contributions of the many persons involved in making the Symposium a success. This review reflects in particular the numerous discussions I have had with Dr. B.F. Gibson and Prof. B.M.K. Nefkens. This work was supported in part by the US Department of Energy (DE-FG05-86ER40285); the Symposium was supported by The George Washington University, the University of California at Los Angeles, the Los Alamos National Laboratory, and the National Science Foundation.

References

1. B.H.J. McKellar, in The Three-Body Force in the Three-Nucleon System, eds. B.L. Berman and B.F. Gibson (Springer-Verlag, Heidelberg, 1986), p. 7. Further references to these Proceedings will be abbreviated as "Symposium Proceedings."

2. H. Primakoff and T. Holstein, Phys. Rev. 55, 1218 (1939).

3. J.L. Friar and B. Frois, Symposium Proceedings, p. 81.

4. C.C. Chang and F.L. Gross, Symposium Proceedings, p. 395.

5. B.F. Gibson, Symposium Proceedings, p. 511.

6. I. Sick, Symposium Proceedings, p. 42.

7. J.L. Friar, private communication referred to in Ref. 6.

8. B.M.K. Nefkens, 2nd Conf. on Intersections between Particle and Nuclear Physics (Lake Louise, Canada, 1986).

9. G.L. Payne, Symposium Proceedings, p. 119.

10. T. Sasakawa, Symposium Proceedings, p. 150.

11. B. Vuaridel et al., Symposium Proceedings, p. 281.

12. C.R. Chen, Symposium Proceedings, p. 222.

13. F. Juster et al., Phys. Rev. Lett. 55, 2261 (1985).

14. B.M.K. Nefkens, Symposium Proceedings, p. 364.

15. B.M.K. Nefkens et al., Symposium Proceedings, p. 385.

16. R.A. Brandenburg, Symposium Proceedings, p. 186.

17. E.L. Tomusiak and H.R. Weller, Symposium Proceedings, p. 177.

18. W. Tornow et al., Symposium Proceedings, p. 251.

19. H.O. Klages, Symposium Proceedings, p. 203.

20. K. Bodek et al., Symposium Proceedings, p. 307.

21. D.R. Lehman, Symposium Proceedings, p. 287.

22. D.D. Faul et al., Phys. Rev. C 24, 849 (1981).

23. B.F. Gibson and D.R. Lehman, Phys. Rev. C 11, 29 (1975) and 13, 477 (1976).

24. M.C. Vetterli et al., Phys. Rev. Lett. 54, 1129 (1985).

25. J.M. Cameron and P.U. Sauer, Symposium Proceedings, p. 319.

26. P.U. Sauer, Symposium Proceedings, p. 107.

27. S. Cierjacks et al., Symposium Proceedings, p. 356.

28. J.M. Cameron, Symposium Proceedings, p. 366.

29. J.M. Cameron et al., Phys. Lett. 103B, 317 (1981) and 10th Int. Conf. on Particles and Nuclei (Heidelberg, 1984).

30. J.M. Laget, Nucl. Phys. A312, 265 (1978) and Second Workshop on Perspectives in Nuclear Physics at Intermediate Energies (Trieste, 1985).

31. J.M. Laget and J.F. Lecolley, 10th Int. Conf. on Particles and Nuclei (Heidelberg, 1984).

32. D.L. Adams, Symposium Proceedings, p. 486.

33. E. Bleszynski, M. Bleszynski, and T. Jaroszewicz, Symposium Proceedings, p. 482.

34. I. Sick, Prog. Part. Nucl. Phys. 13, 165 (1984).

35. D. Day et al., Phys. Rev. Lett. 43, 1143 (1979).

36. A.E.L. Dieperink, T. de Forest, and I. Sick, Phys. Lett. 63B, 261 (1976).

37. L.S. Kisslinger, Symposium Proceedings, p. 432.

38. L.S. Kisslinger, Phys. Lett. 112B, 307 (1982).

39. E.M. Henley, L.S. Kisslinger, and G.A. Miller, Phys. Rev. C 28, 1277 (1983).

HIGHLIGHTS AT FEW BODY XI

Tatuya Sasakawa

Department of Physics, Tohoku University

Sendai 980, Japan

1. INTRODUCTION

The Eleventh International IUPAP Conference on Few Body Systems in Particle and Nuclear Physics (Few Body XI) was held at Tokyo Prince Hotel from 24th to 26th August and at Sendai Civic Auditorium from 27th to 30th August under the auspices of IUPAP, the Science Council of Japan and the Physical Society of Japan. The number of participants was 230, among them 130 from Japan, 28 from USA, 13 from West Germany, 8 from Canada, 5 from Italy and also 5 from Switzerland, etc., altogether from 24 countries. Works by 542 authors were presented in the form of 210 contributed papers, among which 106 articles were discussed in the poster sessions.

This conference was held as a joint conference with the Tenth International Conference on Atomic Physics (ICAP X). At the Tokyo part of this conferece, a participant can freely attend any talk of either conference. As a consequence, any topic related to atomic physics such as H^-, μ-t-d, few-atomic molecules etc., which were discussed in some of this series of conferences, were dismissed from Few Body XI. On the other hand, Few Body XI was devoted to discussions on intermediate and low energy few-nucleon systems. Because the studies in few-nucleon systems have made a great progress since Few Body X at Karlsruhe 1983, we enjoyed hot discussions about many topics.

The contributed papers were divided into six fields; intermediate energy physics, quarks and skyrmions, theoretical techniques, few-body theory and few-body experiments. Six rapporteurs assigned to each of

these items gave review talks, digging out interesting works from contributed papers. Prof. W. Sandhas has given a concluding summary based on
invited talks.

Since all of these review talks will appear in the proceedings of the
conference in a few months, we don't need to have any review of reviews.
Rather, I would pick up some questions, which is crucial to better understanding of nuclear physics and give answers to these questions on the
basis of works presented to this conference. By doing this, I would try
to convey the audience of the present workshop a feeling of the exciting
and successful Few Body XI.

2. NEW FINDINGS, QUESTIONS AND ANSWERS

In this section, we mark "questions and answers" by capital letters.

[A] Which realistic two-nucleon potential is more realistic ?

At present, there are several two-nucleon potentials in the market:
Reid soft core (RSC)[1], Paris (PARIS)[2], super-soft-core (TRS)[3],
Urbana-Argonne (AV)[4], Bonn (BONN)[5] and still several other potentials.
A realistic two-nucleon potential has been derived theoretically, but adjusted to fit two-nucleon experiments, where the energy and momentum in
the two-nucleon system are conserved in the intitial and final states. We
call this situation that the two-nucleon system is on-energy-shell. The
way to answer the question [A] may be to study the behavior of the two-
nucleon system at off-energy-shell. For this purpose, we should take up
phenomena beyond the two-nucleon system. We can consider the following
two phenomena.
(1) To directly measure the p-p bremsstrahlung.
(2) To study three-nucleon systems.
In fact, for many years, one of motivations of studying three-nucleon
problems was to answer this question.

In Few Body XI, a new proton-proton bremsstrahlung experiment at
TRIUMF was reported by P. Kitching. An early experiment at TRIUMF (1980)
agrees with the soft pion approximation than with the prediction of potential models. Now, they used polarized protons and the analysing power
data becomes available for the first time. As reported by H.W. Fearing,
the calculations were done including a large number of corrections. The
results disagree with the soft photon approximation and agree with the

potential model calculations. So far, BONN and PARIS results are approximately correct off-shell as well as on-shell, yielding about the same results.

This is surprising, because PARIS and BONN yield a very different triton binding energy (B_3) and the D-state probability (P_D). The calculation of the triton quantities for BONN was done for the first time by the Sendai group. In Table 1, B_3 are listed for various two-nucleon potentials (2NP), without and with the Tuson-Melbourne (TM) three-nucleon force [6].

Table 1. Triton binding energy

Potential	2NP	2NP+TM
AV	7.68	8.42
PARIS	7.64	8.32
TRS	7.55	8.47
BONN	8.33	9.01
Exp.		8.48

At this moment, we can not say much about the question as to which group of potentials is better, unless we make calculations on the Coulomb energy difference, the (e,e') scattering, the pion absorption etc. However, we point out that the coupling constants of BONN are larger than other potentials. In Table 2, we compare the coupling constants $g_\alpha^2/4\pi$ of BONN and the Brown-Jackson potential (BJ)[7]. The larger coupling con-

	BONN	BJ
π	14.6	14.3
ρ	0.95	0.53
ω	20.0	4.77
δ	3.7064	
η	3.0	
σ	8.0568	6.0

Table 2. The coupling constants $g_\alpha^2/4\pi$ for the r-space version of BONN compared with those given by BJ.

stants makes, first of all, that the central potential due to the σ-exchannnge (representing two- and multi-boson exchange) larger. This is compensated by a large short range central repulsion due to the ω-exchange. Although in the two-body problem this sum of stronger potentials may be compatible with usual sum of weaker potentials, this balance should not be kept in the three-body system, too. As a result, BONN gets larger B_3.

We may conclude thus, (1) when the on-shell properties are almost the same, there is no much difference in off-shell properties in the two-nucleon sustem. (2) On the other hand, the property of a potential that affects most to B_3 is the behavior at two- and multi-pion exchange region.

[B] Dibaryon problem; Does $\pi N\Delta$ account for it ?

At Few Body X, the dibaryon problem was enthusiastically discussed both experimentally and theoretically, and yet no conclusion was led.

The strong energy dependent structure of the NN polarization $\Delta\sigma_L$ suggested the dibaryons in the $^3F_3$ and $^1D_2$ states. The phase shift analysis confirms their existence. Beside a thought that there should be no dibaryon resonance at all and the seeming resonance is due to the threshold effect, there have been two thoughts about the dibaryon:(1) It is a new kind of resonace due to exchanges of a quark. (2) It can be described in terms of $\pi N\Delta$.

In this conference, T.S.H. Lee and T. Ueda discussed this problem by a unitary model of πNN dynamics to treat $NN \to NN$, $NN \to \pi d$, and $NN \to \pi NN$ in a unified way. (The unitary coupled model has been worked also by some other people., e. g., I.R. Afnan and R.J. McLeod. Further references are seen in [8].) These authors reproduced the phase shift including $^1D_2$ and $^3F_3$ very well. However, since a simple πNN model within the framework of conventional meson-exchange model the strong structure of $\Delta\sigma_L$ can not be reproduced, Lee expressed the need for the dibaryon state consisting of six quarks. On the other hand, by taking account of πN and ρN channel coupling in the P_{33} state, the effect of backward going pion and some other effects, Ueda reproduced the energy dependence of $\Delta\sigma_L$ of pp-pp, and concluded that the dibaryon is reproduced by the πNN dynamics including Δ.

Also, it was pointed out by Blankleider that the lack of backward going pion in a similar models by other authors might be responsible to the undersestimate of the pp $\to$ πd cross section.

[C] Where the effect of three-body force is seen and to what extent ?

Before answering this question, we may look back the progress in the triton binding energy calculations after Few Body X.

[C1] Progress after Few Body X

[C11] Five channel calculations in 1983

In 1983, four groups performed five channel calculations of the Faddeev equation for RSC by completely different methods. These groups obtained the triton binding energy of from 7.02 to 7.04 MeV. In view of big computer programs, the difference in these values may not be taken seriously. Rather, this agreement shows that apart from values themselves we are now solving the three-body problem without a mistake.

These groups also calculated the first order perturbation correction due to the TM three-body force. The calculated binding energies are given in Table 4. These values are 1MeV less than the experimental value. As a result, a pessimistic feeling was prevailing in the effect of the three-nucleon force. At the Few Body X Conference in Karlsruhe 1983, people felt that the conservative approach using the potential model is

dying and we might have to recourse to the quark model to get a correct
nuclear force at short distances.

Table 4. Binding energy of triton calculated from
RSC5 and the first order perturbation energy for TM

	Purdue[9]	Bochum[10]	Los Alamos[11]	Sendai-Osaka[12]
RSC5	6.98	7.04	7.02	7.03
$- E_3$	$- 0.12$	0.16	0.41	0.47
B_3	6.86	7.20	7.43	7.50

[C12] Eighteen channel calculations in 1984

After the Karlsruhe conference, the Sendai gruop performed the 18-
channel calculations. The value obtained from calculations with a two-body
potential alone is about 1 MeV less than the experimental value of 8.48
MeV. However, if we calculate the first order perturbation energy for TM,
we obtain the values which are very close to the experimental value as
shown in Table 5.

Table 5. Binding energy of triton calculated for various
two-nucleon potentials. The first order perturbation value
for TM is added to the two-nucleon binding energy [12].

Number of channels	RSC	URG	PARIS
5	7.50	7.55	7.73
18	8.13	8.00	8.23

The NN form factor appeared in TM is assumed to be the monopole form
$[(\Lambda^2 - \mu^2)/(\Lambda^2 + q^2)]$, with cut-off mass of $\Lambda = 800$ MeV. This eighteen
channel calculation with TM as the first order pertturbation first cast
a hope of getting the triton binding energy [12].

[C13] 34-channel caculation in 1985

Stimulated by the result of [12], the Los Alamos-Iowa group has solved the 34-channel problem in 1985 and concluded that we get a convergent
calue at the 34-channel calculation [13]. Alsmost simultaneously, we solved the Faddeev equation with TM for 18- [14], 26- [15] and 34-channels at
the end of 1985 and found that if we take $\Lambda = 700$ MeV, we obtain the experimental triton binding energy, as shown in Table 1 [16].

Here, we have to mention that the Hannover group has made the 34-
channel calculation as early as in 1983 [17]. The three-nucleon force
that they have adoptedis due to the excitation of a nucleon to Δ. Unfortunately, the contribution from the three-nucleon force is only 0.32 MeV,
and the binidng energy of the triton is 7.7 MeV.

[C2] Scaling of physical quantities with B_3

We are thus able to calculate the triton physical quantities such as the D/S ratio of the asymptotic normalization constants [17], charge radius [18], momentum distribution [18], with various two- and three-nucleon potentials for various number of channels. Together with the Los Alamos-Iowa results on trinucleon charge radius and Coulomb energy E_c [19] and the doublet n–d scattering length a_2 [20], these results show that *the calculated values of a triton physical quantity are simply correlated to the calculated binding energy.* This property may be called the scaling of triton physical quantities with B_3. Making use of this scaling property and the triton binding energy, we can either predict the experimental value of this quantity, as in the case of the D/S-ratio of the asymptotic normalization constants, or can point out that something is missing in the calculation, as in the case of E_c, where this prediction gives 652 ± 2 keV against the experimental value of 764 keV. The simple relation between a_2 and B_3 has been known as the Phillips curve [21].

[D] What information the electron scattering experiments can give ?

[D1] The impressive experimental results of trinucleon e.m. form factors were published from Saclay in the last year [22]. The results provided an important test of our understanding of the electromagnetic property of trinucleon systems. Theoretically, all of the Hannover, Quebec, Sendai and Los Alamos groups performed basically similar calculations, taking account of (1) the three-body force contribution in the wave function, (2) the exchange currents [in the Sendai case, $\pi(ps)$-, ρ-, ω-pair currents, pionic $\pi\pi\gamma$, $\rho\pi\gamma$, $\omega\pi\gamma$, and $\pi N\Delta$ currents], and (3) the retardation current that is the difference of the recoil and renormalization currents appeared at the third order (M_N^{-3}) Foldy-Wouthysen transformation. Although the given results are basically consistent with each other, the agreement with experimental results looks slightly better in the Hannover and Quebec results than in the Sendai and Los Alamos results. This might be due to the pv-nature of their coupling.

Although a more careful calculation is yet to be made taking account of the consistency between nuclear forces and the exchange current and the continuity equation, it seems that there is no explicit need for accommodating the quark degrees of freedom, as P.U. Sauer has concluded.

[D2] Electron inclusive scattering

The concept of the y-scaling was introduced by West [23]: The nuclear structure function, which is simply related to the inclusive cross section by a system of non-interacting particle becomes, at a large momentum transfer, only function of the nucleon momentum in the direction of the momentum transfer. This nucleon momentum is denoted by y: $y = (\vec{k}.\vec{q})/q = M(\omega - q^2/2M)/q$, where $\vec{q}$ is the momentum transfer, $\vec{k}$ the nucleon momentum, $\omega = (M^2 + (\vec{k} + \vec{q})^2)^{1/2}$, the energy of outgoing nucleon. C. Ciofi degli Atti has made an extensive review. The following results are worthy of mentioning.

(1) The theory of y-scaling has been based on PWIA. However, an explicit introduction of the final state interactions in the theory is a prerequisite for extracting a reliable information on nucleon dynamics from y-scaling.

(2) Inclusion of the three-nucleon potential in the wave function brings the calculated result closer to the experimental data, as shown by the Sendai resuts.

(3) From the Faddeev calculation, the need for 6q- and 9q-components in the triton was once suggested [24],[25]. This suggestion was made on the basis of the Brandenburg-Kim-Tubis five channel wave function. The 34-channel Sendai wave function suggests no need for quarks.

(4) The Sendai calculation shows that if we increase the cut-off mass of πNN form factor, the agreements between data and the theoretical result becomes better. This means that we need higher momentum components in the triton wave function. However, the increase of the cut-off mass of the πNN form factor makes the triton overbound, as shown in Table 6 [18].

Table 6. Examples showing a strong Λ-dependence of the triton binding energy

[AV+TM(Λ)]26		[RSC+TM(Λ=1000)]		[AV+TM(Λ=1000)]	
	B_3	Number of channels	B_3	Number of channels	B_3
600	7.84				
700	8.33	5	8.87	5	10.27
800	9.16	18	12.16	18	13.86

Therefore, the strategy to mix higher momentum components into the triton is to take account not noly of TM(which is a ππ-exchange three-nucleon force), but also ρπ- and ρρ-exchange three-nucleon force. This is because the ρ-meson is short range and inclusion of it makes increase higher momentum components on the one hand, and the ρπ-exchange three-nucleon force is a little bit repulsive [26], thus rescuing from over-bound on the other.

[D3] EMC effect

The EMC effect has been thought as a realization of the quark-gluon picture in nuclei. This effect was calculated for ^{3}He by using the Sendai 18-channel triton wave function for RSC with TM, but without a quark [27]. The results were compared with the ^{4}He experiment. The calculated results are very close to experimental results. This suggests that the EMC effect is reproduced without recourse to a quark. We urge that the electron experiments for [3]He be done.

[E] How photoreactions are useful for understanding of a few-nucleon systems ?

Together with the electron scattering, we can test the theoretical wave function of a three-nucleon system by photoreactions, because the electromagnetic operators are in principle known, and in some favorable cases, the selection rule makes the transition mechanism clear.

[E1] Low energy

Here, I would mention an interesting work by the Basel group and J. Torre [28], reviewed by D. Lehman.

The D-state probability P_D of the triton ranges from 8.5% to 10% depending on the two-nucleon potential, except for BONN, which gives rather small value of 6.8%. In Table 7, we list P_D for various potentials obtained from 34-channel calculations, with an exception of the 18-channel calculations for RSC [18].

Table 7. The D-state probability of triton.

Potential	2NP	2NP+TM(700)	2NP+TM(800)
AV	8.96	9.23	9.61
PARIS	8.50	8.67	8.99
TRS	8.60	8.81	9.17
BONN	6.81	6.79	
RSC	9.42	9.86	10.29

By the Basel group, it has been noted that the tensor analyzing power A_{yy} of the capture reaction with polarized deuteron beams $H(\vec{d},\gamma)^3He$ shows a strong enhancement of the [3]He D-state effects. At $E_d \simeq 30 - 45$ MeV, this process is strongly dominated by E1. Therefore, the exchange currents need not be calculated. The main part of A_{yy} is due to transitions into D-state components with p-waves both for the interacting pair and the spectator. The experiment at $E_d = 29.9$ MeV yielded $A_{yy} = 0.0282$ for $\theta_{cm} = 96^o$, while the calculation done by Torre with RSC gives $A_{yy} = 0.0339$ for the same angle and energy. This suggests that P_D for RSC is too big. We should calculate A_{yy} for all other potentials.

[E2] Intermediate energy

G. Tamas pointed out an important effect of the three-body absorption in the breakup reaction of $^3He(\gamma,pp)n$. In this reaction the two-body channel (of a quasi-bound pp pair and n) is strongly suppressed for the following reasons.

(1) A pp pair has no dipole moment. As a result, the photon is absorbed in higher multipolarities, and it makes the correlated amplitude decrease. (2) No charged pion can be exchanged, and a $N\Delta$ system in a S-state can not decay into a ^1S_o pp pair.

Therefore, the three-body process takes place in the following manner. First, a photon is absorbed by a neutron and emits a pion. Subsequently, this pion is absorbed by a proton, followed by emission of a pion from it. This pion is absorbed by another proton. This mechanism is a "three-body-current", which is a pair current followed by a three-body force. Since the pair current is the largest among exchange currents, and the three-body force is not negligibly small as we saw in [C], this process is very plausible. A careful experiment and analysis of this process might help getting information about the three-body force. By the way, we should mention that an evidence for a direct three-nucleon pion absorption was obtained at SIN [29].

[F] What was achieved in the nd scattering and what is its problem ?

As stated in [A], the motivation of the nd scattering experiment has been to provide information on the nucleon-nucleon interaction. For this purpose, the pd scattering is of course more favorable. However, since we have no hopeful prospect of having a relyable theory on the pd scattering in a foreseeable future, a great effort was made to have a good polarized neutron beam by several groups up to 50MeV. In parallel, theoretical development has taken a long time even for the nd scattering, beginning from a use of the simple Yamaguchi-type s-wave separable interaction [30], phenomenological separable interactions for higher partial waves [31], a simple local potential for s-wave [32], and recent extention of it to higher partial waves [33].

Recently, Y. Koike and J. Haidenbauer have succeeded in yielding relyable calculations using a realistic two-nucleon interaction like PARIS. They calculated the elastic cross section and the analyzing power A_y up to 20 MeV. For the two-body interaction, partial waves up to the f wave were taken into account.

For the elastic cross section, new Karlsruhe results were in excellent agreement with precise data from UC Davis. In the energy range of 20 to 30 MeV, the old discrepancy that the calculated results were too low than the experimental results in the backward angles has been removed. Theoretically, the importance of the p wave interaction was emphasized. The p wave interaction increases the cross section both in the forward and backward region, while it makes minimum lower more than 15%.

Most accurate experiments were done by TUNL, UC Davis and Karlsruhe up to 50 MeV. The importance of the spin triplet p-wave interaction for A_y has been known for many years [31]. At present moment, Koike's refined calculation reproduced the main feature, but suffers a convergence problem: Increasing number of separable terms makes the agreement with experimental results worse. He conjectured that one or some of following reasons may be responsible to this difficulty; (1) The triplet p-state is not well-represented by PARIS. (2) A_y is quite sensitive not only to the on-shell- but also off-shell-behavior of the p-wave interaction. (3) Since the NN p-state can couple to the NΔ state, the effect of three-nucleon force might be important for the NN triplet p-wave. In this case, we should include up to very higher partial waves as seen in the bound state calculations [13][16] [18].

At Few Body XI, no breakup calculations was reported.

[G] Some advance made in the system of A > 3

[G1] Four nucleon scattering problem

The importance of the p-wave (3+1) amplitude has been pointed out by J. Tjon [34]. However, in his calculation, a full convergence of his separable expansion for the amplitude could not be obtained. Recently, A. Fonseca utilized a convergent energy dependent pole approximation, and made a very important advance in the field of four-nucleon scattering problem. He calculated the elastic scattering of a proton from ^{3}He at E_p = 9.51 MeV, and obtained a result for the differential cross section that enjoyes a good agreement with the experimental results within 10%. This should be compared with the difference of 50% when the p-wave is omitted. His calculated phase shifts of the elastic scattering of a neutron from ^{3}H are in good agreement with the resonating group method calculated by Tang and collaborators, in which the virtual breakup of the deuteron cluster is taken into account.

A calculation of the breakup reaction p +^{3}He $\rightarrow$ p + p + d was performed by Mdlalose et al. The main feature of the experimental results are reproduced without any artificial normalization.

[G2] Few-body reactions

The Osaka group proposed for the first time a practicable yet divergent-free connected kernel theory of nuclear reactions, named the Multi-Three-Cluster Coupling (MTCC) Model and the Multi-Two-and Three-Cluster Coupling (MTCC) Model. For instance, in treating the d + α scattering, the three cluster partition (n,p,α) is coupled to the Faddeev systems of the three-cluster partitions (n,d,^{3}He) and (p,d,^{3}H). The calculated results fairly reproduce the experimental results.

[G3] Faddeev calculation for ^{12}C cluster model

The Noda group (Science University of Tokyo) performed the Faddeev calculation with the Pauli correct alpha-alpha potential for three alpha model of ^{12}C. Low-lying levels were reproduced in correct order, whereby the three-body and Coulomb forces give very important contribution.

[H] There are many other topics which made a great progress after Few Body X. These include the study of two-nucleon force from the quark and skyrmion models, the study of relativistic effects, possible existence of the hyper dibaryon state, the coupled cluster systems and few-body experiments. I regret that no space is left for reviewing these subjects.

3. CONCLUDING REMARKS

In concluding this talk, I would remark a few subjects that I think very impressive.

(1) Where we need quarks ?

From articles discussed in Few Body XI, we came to know that many phenomena can be described by a potential model, without a quark. These phenomena include the triton binding energy, and accordingly, other triton physical quantities, the EMC effect, the y-scaling and the electromagnetic form factors. Then, is there any quantity that needs a quark ? Sure there are: We don't know how to calculate the πNN form factor. Only plausible way should be a quak model. Also, we need a quark interpretation in treating $p\bar{p} \to K\bar{K}$ as A. Green reviewd. However, it seems that for some years to come, we can work with in more safe way of employing a potential model and yet can draw physically meaningful conclusions as we did for the scaling of triton physical quantities this time.

(2) Although the calculation of triton binding energy with the Bonn potential casted a controversy about the necessity of a three-body force, the analysis of the response functions and the y-scaling clearly shows the necessity of a three-nucleon force. Together with interesting experiments of three-body γ- and π-absorption by Saclay and Basel groups, the study of three-body forces including ρ- and ω-exchange should be done urgently.

(3) We need very much to have a continuum wave function of three-nucleon systems not only for the nd and pd scatterings, but for the analysis of γ-absorption, electron scattering, and meson scattering. The use of PWIA should now be out of date.

(4) A new phenomenology of handling scattering as well as bound states, taking both advantage of the cluster structure and the Faddeev calculation

was for the first time presented in this conference. I would expect a fruitful future of this approach.

References

1. Reid, R.V.: Ann. Phys. (N.Y.) 50, 411 (1968)

2. Lacombe, M. et al.: Phys. Rev. C21, 861 (1980)

3. de Tourreil, R. et al.: Nucl. Phys. C31, 2266 (1985)

4. Wiringa, R.B.et al.: Phys. Rev. C29, 1207 (1984)

5. Machleidt, R.: Lecture note at the Los Alamos Workshop, June 1985, TRI-PP-85-68.

6. Coon, S.A., et al.: Nucl. Phys. A317, 242 (1979); Coon,S.A., Glöckle, W : Phys. Rev. C23, 1970 (1981)

7. Brown, G.E., Jackson, A.D.: The Nucleon-Nucleon Interaction, Amsterdam: North Holland. 1979

8. Afnan, I.R., McLeod, R. J.: Phys. Rev. C31, 1821 (1985)

9. Muslim, Kim, Y. E., Ueda, T.: Phys. Lett. 115B, 273 (1982); Nucl. Phys. A393, 399 (1983)

10. Bömelburg, A.: Phys. Rev. C28, 2149(1983)

11. Wiringa, R.B., Friar, J.L., Gibson, B.F., Payne, G.L., Chen, C.R.: Phys. Lett. 143B, 273 (1984)

12. Ishikawa, S., Sasakawa, T., Sawada, T., Ueda, T.: Phys. Rev. Lett. 53 1877 (1984)

13. Chen, C.R., Payne, G.L., Friar, J.L., Gibson, B.F.: Phys. Rev. Lett. 55, 374 (1985); Phys. Rev. C33, 1740 (1986)

14. Sasakawa, T.: Proceedings of the Xth European Symposium on the Dynamics of Rew-Body Systems, Budapest 1986

15. Sasakawa, T., Ishikawa, S.: (Aoki, Y. and Yagi, K. ed.) Deuteron Involving Reactions and Polarization Phenomena,Singapore: World Scientific. 1986

16. Sasakawa, T.,Ishikawa, S.: Few-Body Systems, 1, 3 (1986)

17. Ishikawa, S., Sasakawa, T.: Phys. Rev. Lett. 56, 317 (1986)

18. Ishikawa, S., Sasakawa, T.: Few-Body Systems, 1, 143 (1986)

19. Friar, J.L., Gibson, B.F., Chen, C.R., Payne, G.L.: Phys. Lett. 161B, 241 (1985)

20. Chen, C.R., Payne, G.L., Friar, J.L., Gibson, B.F.: Phys. Rev. C33, 401 (1986)

21. Phillips, A.C.: Rep. Prog. Phys. 40, 905 (1977)

22. Juster, J.-P. et al.: Phys. Rev. Lett. 55, 2261 (1985)

23. West, G.W.: Phys. Rep. 18, 263 (1975)

24. Pirner, H.J., Vary, J.P.: Phys. Rev. Lett. 46, 1376 (1981)

25. Sick, I.: Nucl. Phys. A396, 455c (1983)

26. Coon, S.A., Pena, M.T., Ellis, B.G.: Phys. Rev. C30, 1366 (1984)

27. Uchiyama, T., Saito, K.: Private communication.

28. Jourdan, J. et al.: Phys. Lett. 162B, 269 (1985)

29. Backenstoss, G. et al.: Phys. Rev. Lett. $\underline{55}$, 2782 (1985)

30. Aaron, R., Amado, R.D.: Phys. Rev. $\underline{150}$, 857 (1966)

31. Dolleschall, Nucl. Phys. $\underline{A201}$, 264 (1973)

32. Kloet, W.M., Tjon, J.A.: Ann. Phys. (N.Y.) $\underline{79}$, 407 (1973)

33. Takemiya, T.: (Sasakawa, T. et al. ed.) Proceedings of Few Body XI, Amsterdam: North Holland. 1987

34. Tjon, A.J., Phys. Lett. $\underline{63B}$, 391 (1976)

THE CEBAF PROJECT AND ITS FEW-BODY RESEARCH PROGRAM

Jean Mougey
Continuous Electron Beam Accelerator Facility[*1]
12070 Jefferson Avenue
Newport News, VA 23606 USA

Abstract

The 4 GeV Continuous Electron Beam Accelerator Facility (CEBAF) planned to be built in Newport News, Virginia will allow detailed coincidence studies of electron and photon induced reactions on few body systems, including polarization experiments. The facility and some examples among the planned experimental program are briefly discussed.

I. Introduction

For many years, the electromagnetic probe has been used as a precise microscope to obtain quantitative information on the structure of nucleon and nuclei. Electro- and photonuclear reactions have put strong constraints on the self consistent mean field picture of nuclei. They also gave the cleanest evidences for the role of meson currents, nucleon resonance excitation, relativistic effects and the need for introducing explicitly the internal quark structure of hadrons. It is the goal of new generation high energy and high duty factor electron accelerators like CEBAF to extend theses studies to higher energy and momentum transfer regimes in a variety of specific channels. With photon wave lengths much smaller than 0.3 fm, essential information is expected on the structure of nucleons and their resonances in nuclear matter, and on the nature of multinucleon subsystems and the strong nuclear force.

A $\leq$ 4 nucleon systems provide the best suited "nuclear laboratory" for such studies. Covering the full range of nucleon binding energies and nuclear densities they exhibit all the above mentioned effects without having the full complexity of heavier nuclei. This allows to perform completely exclusive experiments without resorting to extremely high resolution, difficult

---

1   On leave of absence from DPhN/HE, CEN-Saclay, 91191 Gif-sur-Yvette

to achieve at high energy. Consequently, a major part of the CEBAf physics program implies d, [3]He, and [4]He targets. A few examples will be discussed in this talk, after a short description of the CEBAF experimental facilities.

II. The CEBAF Experimental Facilities

The configuration of the CEBAF accelerator is sketched in Fig. 1, and its main characteristics are given in Table 1. It consists of two 0.5 GeV superconducting linacs through which the $200\mu A$ electron beam is circulated 4 times. Microbunch separation using 2495 MHz RF deflection allows to deliver three simultaneous beams at different but correlated energies, and possibly two or more order of magnitude different intensities to the three planned end stations. The installation of a GaAs polarized electron source is being studied. A detailed description of the CEBAF project can be found in Ref. 1.

Table 1
Main CEBAF Design Parameters

Beam Characteristics		
	Energy	0.5-4 GeV
	Average current	$200\mu A$
	Emittance	$2 \cdot 10^{-9}$ πm
	Energy spread ($\delta p/p$)	$< 10^{-4}$
	Duty factor	100%
Linac Parameters		
	Type	Superconducting CW recirculating
	Frequency	1500 MHz
	Cavity design Gradient	5 MeV/m
	Cavity design residual Q	$3 \cdot 10^{9}$
	Number of cavities (klystrons)	418 (418)

Given the high duty factor and the high quality of the electron beam, preliminary designs for the experimental equipment have been performed with emphasis on coincidence experiments, and the possibility to perform high missing mass resolution measurements as well as kinematically complete multiparticle emission experiments. Consequently, two rooms (Hall A and C) will be equipped with several detection arms, mainly spectrometers, around the same pivot, while the third one (Hall B) will house a nearly 4π detector, so-called Large Acceptance Spectrometer (LAS). The characteristics

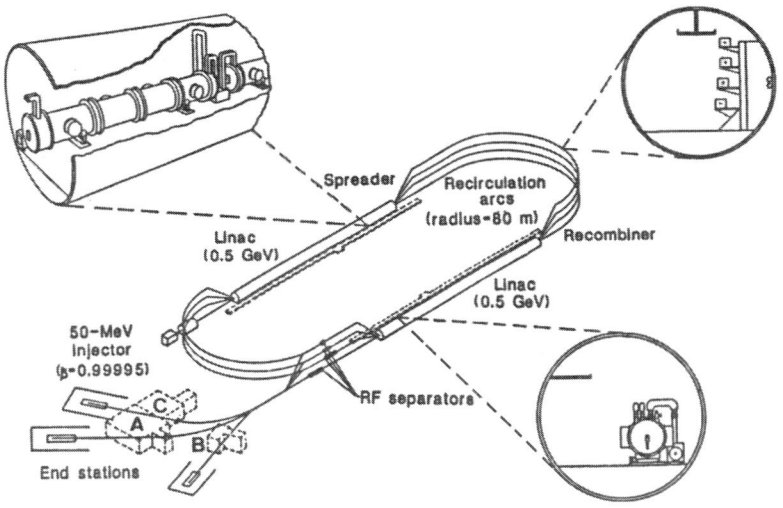

Fig. 1   Sketch of CEBAF accelerator configuration

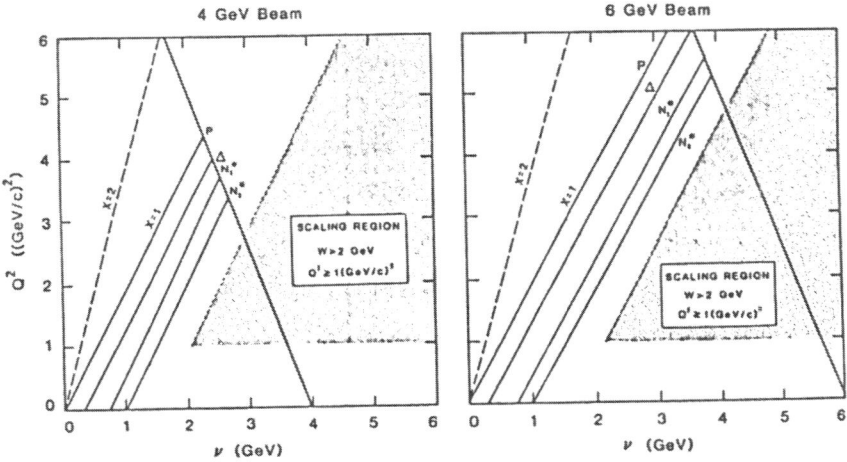

Fig. 2   Accessible regions of the (Q^2,ν) plane. The left part of each figure corresponds to $\sigma_{Mott} > 10^{-2}\ \mu b.sr^{-1}$.

of the detector arrangements follow from the requirements of a set of typical experiments.

The kinematical domain accessible with 4 GeV and 6 GeV incident electrons is shown in Fig. 2 as a function of the squared four momentum transfer $Q^2 = 4 \ EE' \sin^2 \theta/2$ and the energy transfer $\nu = E-E' = (W^2 - M_T^2 + Q^2)/2M_T$, where M_T and W are the target mass and the total invariant mass of the final state. Fig. 3 shows the conditions needed to perform transverse/longitudinal separations by varying the virtual photon polarization parameter $\epsilon = (1. + 2 \ \vec{q}^2/Q^2 \ \tan^2\theta/2)^{-1}$. One notices that the possibility to reach very forward angles, 10° or below, is essential, for both the electron and the hadron arms, the latter being often in the direction of $\vec{q}$ at high momentum transfer.

One of the most stringent requirement is the missing mass resolution needed i) when final bound nuclear states have to be isolated or ii) when signal/noise (true/accidental) ratio has to be optimized in fully exclusive experiments dealing with very low cross sections. Hall A is designed for such experiments -- mainly (e,e'p) and (e,e'K) reactions -- while Hall C will be better suited for moderate (~ 10 MeV) resolution experiments.

The main parameters for the preliminary spectrometer designs are listed in Table 2. The layout of Hall A electron spectrometer is shown in Fig. 4. Like Hall A hadron and Hall C electron spectrometer, it is planned to make use of a few modular magnetic elements (one type of homogeneous field, rectangular, superconducting dipole, and two types of superconducting $\cos2\theta$ quadrupoles, with higher order correcting coils). The configuration chosen for Hall A, with one horizontal and one vertical spectrometer, allow extended target operations while optimizing acceptances and costs. Initial plan for Hall C (Fig. 5) includes two non-focusing hadron spectrometers (each of them consisting in a single dipole). The possibility to use non-magnetic devices in direct view of the target, like arrays of scintillation and lead glass counters allowing to detect neutrals, is being investigated. The Large Acceptance Spectrometer in Hall B (Fig. 6) has been designed for photonuclear and low luminosity ($\lesssim 10^{-33} \ cm^{-2}s^{-1}$) electronuclear studies. Fully instrumented, it will allow multiparticle detection and identification within ~ 80% of 4π (15° minimum forward angle) and 0.1 to 3 GeV/c in momentum.

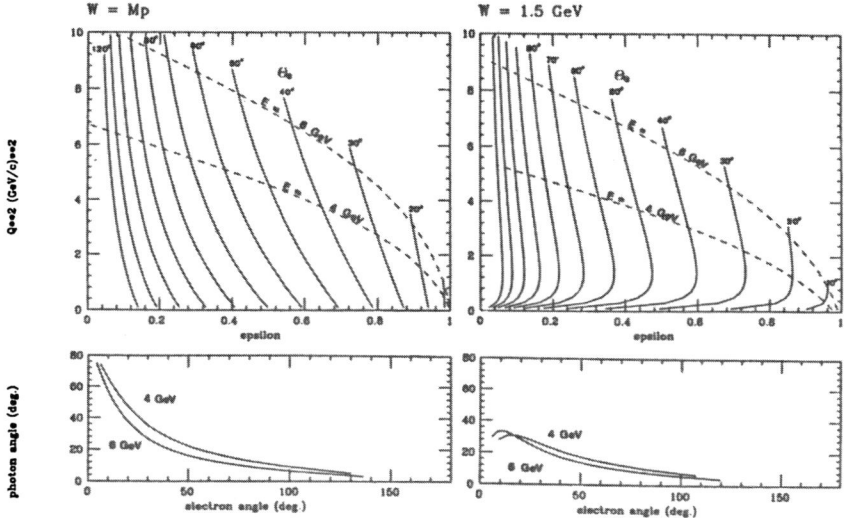

Fig. 3   Kinematical constraints for longitudinal/transverse separations in elastic and inelastic ep scattering.

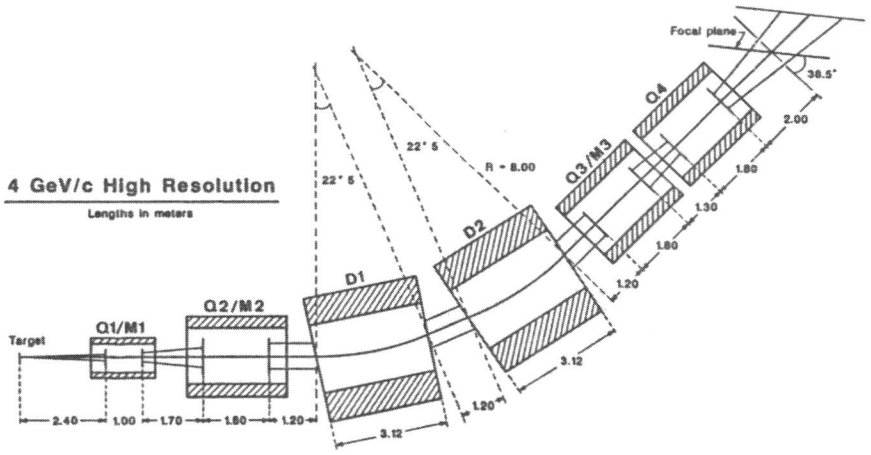

4 GeV/c High Resolution

Lengths in meters

Fig. 4   Sketch of the CEBAF High Resolution Electron Spectrometer (Hall A).

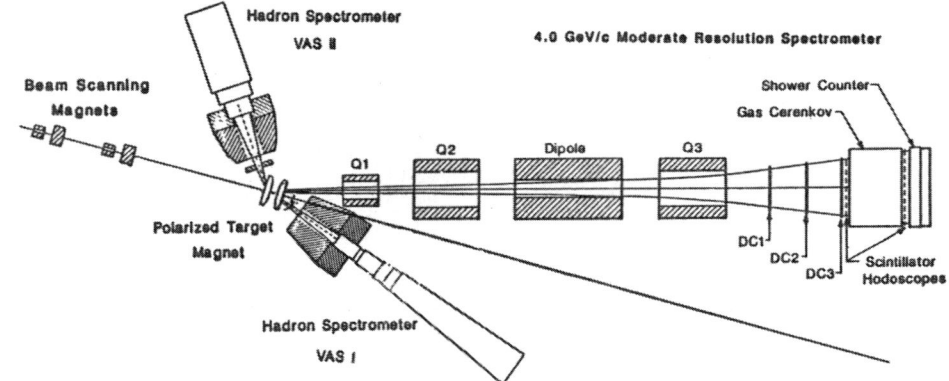

Fig. 5 Spectrometer arrangement in Hall C.

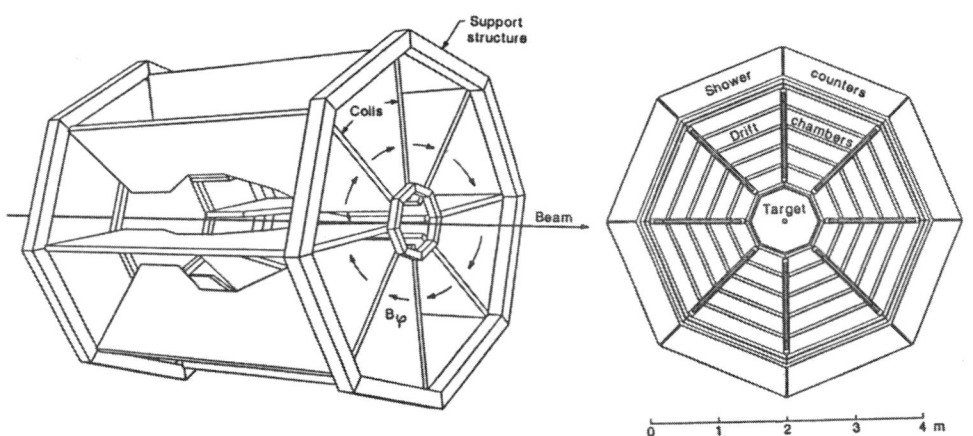

Fig. 6 Sketch of the Large Acceptance Toroidal Spectrometer

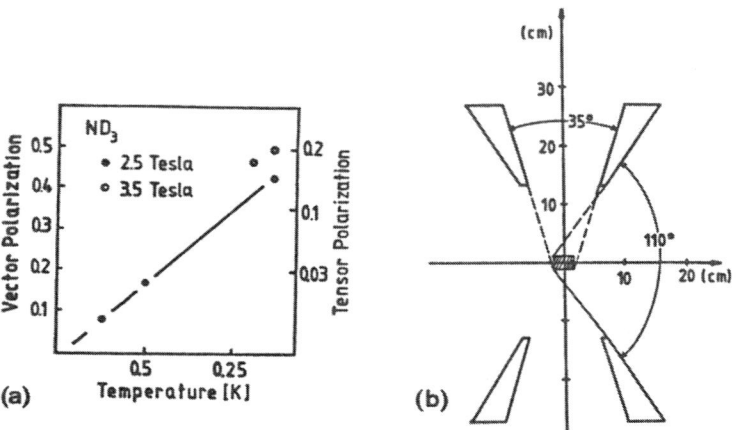

Fig. 7 Solid state polarized ND$_3$ Targets: a) deuteron polarization[2] b) coil geometry for CEBAF beams.

Table 2

Preliminary Designs Parameters for CEBAF Spectrometers

	Type	Maximum Momentum (GeV/c)	Momentum Acceptance (%)	Momentum Resolution	Solid Angle (msr)	Angular Range
Hall A Electron	QQDDQQ horizontal	4. (6)	10	$< 10^{-4}$	11.	$10° \rightarrow 130°$
Hall A Hadron Solution I	QQDD vertical	3.	15	$< 10^{-4}$	11.	$10° \rightarrow 160°$
Hall A Hadron Solution II	QQDMD vertical	1.2	15	$\sim 3 \cdot 10^{-5}$	35.	$20° \rightarrow 150°$
Hall C electron	QQDQ	5.5	30	$\sim 3 \cdot 10^{-4}$	6.6	$10° \rightarrow 160°$
Hall C Hadron I	Dipole (VAS I)	3.5	> 50	$\sim 3 \cdot 10^{-3}$	50.	$16° \rightarrow 150°$
Hall C Hadron II	Dipole (VAS II)	1.7	> 50	$\sim 3 \cdot 10^{-3}$	70.	$25° \rightarrow 150°$
Hall B	Large Acceptance Spectrometer			Toroidal field, max. strength 8 superconducting coils Solid Angle Momentum Resolution		1T 80% of 4π $\sim 10^{-2}$

Based on experiences at Bonn[2] and SLAC[3], the use of polarized solid state targets in all three CEBAF end stations is being discussed. By using NH_3 and ND_3 as polarized target materials, one partially circumvents the low resistivity of most chemical compounds to radiation damage. Fig. 7a shows deuteron vector and tensor polarizations achieved at Bonn in various temperature and magnetic field conditions. A $\sim 80°K$, $\sim 10^{17}$ e/cm^2 preirradiation has to be performed first. Careful study of the geometry of the superconducting coils producing high field is required to achieve minimum dead zone (Fig. 7b). The very good CEBAF beam emittance allows to use small target volumes. With reasonable extrapolation of existing technology, beam current up to 5nA on 10^{23} deuterons/cm^2 ND_3 targets with 0.3 tensor polarization can be achieved, leading to a figure of merit (L x P^2) of $3 \cdot 10^{32}$, even higher than currently expected internal target performances[4].

III. Possible Programs on Few-Body Systems at CEBAF

We describe briefly a few examples of experiments on few nucleon
systems which have been discussed in the CEBAF Workshops and Summer
Study Groups. A more exhaustive description of the physics possibilities can
be found in the corresponding proceedings.

1) (e,e'p) and (e,e'n) reactions on light nuclei

Single nucleon emission experiments, especially on light systems, will
constitute an important part of the CEBAF program. In the general case
of both polarized beam and target, the one-photon-exchange (Fig. 8) cross
section for the (e,e'N) process provides nine invariant structure functions
f_{ij}[5], four of which appear only when the target nucleus is polarized. In
principle, a measurement of these structure functions over a wide kinematical
domain would give complete and basic information on the nuclear current
operator. In practice, some are small, and need out-of-plane experiments to
be extracted.

The simplest case is the d(e,e'p)n process. Data for the full cross
section are available up to $|\vec{n}| \sim$ 350 MeV/c[6] in the quasielastic regime,
and up to 500 MeV/c in a kinematical region where mesonic currents and
isobar configurations give important contributions[7]. Much richer
information could be obtained from the separated structure functions, as they
exhibit different sensitivities to final state interactions and non-nucleonic
degrees of freedom[8]. The longitudinal one, almost not affected by meson
currents and Δ's, is the best suited for studying high momentum components
and short range effects. Its separation should be achievable at CEBAF up
to $|\vec{n}| \sim$ 500 MeV/c and $Q^2 \sim 1$ $(GeV/c)^2$ with better than 10%
accuracy[9]. Few coincidence data are available at higher (np) relative
energies[10], and more systematic and accurate ones would be highly
desirable.

For three-nucleon systems, in which the one-body spectral function can
be computed, the two isospin components T = 0 and T = 1 of the residual
pair can be isolated by measuring the (e,e'p) cross section in both ^{3}He and
^{3}H, or the (e,e'p) and (e,e'n) cross sections on a single isotope. Extending
the existing measurements[11,12] to higher recoil momenta where D-state
contributions dominate may give unique information on 3-body forces,
provided the measurements are performed at large momentum transfer[13].

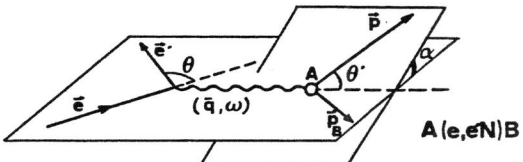

Fig. 8   Kinematics of the A(e,e´N)B process

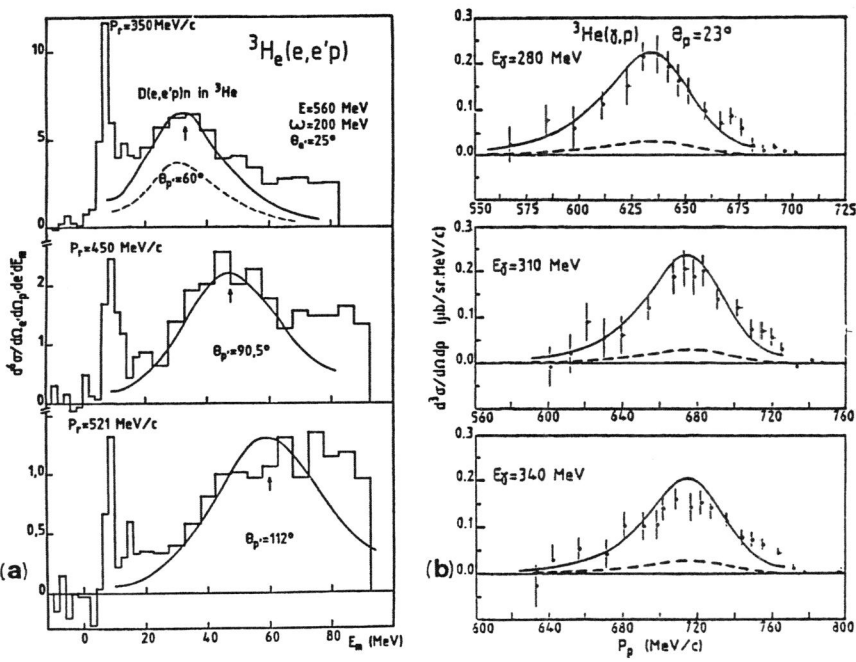

Fig. 9   ³He(e,e´p)np results (a) and ³He(γ,p)np results (b) from Saclay[16,17] compared to Laget's calculation including (full curves) or not (dashed curves) pion exchange and FSI.

The (e,e´p) reaction can also be used to investigate how the elementary eN interaction is modified when the nucleon is embedded in the nuclear medium. Although single arm quasielastic scattering data[14] and recent (e,e´p) experiments[15] may indicate some modification in the nucleon electromagnetic properties in nuclei, this point is still controversial, and more detailed studies are clearly needed. Counting rate estimates for the ^{4}He(e,e´p)^{3}H reaction under appropriate kinematics show that momentum transfers up to $Q^2 = 2(GeV/c)^2$ can possibly be achieved.

2) (e,e´2N) reactions

The cleanest evidence for two-nucleon correlations is, at present, the spectrum of protons emitted in the continuum of the reactions ^{3}He (e,e´p)X[16] and ^{3}He(γ,p)X[17] recently measured at Saclay (Fig. 9). However, separate determination of the longitudinal cross section, allowing to eliminate MEC and Δ contributions, is still lacking. The most promising way to study correlations is the (e,e´2p) reaction[18]. Selection rules strongly suppress two-body currents in the transverse cross section, as the 2p system has no dipole moment to couple with the photon. Possible 3-body currents may then show up in the ^{3}He(e,e´2p)n reaction. The longitudinal part is best suited for a detailed study of the two-body correlations. Counting rate estimates under CEBAF conditions have shown the feasibility of the experiment although a longitudinal/transverse separation would be very difficult to achieve. Moreover, the use of three spectrometers would lead to important constraints on the kinematically choices. Use of non magnetic, large solid angle devices as well as a broad survey of two-nucleon emission processes using the LAS may be an appropriate way of starting this program.

3) Nucleon and deuteron form factors

Precise knowledge of $G_E^n(q^2)$ is of fundamental importance for testing microscopic models of the nucleon, as well as for quantitative analysis of most high Q^2 electron scattering data. Attempts to measure it by Rosenbluth separation in quasielastic e + d scattering, or extract it from elastic e + d scattering, gives inaccurate or uncertain answers beyond $Q^2 \sim 0.5$ $(GeV/c)^2$ (Fig. 10a). Measuring the recoil neutron polarization in quasifree d($\vec{e}$,e´n)p reactions with polarized electrons[19] is hampered by poor and badly known efficiency of neutron polarimeters. Quasielastic scattering of polarized electrons on neutrons with spin oriented in the scattering plane, perpendicular to the direction of $\vec{q}$ (from a vector polarized deuteron target) is being considered for CEBAF. The polarized cross section writes as

$$d\sigma/d\Omega = d\sigma/d\Omega \mid_{unpol} [1 + p_e p_n A^n(Q^2)] \text{ with } A^n(Q^2) \propto 2G_E^n G_M^n$$

p_e and p_n being the electron and neutron polarization. With realistic figures on achievable polarizations and luminosities, measurements of G_E^n for Q^2 up to ~ 1.5 $(GeV/c)^2$ seem feasible, with $\delta A = \pm 0.02$ accuracy (Fig. 10b).

The separation of charge G_C (Q^2) and quadrupole G_Q (Q^2) deuteron form factors requires also polarization experiments. These form factors are very sensitive to short range effects and possible 6-quark contributions in the deuteron wave function (Fig. 11a). It has been proposed[19] to measure the recoil deuteron tensor polarization t_{20} in unpolarized ed scattering, and such experiments are underway[20]. An alternative method being investigated for CEBAF is to use a tensor polarized deuteron (ND_3) target. In combination with $A(Q^2)$ and $B(Q^2)$, the measurement of ratios $R_i = \sigma_i(pol.)/\sigma_0(unpol.)$ for various target orientations enable separation of all 3 form factors. At CEBAF energies, measurements of $R_{\parallel}$ (target spin aligned parallel to the virtual photon) with an accuracy of $\delta R = \pm 0.1$ for Q^2 up to ~ 2$(GeV/c)^2$ appear feasible (Fig. 11b).

4) $(e,e'p\pi^-)$ and $(\gamma,p\pi^-)$ reactions in light nuclei

When the total center of mass energy $W_{N\pi}$ of the final $N\pi$ pair is around the Δ mass, these reactions provide the most direct way of studying Δ production and propagation in nuclear medium, and are expected to play a significant role in the CEBAF physics program. To my knowledge, the only experimental work attempted until now was the study of reactions $^2H(\gamma,p\pi^-)p$ and $^2H(\gamma,pp)\pi^-$ performed at Saclay[22]. Although limited in accuracy by the low (~ 1%) beam duty factor, the data (Fig. 12) show significant deviations from a pure quasifree process[23]. Using virtual photons, both Q^2 and the outgoing Δ kinematics can be varied. Feasibility studies under CEBAF Hall C spectrometer conditions[24] show (Table 3) coincidence rates of ~ 10^3/hr for the reaction $^4He(e,e'p\pi^-)x$ with 20μA beam on 0.2gm/cm^2 target. A possible program could include

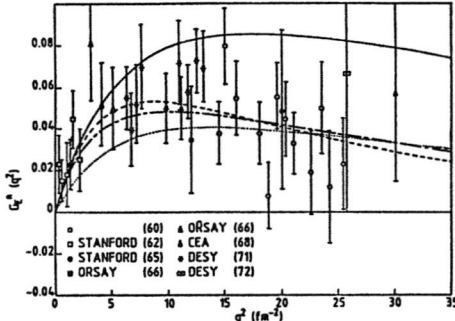 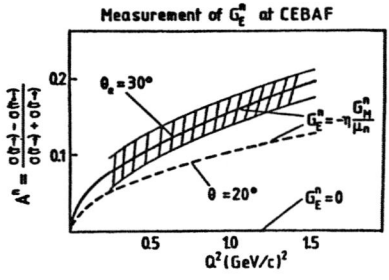

Fig. 10   Neutron electric form factor G_{En} (a) present status (b) achievable accuracy in CEBAF asymmetry measurement.

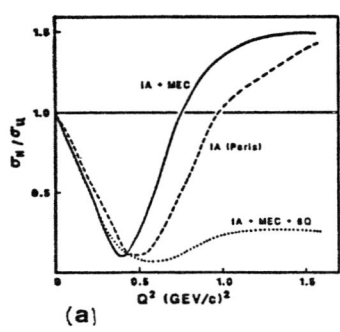 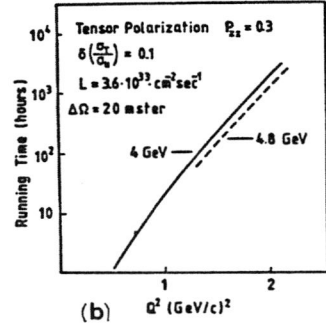

Fig. 11   Ratio $R_{||} = \sigma_{pol}/\sigma_{unpol}$ for elastic ed scattering: (a) theoretical predictions, including 6-quark effect[21] (b) expected running time for 10% accuracy on R.

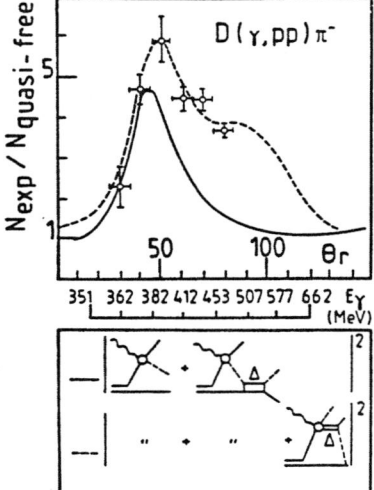

Fig. 12

Ratio of measured yield for the $\gamma d \rightarrow pp\pi^-$ reaction[22] to the one-nucleon quasifree contribution. Solid curve includes πN rescattering, dashed curve includes also 2π production mechanisms.

longitudinal/transverse separation at the Λ peak at fixed $Q^2 = 0.3 \ (\text{GeV}/c)^2$, Λ form factor in the range $Q^2 = 0.3 - 1.$ at fixed $\epsilon \sim 0.95$ $(\theta_e = 15°)$, Λ angular distributions for fixed initial nucleon momentum $\vec{p}_i$ as well as Λ-production at high $\vec{p}_i$. Similar studies can be attempted on higher resonances, although multipion emission, competing strongly above the Λ, could make the LAS in Hall B a better suited detection device. The search for resonances which decouple from the πN-channel (and therefore cannot be detected in inelastic πN scattering) would be an important test for QCD inspired models[25].

Table 3
$^4\text{He}(e,e'p\pi^-)x$ coincidence rates at $Q^2 = 0.3 \ (\text{GeV}/c)^2$
and $W_{p\pi} = 1.23$ GeV (from Ref. 24)

E (GeV)	θ_e	Γ $(\text{GeV}^{-1}\text{sr}^{-1})$	$N(p\pi)$ (hr^{-1})
2	18°	4	0.8×10^3
2.5	14°	6	1.3×10^3
3	12°	10	2×10^3
3.5	10°	14	3×10^3

5) Parity experiments

Among the weak interaction experiments which may be relevant to CEBAF, the measurement of parity-violating asymmetries in elastic and inelastic ep and ed scattering are the most challenging ones. They provide basic tests of the standard model of electroweak interactions, through the predicted relationship between hadronic neutral and charge changing weak currents and electromagnetic currents through a unique quantity, $\sin^2\theta_W$. Experiments on the nucleon[26] would consist of measuring the asymmetry $A = (\sigma_L - \sigma_R)/(\sigma_L + \sigma_R)$ for longitudinally polarized electrons. Like the previous SLAC-Yale experiment[27], both elastic ep $\rightarrow$ ep and inelastic ep $\rightarrow$ eΛ^+ reactions, which have large cross sections in the range $0.1 \leq Q^2 \leq 1 \ (\text{GeV}/c)^2$ relevant to CEBAF, will determine different combinations of the coupling constants of the standard model. In this domain $A \sim 10^{-5} - 10^{-4}$ should be measured with an accuracy of $\delta A \sim 10^{-7}$ to 10^{-5} for a substantial improvement over previous results. Other relationships between the isoscalar currents appearing in hadronic neutral weak currents and the isoscalar part of the electromagnetic current can be tested using ed $\rightarrow$ ed and ed $\rightarrow$ e(np, ^{1}S$_o$) reactions as $\Delta T = 0$ and $\Delta T = 1$

isospin filters[28]. Although extremely difficult and time consuming, these experiments constitute an exciting challenge for CEBAF.

References

1. CEBAF Design Report, May 1986
   C.W. Leemann, 1986 Linear Accelerator Conference, Stanford, June 2-6, 1986.

2. G. Baum, et. al., Phys. Rev. Lett. 45, 2000 (1980).

3. W. Meyer, Nucl. Phys. A446, 381c (1985) and references therein.

4. M.C. Green, ANL Aladdin design, private communication.

5. S. Boffi et. al., Nucl. Phys. A435, 697 (1985) and invited talk to the 1986 CEBAF Summer Workshop, CEBAF, June 23-27, 1986.

6. M. Bernheim et. al., Nucl. Phys. A365, 349 (1981).

7. S. Turck-Chieze et. al., Phys. Lett. 142B, 145 (1984).

8. H. Arenhovel, Nucl. Phys. A384, 287 (1982).
   A. Buchmann et. al., Nucl. Phys. A443, 726 (1985).

9. J. Morgenstern, RPAC Report, CEBAF, January 1986.

10. W. Mehnert, Thesis, Bonn IR-84-28 (1984).

11. E. Jans et. al., Phys. Rev. 49, 974 (1982).
    J. M. Laget, Phys. Lett. 151B, 325 (1985).

12. C. Marchand, Proc. of the Intl. Symposium on Three-Body Forces in the Three-Nucleon System, Ed. by B.L. Berman and B.F. Gibson, Springer-Heidelberg, 1986, pg. 338.

13. I. Sick, ibid, pg. 42

14. Z.E. Meziani et. al., Phys. Rev. Lett. 54, 1233 (1985).

15. G. van der Steenhoven et. al., Phys. Rev. Lett. 57, 182 (1986).
J. Morgenstern, Nucl. Phys. A446, 315c (1985).

16. J. Morgenstern, 2nd Workshop on Perspectives in Nucl. Phys. at Int.
Energies, World Scientific Ed., 1986, p. 355 and C. Marchand, private
communication.

17. N. d'Hose et. al., 8eme Session d'Etude de Physique Nucleaire,
Aussois, February 1985.

18. J.M. Laget, Nucl. Phys. A446, 489c (1985), and submitted to Phys.
Rev. Lett.

19. R. Arnold et. al., Phys. Rev. C23, 363 (1981).

20. M.E. Schulze et. al., Phys. Rev. Lett. 52. 597 (1984).

21. A.P. Kobyshkin, Sov. J. Nucl. Phys. 28, 252 (1978).

22. P. Argan et. al., Phys. Rev. Lett. 41, 86 (1978).

23. J.M. Laget, Phys. Rep. 69, 1 (1981).

24. P. Stoler, RPAC report, CEBAF, January 1986, p. 5-90 and CEBAF
Baryon-Info Note #10, September 1986.

25. N. Isgur, Proc. 1984 CEBAF Summer Workshop, June 1984.

26. P. Souder et. al. in Future Directions in Electromagnetic Physics, P.
Stoler ed., 1981, p. 385 and references therein.

27. C.Y. Prescott et. al., Phys. Lett. B77, 347 (1978); Phys. Lett. B84,
524 (1979).

28. W-Y. Pauchy Hwang and T.W. Donnelly, Phys. Rev. C33, 1381
(1986).

MONOCHROMATIC AND POLARIZED GAMMA RAY BEAMS
FOR THE STUDY OF FEW BODY INTERACTIONS

LADON Collaboration[*], presented by
C. Schaerf

Dipartimento di Fisica, Università di Roma, and INFN, Sezione di Roma

ABSTRACT. - We present the results of the LADON monochromatic and polarized photon beam at Frascati. We discuss the future prospects of Ladon beams at other electron storage rings and we suggest some possible lines of research.

PAST RESULTS AND FUTURE PROSPECTS

Monochromatic and polarized gamma ray beams produced by the back-scattering of Laser light from the high energy electrons circulating in a storage ring have proved to be a valuable tool for the study of photonuclear reactions[1] in the energy region from 5 to 80 MeV. In 1987 a tagging spectrometer should be added to the Ladon beam in Frascati[2] and the LEGS[3] facility should enter into operation, thus extending the gamma ray energy to more then 300 MeV and the available intensity to 10^7 photons/sec.

The use of the 6 GeV storage ring of the European Synchrotron Radiation Facility, will allow the exploration of a new energy domain, up to 1500 MeV, with an intensity of 10^7 photons/sec, and an energy resolution of 10-14 MeV. Table I present a summary of present and future Ladon beams. Moreover a proper optical insertion in the machines magnetic lattice should yield a further improvement in the gamma ray energy resolution by about a factor of three.

(*)D. Babusci, P. Belli, R. Bernabei, L. Casano, S. d'Angelo, M.P. de Pascale, G. Giordano, B. Girolami, A. Incicchitti, G. Matone, M. Mattioli, P. Picozza, M. Preger, D. Prosperi, C. Schaerf and B. Spataro.

Table I

Location	Frascati		Brookhaven		Grenoble	
Storage ring	Adone		NSLS		ESFR	
Project	Ladon	Taladon	LEGS-Labro			
Technique	collimat.	tagging	tagging		tagging	
El. En. E(GeV)	1.5	1.5	2.8		5	6
Laser	Argon		Argon	Nd-Yag x 4	Argon	
λ (nm)	514.5		351.1	266.0	514.5	351.1
k_1 (eV)	2.45		3.59	4.74	2.45	3.59
k_{max} (MeV)	78		370	465	780	1500
Δk (MeV) (FWHM)	9	4	6.4	8	10	14
Intensity (s^{-1})	$2 \cdot 10^5$	10^6	10^7		10^7	

More important is the consideration that the beam should be almost continously available since we are now convinced that it is possible to work without interfering with the normal operation of the machine:

1) the tagging technique, which is now in preparation for the Frascati and Brookhaven facilities, allows the simultaneous use of a broad spectrum of photons, thus overcoming the need to change the energy of the electrons;

2) the technology developed at LEGS[3] to position the Laser beam over the circulating electrons, should be tested and operational by 1988;

3) a circulating current of $1.3 \cdot 10^{12}$ electrons (100 mA) and a beam life time of 10 hours corresponds to a beam loss of $3.6 \cdot 10^7$ electrons/sec, larger of what we can produce with our maximum usefull intensity of 10^7 photons/sec.

At energies above a few tens of MeV, the tagging technique provides a better gamma ray energy resolution and therefore we will limit these considerations to such a case referring to our previous works[4] for a discussion of collimated beams.

KINEMATICS

The kinematics of the reaction is indicated in fig. 1. A low energy, Laser, photon impinges head-on with a high energy electron. In the collision the electron transfers a fraction of its energy to the photon. For backward scattering the energy of the outgoing gamma ray is given by:

$$k_{max} = E\frac{z}{1+z} \tag{1}$$

where:

$$z = \frac{4Ek_1}{m^2} \tag{2}$$

and where:

E is the incident electron energy,
E' is the scattered electron energy,
k_1 is the incident photon energy and
m is the electron mass.

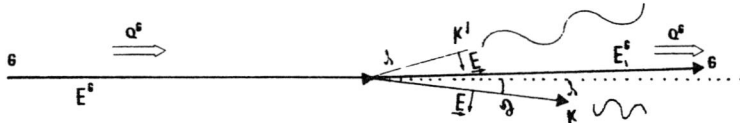

Fig. 1 - Kinematics for Compton scattering on relativistic electrons. The most critical variable is the angle θ between the directions of the incoming electron and the scattered gamma ray.

Neglecting the energy of the incoming photon (a few eV), the energy of the final gamma ray is give by:

$$k = E - E' \tag{3}$$

and its variance[2]:

$$\sigma_k = E\frac{\sigma_{xT}}{d} \tag{4}$$

where:

σ_{xT} is the total radial spread of the electron beam at the position of the tagging detector;
d is the dispersion of the storage ring magnetic lattice (A_{13} or A_{16} of the Transport matrix) from the interaction region to the tagging detector.

EXPERIMENTAL APPARATUS

The typical lay-out for a tagged Ladon apparatus is composed (Fig. 2) of a straight section where the Laser beam is superimposed to the electrons equilibrium orbit and in which the photon-electron interaction takes place and a tagging region. In this last region are located a series of microstrip solid state silicon detectors (μSSSD) which measure the position of the scattered electrons away from the main orbit. The relative position of the interaction region and the tagging region must be selected primarily in such a way as to minimize the result of eq. 4 and therefore optimize the energy resolution of the gamma ray beam.

The gamma ray beam is emitted tangentially to the interaction region and therefore the photonuclear targets and the experimental apparatus should be located along that line in the direction of the circulating electrons.

With the Grenoble machine and the design presently available, ESRP-27/3[5], the best solution appears the one indicated by D/2-W[2]: with the Laser-electron interaction region located in a short straight section (a dispersive one, designated by D) and the tagging spectrometer in the following, wiggler type, straight section, (designated by W).

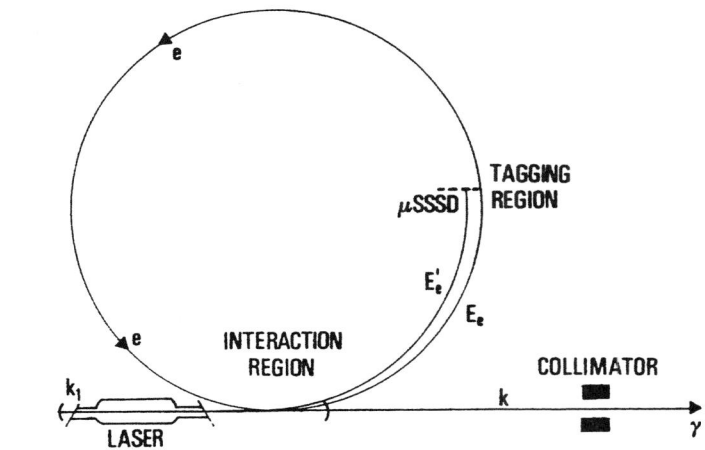

Fig. 2 - Schematic representantion of the experimental apparatus on the storage ring. The electrons which lost energy in the interaction region drift away from the equilibrium orbit till they reach a microstrip silicon solid state detector in a selected position of the storage ring.

Therefore the optical bench for the alignment and monitoring of the Laser beam should be located tangentially to the short straight section, in the same direction where is located the gamma ray beam collimator, the monitoring apparatus and the experimental area.

GAMMA RAY POLARIZATION

For high energy electrons the helicity is a good quantum number. The electron spin does not flip during relativistic Compton scattering. For backward scattering there is also no transfer of orbital angular momentum. For this reason the photons scattered in the very forward direction to not change their angular momentum, and therefore their polarization. This is partly true also for photons scattered at slightly different angles and which have an energy lower than k_{max}. Since the polarization of the gamma rays is the same as that of the Laser beam, we can easely produce gamma ray beams of any polarization changing the polarization of the Laser light. In particular we can produce linearly polarized and circularly polarized gamma ray beams with almost full polarization.

Linear polarization

The usefulness of the linear polarization is widely recognized. Defining the plane in which the electric vector oscillates produces strong azimuthal anysotropies with a typical $\cos(2\varphi)$ distribution in the differential cross section. The simmetry plane of this distribution is different for transitions of different multipolarities thus introducing a new discriminating observable.

Circular polarization

Circularly polarized gamma rays have not been extensively used in the study of photonuclear reactions, and not only for their limited availability. Circularly polarized photons are in a spin eigenstate. Each direction of polarization corresponds to an orientation of the photon spin parallel or antiparallel to its direction of motion ($s_z = \pm 1$). If no other spin variables are observed in the experiment, the simplest observable which can be constructed with the spin of the photon (an axial vector) and the kinematical variables of the reaction is:

$$M = \vec{s} \cdot \vec{p}_1 \times \vec{p}_2 \qquad (5)$$

and this because photonuclear reactions do conserve parity and therefore the simplest way to construct a scalar with $\vec{s}$ is to multiply it by an other axial obtained by the vector product of two indipendent momenta. The photon spin s being in the same direction of the photon momentum $\vec{k}$ we can write more clearly:

$$M= \vec{k} \cdot \vec{p_1} \times \vec{p_2} \tag{6}$$

Any asymmetry in the differential cross section obtained by the inversion of the gamma ray polarization (spin) should be proportional to M.

Let us know consider the value of M in some nuclear processes:

a) single particle processes:

a photon strikes a particle bound in a nucleus with momentum p_i and ejects it into the continuum with momentum p_f:

$$k + p_i = p_f$$

in this case k, p_i and p_f are coplanar and M is equal to zero.

b) two particles processes with two body final state:

the photon strikes a target at rest and two particles come out of the reaction:

$$k + m = p_1 + p_2$$

Also in this case k, p_1 and p_2 are coplanar and therefore M is zero.

If M is zero no asymmetry should be observable if we invert the photon spin. It appears reasonable that the same rule should apply also to the case in which the photon interacts only with a part of the target nucleus and some spectator nucleons are emitted in a direction outside the reaction plane.

c) three body forces and three body final states:

if the photon strikes a system at rest and three or more particles are emitted, then they are not necessarily coplanar:

$$k + m = p_1 + p_2 + \ldots \ldots p_n$$

in this case M can be different from zero for some selected values of the kinematical variables and reversing the gamma rays circular polarization an asymmetry can be observed.

Concluding this chapter we would like to stress that a measurement of the cross section asymmetry

$$A = \frac{\sigma(s_z = +1) - \sigma(s_z = -1)}{\sigma(s_z = +1) + \sigma(s_z = -1)} \tag{7}$$

obtained by reversing the photon circular polarization could present a way to selectively observe the contribution of the three body forces.

A GRASER: IS IT POSSIBLE?

The LADON technique has proved itself valuable to produce monochromatic and polarized gamma rays starting from a Laser beam. Laser light is also coherent. Why our gamma rays are not coherent? The photons scatter on an incoherent bunch of electrons and in this process loose their phase. To conserve the coherence of the Laser light in the scattering process we must introduce some coherence in the electron beam. As an example suppose that the electron bunch is divided in thin slices of electrons all placed at the same distance from one another as indicated in fig. 3. Each electron slice constitute a plane of free electrons and acts therefore as a semitransparent mirror for the Laser light. If these mirrors are separated by a distance which is an integer multiple of half wavelenght (D= n λ/2) of the light, then the amplitudes scattered by different planes add coherently and produce a coherent beam of scattered gamma rays. We have a Graser.

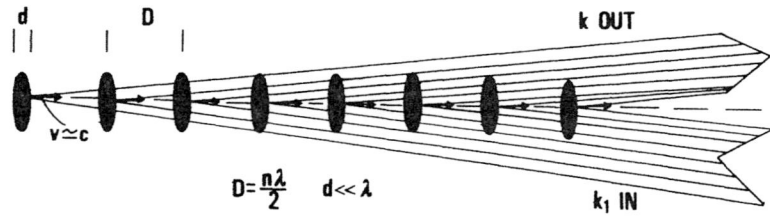

Fig. 3 - Principle of production of a coherent back scattered LADON beam. Photons scattered from evenly spaced electrons bunches add coherently to produce a coherent gamma ray beam. The photon wavelength λ is measured in the electron rest frame (ERF).

A more rigorous three dimensional analysis [6] yelds for the incoherent and coherent scattered intensities:

$$I_{inc} = N\, I_0 \qquad\qquad (8)$$

$$I_{coh} = N\,(N-1)\,|\,G(\vec{q})\,|^2\, I_0 \qquad\qquad (9)$$

where:

$$\vec{q} = \vec{k} - \vec{k}' \tag{10}$$

is the three-momentum transfered from the photon to the electron calculated in the Electron Rest Frame (ERF) and

$$G(\vec{q}) = \int_V d\vec{r}\, e^{i\vec{q}\cdot\vec{r}} \rho(\vec{r}) \tag{11}$$

is the form factor of the electron bunch and $\rho(r)$ its spatial density distribution in the interaction region calculated in the ERF.

$$N \approx 10^{12}$$

is the number of electrons in a bunch.

Therefore the ratio of the coherent to the incoherent intensity is given with good approximation by:

$$\frac{I_{coh}}{I_{inc}} = N \mid G(\vec{q}) \mid^2 \tag{12}$$

Since N is a very large number it is sufficient a small amount of spatial structure in the electron beam to produce a gamma ray beam of useful coherence. Unfortunately the spatial coherence required for the electron bunch is at distances of the order of the photons wave lenght in the ERF. An Angstrom (10^{-10}m) at typical Ladon energies.

As indicated in ref. (6) this tecnique suggests a way to produce coherent x-ray beams at energies of the order of tens of keV using a Free Electron Laser to modulate the electron bunch. For gamma ray energies we must still wait more advanced studies.

REFERENCES

1) M.P. De Pascale, G. Giordano, G. Matone, D. Babusci, R. Bernabei, O.M. Bilaniuk, L. Casano, S. d'Angelo, M. Mattioli, P. Picozza, D. Prosperi, C. Schaerf, S. Frullani and B. Girolami, Phys. Rev. C 32, 1830 (1985).

   R. Bernabei, A. Incicchitti, M. Mattioli, P. Picozza, D. Prosperi, L. Casano, S. d'Angelo, M.P. De Pascale, C. Schaerf, G. Giordano, G. Matone, S. Frullani and B. Girolami, Phys. Rev. Lett. 57, 1542 (1986).

2) M. Preger, B. Spataro, R. Bernabei, M.P. De Pascale and C. Schaerf, Nucl. Instr. & Meth. A 249, 299 (1986).

3) A.M. Sandorfi, M.J. Levine, C.E. Thorn, G. Giordano, G. Matone and C. Schaerf, IEEE Transsactions on Nuclear Sciences, NS-30, n° 4, 30B3 (1983).

4) L. Federici, G. Giordano, G. Matone, G. Pasquariello, P.G. Picozza, R. Caloi, L. Casano, M.P. De Pascale, M. Mattioli, E. Poldi, C. Schaerf, M. Vanni, P. Pelfer, D. Prosperi, S. Frullani, and B. Girolami, Nuovo Cimento 59B, 247 (1980).

   L. Federici, G. Giordano, G. Matone, G. Pasquariello, P.G. Picozza, R. Caloi, L. Casano, M.P. De Pascale, M. Mattioli, E. Poldi, C. Schaerf, P. Pelfer, D. Prosperi, S. Frullani and B. Girolami Lett. Nuovo Cimento 27, 339 (1980).

5) B. Buras and S. Tazzari. Report of the ESRP, second edition, c/o CERN, LEP Division (1985).

6) G. Matone and A. Luccio. Coherent Backscattering in the Soft X-Ray Region. LEGS BNL 38460, June 1986.

THE TRIUMF KAON FACTORY PROPOSAL

D.A. Hutcheon
TRIUMF
4004 Wesbrook Mall
Vancouver, Canada V6T 2A3

Abstract

TRIUMF has recently proposed a kaon factory, consisting of a group of accelerators to produce a 100 µA, 30 GeV proton beam. The science program, the system of accelerators, the experimental area, and the current status of the proposal are described.

Introduction

To provide background to the TRIUMF kaon factory proposal, we begin with a reminder of the properties of the present accelerator and the types of experiment currently performed. TRIUMF is an isochronous synchrotron producing proton beams of 200 to 500 MeV. The protons are accelerated as H^- ions and extracted following passage through a thin stripping foil; it is thus possible to have two beams extracted simultaneously, with the energy and intensity of each beam independently variable (Fig. 1). About two-thirds of the beam time is operation with 140 µA unpolarized protons extracted for production of secondary beams of pions and muons. Typical experiments done with these beams are: elastic scattering of low-energy π^+ and π^- by nuclei, search for neutrinoless $\mu \to e$ conversion, π^+ elastic scattering from a tensor-polarized deuterium target, search for right-handed coupling in the decay of polarized muons, and chemistry/solid state physics studies by muon spin rotation. An applied program includes pion cancer therapy studies and radio-isotope production. The other one-third of the beam time a polarized proton beam of intensity up to 0.75 µA is used (either directly or as a means of producing polarized neutrons) for studies of nuclear structure or of few-body systems. These experiments include: measurement of charge symmetry breaking in np elastic scattering, radiative

596

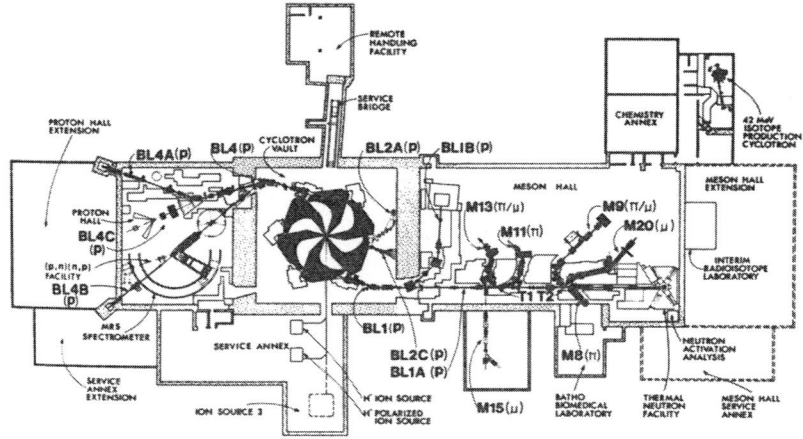

Fig. 1. Layout of the TRIUMF cyclotron and beam lines

capture of polarized neutrons or protons, spin transfer in elastic or inelastic p-nucleus scattering, and studies of (p,n) and (n,p) reactions [1].

With this background of a strong program in particle decays and symmetry laws, it is natural that TRIUMF has sought to extend the "high intensity" frontier from π and μ production to include also kaons, neutrinos and antiprotons. This has resulted in TRIUMF putting forward to the Canadian government a proposal for a 30 GeV, 100 μA proton accelerator [2].

Scientific Program of a Kaon Factory

In what follows, no attempt was made to isolate "few-body" aspects of the program. However, my interests in the current few-body experiments at TRIUMF no doubt has affected the emphasis placed on the various possible experiments. We highlight those experiments which would benefit most by the 100-fold increase of intensity over what is currently available at the Brookhaven AGS.

Rare decay modes of kaons

These are a test of the Standard Model or of extensions to it. For example, $K_L^0 \to \mu e$ or $K^+ \to \pi^+ \mu e$ would reveal mixing of lepton generations, while $K_L^0 \to \mu^+ \mu^-$ and $K \to \pi e^+ e^-$ are sensitive to heavy-quark couplings and we could explore interaction mechanisms in high-statistics measurements. Observation of $K^+ \to \pi^+ \nu \bar{\nu}$ would supply constraints on t-quark mass and mixing. Figure 2 shows the current limit or value, those for proposed experiments at existing labs, and those projected for a kaon factory, for the branching ratios in various kaon decay modes.

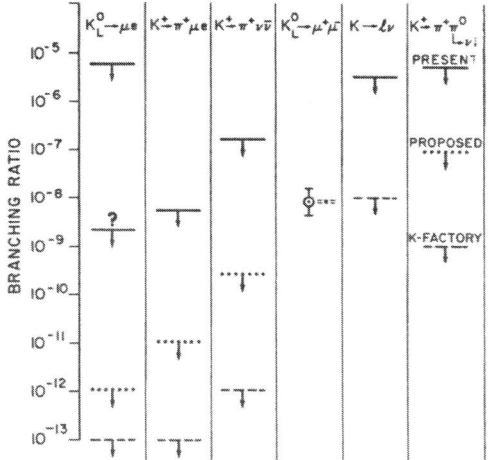

Fig. 2. Branching ratios of some rare kaon decays. Limits achievable with present accelerators are shown by a dotted line, and with a kaon factory by dashed lines.

CP-violation in K decay

This has remained one of the outstanding puzzles in physics of the past two decades. The CP-violating $K_L^0 \to 2\pi$ decays occur with a branching ratio of about 10^{-3} that of the CP-allowed $K_L^0 \to 3\pi$ decay. They have been explained either by "milliweak" models with $\Delta S=1$ and direct CP violation in $K_2^0 \to 2\pi$, or by "superweak" mixing brought about by a new interaction with $\Delta S=2$. One possibility of distinguishing between superweak and milliweak models lies in precise comparison of the branching ratios of $K_L^0 \to \pi^+\pi^-$ and $K_L^0 \to \pi^0\pi^0$. Present data are not sufficiently accurate to discriminate. A proposed arrangement using long, twin beams of K^0's, with a regenerator alternating between them, is shown in Fig. 3.

Neutrino physics

A direct measure of mixing of $\mu-$ and e-neutrinos is obtained by looking for ν_e-initiated events when the detector is placed in a beam initially consisting of ν_μ only. Contaminant ν_e in the beam can be suppressed by "narrowband beam" operation, but at the expense of intensity. Since potential event rates are not high, the high intensity of a kaon factory is essential in looking for neutrino oscillations.

Neutrino-electron elastic scattering ($\nu_\mu e$ vs $\overline{\nu}_\mu e$) affords a very clean measurement of $\sin^2\theta_w$. Kaon factory intensities would provide far better statistics than have been obtained with present facilities.

Elastic scattering of electron neutrinos and electrons could provide unique information on charged current and neutral current amplitudes, and their interference.

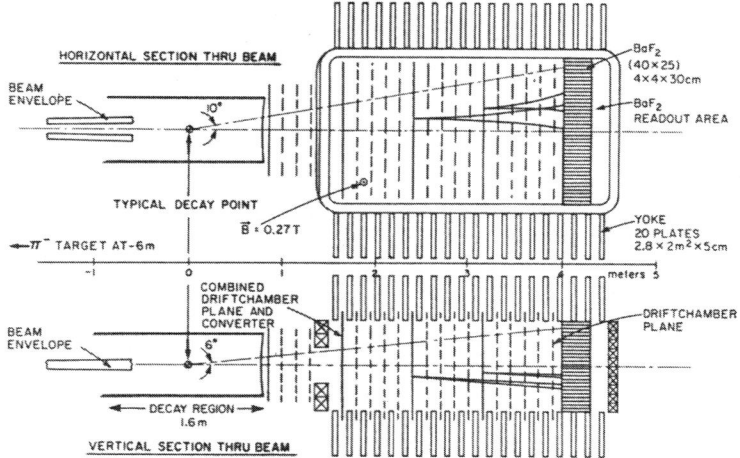

Fig. 3. Spectrometer for a proposed experiment to study CP-violation via $K^0 \to 2\pi$.

Hypernuclear physics

Injection of a strange particle into the nucleus may shed light on the question of quark deconfinement in nuclei. The Λ viewed as a particle is not prevented by Pauli exclusion from seeking the lowest energy level in a nucleus. However, if there is deconfinement, one would expect that the u and d quarks of the Λ would be Pauli-blocked by the u and d quarks of the neutrons and protons. The difference between confined and partially deconfined quarks should appear in hypernuclear binding energies, decay rates, and Ml "Λ spin flip" transition probabilities.

The intensities at a kaon factory would make it feasible to study S=-2 systems, for example through the reaction (K^-, K^+). A postulated $\Lambda\Lambda$ bound system, the "H" dibaryon, might be sought via $^3He(K^-, K^+)Hn$.

Baryon and meson spectroscopy

The identification of calculated light-quark configurations with observed states remains incomplete. With many, overlapping states this is not an easy task. It is proposed that a polarized hydrogen target be used to provide additional constraints to identify missing states of the S=-1 K^-p system (Y^* resonances from 1520 to 2430 MeV). Polarization data in $K^-p \to \bar{K}^0 n$ and $K^-p \to \Sigma^-\pi^+$ would be important constraints on phase-shift analyses.

Antiprotons

Antiprotons of 3 to 7 GeV/c incident on a fixed p target could search for charmonium ($c\bar{c}$) states other than ones readily reached by e^+e^- colliders (1^- directly and $0^+, 1^+, 2^+$ indirectly).

Nuclear structure with K⁺

The K^+ beams are of interest in measuring matter density in nuclei. The long mean free path in nuclear matter of K^+ compared with other hadronic probes (Fig. 4) and lack of resonances in the K^+N cross section make it an ideal probe of the nuclear interior. Present data on ^{12}C and ^{40}Ca elastic scattering do not have sufficiently small error bars for a definitive measurement of neutron distributions.

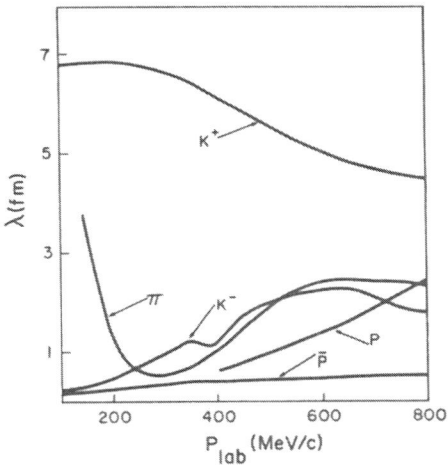

Fig. 4. Mean free path of various hadronic probes in nuclear matter.

Polarized protons

Dramatic spin effects have been seen in the scattering of a polarized proton beam from a polarized proton target at 11.75 GeV/c (Fig. 5). The largest effects occur at large values of $p_\perp$ where cross sections are low.

Clearly, not all of the possibilities cited would be realized in the early experimental program, nor will the list of experiments remain unchanged. Rather, it illustrates the type of experiments most likely constitute the scientific program of the kaon factory.

The System of Accelerators

There are three accelerators in the system: (1) the present TRIUMF cyclotron as injector (0.44 GeV), (2) a Booster synchrotron of 3 GeV, and (3) the main Driver synchrotron of 30 GeV. After each accelerator is a storage ring to provide the appropriate time structure in the beam: (1) the 0.44 GeV Accumulator, (2) the 3 GeV Collector, and (3) the 30 GeV Extender. The five rings would be housed in two tunnels, as shown in Fig. 6.

600

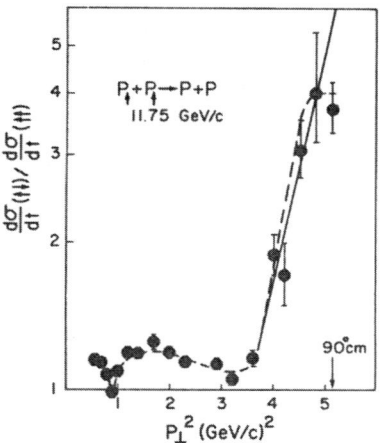

Fig. 5. Dependence of the proton-proton elastic scattering cross section upon relative spin orientations of a polarized beam and a polarized target.

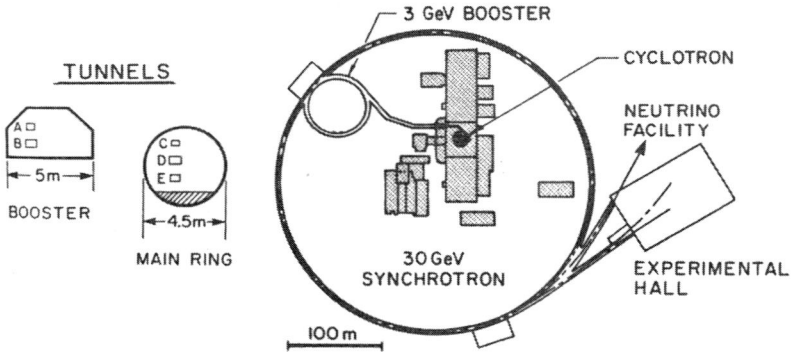

Fig. 6. Proposed layout of the TRIUMF kaon factory accelerators, with cross sections through the tunnels.

The synchrotrons are both fast cycling, with attendant need for high rf voltages. To ease the rf requirements, the magnet cycle is asymmetric, with a rise time three times longer than the fall time. Most of the rf frequency swing occurs in the Booster, largely decoupling the problems of providing high rf voltages and large frequency change. The Booster also allows smaller apertures to be used in the main rings.

In detail, the acceleration cycle is as shown in Fig. 7. The continuous train of pulses from the cyclotron (23 MHz) is extracted as H⁻ ions and stacked in the Accumulator using a stripper. After 20 ms the accumulated beam is dumped into the Booster which, cycling at 50 Hz, accelerates the beam to 3 GeV. The beam is transferred to the Collector, which collects

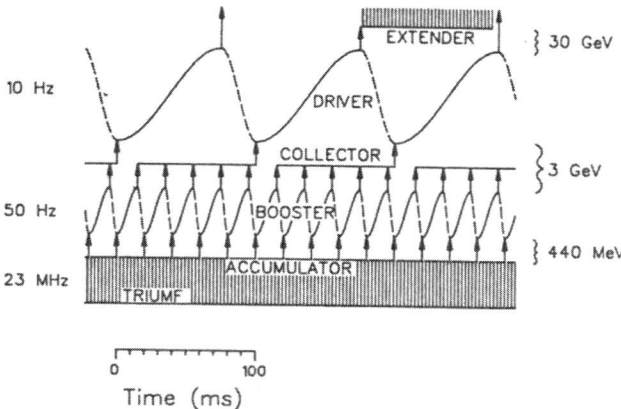

Fig. 7. Time structure of the beam at various stages in acceleration and extraction.

5 Booster pulses and manipulates the longitudinal emittance of the beam. It is then passed to the main 10 Hz Driver synchrotron where it is accelerated to 30 GeV. From here the beam either is extracted in a single turn or transferred to the Extender from which it may be slowly extracted using a one-third integer resonance. The rings in each tunnel have identical lattices and tunes, providing straight-forward matching during beam transfer. Parameters of the Booster and Driver are summarized in Table 1.

Beam losses would be minimized by bucket-to-bucket transfer from one stage to the next. Five out of each 45 buckets would be left empty to allow for turn-on of kicker magnets. It is estimated that ~10% of the beam

Table 1. Synchrotron design parameters

	Booster	Driver
Energy	3 GeV	30 GeV
Radius	$4.5 R_T = 34.11$ m	$22.5 R_T = 170.55$ m
Current	100 μA $= 6 \times 10^{14}$/s	100μA $= 6 \times 10^{14}$/s
Repetition Rate	50 Hz	10 Hz
Charge/Pulse	2 μC $= 1.2 \times 10^{13}$ppp	10 μC $= 6 \times 10^{13}$ppp
No. Superperiods	6	12
Lattice $\{$ Focusing	FODO	FODO
Structure $\}$ Bending	OBOBBOBO	BBBBBOBO
No. Focusing Cells	24	48
Maximum $\beta_x \times \beta_y$	15.8 m $\times$ 15.2 m	38.1 m $\times$ 37.5 m
Dispersion η_{max}	4.0 m	9.09 m
Transition $\gamma_t = 1/\sqrt{\eta}$	9.2	∞
Tunes $\nu_x \times \nu_y$	5.23 $\times$ 6.22	11.22 $\times$ 12.18
Space Charge $\Delta \nu_y$	-0.15	-0.09
Emittances $\{$ $\epsilon_x \times \epsilon_y$	139π $\times$ 62$\pi(\mu$m$)$	37π $\times$ 16$\pi(\mu$m$)$
at Injection $\}$ ϵ_{long}	0.064 eV-s	0.192 eV-s
Harmonic	45	225
Radiofrequency	46.1 $\rightarrow$ 61.1 MHz	61.1 $\rightarrow$ 62.9 MHz
Energy gain/turn	210 keV	2000 keV
Maximum RF Voltage	576 kV	2400 kV
RF cavities	12 $\times$ 50 kV	18 $\times$ 135 kV

intensity would be lost in the cyclotron at the pre-stripper (but this "lost" beam could be extracted down an existing beamline), and that roughly 10 µA would be lost upon injection/extraction in the rings.

It is to be noted that a pulsed H⁻ linac as an injector, instead of the TRIUMF cyclotron, would eliminate the need for the Accumulator ring. However, the cost of such a machine is estimated to be about \$50 M Canadian vs a cost of \$5 M Canadian for the ring. The (1985) cost estimates for the other rings are \$25.4 M for B, \$13.5 M for C, \$83.3 M for D and \$18.2 M for E, excluding the cost of the tunnels.

Experimental Area

We have chosen one large experimental hall, rather than several separate, smaller halls. This gives greater flexibility and possibility of shared services, albeit with more of the area used for shielding. The hall layout is shown in Fig. 8: the floor is 8 m below grade and is 75 m by 120 m, with a further 75 m by 20 m staging area at grade level.

The fast-extraction beam enters one corner of the hall, where the target and horn for neutrino production are located. The neutrino detector will be in a separate building at the end of a decay tunnel and shielding. The slow-extraction beam will be diverted in a switchyard before the hall into one of 3 lines, A,B or C. Inside the hall, line D splits off from line C. Generally, each line has a high-momentum secondary channel at small take-off angle and a lower-momentum channel at a larger angle. Extensive

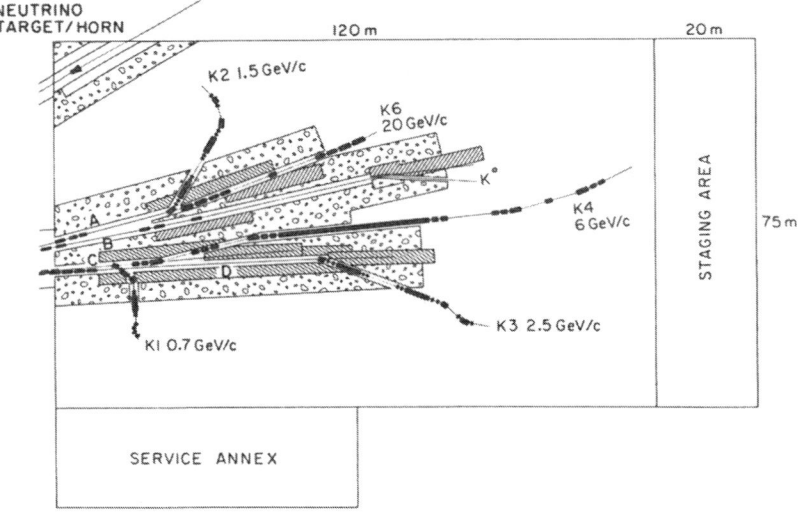

Fig. 8. Proposed layout of the experimental area.

iron and concrete shielding is required to contain the hadrons and muons produced in the target and beam dumps.

Characteristics of the secondary channels are given in Table 2. These span a wide range in momenta, and (except for K6) have one or two separators each. Estimated fluxes of secondary beams are given in Table 3.

Again, it is clear that not all of the facilities described would be available in the initial stages of operation, but the intent is to indicate the range of experiments that it should be possible to accommodate.

Table 2. Properties of the secondary beam lines

P GeV/c	K^- $\times 10^6/s$	K^+	π^- $\times 10^6/s$	π^+	$\bar{p}$ $\times 10^6/s$	Channel
6.0	14.6	33.5	1.91	3.60	23.4	
5.0	16.5	32.3	3.27	5.71	44.0	K4
4.0	9.8	17.3	3.80	6.20	56.2	
3.0	2.5	4.5	3.23	5.02	43.1	
2.5	66.4	119	15.8	23.8	10.6	
2.25	52.6	97.6	18.2	27.1	10.3	
2.00	39.1	75.6	20.8	30.4	91.0	K3
1.75	25.8	51.6	23.2	33.4	78.4	
1.50	14.3	27.0	25.4	35.7	51.7	
1.25	5.4	9.7	26.9	36.8	25.5	
1.5	146.7	278	37.1	52.3	61.4	
1.4	86.4	161	33.9	47.5	40.7	
1.3	60.1	110	30.8	42.4	29.5	
1.2	39.5	70.3	27.2	37.4	19.2	
1.1	24.2	42.1	23.7	32.1	11.4	K2
1.0	13.5	23.6	20.3	27.1	6.3	
0.9	6.7	11.6	17.0	22.2	3.2	
0.8	2.8	4.8	13.5	17.6	1.45	
0.7	.94	11.7	10.4	13.2	0.67	
.70	53.3	95.9	69.3	88.0	4.4	
.65	32.4	59.4	63.4	79.6		
.66	19.1	35.5	51.9	65.6	1.55	
.55	10.0	19.4	44.3	55.4		K1
.50	4.6	9.7	36.4	44.8	0.58	
.45	1.8	4.2	26.9	32.6		
.40	.65	1.6	17.1	23.0	0.19	

Table 3. Calculated fluxes for the secondary beam lines

Beam Line	Momentum Range (GeV/c)	Solid Angle (msr)	Takeoff Angle (Deg.)	Momentum Acceptance (%)	Length (m)	Length of Separator
K1	0.4 - 0.7	6	10	5	17.9	1×3.0 m
K2	0.7 - 1.5	1.6	0	3.8	29.7	2×3.0 m
K4	2.0 - 6.0	.08 (6 GeV/c) .30 (3 GeV/c)	0	3	115	1×30 m
K0	0.5 - 10.0	0.03	6	Wide	20	
K3	1.25 - 2.5	.5 (2.5 GeV/c) 2.0 (1.5 GeV/c)	0	4	54	2×7.5 m dc
K6	up to 20 GeV/c	0.16	0	8	46.3	unseparated
Muon	30-200 MeV/c	35	135	10	18	1×1.5 m rf or 3 m dc

Status of the Proposal

When TRIUMF's proposal was submitted in the fall of 1985, two review panels were created by the two agencies involved in funding TRIUMF and TRIUMF experiments (the National Research Council and the Natural Sciences and Engineering Research Council). The first, a Technical Review panel, examined the scientific merit and technical feasibility of the proposal. It met in February 1986 and produced a report which was very favorable concerning the scientific case for building a kaon factory, the advantages of TRIUMF as a Canadian site, the technical feasibility, and the soundness of the cost estimates. The panel recommended prompt action, and suggested that the construction period be shortened from 7 to 5 years. The second panel, an overview panel, was to assess the impact of a kaon factory on other parts of the science program in Canada, and upon the economic and technological benefits for local and national industries. The overview panel met in April and has just issued its report. It too is favorable to the proposal, but with caveats about the need to protect existing programs. With the reviews completed, the next stages will be consideration of the proposal by the Minister of Science and Technology and eventually by the full cabinet of the federal government. In parallel, we would expect to discuss with foreign groups the possibility of joint funding, especially of detector systems.

Design calculations have continued on: slow-extraction efficiency of the Extender, acceleration of polarized protons, and filling of the Accumulator ring [3]. Tests of H^- extraction from the cyclotron using rf and dc deflectors have been made during the March 1986 shutdown. Despite some mechanical problems, an 85% extraction efficiency was demonstrated and there appears no problem in achieving the design goal of 90% or better.

References

1. TRIUMF Annual Report, 1985.
2. TRIUMF KAON Factory Proposal, September 1985.
3. TRIUMF Progress Summary, First Quarter 1986.

THE EHF PROJECT AND ITS RESEARCH PROGRAM*

Tullio Bressani

Istituto di Fisica Superiore dell'Università, 10125 Torino (Italy)

Istituto Nazionale di Fisica Nucleare, Sezione di Torino

Abstract

The European Hadron Facility (EHF) project is briefly described. It consists of an accelerator complex designed to supply a 100 μA primary proton beam of 30 GeV, with a wide spectrum of intense secondary beams of high quality. The outlines of the research program, centered on the physics in the Confinement region, a yet unresolved problem, and on the search of signals of new physics by precision measurements, are discussed.

1. Introduction

The European Hadron Facility project is a proposal, still under final study, for the construction in Western Europe of a new facility for the basic research in Intermediate Energy Physics in the 1990's and beyond. The initiative grown up by the consciousness of many european physicists, active in the field and that contributed strongly to the experimental and theoretical progress, that their discipline would be condamned to a fast decline unless some new action was started. An EHF Study Group, under the chairmanship of Prof. F. Scheck from Mainz, was created in March 1983 and an EHF Design Group, under the chairmanship of Prof. F. Bradamante from Trieste, in December 1985. Details on the activity of the EHF Study Group can be found in Ref. 1). The EHF Design Group, an international team of accelerator physicists and engineers, was asked by the EHF Study Group to develop a technical project for an accelerator complex to produce a

---

* On behalf of the EHF Study Group

100 μA proton beam (6 x 10^{14} protons/sec) to an energy of 30 GeV, with the further constraints:

1) capability of producing polarized proton beams

2) fast and slow extraction systems

3) easy upgrade of the designed maximum energy to ~40 GeV

4) 960 m for the length of the main ring (it could fit into the CERN ISR tunnel)

    Regarding the possible sites for EHF, it seems at moment that there are the following realistic options:

a) a new site in Italy

b) CERN, in the tunnel of the dismantled ISR

c) SIN, Villigen

More details may be found in Ref.1), in a technical document on the feasibility Study for an EHF[2] and in Ref.3) and Ref. 4).

2. The conceptual design of the Accelerator Complex

    A substantial difference of EHF with respect to similar projects (LAMPF II, KAON at TRIUMF) is that no existing machine was suggested as injector.

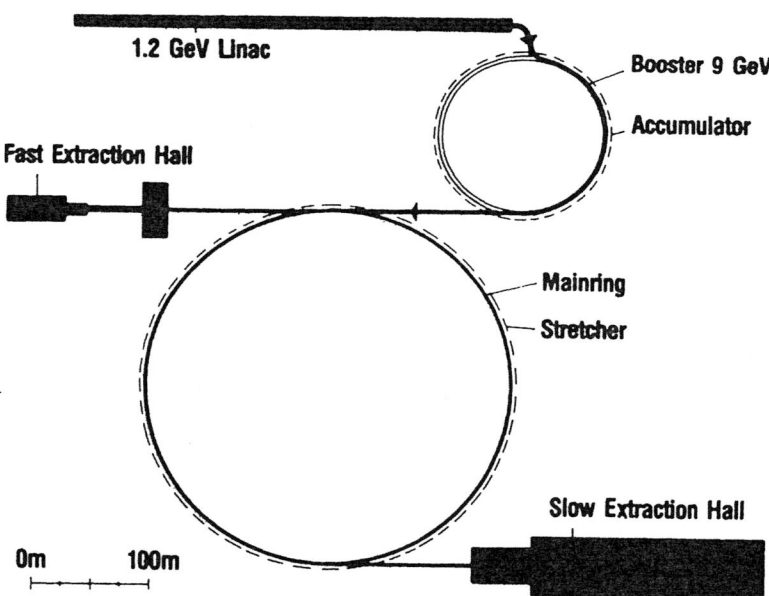

Fig. 1 - Schematic layout of the European Hadron Facility

Then all the parameters of the different machines could be optimized without constraints. An option using the 590 MeV SIN isochronous cyclotron as injector has been kept in mind, but has not been studied by the EHF Study Group.

The proposed EHF is the complex of accelerators schematically illustrated in Fig. 1, whose main components are a high-energy LINAC, accelerating a H⁻ beam to 1.2 GeV, and two fast-cycling synchrotrons, a 9 GeV Booster Ring and a 30 GeV Main Ring, with radii and repetition rates of ratios 1:2 and 2:1 respectively. The repetition rates of the LINAC and of the Booster are the same, 25 Hz. The H⁻ beam pulse coming from the LINAC is stripped into a proton beam by passing through a thin foil and injected directly into the Booster over 200 turns.

Two more rings complement the system, a 9 GeV Accumulator, with the same radius as the Booster, where the Booster pulses are stored before being transferred to the Main Ring, and a 30 GeV Stretcher Ring, having the same radius as the Main Ring synchrotron, where the fast extracted 30 GeV beam from the Main Ring is stored and then slowly extracted to produce 100% duty factor secondary beams. The 1:2 ratio between the repetition rates of the Main Ring and the Booster allows to have an Accumulator of the same size as the Booster, rather than of the same size as the Main Ring. Only one Booster pulse is thus stored in the Accumulator, the second one passing through the just emptied Accumulator and going directly to the Main Ring.

TABLE I

	LINAC	BOOSTER	MAIN RING
Energy	1.2	1.2 – 9	9 – 30
RFQ (MHz)	50/400	–	–
Chopper (MHz)	0.55	–	–
Alvarez (MHz)	400	–	–
Side Coupled Linac (MHz)	1200	–	–
Circumference (m)	–	480	960
Rep. Rate (Hz)	25	25	12.5
Protons (10^{13})	–	2.5	5
RF (MHz)	–	50.5 – 56	56 – 56.2
Frequency swing	–	11%	0.4%
Peak RF Voltage (MV)	–	1.2	2

Table I – Summary of the EHF Parameters

The use of these two relatively low cost storage rings allows to continuously run acceleration cycles in the Booster and Main Ring without the need to "flat-top" of "flat-bottom" the magnetic cycles. Table I summarizes the parameters of the proposed EHF.

The operation of the complex can be understood by looking at the time diagram in Fig. 2.

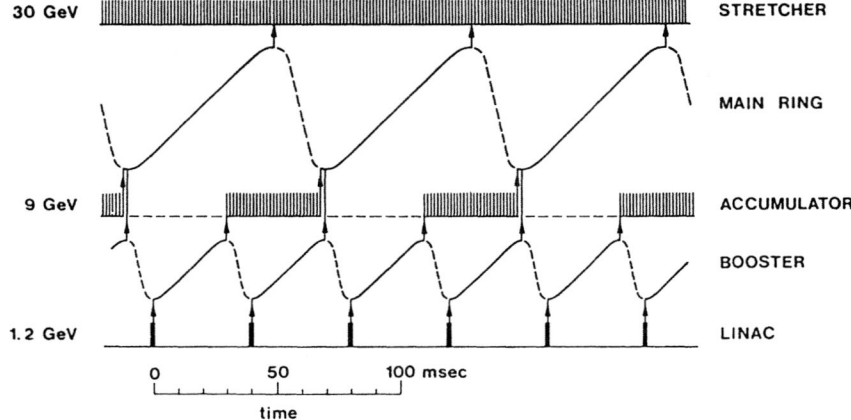

Fig. 2 — Time diagram for the EHF, showing the acceleration cycles of the Booster and of the Main Ring, and the beam transfers in the various stages.

Acceleration cycles of the Booster and of the Main Ring are shown, as well as the beam transfers between the various rings. Bucket-to-bucket transfer of the beam from one machine to the next minimizes beam losses. The ramping of the magnets is done with a dual resonance circuit in which the up and down ramp frequencies are in a 1:3 ratio, so that the ramp frequency of the two synchrotrons is 2/3 of the repetition rate, with a corresponding reduction of the peak RF voltage.

A crucial issue in the design of every single part of the EHF has been the minimization of beam losses.

A 1% beam loss at EHF would correspond to the loss of the full beam at the CERN PS or at the Brookhaven AGS, a perspective which is clearly unacceptable and would put excessive requirements on shielding and remote control of the machine components and seriously question the reliability of the entire complex. It turns out, however, that by careful design and parameter optimization beam handling can be effectively lossless, so that general radioactivity level will be equivalent to that of an existing accelerator of the same energy. More details may be found in Ref. 3.

A characteristic feature of the proposed complex compared to similar projects is the high injection energy of the Main Ring (9 GeV), which in turn demands a high energy LINAC (1.2 GeV). Such a choice presents several advantages, in particular:

1) the possibility of using "Siberian Snakes" in the Main Ring in order to preserve the beam polarization

2) the possibility for a staged construction where the LINAC, Booster and Accumulator are built first and the Main Ring and Stretcher added later. A useful physics program with low energy Kaons and neutrinos could be started soon; for that reason the Fast Extraction Hall can be fed also directly from the Booster.

Concerning the intensity gain of the secondary beams, an increase of a factor even larger than that of the primary proton beam (10^2) may be expected by substantial improvements in the target and secondary beams design. A nice system for Multiple Achromatic Extraction of Independent Momentum beams (MAXIM) was proposed[5]. Such a system allows the extraction of several secondary high-energy charged-particle beams from a production target at zero angle. The system consists of a pair of sectors of concentric circular bending magnets of opposite polarities, centred on the production target. As the magnetic field is rotationally symmetric with respect to a vertical axis through the target, any charged particle emitted from the target centre will emerge from the system travelling on a radial plane through the axis. Although all momenta are focused into any radial direction, there is a one-to-one correspondence between any given direction and the sign and momentum of the forward emitted secondaries.

A possible beam layout for the Slow Extraction Area using two MAXIM systems and three production target was studied, with about ten beam lines for experiments. The intensities of $\pi^-(\sim 11^{11}/s)$, $K^-(\sim 10^9/s)$ and $\bar{p}(\sim 10^9/s)$ which can be obtained are impressive and would allow a real breakthrough in the Intermediate Energy Physics.

3. Outlines of the research program

The research program at EHF may be roughly divided into two parts:
i) Physics in the Confinement region
ii) Physics at the Precision frontier.

In my personal opinion, the study of the physics in the Confinement region would be the characterizing and unique aspect of EHF. The Confinement is a problem which is not yet solved in the Hadronic Physics[6] and is a central point of the Modern Physics. There is a lot of work that still has to be done in the few-quark systems (Particle Physics) and practically everything has to be done for the study of many-quark systems (Nuclear Physics), where possible collective or coherent effects may appear as a unique prime of EHF. I will list in the following some relevant and unique experiments that could be performed.

1) Quark clusters in the structure of light mesons and baryons

Taking as example the pion, it is very likely not only a simple $(q\bar{q})$ system, but has also $3q$-$3\bar{q}$, $5q$-$5\bar{q}$ and so on components. The diffractive dissociation by nuclei would allow the decomposition into these substructures[7].

2) Spectroscopy of the mesons (and hadrons) with heavy flavours: the quantitative test of the QCD.

The $(q\bar{q})$ system is the simplest one that can be treated with QCD. Taking into account the quark masses, only for the $(c\bar{c})$ and $(b\bar{b})$ systems we may believe in perturbative, non relativistic two-body approaches. There are many states of these systems, predicted theoretically, that have still to be discovered and whose main parameters (energy and width) have then to be measured. Whereas the spectroscopy of the $(c\bar{c})$ system will be in an advanced stage when an accelerator like EHF could operate, that of the $(b\bar{b})$ system would still have to be done. The production of $(b\bar{b})$ in $\bar{p}$-p collisions may be favourably compared with that obtained in e^+-e^- collisions, in spite of the huge background[8]. A $\bar{p}$-p collider like SUPERLEAR[9] would be obviously necessary in order to benefit from the high flux of $\bar{p}$'s from EHF. If the energy of such a collider would be increased to 20 GeV (10 GeV for each beam), it could also be considered as a serious alternative[10] to the proposed B-factories based on e^+-e^- colliders.

3) Search and study of the glueballs

Following the first indications, not yet confirmed and universally accepted, the existence of these (gg) states would be one of the most spectacular experimental results of EHF. Production of glueballs in $\bar{p}$-p collisions seems one of the most promising methods.

4) Baryonium $(qq\bar{q}\bar{q}, qqq\bar{q}\bar{q}\bar{q})$ states

Even if not yet observed at LEAR, simple arguments based on duality diagrams strongly suggest their existence. More complete and precise experiments are necessary.

5) Dibaryons $(qqqqqq)$

Many experiments performed at several Laboratories have not yet produced a clear-cut evidence for these object that would constitute the interface between the few-and many-quark systems. They would in fact demonstrate the existence of color degrees of freedom in nuclei. The most spectacular dibaryon is the H particle, a (2s; 2u;2d) configuration, with a predicted mass lower than $2M_\Lambda$ and than stable.

6) Spin physics

Measurement of spin observables with polarized proton and antiproton* beams and targets would allow the understanding of unexpected features observed in recent experiments[11] and which are a serious problem for QCD. On the contrary such effects are predicted by an alternative, even if less popular, theory of strong interactions, the Anysothropic Chromo-Dynamics[6] (ACD).

7) Insertion experiments

The general approach of these measurements is that of substituting one (possibly two) of the u- and d-quarks of a nucleus with a "marked" s-quark. The quark structure of a nucleus is then studied at $Q^2 \cong 0$, and the method is similar to the "radioactive marked nucleus" technique of the molecular physics. In other words, this class of experiments belongs to the hypernuclear physics, where an impressive amount of first-class experiments can be performed.

8) Scattering experiments

By using different probes (nucleons, antinucleons, pions, Kaons) one could change in a controlled way the quark content of the probe and then explore, in the same kinematical conditions, the underlying quark structure of a nucleus. Among all, K^+ scattering is of first importance. The reason is that the K^+ is the only hadronic probe containing an $\bar{s}$-quark, that cannot annihilate with a real s-quark, not contained in a nucleus. It is then able to penetrate in the interior of the nucleus and explore hopefully the quark substructure.

9) Deposition experiments

By means of $\bar{p}$ annihilations in the interior of a heavy nucleus it is possible to reach, in a portion of the nuclear matter, the same density and temperature conditions that could be obtained in relativistic nucleus-nucleus collisions and then search for new phases of the hadronic matter (the quark-gluon plasma).

The Physics at the Precision frontier may be considered as complementary to the Physics at the extreme high energy. The final output would be the identification of signals for new physics, not covered by the Standard Model. An extreme precision and the search for very rare events are an alternative approach to the multi-TeV energy.

* obtained by the classical technique of production from Λ decay.

612

We could mention the following outstanding experiments:

10) Rare Kaon decays

With the intense K beams from the EHF, and with a dedicated apparatus, up to seven orders of magnitude may be gained on the existing limits for the decays: $K_L \to \mu e$, $K_L \to \mu^+ \mu^-$, $K^+ \to \pi^+ \mu e$, $K^+ \to \pi^+ \pi^\circ$, $K \to l \nu_H$, $K^+ \to \pi^+ \nu \nu$. The last process is perhaps the most interesting one, since it can be related also to the mass of the t-quark, to that of the the heavy bosons and so on. More details may be found in Ref. 12).

11) Neutrino oscillations

The improvement of the present limits is an experimental task that will be carried out with different techniques. A unique advantage of EHF compared to other facilities with similar intensities is that it will be located in Western Europe, where, at the Gran Sasso Laboratory, the most powerful detectors and facilities for underground physics will be into operation.

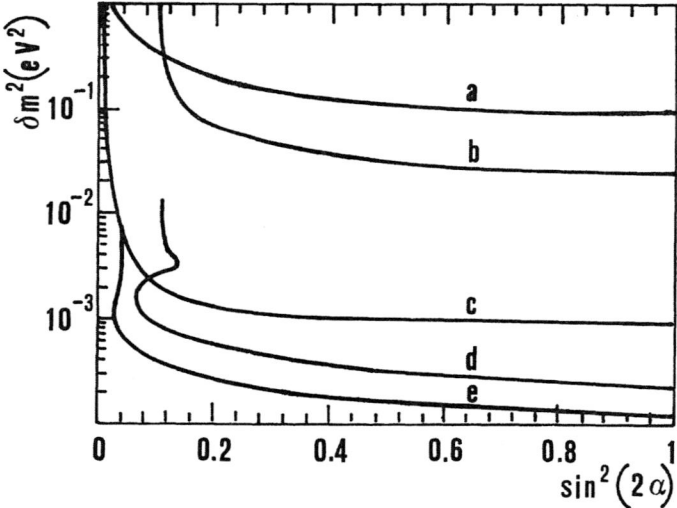

Fig. 3 - 90% C.L. limits on the oscillation parameters
    a) from accelerators
    b) from reactors
    c) from the TRIUMF KAON facility
    d) from EHF + Gran Sasso and a CHARM-like detection apparatus
    e) from EHF + Gran Sasso and an ideal apparatus.

The main advantage in the use of a high intensity proton beam combined with an underground detector consists in the possibility to exploit the large distance between the detector and the accelerator while maintaining a good neutrino flux in order to perform the experiment in a reasonable time. De Vincenzi and Pistilli[13] have studied the approach, and they found the results shown by Fig. 3, where the limit foreseen at the TRIUMF KAON machine[14] is also shown for comparison. The advantage of the EHF + Gran Sasso approach is evident.

4. Conclusion

The large amount of unique, first-class experiments and the convincing technical project ensure from the beginning the success of EHF. On the other hand the estimated cost (850 Millions of Deutsch Marks) put directly EHF at the same scale and then on a collision trajectory with other accelerator projects, under study and discussion mainly among the High Energy community. The circumstance that the EHF community interfaces the Nuclear and Particle Physics fields, acknowledged in lofty academic circles as an example of the right direction towards an essential unification, is often contradicted when the discussions focus on substantial arguments, like the possible sources of funds. There are also attempts to divide the wide experimental program of EHF into different slices, and then compare each of them to what one could obtain with dedicated (and less expensive!) facilities. In brief, an approach of the type "DIVIDE ET IMPERA", well established by an eminent Roman more than 2000 years ago[15], is being tried on our community. Just educated by History, we will try to overcome these attempts and hope to succeed in our enterprise.

ACKNOWLEDGEMENT

I would like to thank Miss L. Ceretta for the careful preparation of the typed version of the manuscript.

REFERENCES

1) EHF, The European Hadron Facility, Letter of Intent (1986)

2) Feasibility Study for a European Hadron Facility, Int. Document EHF-86-33 (1986)

3) F. Bradamante, in Proc. of the Int. Conference on a European Hadron Facility, Mainz, March 10-14, 1986, ed. Th. Walcher, to be published as a special issue of Nuclear Physics B (quoted in the following as Mainz 1986)

4) T.Bressani,in Perspectives on Theoretical Nuclear Physics, Proceedings of the Workshop held in Cortona (Italy), September 16-18, 1985, eds. L. Bracci et al. (ETS Editrice, Pisa, 1986), pp. 75-83

5) C. Tschalär, "Multiple Achromatic Extraction System", to be published in Nucl. Instr. and Meth.

6) G. Preparata, in Proc. of the Winter School on Hadronic Physics at Intermediate Energy, 1st Course, Folgaria (Italy), February 17-22, 1986, eds. T.Bressani and R.A. Ricci, to be published by North-Holland Physics Publishing, (quoted in the following as Folgaria 1986); G. Preparata, in Mainz 1986

7) B. Povh, in Proc. of the Workshop on Nuclear and Particle Physics at Intermediate Energies with Hadrons, Trieste, 1-3 April, 1985, eds. T. Bressani and G. Pauli, Conf. Proceedings Vol. 3 (Italian Physical Society, Bologna, 1986), p. 43 (quoted in the following as Trieste 1985)

8) P. Dalpiaz, in Mainz 1986 and Folgaria 1986

9) See e. g. L. Tecchio, in Folgaria 1986

10) P. Pistilli, in Folgaria 1986

11) G. Preparata, in Proceedings of the 6th International Symposium on High Energy Spin Physics Marseille, Sept. 1984, ed. J. Soffer, Journal de Physique 46C2 (1985), 128

12) N. Paver, in Trieste 1985, p. 51

13) M. De Vincenzi and P. Pistilli, in Folgaria 1986

14) KAON Factory Proposal, prepared by TRIUMF (Sept. 1985)

15) C. Julius Caesar, in "De Bello Civili" (49 b.C.)

LIST of PARTICIPANTS

Y. Akaishi
Department of Physics,
Hokkaido University
Kita-ku Kita-10 Nishi-8
Sapporo 060, Japan

E.O. Alt
Institut fur Physik
Universitat Mainz
Staudinger Weg 7
D-6500 Mainz
W. Germany

L. E. Antonuk
Nuclear Research Center
University of Alberta
Edmonton, Alberta T6G2N5
Canada

H. Arenhoevel
Institut fur Kernphysik
Universitat Mainz
J.J. Becher Weg 33
D-6500 Mainz
W. Germany

G. Anton
Physikalisches Institut
Universitat Bonn
Nussallee 12, D-5300 Bonn-1
W. Germany

R.G. Arnold
The American University
SLAC P.O.B. 4349
Stanford, CA 94305, U.S.A.

G. Backenstoss
Institut fur Physik
Experimentalphysik
Klingelbergstrasse 82
CH 4056 Basel
Switzerland

B.L.G. Bakker
Nat. Lab. VU
P.O.Box 7161
1007 MC Amsterdam
Holland

J.L. Ballot
Institut de Physique Nucleaire
Universite de Paris-Sud
B.P. 1, 91406 Orsay Cedex
France

O. Benhar
I.N.F.N. - Sezione Sanita'
V.le Regina Elena 299
00161 Roma Italy

B.L. Berman
The George Washington University
Dept. of Physics
Washington, D.C. 20052
U.S.A.

G. Berthold
Inst. fur Theoretische Physik
Universitat Graz
Universitatsplatz 5
A-8010 Graz
Austria

N. Bianchi
I.N.F.N. Lab. Nazionali di
Frascati
Via E. Fermi
00044 Frascati Italy

A. Boemelburg
Ruhr Universitat Bochum
Institut fur Theoret. Physik II
Universitatstrasse 150
D-4630 Bochum,
W. Germany

S. Boffi
Dipartimento di Fisica
Universita' di Pavia
Via Bassi 6
27100 Pavia Italy

T. Bressani
Direttore della Sezione I.N.F.N
Corso M. D'Azeglio 46
10125 Torino Italy

W.H. Breunlich
Austrian Academy of Sciences
Boltzmanng. 3
A-1090 Wien
Austria

M. Bruno
Dipartimento di Fisica
Universita' di Bologna
Via Irnerio 46
40126 Bologna Italy

R. Buettgen
K. F. A. Julich GmbH
Institut fur Kernphysik
Postfach 1913
D-5170 Julich 1
W. Germany

F. Cannata
Dipartimento di Fisica
Universita' di Bologna
Via Irnerio 46
40126 Bologna Italy

L. Canton
Dipartimento di Fisica
Universita' di Padova
Via F. Marzolo, 8
35131 Padova Italy

C.E. Carlson
Physics Dept.,
College of William and Mary
Williamsburg, VA 23185
U.S.A.

. Cattapan
Dipartimento di Fisica
Universita' di Padova
Via Marzolo 8
35131 Padova Italy

C. Ciofi degli Atti
I.N.F.N. - Sezione Sanita'
V.le Regina Elena 299
00161 Roma Italy

S.A. Coon
Dept. of Physics
University of Arizona
Tucson, AZ 85721 U.S.A.

E. Cravo
Centro de Fisica Nuclear
Universidade de Lisboa
Av. Prof. Gama Pinto 2
1699 Lisboa Codex
Portugal

M. D'Agostino
Dipartimento di Fisica
Universita' di Bologna
Via Irnerio 46
40126 Bologna Italy

S. D'Angelo
Dipartimento di Fisica
II Universita' degli Studi
Via O. Raimondo
00173 Roma Italy

N.E. Davison
Dept. of Physics
University of Manitoba
Winnipeg, Manitoba R3T 2N2
Canada

E. De Sanctis
I.N.F.N. Lab. Naz. di Frascati
Via E. Fermi, 13
00044 Frascati Italy

J.J. de Swart
Institute of Theor. Physics
University of Nijmegen
Toernooiveld
6525 ED Nijmegen
Holland

C. de Vries
NIKHEF-K
Postbus 4395
1009 AJ Amsterdam, Holland

P.K.A. de Witt Huberts
NIKHEF-K
P.O. BOX 4395
1009 AJ-Amsterdam Holland

G. Dillon
Dipartimento di Fisica
Universita' di Genova
Via Dodecanneso 33
16146 Genova Italy

P. Doleschall
Central Research Inst. for Physics
H-1525 Budapest, Pf. 49
Hungary

M. Fabre de la Ripelle
I.P.N.
B.P.1
91406 Orsay
France

J.L. Friar
Los Alamos Nat. Lab.
Theory Division
M.S. B293
Los Alamos, NM 87545, U.S.A.

S. Fantoni
Dipartimento di Fisica
Universita' degli Studi
Piazza Torricelli, 2
56100 Pisa Italy

M. Frisoni
Dipartimento di Fisica
Universita' di Bologna
Via Irnerio 46
40126 Bologna Italy

B. Frois
CEN-Saclay
DPhN-HE
91191 Gif-sur-Yvette, Cedex
France

S. Frullani
Istituto Superiore di Sanita'
Laboratorio di Fisica
V.le Regina Elena 299
00161 Roma Italy

H. Garcilazo
Institut fur Kernphysik des
Kernforschungszentrums Karlsruhe
Postfach 3640
7500 Karlsruhe 1
W. Germany

F. Garibaldi
Istituto Superiore di Sanita'
Laboratorio di Fisica
V.le Regina Elena 299
00161 Roma Italy

K. Geissdorfer
Physikalisches Institut der
Universitat Erlangen-Nurnberg
D-8520 Erlangen
Erwin-Rommel-str.1
W. Germany

M. Giannini
Dipartimento di Fisica
Universita' di Genova
Via Dodecanneso 33
16146 Genova Italy

G. Giardina
Dipartimento di Fisica
Universita' di Salerno
84100 Salerno
Italy

B. Girolami
Istituto Superiore di Sanita'
Laboratorio di Fisica
V.le Regina Elena 299
00161 Roma Italy

H. Grimm
Max-Planck Inst. fur Chemie
Saarst. 23
D-6500 Mainz
W. Germany

F. Gross
CEBAF
12070 Jefferson Avenue
Newport News, VA 23606
U.S.A.

W. Heinrich
Inst. fur Theoretische Kernphysik
Endenicher Allee 11-13
D-5300 Bonn 1
W. Germany

M. Hermann
I.N.F.N. Sezione di Bologna
Via Irnerio 46
40126 Bologna Italy

T. Hippchen
K. F. A. Julich GmbH
Institut fur Kernphysik
Postfach 1913
D-5170 Julich 1
W. Germany

K. Holinde
K. F. A. Julich GmbH
Institut fur Kernphysik
Postfach 1913
D-5170 Julich 1
W. Germany

B. Holzenkamp
K. F. A. Julich GmbH
Institut fur Kernphysik
Postfach 1913
D-5170 Julich 1
W. Germany

D.A. Hutcheon
TRIUMF
4004 Wesbrook Mall
Vancouver BC V6T 2A3
Canada

A. Incicchitti
Dipartimento di Fisica
Universita' di Roma
Piazzale Aldo Moro
00185 Roma Italy

M. Jodice
Istituto Superiore di Sanita'
Laboratorio di Fisica
V.le Regina Elena 299
00161 Roma Italy

B. Karlsson
TANDEM ACC. LAB.
BOX 533
S-75121 Uppsala
Sweden

K.H. Krause
Max Planck Institut
Saarstr. 23
D-6500 Mainz
W. Germany

C. Kuhl
Inst. fur Theoretische Kernphysik
Endenicher Allee 11-13
D-5300 Bonn
W. Germany

J. M. Laget
CEN-Saclay
DPhN HE
91191 Gif-sur-Yvette, Cedex
France

P. Leleux
Universite' Catholique
Chemin du Cyclotron, 2
B-1348 Louvain-la-Neve
Belgium

P. Levi-Sandri
I.N.F.N. Lab. Nazionali di
Frascati
Via E. Fermi
00044 Frascati Italy

S. Liuti
Dipartimento di Fisica
Universita' di Perugia
Via Elce di sotto
06100 Perugia Italy

L. Lovitch
Dipartimento di Fisica
Universita' di Ferrara
Via Paradiso 12
44100 Ferrara Italy

C. Marchand
DPHN-HE CEN Saclay
F-91191 Gif-sur Yvette
France

M. Martini
Dipartimento di Fisica
Universita' di Ferrara
Via Paradiso 12
44100 Ferrara Italy

P.A. Massaro
Dipartimento di Fisica
Universita' di Bari
Via Amendola 173
70126 Bari Italy

W. J. McDonald
Laboratoire National Saturne
BP n.2
91190 Gif-sur-Yvette Cedex
France

Z.E. Meziani
Stanford University
Physics Department
Varian Bldg., room 152
Stanford, CA 94305 U.S.A.

J. Morgenstern
CEN Saclay
DPhN-HE B.P.2
91191 Gif-sur-Yvette Cedex
France

B. Mosconi
Dipartimento di Fisica
Universita' di Firenze
Largo E. Fermi 2
50125 Firenze, Italy

J. Mougey
CEBAF 12070 Jefferson Ave.
Newport News, VA 23606
U.S.A.

J. Neubauer
Inst. fur Theoretische Kernphysik
Endenicher Allee 11-13
D-5300 Bonn 1
W. Germany

R. Neuhausen
Inst. fur Kernphysik
University of Mainz
Postfach 3980
D-6500 Mainz
W. Germany

P. Niessen
Institut fur Kernphysik
Zulpicher Str. 77
5000 Koln 41
W. Germany

P. Obersteiner
Institute for Theoretical Physics
University of Graz
Universitatsplatz 5
A-8010 Graz
Austria

S. Oryu
Department of Physics
Science University of Tokio
Noda-City Chiba-ken, 278
2641 Yamazaki
Japan

E. Pace
Dipartimento di Fisica
Universita' "La Sapienza"
P.le A. Moro 2
00185 Roma Italy

J. Pauschenwein
Inst. for Theoretical Physics
University of Graz
Universitatsplatz 5
A-8010 Graz
Austria

G.L. Payne
Dept. of Physics and Astronomy
The University of Iowa
Iowa City, Iowa 52242
U.S.A.

P. Picozza
Dipartimento di Fisica
Universita' "La Sapienza"
P.le A. Moro 2
00185 Roma Italy

G. Pisent
Dipartimento di Fisica
Universita' di Padova
Via Marzolo, 8
35131 Padova Italy

S. Platchkov
DPHN-HE CEN Saclay
F-91191 Gif-sur-Yvette
France

622

W. Plessas
Institut Fur Kernphysik
K F A Julich
Postfach 1913
D-5170 Julich
W. Germany

B. Polke
Institut fur Kernphysik
Zulpicher Str. 77
5000 Koln 41
W. Germany

C. Preuss
Max Planck Institut fuer Chemie
Saarstrasse 23
D-6500 Mainz
W. Germany

D. Prosperi
Dipartimento di Fisica
Universita' di Roma
Piazzale Aldo Moro
00185 Roma Italy

G. Rauprich
Institut fur Kernphysik
Zulpicherstr. 77
D-5000 Koln 41
W. Germany

M. Rosa-Clot
Dipartimento di Fisica
Universita' degli Studi
Piazza Torricelli, 2
56100 Pisa Italy

S. Rosati
Dipartimento di Fisica
Universita' degli Studi
Piazza Torricelli, 2
56100 Pisa Italy

A. Rotondi
Dipartimento di Fisica
Universita' di Pavia
Via Bassi 6
27100 Pavia Italy

G. Salme'
I.N.F.N. - Sezione Sanita'
V.le Regina Elena 299
00161 Roma Italy

P. Salvisberg
Institut fur Physik
Universitat Basel
Experimentelle Kernphysik
Klingelbergstrasse 82
CH 4056 Basel
Switzerland

W. Sandhas
Inst. fur Theoretische Kernphysik
Endenicher Allee 11-13
D-5300 Bonn 1
W. Germany

M. Sanzone
Dipartimento di Fisica
Universita' di Genova
Via Dodecanneso 33
16146 Genova Italy

A.M. Saruis
ENEA Centro E. Clementel
Via Mazzini 2 Italy
40138 Bologna Italy

T. Sasakawa
Department of Physics
Tohoku University
Aramaki Aoba
980 Sendai, Japan

P.U. Sauer
Institut fur Theoretische Physik
Appelstrasse 2
D-3000 Hannover 1,
W. Germany

M. Sawicki
Theoretical Physics
Warsaw University
Hoza 69
00-681 Warszawa
Poland

C. Schaerf
Dipartimento di Fisica
Universita' di Roma
Piazzale Aldo Moro
00185 Roma Italy

H. P. gen. Schieck
Institut fur Kernphysik
Zulpicher str. 77
D-5000 Koln 41
W. Germany

E. W. Schmid
Institut fur Theoretische Physik
Auf de Morgenstelle 14
D-7400 Tubingen
W. Germany

W. Schuster
Physikalisches Institut der
Universitat Erlangen-Nurnberg
D-8520 Erlangen
Erwin-Rommel-str.1
W. Germany

S. Simula
Istituto Superiore Sanita'
V.le Regina Elena 299
00161 Roma Italy

I. Slaus
R. Boskovic Institute
P.O.B. 1016
41001 Zagreb
Yugoslavia

J. Sobolewski
Max-Planck-Institut
Saarstr. 23
D-6500 Mainz
W. Germany

M. Sotona
Institut for Nuclear Physics
250 68 Rez near Prague
Czechoslovakia

A. Stadler
Institut fur Theoretische Physik
Universitat Graz
Universitatplatz 5
A-8010 Graz
Austria

M. Steinacher
Institut fur Physik
Universitat Basel
Experimentelle Kernphysik
Klingelbergstrasse 82
CH 4056 Basel
Switzerland

J. Strate
Physikalisches Institut der
Universitat Erlangen-Nurnberg
D-8520 Erlangen
Erwin-Rommel-str.1
W. Germany

J.A. Tjon
Fysisch Laboratorium
Rijksuniversiteit Utrecht
P.O. Box 80000
NL-3308 TA Utrecht
Holland

E. Truhlik
Nucl. Physics Institute
Czechoslovak Academy of Scien
250 68 Rez
Czechoslovakia

W. Turchinetz
MIT-Bates Accelerator
P.O.B. 846
Middleton, Ma 01949, U.S.A.

G.M. Urcioli
I.N.F.N. - Sezione Sanita'
V.le Regina Elena 299
00161 Roma Italy

E. van Meijgaard
Fysisch Laboratorium
Rijsuniversiteit Utrecht
P.O. Box 8000
NL-3308 TA Utrecht
Holland

R. Vinh Mau
Division de Physique Theorique
Institut de Physique Nucleair
Universite' Paris Sud
B.P. 1-91406 Orsay Cedex
France

M. Viviani
Dipartimento di Fisica
Universita' degli Studi
Piazza Torricelli, 2
56100 Pisa Italy

C. von Ferber
Physikalisches Institut
der Universitat Bonn
Nussallee 12
D-5300 Bonn 1
W. Germany

W. von Witsch
Institut fur Strahlen und
Kernphysik
Universitat Bonn
Nussallee 14-16
D-5300 Bonn
W. Germany

M.F. Wagner
Institut fur Physik
University of Mainz
D-6500 Mainz
W. Germany

H. Walliser
Inst. fur Theoretische Physik
Universitat Tubingen
Auf der Morgenstelle 14
D-740 Tubingen
W. Germany

H. Walter
Inst. fur Theoretische Kernphysik
Endenicher Allee 11-13
D-5300 Bonn 1
W. Germany

P. Weber
Institut fur Physik
Universitat Basel
Experimentelle Kernphysik
Klingelbergstrasse 82
CH 4056 Basel
Switzerland

H.R. Weller
Duke University
Dept. of Physics
Science Drive
Durham, NC 27706
U.S.A.

R.B. Wiringa
Physics Division
Argonne Nat. Lab.
9700 South Cass Ave.
Argonne, IL 60439, U.S.A.

C.W. Wong
Physics Dept., UCLA
405 Hilgard Ave.
Los Angeles, CA 90024
U.S.A.

A. Zieger
Max Planck Inst. fur Chemie
Saarstrasse 23
D-6500 Mainz
W. Germany

ELECTRONIC MAIL DIRECTORY

USER NAME	USERID	NODE	NETWORK
L.E. Antonuk	TMP3	UALTAMTS	BITNET
R. G. Arnold	Arnold	SLACVM	BITNET
J.L. Ballot	Ballot	FRCPN11	EARN
O. Benhar	SANTEO	ICNUCEVM	EARN
S. Boffi	Boffi	VAXPV	INFNET[*]
M. Bruno	Bologna	IPIVAXIN	BITNET
F. Cannata	Bologna	IPIVAXIN	BITNET
C. Ciofi degli Atti	SANTEO	ICNUCEVM	EARN
S.A. Coon	Coon	ARIZJVAX	BITNET
M. D'Agostino	Bologna	IPIVAXIN	BITNET
N.E. Davison	Vanoers	UOFMCC	BITNET
	NED	TRIUMFCL	BITNET
J.J. de Swart	U634999	HNYKUN11	EARN
S. Fantoni	Many	ICNUCEVM	EARN
J.L. Friar	Friar%E	LLL-MFE	ARPANET
(Generic address	GJJ	LANL	ARPANET
at Los Alamos)			
M.Frisoni	Bologna	IPIVAXIN	EARN
B. Frois	B33426	ANLVM	BITNET
H. Garcilazo	IKP033	DKAKFK3	EARN
M.M. Giannini	Giannini	VAXGE	INFNET[*]
F. Gross	Gross	CEBAFVAX	BITNET
B. Goulard	578	CDECUDEM	CDN
D.A. Hutcheon	SMURF	TRIUMFCL	BITNET
S. Liuti	SANTEO	ICNUCEVM	EARN
L. Lovitch	Lovitch	VAXFE	INFNET[*]

C. Marchand	GNN.ZF	GEN	EARN
W.J. McDonald	TMP3	UALTAMTS	BITNET
Z.E. Meziani	Meziani	SLACVM	BITNET
S. Oryu	PH5290	JPNSUT30	BITNET
E. Pace	SANTEO	ICNUCEVM	EARN
G.L. Payne	BLABFGVA	UIAMVS	BITNET
W. Plessas	TPH127	DJUKFA11	EARN
M. Rosa-Clot	Many	ICNUCEVM	EARN
S. Rosati	THEO	ICNUCEVM	EARN
C. Schaerf	Schaerf	VAXROM	INFNET[*]
G. Salmé	SANTEO	ICNUCEVM	EARN
W. Sandhas	UNPØ21	DBNRHRZ1	EARN
P.U. Sauer	BBBG8	DHVRRZ01	EARN
H.P. gen.Schieck	ABK14	DKØRRZKØ	EARN
J.A. Tjon	Tjon	HUTRUU51	EARN
E. Van Meijgaard	VanMeijgaard	HUTRUU51	EARN
R. Vinh Mau	Lacombe	FRCPN11	EARN
C. Von Ferber	UNPØ32	DBNRHRZ1	EARN
H. R. Weller	DTUNL	TUCC	BITNET
R. B. Wiringa	B30916	ANLVM	BITNET

---

*The Bitnet address of the BITNET-INFNET Gateway is : INFNGW at IPIVAXIN